N

ALCOHOL INTOXICATION AND WITHDRAWAL

Experimental Studies II

ADVANCES IN EXPERIMENTAL MEDICINE AND BIOLOGY

Recent Volumes in this Series

Volume 50
ION-SELECTIVE MICROELECTRODES
Edited by Herbert J. Berman and Normand C. Hebert • 1974

Volume 51
THE CELL SURFACE: Immunological and Chemical Approaches
Edited by Barry D. Kahan and Ralph A. Reisfeld • 1974

Volume 52
HEPARIN: Structure, Function, and Clinical Implications
Edited by Ralph A. Bradshaw and Stanford Wessler • 1975

Volume 53
CELL IMPAIRMENT IN AGING AND DEVELOPMENT
Edited by Vincent J. Cristofalo and Emma Holečková • 1975

Volume 54
BIOLOGICAL RHYTHMS AND ENDOCRINE FUNCTION
Edited by Laurence W. Hedlund, John M. Franz, and Alexander D. Kenny • 1975

Volume 55
CONCANAVALIN A
Edited by Tushar K. Chowdhury and A. Kurt Weiss • 1975

Volume 56
BIOCHEMICAL PHARMACOLOGY OF ETHANOL
Edited by Edward Majchrowicz • 1975

Volume 57
THE SMOOTH MUSCLE OF THE ARTERY
Edited by Stewart Wolf and Nicholas T. Werthessen • 1975

Volume 58
CYTOCHROMES P-450 and b_5: Structure, Function, and Interaction
Edited by David Y. Cooper, Otto Rosenthal, Robert Snyder, and Charlotte Witmer • 1975

Volume 59
ALCOHOL INTOXICATION AND WITHDRAWAL: Experimental Studies II
Edited by Milton M. Gross • 1975

Volume 60
DIET AND ATHEROSCLEROSIS
Edited by Cesare Sirtori, Giorgio Ricci, and Sergio Gorini • 1975

ALCOHOL INTOXICATION AND WITHDRAWAL

Experimental Studies II

Edited by

Milton M. Gross

Downstate Medical Center
Brooklyn, New York

PLENUM PRESS • NEW YORK AND LONDON

Library of Congress Cataloging in Publication Data

Main entry under title:

Alcohol intoxication and withdrawal.

(Advances in experimental medicine and biology; v. 59)
"Proceedings of a symposium on Alcohol Intoxication and Withdrawal: Experimental Studies, held in Manchester, England, June 24-28, 1974, as part of the 20th International Institute on the Prevention and Treatment of Alcoholism, International Council of Alcoholism and Addictions."
Includes bibliographies and index.
1. Alcoholism—Congresses. 2 Alcohol—Toxicology—Congresses. 3. Alcoholism—Treatment—Congresses. I. Gross, Milton M. II. Series. [DNLM: 1. Alcohol, Ethyl—Toxcity—Congresses. 2. Alcoholic intoxication—Congresses. 3. Alcoholism—Therapy—Congresses. 4. Drug withdrawal symptoms—Congresses. W1 AD559 v. 59 1974 / WM274 A3525 1974]
RC565.A443 616.8'61 75-16174
ISBN 0-306-39059-0

Proceedings of a symposium on Alcohol Intoxication and Withdrawal: Experimental Studies, held in Manchester, England, June 24–28, 1974 as part of the 20th International Institute on the Prevention and Treatment of Alcoholism, International Council of Alcoholism and Addictions

A Division of Plenum Publishing Corporation
227 West 17th Street, New York, N.Y. 10011

United Kingdom edition published by Plenum Press, London
A Division of Plenum Publishing Company, Ltd.
Davis House (4th floor), 8 Scrubs Lane, Harlesden, London, NW10 6SE, England

Printed in the United States of America

Introduction

The acquisition of new knowledge is a continuing process. As part of that process, it is helpful to take stock periodically. It is for this reason that these international symposia on experimental studies of alcohol intoxication and withdrawal will be held every two years. The first was held in Amsterdam in September, 1972. The proceedings were published by this press as Volume 35 in the series Experimental Medicine and Biology and was entitled *Alcohol Intoxication and Withdrawal: Experimental Studies* I. This volume contains the proceedings of the symposium in Manchester held in 1974. The next symposium is planned for 1976.

There are several useful ways of taking stock. The one chosen for these symposia is to present the most recent findings in order to sharpen the focus on where we are now. It is for this reason that the papers in this publication present new data. The topics range from molecular to clinical frames of reference in the conviction that, in studies of alcohol intoxication and withdrawal, ultimately it will all have to come together. In the meantime, each vantage point can challenge and stimulate the others.

That the symposium took place, and the high quality of its content, are both tributes to the dedication, enthusiasm and caliber of the participants. Those who chaired each section, Drs. Ernest Noble, Sir Hans Krebs and his co-chairmen Charles Lieber and Richard Veech, Henri Begleiter, Ian Oswald and Arnold Ludwig, played critical roles in the selection of participants and conducting the proceedings. In addition, Drs. Charles Lieber and Richard Veech edited the papers in the biochemistry section. To them and to all the other participants goes my deepest appreciation and the credit for once again making the symposium something special.

It is essential to note that these proceedings are activities of the International Council of Alcoholism and Addictions. It is to the credit of its directors, Archer and Eva Tongue, that they had the vision to recognize the value of these symposia. Without their support and encouragement, neither the symposia, nor my own task in organizing them would have been possible. For their friendship and help I am very grateful. I would also like to thank Drs. Brian Hore and Myrrdin Evans for the important parts they played in arranging for the symposium to take place in Manchester.

Obviously a great deal of activity is involved before and after such a symposium. Mrs. Lee Davis played a central role in the preparations for the meeting and of this volume. I am deeply appreciative of her help.

Before all else, for myself, I could never have carried out this task without the continuing love, encouragement and patience of my wife and children.

Milton M. Gross
Chairman
Section on Biomedical Research in Alcoholism
ICAA

Contents

Section 1. NEUROCHEMISTRY

Ernest Noble, Chairman

Section II. BIOCHEMISTRY

Sir Hans Krebs, Chairman

Charles S. Lieber and R. L. Veech, Co-chairmen

Section III. EXPERIMENTAL STUDIES IN ANIMALS

Henri Begleiter, Chairman

GENERAL DEPRESSANT DRUG DEPENDENCY : A BIOPHYSICAL HYPOTHESIS

Martyn W. Hill and A.D. Bangham

Biophysics Unit, A.R.C. Institute of Animal Physiology

Babraham, Cambridge (U.K.)

INTRODUCTION

We wish to propose a general hypothesis concerning the underlying physical effects and physiological changes that might lead to general-depressant drug dependency and to the manifestations of a general-depressant, withdrawal syndrome. To us it seems more important at this stage, and more in line with clinical observation, to accent the similarities between the wide range of C.N.S. depressants that are drugs of dependence, than to point to their differences. The hypothesis can be summarized as follows: Dependency occurs when the lipid composition of some critical membrane or membranes are modified so as to return the physical state (fluidity) of that membrane to its pre-affected value. The word 'general' is used here in the same sense as general depressant and general anaesthetic, it being now widely accepted that general anaesthesia is a non-specific effect induced by a wide range of chemically dissimilar compounds including even the noble gases, furthermore increased pressure antagonises anaesthesia (Miller, 1972) and increased temperature acts synergistically (Hill and Bangham, to be published). The recent proposed Gibbs free energy hypothesis of general anaesthesia (Hill, 1974) attempts to explain these effects at a fundamental thermodynamic level; thus the word 'general' implies a fundamental thermodynamic effect brought about by a wide range of different chemical compounds, which is neither fundamentally chemical nor biochemical, although the induced changes will result in biochemical modifications.

The proposed hypothesis depends on two premises, the first is that the effects are general in the above sense, and the second

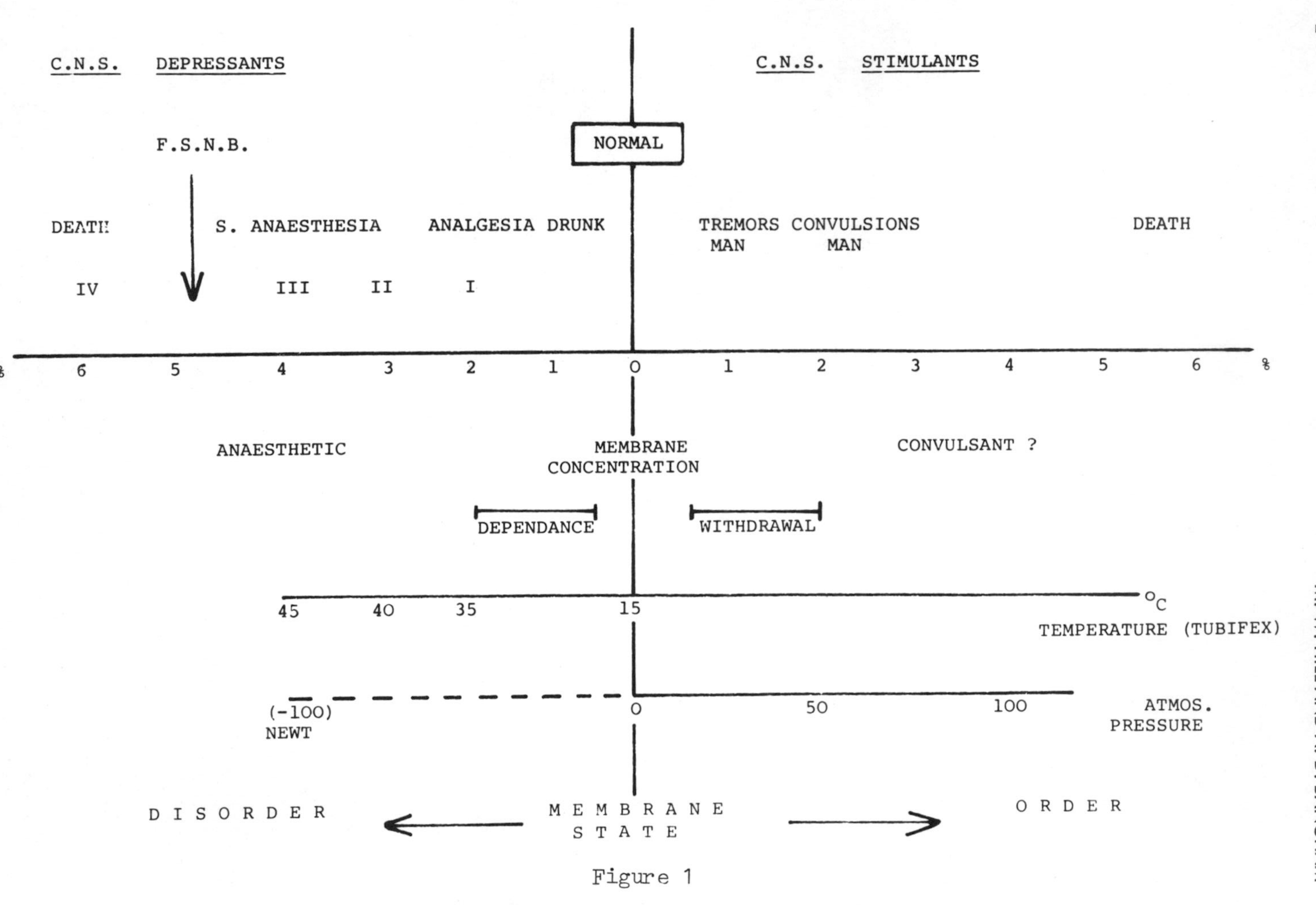
C.N.S. DEPRESSANTS
C.N.S. STIMULANTS
F.S.N.B.
NORMAL
DEATH
S. ANAESTHESIA
ANALGESIA DRUNK
TREMORS MAN
CONVULSIONS MAN
DEATH
IV
III
II
I
%
6
5
4
3
2
1
0
1
2
3
4
5
6
%
ANAESTHETIC
MEMBRANE CONCENTRATION
CONVULSANT ?
DEPENDANCE
WITHDRAWAL
45
40
35
15
°C
TEMPERATURE (TUBIFEX)
(-100) NEWT
0
50
100
ATMOS. PRESSURE
DISORDER
MEMBRANE STATE
ORDER

Figure 1

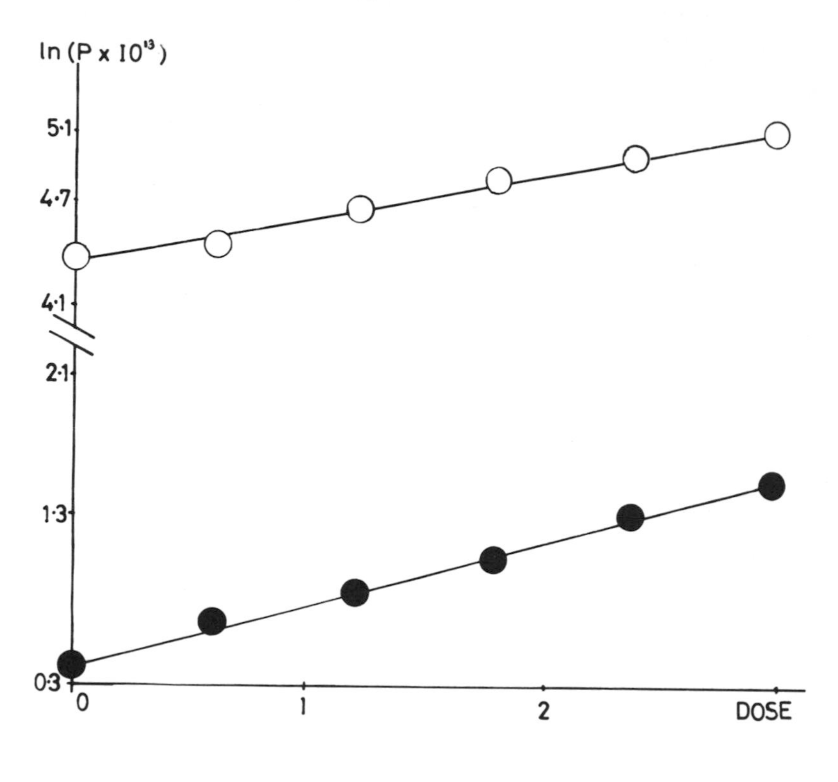

Figure 2

The effect of increasing concentrations of diethyl ether (1 - 25 mM) on the alkali metal cation permeability in liposomes at 37° C in the presence of valinomycin 1 μmole/mole phospholipids, O-O K permeability; ●-● Na permeability (from Johnson et al. 1973).

that organisms are capable of adapting and indeed do adapt to changes in environment, e.g. compositional changes of cell membranes related to body temperature or prevailing pressure.

Evidence for the first premise can be gained from clinical observations and from a model system, the liposome, which will be discussed as an illustration of the meaning of the term, 'general'. The second premise depends in part on the first, and is the fact that adding foreign molecules is equivalent to an increase in temperature and the best examples of animal adaptation are drawn from those adapted to changes in temperature.

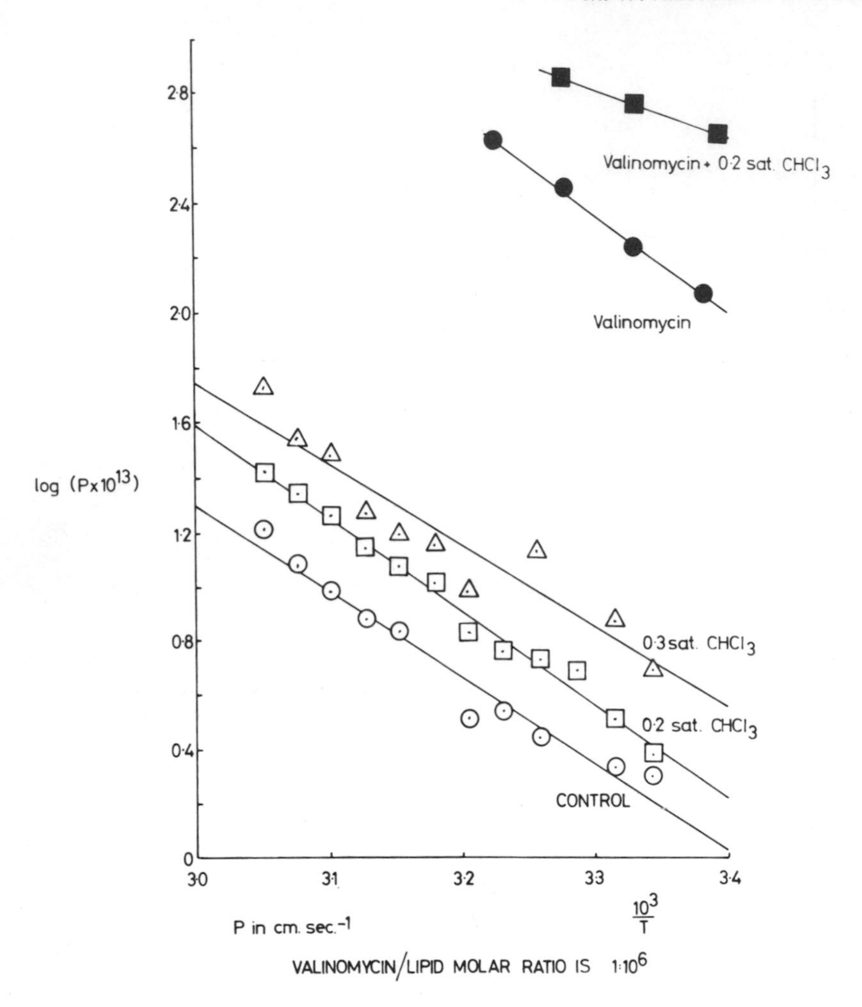

Figure 3

The effect of temperature and anaesthetic (chloroform) on the permeability of liposomes to potassium ions (From Johnson and Bangham, 1969).

GENERAL EFFECTS ON MEMBRANES

Figure 1 illustrates the main points of the relationships that occur between temperature, pressure and added depressants or anaesthetics to the left and added stimulants to right; the central vertical line denoting a normal environment. From normal to the right we notice that the membrane increases in

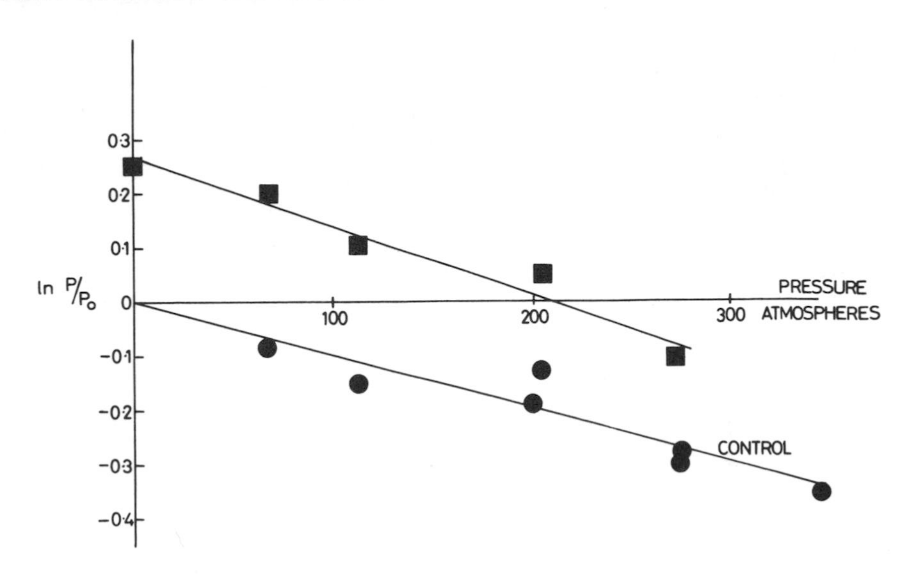

Figure 4

The effect of increasing the pressure of the non-anaesthetizing gas, helium, on the permeability of liposomes to potassium at 37°C ●-● without ether; ■-■ with 27.3 mM ether (From Johnson et al. 1973).

order and from normal to the left it decreases (bottom of diagram); we can proceed to the left by adding stimulant, withdrawing drugs or increasing the pressure, in all cases tremors occur, followed by convulsions and sometimes by death. In the opposite direction, we can add anaesthetics and go through the stages of intoxication, analgesia, anaesthesia and ultimately death. "F.S.N.B." refers to the membrane concentration that produced 50% inhibition of the nerve transmition in the frog sciatic nerve. The number on the top horizontal line refers to approximate values of the membrane concentration of a drug in moles per cent. The second line indicates the effects of temperature on a mud worm, the tubifex (Hill and Bangham, to be published) and the bottom line delineates the effects of pressure, 100 Atmos reversing anaesthesia in the newt (Miller, 1972) and is thus equivalent to 4 moles per cent membrane concentration of stimulant. We propose that tolerance is induced at the sort of concentrations indicated and the membranes adjust their order so that in the presence of the compound the order is back to its pre-affected value. On withdrawal of a

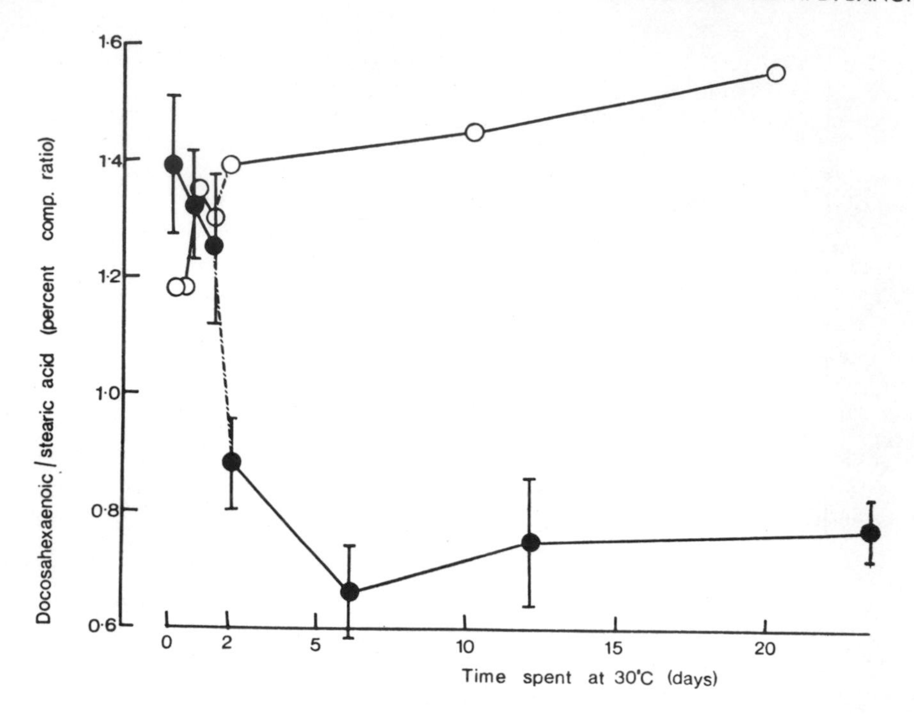

Figure 5

Time course of temperature induced changes in the fatty acid composition of goldfish mucosal membranes, and amino acid transport by the goldfish intestine. ●-● ratio of docosahexanoic to stearic acid in total lipids. O-O ratio of phenylalanine to valine transport at various times after fish adapted to 16°C were placed in water at 30°C (From Smith and Kemp, 1971).

depressant, initially the membrane will be more ordered than normal and hence be to the right of the normal line; this results in the organism reacting as if it were under hydraulic pressure or being affected by CNS stimulants. If the membrane state of a dependent organism is more ordered upon withdrawal it will obviously take more anaesthetic to produce general anaesthesia as is indeed the case with alcoholics (Jaffe, 1970).

THE LIPOSOME MODEL

The liposome model and its methodology has been fully described elsewhere (Bangham et al., 1974). It consists of a suspension of pure lipid bilayers in an aqueous environment, the bilayers form closed vesicles which entrap solutes. By using isotopically labelled solutes, membrane permeability, (P), can be measured. Figure 2 (Johnson et al., 1971) shows that the log of permeability of liposomes to potassium ions increases linearly with an anaesthetic concentration, in this case, ether. Figure 3 (Johnson and Bangham, 1969) shows the effect of temperature and anaesthetic on the permeability of liposomes to potassium and here the log of the permeability P is proportional to 1/T, an increase in temperature producing the same effect as an increase in concentration of anaesthetic. Figure 4 (Johnson et al., 1973) shows the antagonistic effects of pressure and anaesthesia. These effects may all be explained in terms of an activation energy ΔG^* (Glasstone et al., 1941).

$$\ln P = -\frac{\Delta G^*}{RT} = -\frac{(\Delta E^* - P\Delta V^*)}{RT} + \frac{\Delta S^*}{R} + nC_m + K$$

where ΔE^* ΔV^* and ΔS^* are the internal energy, volume, and entropy of the activated state for permeation; P, T, the pressure and temperature, C_m the membrane concentration of anaesthetic, n allows for the fact that C_m is in terms of moles of membrane and ΔG^* in moles of a permeant ion, K is constant. Full details of the analysis are given in (Hill, 1974). Thus, if a function is dependent on the state of the membrane it will be modified by alterations in anaesthetic concentration, temperature or pressure.

ADAPTATION OF MEMBRANES

Low cholesterol-containing, natural, membranes usually exhibit a phase transition due to the melting-freezing of the lipid hydrocarbon chains. In *Mycoplasma laidlawii*, for example, studied by Melchior et al. (1970) the natural membranes and the lipid extract from them exhibited a transition, and the temperature of the transition was associated with the prevailing growth temperature of the cells. The cells adapt their lipids so as to be in the same liquid-crystalline state if and when the growth temperature has changed. This fact argues strongly that the cells in some way sense the state of their membranes and adjust their lipid composition so that some membrane dependent function is kept to a value independent of the environment. We extend this mechanism to make the prediction that any membrane disordering mechanism, like addition of general depressants, may have the same effect.

A second example refers to a temperature adaptation in the goldfish (Smith and Kemp, 1971). The passage of some amino-acids across the intestine is affected by the temperature at which the goldfish are kept (see Figure 5). The permeability changes that occur on abrupt change of temperature follows the same time course as a change in the composition of the membrane lipid hydrocarbon chains, such that the hotter the temperature of the environment, the less docosohexanoic acid and the greater the stearic acid esterified to the lipids (also Figure 5). Klein et al. (1971) showed in the liposome model system that the permeability of the amino-acids was indeed influenced by the hydrocarbon chain of the lipid. The higher the degree of unsaturation, the higher the permeability. The more hydrophilic and smaller amino-acids being affected the most. Again we can draw the conclusion that the state of the membrane is modified so as to accommodate to the change in environment.

CONCLUSION

Evidence is presented to suggest that dependency on general depressants is a membrane phenomenon, and is a consequence of a natural process of adaptation by the membrane to a new environment, i.e. presence of a drug. The hypothesis attempts to draw together the effects of temperature, pressure, and the presence of foreign molecules into one picture, and to accent their similarities. Abrupt and total withdrawal of a drug from a subject that has become adapted will result in an immediate change in the 'fluidity' of the membrane in a direction opposite to that of characteristics of depression or anaesthesia. Differences between drugs are bound to exist as sensitivities of various membranes to the three environmental parameters will vary, but we consider these to be second order effects and not of theoretical importance to the understanding of the basis of the phenomena under discussion.

REFERENCES

Miller, K.W. Inert gas narcosis and animals under high pressure. In (Eds.) M.A. Sleigh and A.G. MacDonald) The Effects of Pressure on Organisms, Cambridge University Press, p. 363-378, 1972.

Hill, M.W. The Gibbs free energy hypothesis of General Anaesthesia. In (Eds.) M.J. Halsey, J.A. Sutton and R.A. Miller. Molecular Mechanisms of General Anaesthesia. Churchill, Livingstone, p. 132-144, 1974 (in press)

Jaffe, J.H. Drug addiction and abuse. In (Eds.) L.S. Goodman and A. Gilman) The Pharmacological Basis of Therapeutics 4th Edition, Macmillan, New York, 1970.

Bangham, A.D., Hill, M.W. and Miller, N.G.A. Preparation and use of liposomes as models of biological membrane. In (Ed.) E.D. Korn. Methods in Membrane Biology Vol. 1. p. 1-68. Plenum press, New York. 1974.

Johnson, S.M. and Bangham, A.D. The action of anaesthetics on phospholipid membranes. Biochim. Biophys. Acta. Vol. 193, 92-104, 1971.

Johnson, S.M., Miller, K.W. and Bangham, A.D. The opposing effect of pressure and general anaesthetics on the cation permeability of liposomes of varying lipid composition. Biochim. Biophys. Acta. Vol. 307, 42-57, 1973.

Glasstone, S., Laidler, K.J. and Eyring, H. The Theory of Rate Processes. McGraw-Hill, New York. 1941.

Melchior, D.L., Morowitz, H.J., Sturtevant, J.M. and Tsong, T.Y. Characterization of the plasma membrane of Mycoplasma laidlawii. Biochim. Biophys. Acta. Vol. 219, 114-122, 1970.

Smith, M.W. and Kemp, P. Parallel temperature-induced changes in membrane fatty acids and in the transport of amino acids by the intestine of goldfish (Carassius Auratus L.) Comp. Biochem. Physiol. Vol. 39B, 357-365, 1971.

Klein, R.A., Moore, M.J. and Smith, M.W. Selective diffusion of neutral amino acids across lipid bilayers. Biochim. Biophys. Acta. Vol. 233, 420-433, 1971.

THE BINDING OF ALCOHOL TO BRAIN MEMBRANES

Robert G. Grenell,
Professor of Neurobiology
Institute of Psychiatry & Human Behavior
University of Maryland, School of Medicine
Baltimore, Maryland

INTRODUCTION

It has been our thesis that the primary action of alcohol in the central nervous system is on cell membranes. The conceptual and experimental background for such a contention has been reviewed to a large extent by Grenell (1957, 1959, 1971), Kalant (1971), Seeman (1972) and, in certain related respects by Mullins (1954).

The consequences of the presence of alcohol to the membrane state and function are considered as basic to the more macroscopic considerations of disturbances in organ function and behavior. Indirect primary effects are further distinguished from direct effects. The former include the consequences to ion conductance and flux, excitability, etc., when alcohol causes either reversible or irreversible alteration of the membrane's water, protein, and lipid states. Direct effects refer to the affinity of an alcohol molecule for the membrane complex. They include whatever binding might occur, alteration to the shell of "structured water" adjacent to the membrane or trapped in its core, and whatever structural re-configurations might exist.

There is considerable evidence, both in our own studies and in the literature, to show that ethanol behaves physiologically as an additive molecule in the nervous system. Some of this evidence stems from gross behavioral study of the alcoholic.

More to the point are the consequences of local perfusion of ethanol in the brain cortex of animals. Our own studies show that, when varying concentrations of ethanol are so perfused, action potentials recorded from the perfused area undergo precisely the same concentration-dependent excitatory-inhibitory cycle which is shown for nembutal, barbital and other well-known narcotics. Single-unit observations made in this laboratory would tend to confirm that these phenomena exist also at the unit level.

At the molecular level, we had noticed previously that ethanol binding to suspended brain membrane preparations appeared to resembly closely the binding activity of sodium barbital to the same structures. An indirect route to learn more about a probably molecular behavior of ethanol therefore seemed open to us.

For certain considerations of effects of alcohols -- particularly on membranes -- it is helpful to look at alcohol from the point of view of its action as an anesthetic molecule. As one of this class of substances, it has been suggested that alcohol can induce ice, clathrates, or mixed-clathrate formation of the membrane-associated water, in cooperation with protein side-chains. There has been virtually no direct experiemental evidence to support such a suggestion. What evidence there is, suggests that such anesthetics 'melt' or fluidize the membrane-associated water. (Now, our evidence shows that there is an increase in 'free water.') Seeman et al. found that neutral or charged anesthetics increased the hydraulic flow of water molecules through the erythrocyte membrane. Schoenborn et al calculated that xenon increased by 15% the irrotationally bound water associated with hemoglobin. It has been found that procaine increases the proton magnetic resonance (PMR) line-width of the water protons. Although this broadening is compatible with an increase in water viscosity, it is impossible, from these results alone, to decide whether this has anything to do with the state of membrane-associated water since only a small fraction of the cell's organized water can be associated with the membrane (further studies of procaine effects on isolated membrane systems will have to be done). It is clear that methods such as those used in the present and proposed investigation, are, so far, the only ones that can give some of the direct evidence necessary to obtain.

METHODS

The basic, technical, details of the microwave absorption procedure are complex and lengthy. They can be found in the NASA Progress Report, "Molecular Binding in the Cell Surface," by McCulloch and Grenell (Report on Grant #NGR 21-002-040), as well as summarized in the Proceedings of the Seventh International Conference on Medical and Biological Engineering, Stockholm, Sweden, 1967. Only a brief outline will be given here.

An Ultrastable Microwave Oscillator, LFE # 814A-X-21, is used as the source. Its output is fed to the cavity in waveguide through a two-directional coupler. The reflected signal is terminated at the input of a Tektronix Spectrum Analyser, # 1L-30, on a parent oscilloscope frame. Attenuation is used in the guide for low SWR, and a coupling iris at the flange to the cavity reduces perturbations due to cavity tuning. Temperature and humidity control is effected by surrounding the cavity and most of the waveguide with a controlled-environment chamber which can be virtually sealed from the room ambient. The chamber contains a heating system normally moderated by an electronic phase-control circuit with a thermistor as the sensing element but which may be adjusted independently to a fixed output for short-term measurements. Sample solutions are circulated into the chamber, through the sample tube in the cavity, and to an external collecting reservoir by connecting flexible tubing to both ends of the receptacle at the slotted section. Solutions are kept in a temperature holding bath prior to circulation, then added in similar volumes to a graduated vessel attached to the outer chamber wall. Releasing a clamped section of the PVC tubing permits gravity circulation of the solution until a small amount, 2 ml, remains in the upper vessel. Circulation is then checked by re-clamping the tubing. Measurements of Q are then possible as a function of the time that materials remain standing in the sample tube.

The spectrum analyser offers a wide choice of dispersion settings proportional to its sweep. This in effect means that cavity Q's with half-power bandwidths of the order of 2 MHz may be displayed in their entirety or that smaller portions of the trace may be amplified to resolve small differences in the frequency component of the trace for nearly similar solutions. In the Tektronix # 1L-30 plug-in, the provision for stepped IF attenuation facilitates a similar stepwise power amplification over the vertical output of the full-scale trace, since the steps are fixed, with an accuracy of 0.1 db per step. The sweep may be phase-locked to an internal reference crystal, from which calibration markers, 1 MHz $\pm$ 100 Hz, are derived. By referencing the dispersion settings to these markers, stepping errors in the "Dispersion" control may be readily discerned and adjusted to a magnitude which is negligible for intermediate and broad band settings of the control. The high order of accuracy in the control of both vertical and horizontal circuits means that portions of the full-scale display of the trace may be referenced to the absorption parameters as a whole with considerable accuracy.

The water molecule is characterized by a high electric dipole moment, associated in the liquid phase with a single-valued relaxation time, C., and with frequencies in the microwave region. Its dispersion is anomalous, with the real component, k, of its complex dielectric constant, k*, falling from a maximum static value, k'_s, of nearly 85 (relative to the permitting of free space) to a high frequency value, k', of approximately 5 at optical frequencies and beyond. Its loss increases to a maximum, $k''_{max}=k'_s$ (1) in the microwave region (10 - 20 GHz), becoming null at very low and very high frequencies. Values of k* are temperature dependent, and are affected by purity and pH. This dispersion characteristic is well-known and has been fully discussed in the literature. The summary of von Hippel (1954) includes a normalized plot of the frequency dependence of k', k'' and tan Δ vs. Ln (f) as well as an arc plot of k* in the complex plane, based on the treatment of Cole and Cole (1941).

The Debye (1929) relaxation expressions form the basis for the normalized frequency plot of von Hippel. They have been shown to have inadequacies, -- in regard to relaxation for a), non-associated polar liquids (Onsager, 1936) and for b), liquids characterized by some degree of short range order (Kirkwood, 1939). But they fit the observed relaxation characteristics of water rather well, and have been used to extrapolate k_s and k_∞ from data in the microwave region (Smyth, 1955).

We have utilized an expanded normalized Debye plot to determine a standardized value for k' and k'' at frequencies between 9.0 and 9.5 GHz from measurements at 10 GHz summarized in von Hippel. Within this range, any inaccuracies in the Debye plots are small. The standardized k* values have in turn formed a measurements standard for the computation of the relationship between k' and resonant frequency changes in the cavity, and between k'' and Q changes, when water is located in a thin-walled sample tube at a cavity node.

Bound water is characterized by relaxation at frequencies between 0.5 and 1 GHz, an order of magnitude below relaxation for free water (Pennock and Schwan, 1969). Proteins, lipids, and other macromolecules normally found in biological systems characteristically undergo relaxation at considerably lower frequencies, several orders of magnitude below free water; and this relaxation is of the Maxwell Wagner form, as discussed in Schwan (1957). When suspended in water, the macromolecules and their structural aggregations may be viewed as cavity-like spaces distributed within a polar medium of given total volume (Haggis, 1951). Any water which is "irrotationally bound" to such structures can contribute only atomic and electronic polarization terms to the microwave measurement, since molecular orientation is precluded. Similar arguments apply to macromolecules.

An intermediate effect would take place when the rotational freedom of the water molecule is hindered by binding at one or more sites, as in the hydration shell around suspended macromolecules or possibly the "ice-like" structured water (Pauling, 1961) adjacent to the excitable membranes for nerve and muscle. This binding would produce a lesser but nonetheless measurable change in the microwave dielectric constant and loss for a given preparation volume within the cavity. Indeed, any change in the structures themselves due to the microwave radiation would be evidenced by a bound-water change. Alteration in the hydration layers, and core-associated water would occur for even slight structural stretchings, rearrangements, or fragmentations, in the most complicated biological suspensions as well as the simplest ones.

We have observed distinct water movements when membranes and membrane fragments were exposed to microwave irradiation at 20 mW/cm^2 for periods of less than one minute. Improving the sensitivity of our instrumentation would not be difficult. We would then expect to detect water movements under short-term and long-term exposure to weak fields, if tissues are even slightly affected. Thus the cavity measurement technique, due to its unusual sensitivity, may well be unique in its potential for assessing very subtle structural weak-field effects on cellular and sub-cellular suspensions in water and saline.

As recognized by Haggis (1951), Buchanan (1962), Vogelhut (1962), and others, the unusually high dielectric constant and microwave loss peak for water affords us an opportunity to study the binding affinity of other solvent molecules to cellular and sub-cellular biological systems. The microwave dielectric effects of the solvents and of the biological suspensions can be independently determined. Then the effect of their combination is assessed, adjusting for the cavity sample volume. Any discrepancy from the superposition of the independent effects must then relate to either hierarchic displacement of water by a binding solvent ion or to an increase in the hydration for the system.

Changes in the binding activity between the solvent molecule and the suspended biological system may occur as a consequence of weak-field microwave exposure. These changes may relate to cell metabolism, to the effects of metabolic poisons on cellular systems, to cell respiration, or to the binding affinity of stimulants or narcotics for the cell membrane. Using the cavity method, we have studied the binding of ethanol and sodium barbitol to whole membranes and refined preparations of rat brain synaptic membranes. There is evidence for unusual susceptibility to microwave irradiation in the latter, not only in terms of fragmentation, as shown elsewhere, but in terms of the membrane binding propensity to the barbital anion. Obviously, a study of binding in terms of related water movements affords us a rich opportunity for the assessment of subtle microwave irradiation effects on delicate cell membranes.

Three different types of preparation were centrifuged out from the rat brain cortex: a) MF, material consisting chiefly of membrane fragments at a suspension of 2% packed fragments in water; b) MSP, material additionally refined to a larger percentage of membrane synaptic processes in a suspension of 2% packed fragments in water; c) MSP/S, material refined as above to include a larger percentage of synaptic processes but suspended in 0.1M sucrose, 0.4% packed fragments. These stock suspensions were measured alone and in the presence of varying amounts of saline, 0.856 M Na Cl and 0.856 M Na $C_8H_{11}N_2O_3$, and 1.4% and 2% ethanol as were the erythrocyte ghosts, to determine k* changes not due to the superposed correction factors.

The only other method of sufficient importance to present here, concerns the procedure for obtaining synaptic membranes. Rat brains were removed, whole cerebral cortices dissected out, rinsed in ice cold distilled water and placed in cold 0.32 M sucrose (pH 6.2) until homogenization. From this stage on, all operations were performed at 4° C. The tissue was homogenized using an all-glass hand homogenizer (Arthur H. Thomas Co., Phila., Pa.). The final homogenate was prepared as a 10% suspension in sucrose. The homogenate was centrifuged at 1000 g for 20 minutes in an International PR-1 centrifuge and the pellet was washed with one-half of the original sucrose volume, and then spun down at the same rate. Combined supernatant and washings were centrifuged at 11,000 g for 20 minutes in the number 40 angle head rotor (Spinco Division, Beckman Instruments Co., Palo Alto, Ca.) placed in an Arden Ultracentrifuge (Arden Instrument Co., Rockville, Md.) and run. The supernatant was discarded and the pellet washed with 0.32 M sucrose (10 ml/gm original weight of tissue). The suspension was recentrifuged at 11,000 g for 20 minutes. The supernatant was discarded and the pellet osmotically shocked using distilled water (10 ml/gm original weight of tissue). This aqueous suspension was spun at 15,000 g for 30 minutes. The resulting pellet contained myelin, synaptic plasma membranes, mitochondria and the aqueous media containing synaptic vesicles along with small particles of broken membranes.

The osmotically shocked pellet was diluted in distilled water and 1.5 ml suspension equivalent to one-half rat brain cortex was charged on top of a SW 40 Titanium rotor tube containing a discontinuous sucrose density gradient of 3 ml each of 0.8/0.9/1.0/1.2 M sucrose. The rotor was centrifuged for a period of 135 minutes. After centrifugation was completed the bands were eluted using the Autodensiflow coupled to a peristaltic pump (Buchler Instrument Company, Springfield, N.J.). After elution, each sample was diluted slowly with 0.32 M sucrose to a total volume of 5 ml and pelleted down at 20,000 g for 20 minutes.

The pellets were suspended in 5 ml of a 0.6% potassium permanganate barbital-acetate buffer (pH 7.4-Luft's reagent). The samples were allowed to stand for 20 minutes for complete fixation prior to centrifugation of the stained particulate suspension at 25,000 g for a period of 30 minutes. The fixed pellets were then dehydrated in upgrading concentrations of ehtanol for final dehydration. Propylene oxide was then used for the final removal of alcohol. Small samples of pellets were embedded in Araldite 502. Thin sections were cut on a Porter-Blum MT-2 Ultramicrotome and mounted directly on 400 mesh copper grids. The grids were stained with lead acetate. The samples were then examined in a Siemens IA Electron Microscope at 60 kV using a 20 U objective aperture.

TABLE I

Proportional dielectric constant correction due to: a) dilution of 2% MF in water, b) dilution of 0.0856 NaCl and $NaC_8H_{11}N_2O_3$ in water.
$T_c = 38^o$, $f_o = 9.10$ GHz, $\pm$ 0.05 GHz

Mixture (ratio)	MF corrections k''_M	k'_M	NaCl corrections k''_s	k'_s	$NaC_8H_{11}N_2O_2$ corrections k''_s	k'_s
15:10	-0.284	0.657*	0.25	-1.32	0.22	-1.25
20:5			0.125	-0.66	0.11	-0.625
saline alone			0.025	-0.13	0.022	-0.125

TABLE II

Net $\Delta \bar{k}^*_c$ for three preparations of rat brain membrane material, mixed with 0.0856 M Na Cl and with 0.0856 M Na $C_8H_{11}N_2O_3$.

Preparation	Mixture	$\Delta \bar{k}''_c$	$\Delta \bar{k}'_c$
(MSP/S)	20:5 B	-0.10	0.87
MF	15:10 B	-0.087	0.586
"	15:10 Cl	-0.057	0.336
(MSP/W)	7.5:5 B	-0.045	0.515
"	7.5:5 Cl	-0.015	0.585

RESULTS

A correction factor (k^*_c) was used which takes into account the temperature and saline components, the separate effect of the membranes and fragments, and where applicable, effect of sucrose on k*. Proportional corrections which result from the mixture ratios are computed, and all values superposed to compute the magnitude of k^*_c.

A non-zero result in the difference, $\Delta k^*_c = k^* - k^*_c$, becomes a measure of the change in water associated with the suspended structure in an irrotational way, insofar as the approximation to an adequate correctional value holds for the superposition of the separate results. For the 0.0856 M salines and their mixtures with the membranes, the results indicate an increase in the number of free water dipoles for a given volume. Net k* shifts for the three categories of rat brain membrane material are summarized in Table I, which illustrates the higher order discrepancy for preparations mixed with the barbital salt in all but one case. Table II demonstrates evident increase in free water molecules when suspensions of brain membrane fragments (and of erythrocyte ghosts) are mixed with saline including an equal quantity of dissociated barbital and chloride anions.

TABLE III

Net Binding by Suspended Membranes

+ = Shift in direction of increased free water
Δ_1 = (Preparation) (Solution alone, less membranes), m kHz
Δ_2 = (") (" " " " , m *cm

Line	Preparation	M	Δ_1, kHz	Δ_2, *cm,	$\Delta k'$	$\Delta k''$
a)	Erythrocyte ghosts					
1	G/ETOH/W	0.85	21	0.01	0.33	0.06
2	G/ETOH/S	"	70	0.01	1.11	0.06
3	"	0.34	63	-0.01	0.99	-0.06
4	"	0.24	34	-0.01	0.54	-0.06
5	G/MTOH/S	1.24	24	0	0.38	0
1	G/B	0.086	8	--	0.13	--
2	G/S/B	0.084/0.002	18	-0.03	0.29	-0.18
3	"	0.077/0.009	5	0.04	0.08	0.24
4	"	0.069/0.017	9	-0.09	0.14	-0.51
5	"	0.055/0.031	1	-0.09	0.02	-0.51
6	"	0.043/0.043	6	0.08	0.10	0.45
b)	Refined cortical membranes					
1	RCM/ETOH/W	0.034	6	--	0.10	--
2	RCM/B	0.077	26	-0.09	0.41	-0.51
3	"	0.061	10	0.12	0.16	0.69
4	"	0.034	17	0.03	0.27	0.24
5	"	0.02	11	0.01	0.17	0.06
6	"	0.01	17	-0.01	0.27	-0.06
7	"	0.003	16	0.02	0.25	0.12
c)	Refined synaptic membranes					
1	RSM/B	0.086	26	0.12	0.41	0.69
2	"	0.061	-26	--	-0.41	--
3	"	0.034	32	-0.01	0.51	-0.06
4	"	0.02	--	0.08	--	0.45
5	"	0.01	--	0.01	--	0.06
6	"	0.003	4	--	0.06	--

Table III shows the computed shift in the dielectric constant (Δk'), with brain membranes and red cell ghosts plus either barbital or ethanol. In all but one case, results are consistent with an increase in free water, indicating displacement of H^+ and OH^- by the binding ions dissociated in solution.

In effect, calculation shows that with membranes in 2% ethanol, approximately 20% of the ethanol is bound in such a way as to release irrotationally bound water. This represents a most significant change in dielectric constant.

DISCUSSION

The approach presented demonstrates that alcohol (and apparently some other molecules with certain similar effects) appears to bind to membranes. The type of binding that occurs is such that it is associated with a change in the state of membrane water. This brief discussion precludes a lengthy consideration relative to the importance and problems involved in dealing with the state, structure and movements of water in cells (see the review of Cooke and Kuntz, 1974). The major point here is that such water changes allow for the inference that alcohol alters the molecular structure of the membrane in a particular way.

Such an inference leads to the obvious suggestion that membrane changes of this type will, of necessity, have to be associated with changes in permeability, in ion concentrations and fluxes, in reaction rates, etc., ultimately leading to significantly altered functional activity involving neurotransmitters and information processing in the brain.

It has been apparent that neither decreased absorption nor increased oxidation of alcohol seem to be the answer to the production of tolerance to it. The altered response of the individual can be explained thus far only by the suggestion of physicochemical changes in neurons, associated with their excitability cycle. Action currents develop subsequent to changes in structural orientation at the neuron surface. The electrochemical processes involved are self-limiting and cyclic, being completed by cellular processes that restore the original excitable state. The molecular arrangements in the resting cell that are capable of such a transition from one state to another have been referred to as a metastable structure in contrast to the excitable structure. The metastable structure responds to certain stimuli and controls ionic fluxes which give rise to potential changes. The stability of the membrane would depend on the molecular organization of this structure, and consequently, one would propose that the initial, fundamental effects of alcohols would be at this locus.

Perhaps the most provocative result of the observations presented is the fact that they point to so much that remains to be done.

REFERENCES

Cooke, R. and Kuntz, I.D. 1974 Ann. Rev. Biophysics and Bioeng.

Grenell, R.G. 1957 Paper in 'Alcoholism', (H.Himwich, ed.), AAAS Publ. No. 47.

Grenell, R.G. 1959 Quart. J. Stud. Alc., 20:421.

Grenell, R.G. 1971 Chapter 1, Vol. II of 'The Biology of Alcohol', ed. by Kissin and Begleiter, Plenum Press, N.Y.

Kalant, H. 1971 Chapter in Vol. II of 'The Biology of Alcohol' ed, by Kissin and Begleiter, Plenum Press, N.Y.

Mullins, L.J. 1954 Chem. Rev., 54:289.

Footnote: The work reported in this paper was partially supported by a grant to the University of Maryland from the National Institute for The Control of Alcoholism and Alcohol Abuse.

ETHANOL-INDUCED CHANGES IN CATION-STIMULATED ADENOSINE TRIPHOSPHATASE ACTIVITY AND LIPID-PROTEOLIPID LABELING OF BRAIN MICROSOMES

Henrik Wallgren, Pirkko Nikander, and Pekka Virtanen

Research Laboratories of the State Alcohol Monopoly (Alko), and Department of Physiological Zoology

University of Helsinki, Helsinki, Finland

I. INTRODUCTION

Our working hypothesis was presented at the previous ICAA symposium on alcohol intoxication and withdrawal in Amsterdam 1972 (Wallgren, 1973). We assume that prolongation of the depressant action of alcohol causes compensatory changes particularly in those neuronal membrane structures which are involved in control of the excitation cycle of the conducting membranes. Adaptive changes in the synaptic region are presumably also involved and actually form an important focussing point for current alcohol research.

Our work is in part based on a technique for producing severe signs of withdrawal excitability including spontaneous convulsions in the rat by forced feeding of alcohol. The technique was briefly presented at the Amsterdam symposium (Wallgren, 1973), and reported in more detail elsewhere (Wallgren, Kosunen and Ahtee, 1972, 1973). Severe withdrawal signs in the rat have also been reported by Falk, Samson and Winger (1972) who found audiogenic seizures even ending in death in rats induced to consume large amounts of alcohol by a feeding schedule causing polydipsia.

Since the work presented here is previously unpublished and reported in a preliminary form, comment should be made on the contributions made by different members of our group. The data presented in Section II of this paper add to the information on the relationship between dosage regimen and induced tolerance to ethanol.

The experiments were outlined by H. W. and mainly performed by Miss Aino Malila as a part of the work for her Master's thesis. The experiments on cation-stimulated ATPase activity, regarded as a marker enzyme for outer cell membranes (Section III), were planned and carried out by P. N. The experiments involving labeling of the microsomal fraction with ^{14}C-serine, presented in Sections IV and V, were jointly planned by H. W. and P. V. The laboratory work was mainly carried out or supervised by P. V.

II. TOLERANCE INDUCED BY SPACED DOSES VERSUS MAINTAINED ETHANOL INTOXICATION

Sprague-Dawley rats obtained from Oy Orion Ab, Helsinki, 4-5 months of age, were maintained on a liquid diet, the composition of which has been described by Wallgren, Kosunen and Ahtee (1973), with the modification that the casein hydrolysate was replaced by milk powder (fatt-free) and the carragenate by lecithin, keeping the protein content unchanged. Because of the limitations of the available animal material, males and females were used and distributed evenly between the groups. Behavioral tolerance was determined by means of the tilted plane test (Arvola, Sammalisto and Wallgren, 1958), injecting ethanol i.p. as a solution of 10 g ethanol filled to 100 ml with isotonic saline. The doses used are indicated below. Performance on the tilted plane was tested eight times with 20 min intervals after the injection, comparing the sliding angle with that obtained before alcohol administration. All groups were tested as described before beginning of the regimen involving prolonged alcohol administration, the ethanol dose being 2.5 g/kg. No differences in performance were found between the groups.

In an experiment of 20 days duration (Expt. 1), 18 rats were given alcohol by stomach tube twice daily (20 % w/v in water), starting with an initial intoxicating dose and then maintaining intoxication as even as possible for 19 days. Dosage was initially 6 g/kg/day and was soon increased to about 9 g/kg/day. Eighteen other rats received single ethanol doses of 5 g/kg every other day. Both these animals and 15 controls were given water by stomach tube according to the schedule used for the animals maintained intoxicated.

After elimination of the terminal alcohol dose given on day 19, the animals were tested for tolerance on day 20 giving 2.5 g ethanol/kg i.p. The result is shown in Fig. 1. Tolerance developed in both the experimental groups. The animals maintained in a state of continuous intoxication clearly had the highest degree of tolerance. These animals also had obvious although not very severe withdrawal symptoms, and were used in the experiments described in section V of this paper.

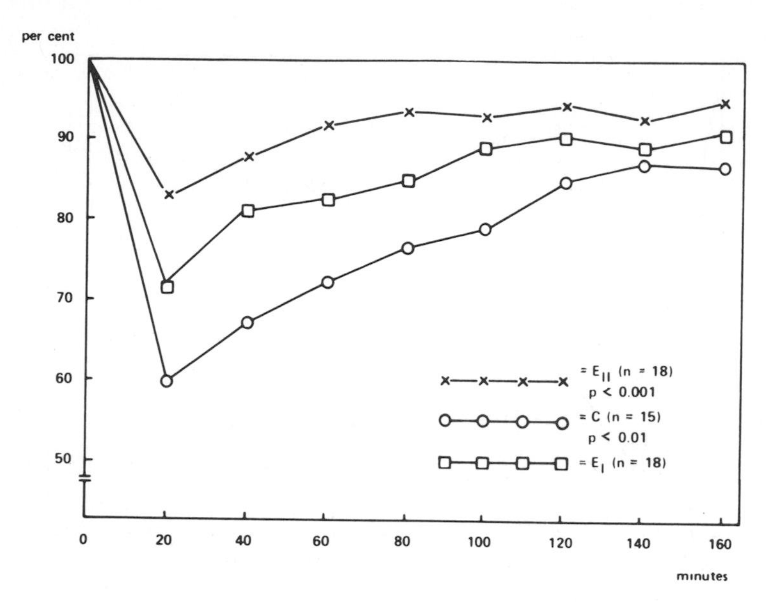

Fig. 1. Performance of rats in the tilted plane test. The animals were injected i.p. with ethanol (10 % w/v in saline) 2.5 g/kg body weight. Performance is expressed as a percentage of the sliding angle before injection. C = controls, E_I = ethanol given every other day ten times in a dose of 5 g/kg, E_{II} = ethanol given twice daily in amounts increasing from 6 to 9 g a day during 19 days. The prolonged alcohol administration was terminated on the morning of the day before testing.

In another experiment of 30 days duration (Expt. 2), 10 controls and 12 experimental rats were fed a liquid diet *ad lib.* for 30 days, the experimental animals receiving 35 % of their calories (5 % v/v in the diet) as ethanol. Their average daily ethanol intake was 9.3 g/kg. The rats were then tested on the tilted plane after injection of 2.7 g ethanol/kg. As shown in Fig. 2, no tolerance developed, neither could signs of withdrawal hyperexcitability be detected. This confirms the experience at the Alko laboratories that our rats even on regimens involving high daily intake of ethanol do not build up significant blood levels and consequently fail to develop increased tolerance as long as they regulate their consumption freely (Wallgren and Forsander, 1963; Wallgren *et al.*, 1967). The finding also provides a parallel to Goldstein's (1972) clear demonstration of the quantitative relationship between alcohol dosage and severity of the withdrawal signs in mice. Some author's (Cicero *et al.*, 1971; Lieber and De Carli, 1973) detection of increased seizure susceptibility in rats drinking alcohol solutions or liquid diets of a high ethanol content may be due to their use of young, growing animals.

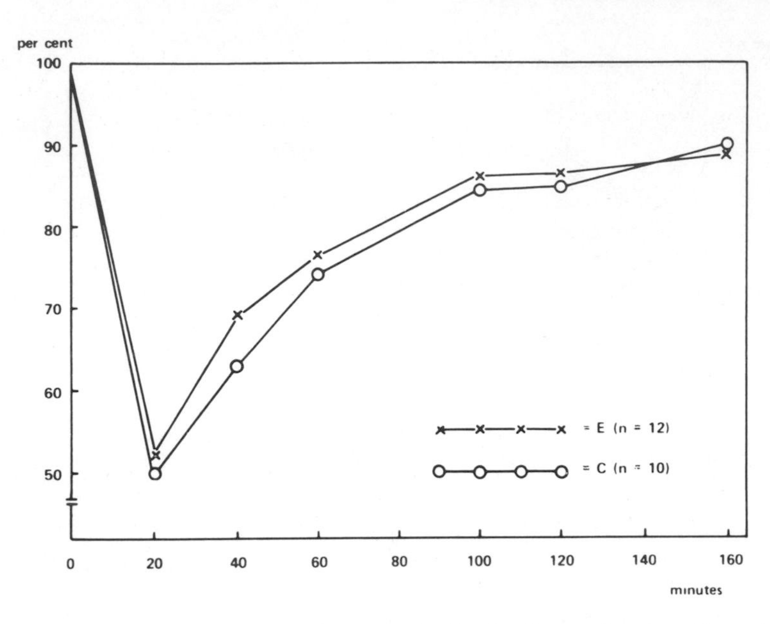

Fig. 2. Performance of rats in the tilted plane test after i.p. administration of 2.7 g ethanol/kg body weight. C = controls, E = animals maintained for 30 days on a liquid diet containing 35 % of the calories as ethanol. The average daily intake of the latter animals was 9.3 g ethanol/kg/day.

III. RELATION BETWEEN CATION-STIMULATED ATP-ASE ACTIVITY, TOLERANCE, AND WITHDRAWAL

Fortyone rats (Oy Orion Ab, Helsinki) were used to induce a withdrawal syndrome by the procedure described by Wallgren et al. (1973). After continuing the regimen for 21 days, administration of alcohol was terminated. During the period of withdrawal hyperexcitability, the animals were tested for the intensity of the withdrawal state by fixing an accelerometer transducer to the base of their tail, enclosing them in a box, provoking a reaction by touching them in the neck region with a rod, and registering the ensuing tail twitches by means of a polygraphic device. Frequences higher than 15 Hz were filtrated. The amplitude and duration of the recorded vibrations were classified by a person who had no knowledge about the treatment of the animals. Controls and experimental animals with minimal reactions were given 0 points. With increasing intensity of the reactions, scores of 1 - 3 were given.

Immediately after testing, the rats were decapitated, the entire brain homogenized, the microsomal fraction isolated by means of fractional centrifugation, the microsome fraction adjusted to

100 µg protein/ml, and the $Mg^{2+}(Na^+, K^+)$-ATPase activity determined using 500 µl enzyme preparation in a final volume of 675 µl according to Bowler and Duncan (1968) and Peacock, Bowler and Anstee (1972).

Fig. 3 summarizes the findings. With increasing degree of withdrawal excitability, there is a proportional increase in Na^+, K^+-ATPase activity and decrease in Mg^{2+}-ATPase activity. The changes in comparison with the microsomal ATPase activities of animals given 0 scores are highly significant. Consequently, the Na^+, K^+-ATPase activity expressed as a percentage of the total activity also increases significantly.

There are conflicting reports concerning changes in cation-stimulated ATPase activity in animals tolerant to ethanol. The discrepancies may in part be due to differences in the fractions used and the conditions of assay. Combined $Mg^{2+}(Na^+,K^+)$-ATPase activity has been reported to increase in whole brain homogenate from rats (Israel _et al._, 1970), homogenates of the cerebral cortex and hippocampus of tolerant cats (Knox _et al._, 1972), and the

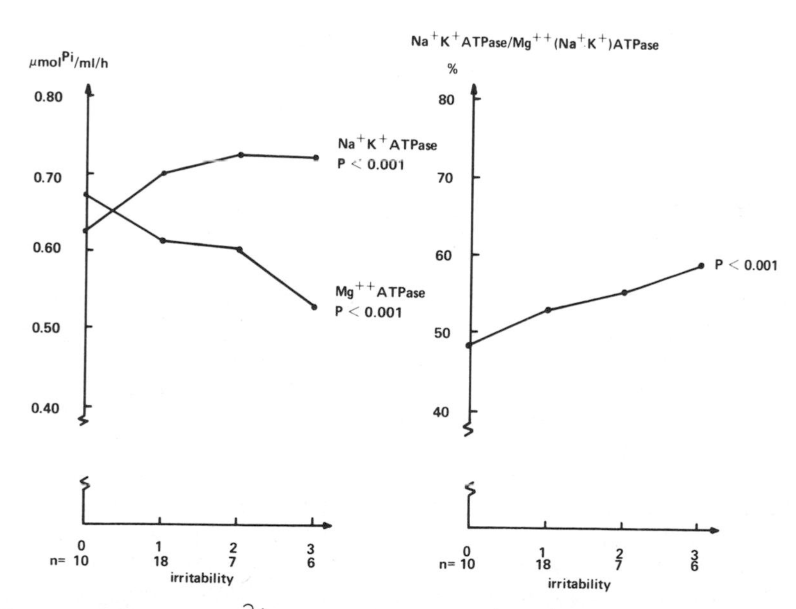

Fig. 3. Activity of Mg^{2+}-ATPase and (Na^+, K^+)-ATPase in cerebral microsomes. The data give activities for rats with different scores for withdrawal irritability, the number of animals being 10 (score 0), 18 (score 1), 7 (score 2), and 6 (score 3). Intoxication had been maintained for 21 days before withdrawal. The left part of the figure shows the enzyme activity, the right part per cent (Na^+, K^+)-ATPase activity of total activity.

microsomal and synaptosomal fractions from rat brain (Roach et al., 1973). No change has been reported in the microsomal fraction of mouse brain (Israel and Kuriyama, 1971), homogenated cerebral cortex from mice (Goldstein and Israel, 1972), rat brain microsomes (Akera et al., 1973), and homogenated reticular formation or amygdala from cats (Knoz et al., 1972). One of us (P.N.) has compared tolerant and non-tolerant rats of the Alko AA and ANA strains and found no change in the microsomal enzyme activity (Table 1).

No clear pattern in terms of regional or fractional differences is yet apparent in the published reports. Roach et al. (1973) have recently reported an increase in Na^+, K^+-ATPase with increasing excitability during withdrawal illness in rats, similar to our finding, but they did not find any changes in Mg^{2+}-ATPase. We are not able to give an interpreparation in functional terms of the observed change. It is, however, compatible with the hypothesis of structural changes in membranes systems as a possible basis of the withdrawal illness.

Control rats

	$Mg^{++}(Na^+K^+)$ATPase	Mg^{++}ATPase	(Na^+K^+)ATPase
AA	25.2	12.6	12.6
n = 10	±1.6	± 1.2	± 1.5
ANA	20.1	9.9	10.2
n = 8	± 2.7	± 1.4	± 2.7

Tolerant rats

	$Mg^{++}(Na^+K^+)$ATPase	Mg^{++}ATPase	(Na^+K^+)ATPase
AA	22.0	10.4	11.6
n = 10	±3.0	± 1.4	± 2.0
ANA	20.3	8.3	12.0
n = 8	± 2.3	± 1.9	± 1.6

Table 1. Microsomal cation-stimulated ATPase activity from brains of two strains of rats. Animals of both strains were made tolerant to ethanol by prolonged intoxication maintained for 19 days according to the technique of Wallgren et al. (1973). The amounts of alcohol given caused marked behavioral tolerance as measured by the tilted plane but was not sufficient to cause withdrawal illness. AA = the high-drinking Alko-Alcohol strain, ANA = the low-drinking Alko-Non-Alcohol strain. The assays were made at different occasions with the different strains and therefore cannot be regarded as indicating a strain difference. The results are expressed as μmoles P_i released per mg protein per hour, averages ± s.d.

IV. MICROSOMAL LIPID AND PROTEOLIPID LABELING *IN VIVO* AFTER ACUTE ETHANOL ADMINISTRATION

Albino female rats of the colony of the Department of Physiological Zoology, Helsinki University, weighing 200-250 g, were anesthetized with sodium pentobarbital, and an adapter for an injection cannula implanted stereotactically into the right lateral brain ventricle. Six days after the operation, the animals were used in experiments. The cannulated animals were symptom-free.

The experimental group received 6 g ethanol/kg b.w. as 15 % (w/v) solution in tap water by stomach tube. The control animals received the corresponding volume of water. In the intoxication group, 5 µc L-serine-^{14}C(U) (sp. a. 156 mc/mM, New England Nuclear Corporation) was injected into the lateral ventricle through the implanted cannula one hour after the administration of alcohol. In the "hangover" group, the labeled precursor was injected 13 hours after administration of ethanol. The contact time with the isotope was 12 hours.

After the 12 hour exposure, the animals were decapitated and the brains were split. The left sides without cerebellum were homogenized in a glass-teflon homogenizer in ten volumes of 0.25 M ice cold sucrose. The right halves of the brains were fixed in 10 % phosphate buffered formaldehyde for histological and autoradiographical examination.

The homogenates were centrifuged at 200 x g for 10 min. The supernatants were recentrifuged for 30 min at 12500 x g. The microsomal fractions were obtained from the recentrifuged supernatants by centrifugation for 60 min at 107000 x g. The microsomal pellets were homogenized in ten volumes of 0.25 M sucrose, and the radioactivities of the total microsomal fractions measured in a Packard Tri Carb 2420 liquid scintillation spectrometer from 10 µl samples dissolved in 10 ml of Insta-Gel (Packard) scintillation liquid. The radioactivities were measured by an external standard method with a counting efficiency of approximately 88 %.

The lipids and proteolipids of the microsomal fraction were prepared by homogenizing with a chloroform-methanol mixture (2:1) using 20 ml per g of rehomogenized fraction (Folch *et al.*, 1951). The extracts were filtered and washed free of non-lipid contaminants by being placed in contact with five volumes of water. When equilibration was reached, the upper water-methanol phases were removed without disturbing the fluff in the interphase. The flasks were then placed at -15°C. After 4 hours, the fluff had frozen, while the chloroform remained liquid. The icy proteolipid fractions were then separated from the solid lipid fractions by filtration through filter paper at -15° (Folch and Lees, 1951). After separation, the solvents were evaporated at +70°C and in vacuum desiccator.

The total dry fractions were dissolved in 10 ml Insta-Gel and the radioactivities counted. The ^{14}C-activity in the proteolipids and lipids was calculated as a percentage of the total radioactivity found in the microsomal fraction.

The results are shown in Fig. 4. Lipid labeling had decreased significantly and to about the same extent in both acute intoxication and the "hangover" phase. Proteolipid labeling decreased markedly during intoxication but had risen to the control level in "hangover". Measurement of the blood alcohols showed that the alcohol had not been entirely eliminated 13 hours after administration.

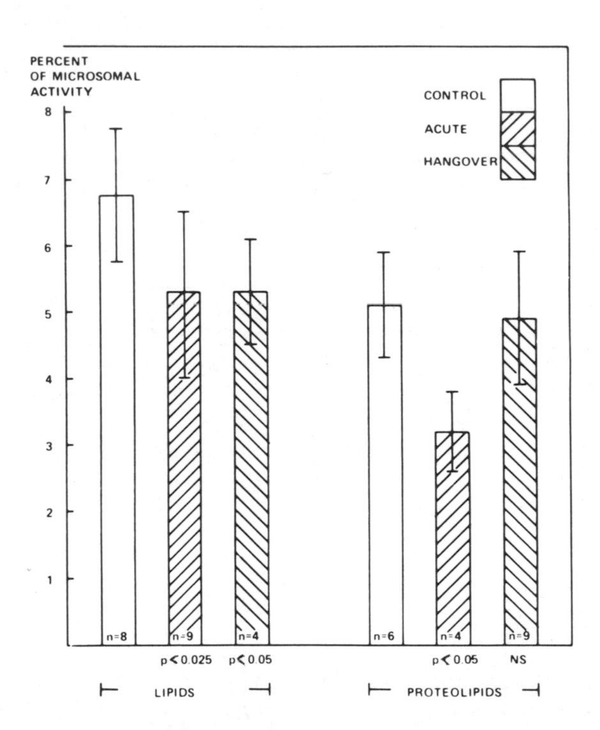

Fig. 4. Effects of acute intoxication and "hangover" on labeling of microsomal lipids and proteolipids *in vivo*. ^{14}C-serine was injected into the right cerebral ventricle of permanently cannulated rats either 1 h (intoxication) or 13 h ("hangover") after administration of 6 g ethanol p.o. Incorporation is expressed as per cent of total microsomal label present in the fractions. Contact time with the precursor was 12 h.

The contact time with the isotope was chosen on the basis of pilot experiments in which i.a. the time needed for uniform distribution of the label through the brain was measured. The serine was injected in tracer quantities only, and as yet we have not adequate control of changes in pool size which are not very likely but cannot be ruled out. The histological and autoradiographical analyses which should give information on possible regional changes in labeling have not yet been performed. In any case, detection of a change in labeling is encouragning in view of our working hypothesis. Currently, we are performing experiments in which similar determinations are made with rats which are at the end of a period of prolonged intoxication or in a state of withdrawal excitability.

V. MICROSOMAL LIPID LABELING IN VITRO AFTER CHRONIC ETHANOL ADMINISTRATION

The animals used were male rats obtained from Expt. 1, described in Section II of this paper. Eight of them were controls and 8 were killed in the withdrawal state after the period of prolonged intoxication, about 24 hours after the last ethanol dose. Tissue obtained from 4 controls and 4 experimental animals was used for incubation without electrical stimulation, and similarly tissue obtained from 4 animals of each group was used for electrical stimulation.

The slices were prepared in a cold room at +5°C according to the procedure described by Wallgren et al. (1974). Two slices were taken from each cerebral hemisphere after removal of the meninges, and transferred into oxygenated, ice cold medium. From each animal, one slice was placed in each of a series of four flasks, combining slices from two animals so that those from the second animal were placed in reversed order in the flasks. In this way, each vessel received a first cut and a second (deeper) cut slice, obtained from two different animals. The time elapsing from decapitation until the start of incubation was 10-15 min.

The incubation medium had the following composition: 30 mM glycyl-glycine; 124 mM NaCl; 5 mM KCl; 1.24 mM KH_2PO_4; 1.30 mM $MgSO_4$; 0.75 mM $CaCl_2$; 10 mM glucose; pH adjusted to 7.4 with 1 N NaOH; gas phase pure oxygen. L-serine-^{14}C (U) of the same quality as described in the preceding section was used as precursor. The incubation flasks contained 2.5 ml of the medium and were flushed in a shaking bath for 5 min at 37°, placed in ice and transferred to the cold room. After preparation of the slices, the flasks were placed in the shaking incubator, flushed with oxygen for 5 min and closed. A preincubation period of 30 min was employed in order to stabilize the conditions of the slices. After that, 1 μc of ^{14}C-serine was injected into the incubation medium of each flask. At the same time, electrical stimulation began in the stimulation groups. The contact time of isotope and the duration of stimulation were 15, 30, 45, and 60 min. Carbon electrodes (Nikander, 1972) were used for the stimulation which was carried out as described by Wallgren et al. (1974). At the end of the incubation, the slices were picked up on an aluminium rider, drained by pulling them up along the wall of the flask, weighed, and transferred to an all-glass homogenizer. Four slices from two incubation flasks of the same animal group with similar conditions of manipulation were pooled and homogenized by hand in ten volumes of 0.25 M ice cold sucrose.

The techniques for preparing microsomes and for counting the radioactivity were similar as described in the preceding section. Because of the small amount of material, it was not possible to separate the lipid and proteolipid fractions. The total lipids were prepared by homogenizing the fractions with chloroform-methanol (2:1),

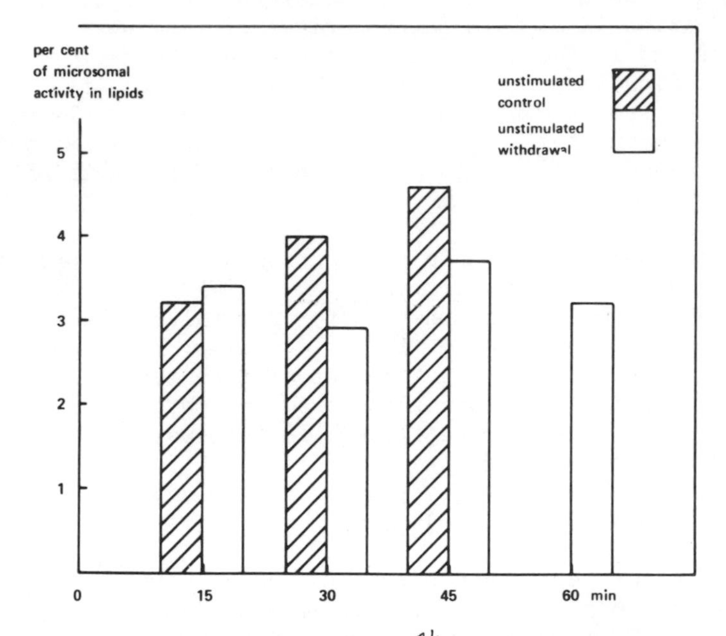

Fig. 5. Incorporation of label from ^{14}C-serine *in vitro* into microsomal lipids obtained from unstimulated sections of cerebral cortex tissue from rats. The withdrawal animals (white bars) were killed during withdrawal excitation after termination of three weeks continuous ethanol intoxication. Each bar represents the activity of 4 pooled slices obtained from different animals. From every animal, four slices were obtained and distributed for incubation in presence of precursor for the time intervals indicated, following preincubation for 30 min. The radioactivity is expressed as a percentage of the total label in the microsomal fraction.

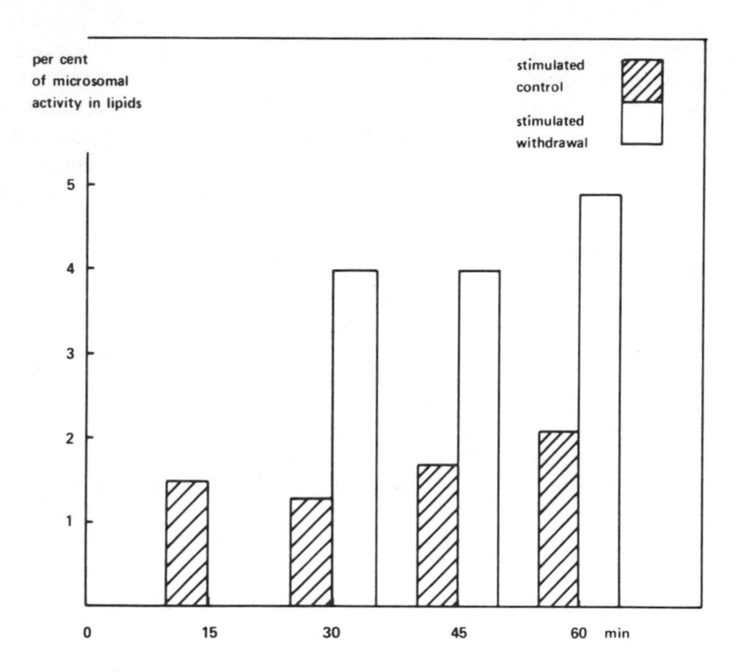

Fig. 6. Incorporation of label from ^{14}C-serine into the microsomal lipids of slices of rat cerebral cortex tissue stimulated electrically *in vitro*. The white bars refer to tissue obtained from animals in a withdrawal state after three weeks of continuous ethanol intoxication. Pooling of slices and distribution of the tissue sections was as explained in the text of Fig. 5. The radioactivity is expressed as a percentage of the total label in the microsomal fraction.

using 20 ml per g of fraction. The extracts were filtered and washed free of non-lipid contaminants by being placed in contact with five volumes of water (Folch *et al.*, 1951). When equilibration was reached, the upper water-methanol phase was removed and the lower chloroform phase taken to dryness under nitrogen. The total dry lipid fractions were dissolved in 10 ml Insta-Gel and counted. The ^{14}C-activity in the total lipids of the microsomal fraction was calculated in per cent of the activity found in the entire fraction. The results are shown in Fig. 5, showing the incorporation found in unstimulated tissue, and Fig. 6, showing the data for stimulated tissue.

The pattern shown by the unstimulated tissue is somewhat irregular. Unfortunately, the sample from the control animals at 60 min was lost. However, the differences are not sufficiently consistent or large to be taken as evidence of any effect of the previous treatment of the animals on precursor incorporation. By contrast, there is a marked and consistent increase in labeling when electrical stimulation is employed. In this instance, change in pool size is highly unlikely and a genuine change in the response of the tissue is indicated. In view of the evidence that processes linked with excitation of nerve cells are particularly sensitive to ethanol (Wallgren et al., 1974) and the hypothesis of compensatory changes in those structures which are directly affected by ethanol during intoxication, the observation is of great interest and indicates that continued work on similar lines is worthwhile.

VI. SUMMARY

The experiments described are based on the hypothesis that prolongation of the depressant action of ethanol leads to compensatory changes in neuronal membrane structures involved in impulse conduction and transmission, and that these become manifest as increased tolerance and withdrawal hyperexcitability.

Behavioral tolerance was tested by means of the tilted plane test in rats consuming 9-10 g ethanol/kg/day in a liquid diet fed ad lib., or given 5 g/kg every other day by stomach tube, or doses rising from 6 to 9 g/kg/day maintaining continuous intoxication. All treatments were continued for about three or four weeks before testing. Rats consuming ethanol at a self-regulated rate did not develop tolerance, evidently because sufficient alcohol levels were not built up. Prolonged intoxication induced a high degree of tolerance and withdrawal symptoms, whereas intoxication every other day induced an intermediate degree of tolerance.

When no definite abstinence symptoms were associated with the behavioral tolerance, cation stimulated ATPase activity of the brain microsomal fraction was not changed. With increasing withdrawal excitability, there was a relative increase in Na^+, K^+-stimulated ATPase and an decrease in Mg^{2+}-stimulated ATPase whereas total activity of the enzyme system was not altered.

^{14}C-serine was used as a precursor in order to detect changes in the metabolism of membrane components. So far, only acute experiments have been carried out in vivo. Heavy intoxication (6 g ethanol/kg by stomach tube) inhibited labeling of brain microsomal lipids and proteolipids. In "hangover", proteolipid labeling had returned to the control level whereas lipid labeling was still depressed. Cerebral cortex slices from rats in a withdrawal state after prolonged intoxication, and from control rats, were incubated in vitro with ^{14}C-serine. Unstimulated tissue showed no effect of the prior treatment. When electrical stimulation was applied, much more activity was recovered in microsomal lipids of slices from withdrawal animals than from controls.

REFERENCES

Akera, T., Rech, R.H., Marquis, W.J., Tobin, T., and Brody, T.M., 1973. Lack of relationship between brain (Na^+, K^+)-activated adenosine triphosphatase and the development of tolerance to ethanol in rats, J. Pharmacol. Exp. Ther. 185: 594 - 601.

Arvola, A., Sammalisto, L., and Wallgren, H., 1958. A test for level of alcohol intoxication in the rat, Quart. J. Stud. Alc. 19: 563 - 572.

Bowler, K. and Duncan, C.J., 1968. Effect of temperature on the Mg^{2+} dependent and Na^+-K^+ATPases of a rat brain microsomal preparation, Comp. Biochem. Physiol. 24: 1043 - 1054.

Cicero, T.J., Snider, S.R., Perez, V.J., and Swanson, L.W., 1971. Physical dependence on and tolerance to alcohol in the rat, Physiol. Behav. 6: 191 - 198.

Falk, J.L., Samson, H.H., and Winger, G., 1972. Behavioral maintenance of high concentrations of blood ethanol and physical dependence in the rat, Science 177: 811 - 813.

Folch, J., Ascoli, I., Lees, M., Meath, J.A., and LeBaron, F.N., 1951. Preparation of lipide extracts from brain tissue, J. Biol. Chem. 191: 833 - 841.

Folch, J. and Lees, M., 1951. Proteolipides, a new type of tissue lipoproteins, J. Biol. Chem. 191: 807 - 817.

Goldstein, D.B., 1972. Relation of alcohol dose intensity of Withdrawal signs in mice, J. Pharmacol. Exp. Ther. 180: 203 - 215.

Goldstein, D.B. and Israel, Y., 1972. Effects of ethanol on mouse brain (Na^+, K^+)-activated adenosine triphosphatase, Life Sci. 11: Part II; 957 - 963.

Israel, Y., Kalant, H., LeBlanc, A.E., Bernstein, J.C., and Salazar, J., 1970. Changes in cation transport and (Na^+, K^+)-activated adenosine triphosphatase produced by chronic administration of ethanol, J. Pharmacol. Exp. Ther. 174: 330 - 336.

Israel, Y. and Kuriyama, K., 1971. Effects of in vivo ethanol administration on adenosine triphosphatase activity of subcellular fractions of mouse brain and liver, Life Sci. 10: 591 - 599.

Knox, W.H., Perrin, R.G., and Sen, A.K., 1972. Effect of chronic administration of ethanol on (Na^+, K^+)-activated ATPase activity in six areas of the cat brain, J. Neurochem. 19: 2881 - 2884.

Lieber, C.S. and De Carli, L.M., 1973. Ethanol dependence and tolerance: a nutritionally controlled experimental model in the rat, Res. Commun. Chem. Pathol. Pharmacol. 6: 983 - 991.

Nikander, P., 1972. Carbon electrodes for stimulation of brain tissue in vitro, J. Neurochem. 19: 535 - 537.

Peacock, A.J., Bowler, K., and Anstee, J.H., 1972. Demonstration of a Na^+-K^+-Mg^{2+} dependent ATPase in a preparation from hindgut and Malpighian tubules of two species of insect, Experientia 28: 901 - 902.

Roach, M.K., Khan, M.M., Coffman, R., Pennington, W., and Davis, D.L., 1973. Brain (Na^+,K^+)-activated adenosine triphosphatase activity and neurotransmitter uptake in alcohol-dependent rats, Brain Res. 63: 323 - 329.

Wallgren, H., 1973. Neurochemical aspects of tolerance to and dependence on ethanol, in Alcohol Intoxication and Withdrawal, Experimental Studies I, (M.M. Gross, ed.), pp. 15 - 31, Plenum Press, New York.

Wallgren, H., Ahlqvist, J., Åhman, K., and Suomalainen, H., 1967. Repeated alcoholic intoxication compared with continued consumption of dilute ethanol in experiments with rats on a marginal diet, Brit. J. Nutr. 21: 643 - 660.

Wallgren, H. and Forsander, O., 1963. Effect of adaptation to alcohol and of age on voluntary consumption of alcohol by rats, Brit. J. Nutr. 17: 453 - 457.

Wallgren, H., Kosunen, A.-L., and Ahtee, L., 1972. Technique for producing an alcohol withdrawal syndrome in rats, Brain Res. 42: 550.

Wallgren, H., Kosunen, A.-L., and Ahtee, L., 1973. Technique for producing an alcohol withdrawal syndrome in rats, Isr. J. Med. Sci. 9: Suppl. 63 - 71.

Wallgren, H., Nikander, P., von Boguslawsky, P., and Linkola, J., 1974. Effects of ethanol, tert.butanol and clomethiazole on net movements of sodium and potassium in electrically stimulated cerebral tissue, Acta Physiol. Scand. 91: 83 - 93.

ALTERATION IN CEREBRAL POLYNUCLEOTIDE METABOLISM FOLLOWING CHRONIC ETHANOL INGESTION

Sujata Tewari and Ernest P. Noble

Section of Neurochemistry
Department of Psychiatry and Human Behavior
University of California, Irvine, California

INTRODUCTION

Ribonucleic acids (RNA) and proteins are known to have unique functions (1-4) in the central nervous system in addition to their general role in cellular metabolism. For example, alterations in RNA of neural tissue have been demonstrated following learning (5) or subsequent to hormonal treatment (6,7).

Our laboratory over the past several years has been engaged in studying the chronic effects of ethanol on brain macromolecules. *In vitro* and *in vivo* studies have shown that ethanol ingestion by rodents adversely affects the protein synthesizing capacity of cerebral tissue (8,9). In addition, under these conditions, ethanol was also found to alter brain RNA metabolism. Thus, brief pulse labeling with (5-^{3}H)orotic acid caused an increase in the incorporation of the precursor into a rapidly labeled brain nuclear RNA fraction obtained from 10% ethanol drinking C57BL/6J mice (10). With an extended pulse labeling period, the initial stimulation disappeared and was replaced by an inhibited incorporation of label into this RNA fraction. Furthermore, in the ethanol drinking animals, the incorporation of RNA precursor into transfer(t) RNA, ribosomal(r) RNA and polysomal RNA fractions were markedly reduced following long-term ingestion of ethanol (11).

The mechanisms by which ethanol causes the abnormal RNA metabolism are not clear. However, one possibility involves the synthesis and processing of RNA with special emphasis on poly-adenylic acid (poly A). These adenine rich polynucleotide sequences have been identified in rapidly labeled mammalian polysomal RNA fraction as well as in the messenger(m) RNA chain and are different from the nucleotide sequences required for the coding of amino acids into protein (12-14).

The present report deals with the chronic effects of ethanol on precursor label incorporation, under both in vivo and in vitro conditions, into various polynucleotides of rat brain. Specifically, comparisons are made among the RNA species obtained from the various ribosomal fractions. Studies are also presented on the influence of long-term ethanol ingestion on poly A synthesis by isolated brain nuclei.

MATERIALS AND METHODS

Two month old male Sprague-Dawley rats were used for the experiments. For at least six weeks, half of the animals drank a 10% ethanol/water solution (v/v) while age-matched controls received water only. Both groups were provided with laboratory chow ad libitum and were maintained in the animal quarters on a 12-hour dark/12-hour light cycle. Twenty-four hours prior to experimentation, ethanol was replaced by water to ensure the absence of ethanol in these animals.

Intraventricular injection of a radioactive precursor of RNA into rat brain was performed according to the method of Noble et al. (15). Techniques for isolation of ribosomal and nuclear fractions have been described elsewhere (16,17). Isolation of free and bound polysomes in the presence of brain RNase inhibitor were carried out according to the procedures described by Blobel and Potter (18). Phenol extraction of RNA was according to Bloemendal et al. (19). Protein and RNA determinations were carried out according to the conventional procedures of Lowry et al. (20) and Mejbaum (21), respectively.

RESULTS

RNA Synthesis

Polysomes in mammalian cells exist in the free state as well as in association with the membranes of the endoplasmic reticulum (ER), and functionally are known to synthesize different types of proteins (22-24). Since brain tissue has been shown to be rich in both free and bound ribosomes (25,26), it was of interest to determine what effects, if any, ethanol might have on the RNA metabolism of these two ribosomal fractions.

In vivo labeling of RNA was carried out first in a mixed

population of ribosomes and later in ribosomes separated into free and membrane bound polysomes. Comparisons were made between chronic ethanol-drinking rats and water-drinking controls.

Incorporation of (5-^{3}H)orotic acid into RNA of mixed population of ribosomes. Cerebral ribosomes from control and ethanol-drinking rats were labeled by intraventricular injection of (5-^{3}H) orotic acid. Ribosomes were isolated at 10, 20 and 120 minutes after exposure to the radioactive precursor and RNA was obtained from these particles. Data presented in Fig. 1 show the presence of relatively small amounts of radioactivity in the RNA fraction at the 10 and 20 minute time points in both control and ethanol groups. Moreover, incorporation of (5-^{3}H)orotic acid into rapidly labeled RNA was significantly inhibited in the ethanol group

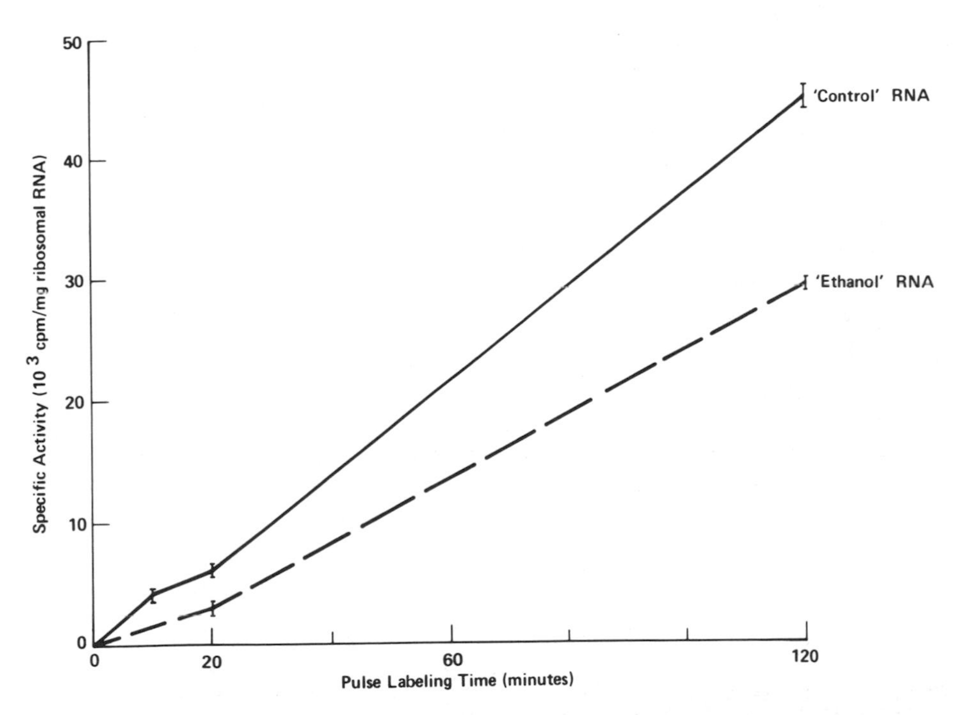

FIGURE 1. Effects of chronic ethanol consumption on the incorporation of (5-^{3}H)orotic acid into RNA of ribosomal fraction. Ten µc of (5-^{3}H)orotic acid (specific activity 15c/mM) were administered intraventricularly into the brains of control and ethanol-drinking rats. Mixed population of labeled brain ribosomes were isolated and phenol extraction of RNA was performed as described in the text. Animals were 167 days old and drank ethanol for 76 days.

where the specific activities at the early time points of 10 and 20 minutes were 28% and 51%, respectively, that of the controls. At the two hour time point, incorporation of (5-^{3}H)orotic acid into the rRNA fraction was still found to be depressed in the ethanol group and its specific activity was 66% that of the control fraction.

Free and bound polysomal RNA.

a. In vivo labeling: These experiments were performed following an intraventricular injection of (5-^{3}H)uridine into the brains of control and ethanol-ingesting animals. Twenty-four hours after the administration of the labeled precursor, free and bound polysomes were isolated in the presence of brain RNase inhibitor (18) and the radioactivity was determined by measuring incorporation into the cold TCA insoluble residue. Labeling of bound polysomes was examined both while they were still attached to the membranes of the ER and after dissociation from the ER. Results given in Table 1 show a greater incorporation of (5-^{3}H) uridine into RNA of free polysomes than bound polysomes in the presence of ER in both control and ethanol groups. However, increased radioactivity in the bound polysomes was obtained after removal of the ER.

It is interesting to note that ethanol had a differential effect on the labeling of the RNA of free and bound polysomes; in the latter group this effect was dependent on the presence of the membranes of the ER. When the labeling pattern was compared between animals ingesting ethanol and water, a decrease in the incorporation of (5-^{3}H)uridine was found in both the free and bound rRNA in the presence of the ER. However, when the ER was removed, increased incorporation of ^{3}H was found in the rRNA of bound ribosomes of the ethanol group.

b. RNA content of polysomes: The RNA content of the two populations of polysomes was determined and the results are given in Table 2. RNA content of the free polysomes was higher than that of bound polysomes (ER absent) in both the control and the ethanol groups. Chronic ethanol administration led to a 17% inhibition in the total yield of free polysomal RNA. On the other hand, increases of 26% and 42% in the RNA content of bound polysomes of the ethanol group were obtained both in the presence and absence of ER, respectively. Thus, the data indicate a differential effect of ethanol on the ribosomal population of brain, an organ characterized by a greater amount of free ribosomes than bound ribosomes (26).

Incorporation of (8-^{3}H)GTP into RNA by isolated brain cell nuclei. The effect of long-term ethanol ingestion was next deter-

TABLE 1

((5-^3H)Uridine Incorporation into RNA by Free and Membrane Bound Polysomes

Fraction	Control (c.p.m./mg RNA)	Ethanol (c.p.m./mg RNA)	% of Control
Free Polysomes	2040	1550	76
Bound Polysomes (ER Present)	1390	1250	90
Bound Polysomes (ER Absent)	4490	6080	135

Animals were pulse labeled by injecting into their brain ventricles 14μc of (5-^3H)uridine (specific activity 24.3c/mM). Free and bound polysomes were obtained 24 hours later and their radioactive content was determined in the cold TCA insoluble residue as described in the text. Mice were 32 weeks old and drank 10% ethanol or water for 27 weeks.

TABLE 2

Effects of Long-Term Ethanol Drinking on RNA Content of Free and Membrane Bound Polysomes

Fraction	Control (mg RNA/g Brain Weight)	Ethanol (mg RNA/g Brain Weight)	% of Control
Free Polysomes	0.350	0.290	83
Bound Polysomes (ER Present)	0.538	0.680	126
Bound Polysomes (ER Absent)	0.026	0.037	142

Free and membrane bound brain polysomes were isolated as described in the text. Mice were 32 weeks old and drank 10% ethanol or water for 27 weeks.

mined on RNA synthesis under *in vitro* conditions in purified brain cell nuclei. Maximum incorporation of (8-^{3}H)GTP into the cold TCA insoluble residue occurred at the 15 minute time point in both the control and ethanol groups (Fig. 2). Incorporation of ^{3}H into nuclear RNA by the ethanol group was inhibited at all time-points examined when compared to the control nuclei. Maximum inhibition was present at 20 minutes at which time the specific activity of RNA in the ethanol nuclei was 66% that of control nuclei.

Polyadenylic Acid Synthesis

Poly A has been found to be naturally present in nuclei (27) and in free and membrane bound polysomes (28) of mammalian cells.

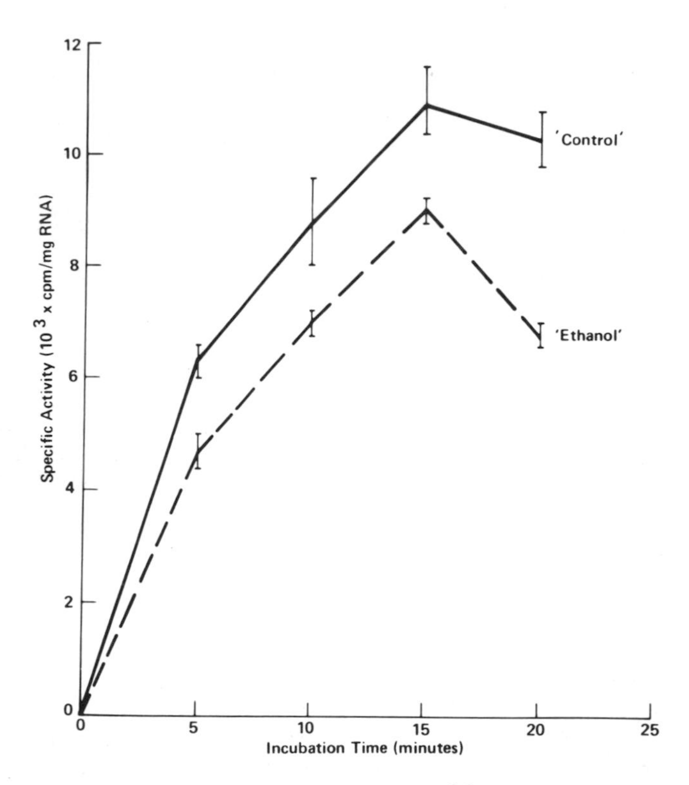

FIGURE 2. *In vitro* incorporation of (8-^{14}C)GTP into TCA insoluble residue by nuclei isolated from control and ethanol drinking rats. Approximately 200µg of nuclear RNA fraction from either group were incubated at 37°C in a final volume of 1ml of a medium containing 0.25M sucrose, 40mM tris buffer (pH 8.6), 70mM KCl, 8mM $MgCl_2$, 1.6mM spermidine, 0.1mM each of ATP, UTP and CTP and 0.5µc of (8-^{14}C)GTP (specific activity 3.9mc/mM). Animals were 160 days old and drank ethanol for 69 days.

Ribosomes and the membranes of the ER were found to contain a DNA like RNA with a 5% poly A content (13). Recently, an ATP polymerizing enzyme system has been described in rat brain nuclei which catalyzes the synthesis of an acid insoluble material containing a linear sequence of AMP linked through the 3'5'phosphodiester bond (17). In the present section the characteristics of poly A synthesis in brain have been examined and the effects of ethanol on the formation of this homopolynucleotide are presented.

In vitro incorporation of (8-^{3}H)ATP into RNA and poly A. Intact brain cell nuclei were isolated from control rats and their capacity to synthesize both RNA as well as poly A was determined. In this study, synthesis of RNA or poly A was measured by determining the incorporation of ^{3}H from (5-^{3}H)ATP into cold TCA insoluble residue. Table 3 compares the results obtained under conditions which favor either RNA or poly A synthesis. As the data indicate, in the presence of (8-^{3}H)ATP alone a large incorporation could be seen in the poly A synthesizing system with little activity in the RNA synthesizing system. However, in the presence

TABLE 3

Characteristics of the RNA and poly A Synthesizing Systems of Rat Brain Nuclei

Addition	(8-^{3}H)ATP Incorporation into	
	RNA	poly A
	c.p.m. X 10^2/mg protein	
(8-^{3}H)ATP	9.50	235.0
+UGC	45.00	23.0
+UGC + DNA	76.10	46.8

Two hundred μg of nuclear protein were incubated in the presence of 0.5μc (8-^{3}H)ATP (specific activity 20.9c/mM) for 15 minutes at 37°C in a final volume of 1 ml of a buffer containing for

1) RNA synthesis: 6mM $MgCl_2$, 3mM $MnCl_2$, 10mM β-mercaptoethanol and 30mM tris buffer (pH 8.0);
2) poly A synthesis: 4mM $MnSO_4$, 60mM KCl, 20mM NaF, 160mM tris buffer (pH 8.5).

Where indicated 0.6mM each of UTP, GTP and CTP (UGC), and 10μg calf thymus DNA were added to the incubation medium. Rats were 203 days old and drank ethanol for 80 days. Reactions were carried out in triplicate and variations among individual values were within 5%.

of the nucleotide mixtures of UTP, GTP and CTP (collectively called UGC) incorporation was stimulated significantly in the RNA system while a large decrease occurred in the poly A system. Furthermore, while the addition of the DNA in the presence of UGC was stimulatory to both the RNA and the poly A system, in the former it was additive while in the latter it failed to relieve substantially the marked inhibition produced by the addition of the UGC mixture.

Effects of ethanol on (8-^{3}H)ATP incorporation into poly A. Addition of exogenous poly A and other factors affecting poly A synthesis were compared between nuclei of control and ethanol drinking rats. As Table 4 demonstrates, deletion of poly A resulted in marked inhibition of its own synthesis under all conditions studied in both groups of nuclei. The addition of a UGC mixture in the presence of poly A caused a decrease in the synthesis of poly A in both groups of control and ethanol nuclei with greater inhibition in the former group. However, the decrease caused by UGC was found to be independent of poly A addition. DNA substituted for poly A could not act as a primer of poly A synthesis in both groups of nuclei; furthermore, the addition of UGC to this mixture failed to effectively change ^{3}H incorporation into the homopolynucleotide.

In another group of nuclei obtained from control and ethanol-drinking rats, additional properties of the ATP polymerizing system were studied. Table 5 compares the effects of the addition of exogenous polyuridylic acid (poly U), UGC, actinomycin-D and $(NH_4)_2SO_4$ on this system. In the complete system, as previously noted (Table 4), synthesis of poly A was markedly inhibited in the ethanol nuclei when compared to the control nuclei. In the absence of primer poly A, the incorporation of (8-^{3}H)ATP into the cold TCA insoluble residue was again reduced in both groups of nuclei. The substitution of poly U for poly A failed to restore the poly A synthesizing activity in the two nuclear preparations. Addition of UGC to the complete system inhibited poly A synthesis (Table 4) while the addition of actinomycin-D, a known inhibitor of RNA synthesis, had no effect on the poly A dependent incorporation of (8-^{3}H)ATP in the control and ethanol group. The inclusion of $(NH_4)_2SO_4$ in the complete system resulted in a marked inhibition of poly A synthesis in both groups of nuclei. The findings with actinomycin-D and $(NH_4)_2SO_4$ are in accord with previously described effects of these compounds on poly A synthesis in other systems (17).

Effects of nucleotide triphosphates on poly A synthesis. The reduction in the activity of the poly A synthesizing system in the presence of UGC mixture in the above studies prompted an examination of the effects of specific nucleotide triphosphates on

TABLE 4

Requirements for the Incorporation of (8-^{3}H)ATP into poly A by Brain Nuclei Obtained from Control and Ethanol-Treated Rats.

Medium	Control Nuclei		Ethanol Nuclei	
	pmol AMP/ mg Protein	% Control	pmol AMP/ mg Protein	% Control
Complete System	6.47	100.0	1.27	19.7
-Poly A	2.73	42.2	0.40	6.4
+UGC	0.56	8.5	0.60	9.5
-Poly A + UGC	0.85	13.2	0.79	12.3
-Poly A + DNA	0.52	8.0	0.52	8.0
-Poly A + DNA + UGC	0.64	9.9	0.69	10.5

The complete system in a final volume of 1ml contained: 100μg nuclear protein, 160mM tris buffer (pH 8.5), 4mM $MnSO_4$, 60mM KCl, 20mM NaF, 0.2mM ATP, 0.5μc (8-^{3}H)ATP (specific activity 20.9c/mM), 50μg poly A and 200μg BSA. Where indicated, 0.2mM each of UTP, GTP and CTP (UGC) and 50μg of calf thymus DNA were added. The mixtures were incubated at 37°C for 15 minutes. Animals were 180 days old and consumed ethanol for 70 days.

this system. Again, comparisons were made between brain nuclei obtained from water and ethanol-drinking rats. Using the complete system, incorporation of (8-^{3}H)ATP into the cold TCA insoluble residue was markedly decreased in the ethanol nuclei (Table 6). With respect to the individual nucleotide triphosphates, the addition of UTP resulted in the largest inhibition of ^{3}H incorporation in both groups of nuclei. Furthermore, inhibition of label incorporation were also observed when GTP and CTP were added individually to each of the two groups of nuclei.

DISCUSSION

Incorporation of precursors into brain RNA and protein have been reported to be adversely affected following chronic ingestion of ethanol (11,29). Detailed examination of RNA metabolism showed

TABLE 5

Comparison of the Requirements for the Incorporation of (8-^{3}H)ATP into poly A Between Control and Ethanol-Treated Rat Brain Nuclei

Medium	Control Nuclei		Ethanol Nuclei	
	pmol AMP/ mg Protein	% Control	pmol AMP/ mg Protein	% Control
Complete System	7.09	100.0	2.10	29.0
-Poly A	1.68	23.8	0.72	10.7
-Poly A + Poly U	1.19	16.6	0.87	12.4
+UGC	2.90	41.0	1.14	16.2
+Actinomycin-D	7.03	99.0	2.29	32.0
+$(NH_4)_2SO_4$	0.50	7.2	0.77	10.8

Reaction conditions were as described in Table 4. Seventy µg of nuclear protein fraction from each group were used. Where indicated 0.2mM UGC mixture, 50µg poly A or poly U, 20µg of actinomycin-D or 0.4M $(NH_4)_2SO_4$ were added. Animals were 180 days old and drank ethanol for 70 days.

TABLE 6

Effect of Various Nucleoside Triphosphates on the Formation of poly A by Rat Brain Nuclei

Incubation Conditions	Control Nuclei		Ethanol Nuclei	
	c.p.m. X 10^3/ mg Protein	% Control	c.p.m. X 10^3/ mg Protein	% Control
Complete System	16.75	100 ± 2	5.87	35 ± 2
+UTP	1.89	11 ± 2	1.73	11 ± 2
+GTP	3.68	22 ± 7	2.24	14 ± 2
+CTP	4.01	24 ± 1	2.16	13 ± 1

Incubation conditions were as described in Table 4. Sixty-four µg of control and 82ug of ethanol nuclear protein fractions were used for the experiments. Rats were 92 days old and drank ethanol for 60 days.

that the decreased incorporation of labeled orotic acid into cytoplasmic RNA fractions were not a function of reduced availability of nucleotides in the brains of ethanol-treated animals (30). The evidence suggests that the observed changes in proteins and RNA most likely are a product of reduced macromolecular biosynthesis and/or possibly, in the case of RNA, of diminished availability of this molecular species in, or its delivery to, its cytoplasmic destination.

In this report, we have presented results which point to a reduction in precursor incorporation into the messenger(m) RNA and the rRNA fractions of the free polysomal unit obtained from the brains of ethanol-drinking animals. In the case of mRNA the evidence is somewhat tentative and was derived from the lower specific activity of polysomal RNA obtained from ethanol-drinking than water-drinking rats in the first 20 minutes of the study (Fig. 1). Since Penman et al. (31) and Girard et al. (32) have shown that only the mRNA was labeled during brief exposure to radioactive precursors, and because rRNA requires at least 60 minutes after nuclear synthesis to migrate into the cytoplasm (33), the decreased incorporation of label into RNA during the first 20 minutes of our study most probably represents radioactive mRNA. On the other hand, the decreased specific activity of ethanol RNA two hours following introduction of precursor reflects the contribution of both mRNA and rRNA.

In view of the relatively short half-life of mRNA (3-4 hours) as compared to rRNA (6-12 days) (31,34), synthesis of rRNA is indicated in experiments where polysomal labeling periods have been extended up to 24 hours in both control and ethanol-drinking rats. The longer incorporation study was conducted to ascertain more specifically the effects of ethanol on rRNA of free and bound polysomes. As Table 1 shows, 24 hours following the introduction of (5-^{3}H)uridine, incorporation of label into ethanol-treated animals was decreased in RNA, predominately rRNA, of both free and bound polysomes still attached to the ER. However, when the ER was removed, the specific activity of RNA was higher in the bound polysomes of ethanol-treated animals than controls. The increased labeling of bound rRNA in the ethanol group paralleled the total RNA yield of the bound ribosomes irrespective of the presence of ER (Table 2). In fact there was further enhancement with removal of the membranes.

In the case of polysomes still attached to the ER, the effect of ethanol on the incorporation of label into RNA differed from the results obtained with polysomes which were dissociated from their ER. The lower incorporation of label into the RNA of the polysome-ER complex could be due to the association of other types of RNA with the ER during the isolation of bound polysomes (35).

The differential effects of ethanol on the various populations of polysomes could influence the nature of proteins they synthesize which may then alter brain function. What the biochemical mechanisms are that subserve these effects are unknown. The decreased labeling of the free polysomes obtained from ethanol-treated animals suggests a more rapid turnover of rRNA, possibly due to increased susceptibility of free polysomes to breakdown. In contrast, the labeling pattern in bound polysomes, freed of their attachment to the ER, would suggest that ethanol treatment diminished the turnover rate of their rRNA. How ethanol produces this effect is unclear, although previous studies have shown that normally the RNA of bound polysomes exhibit a slower turnover rate than the free polysomes (32,36). It may be presumed that the administration of chronic ethanol could, by proliferating the ER, further protect the bound polysomes thus rendering them less susceptible to attack by degradative processes. Other explanations, of course, are possible but further experimentations are necessary to shed light on this question.

The above described changes in mRNA and rRNA of the polysomal fraction raises the possibility that ethanol might alter the transcription of RNA. Therefore, the ethanol inhibited RNA synthesis could be a reflection of subnormal RNA polymerase activity as shown by the *in vitro* results of $(8-^{14}C)$GTP incorporation (Fig. 2). The labeling pattern of the *in vitro* RNA products have been shown to be similar to those found under *in vivo* conditions (37). Recent findings indicate that most eukaryotic mRNAs contain repet itive poly A tracts, thereby suggesting an important functional role for these nucleotides in the organism (38 40). To assess the effects of ethanol on this homopolynucleotide, *in vitro* experiments were conducted.

As Tables 3 and 4 indicate, long-term ethanol administration markedly reduced poly A synthesis of brain nuclei. That this system, with its own endogenous poly A, was partially dependent on exogenous poly A is shown by the reduced incorporation of $(8-^{3}H)$ATP into cold TCA insoluble residue in the absence of primer poly A. It is also observed that this system could not be activated either by the addition of DNA (Table 4) or poly U (Table 5) in both the control or ethanol group. In addition, the properties of the single nucleotide triphosphate (ATP)-dependent reaction was strikingly different from the reaction requiring the four nucleotide triphosphates (NTP). The four NTP-dependent reaction is known to be inhibited by actinomycin-D, while incorporation of the single NTP $[(8-^{3}H)ATP]$ into cold TCA insoluble residue remained unaffected. Furthermore, UTP, GTP and CTP added either together (Table 3) or separately (Table 6) drastically inhibited the conversion of $(8-^{3}H)$ATP into (^{3}H)poly A.

The data presented indicate not only an active ATP polymerizing system in rat brain primed by poly A but also the susceptibility of this system to chronic ethanol exposure. These findings could have an important bearing on the results obtained from the previous RNA labeling studies. As poly A is thought to be an essential part of the mRNA processing mechanism, the reduced availability of mRNA in the ethanol-treated brains is probably a function of decreased synthesis and/or defective modification of this RNA. Since earlier studies in this laboratory have demonstrated reduced protein synthesis by both a mixed population of ribosomes as well as free and bound polysomes in brains of ethanol-drinking animals (30), the present findings suggest that this reduction could involve mRNA by affecting the poly A synthesizing system at the prepolysomal level. Since the presence of the poly A tail in mRNA molecules is also a prerequisite for successful production of this RNA (28,39,40), the severe inhibition of poly A synthesis induced by ethanol could cause an imbalance in the brain cell's protein synthetic capacity.

In conclusion, although the locus of protein synthesis is primarily in the cytoplasm, the interaction among different RNA species plays a regulatory role in this process. An important factor to be considered in this process is the stabilization of mRNA which is largely dependent on poly A (41). Inhibition of polyadenylation by ethanol, thus, will directly interfere with the stabilization process and hence lead to decreased protein synthesis.

ACKNOWLEDGMENTS

The technical assistance of Michael E. Arquilla and Charles H. Boniske is gratefully acknowledged. This work was supported by research grant (AA-00252) from the National Institute on Alcohol Abuse and Alcoholism (NIAAA), ADAMHA. One of us (S.T.) is a recipient of a Research Scientist Development Award from the NIAAA while the other (E.P.N.) is a Guggenheim Fellow. Figures 3 and 4 and Tables 1 and 2 have been reprinted by permission from Alcohol and Abnormal Protein Biosynthesis, edited by M. A. Rothschild, M. Oratz, and S. S. Schreiber, Pergamon Press Inc., 1975.

REFERENCES

1. Moore, B. W., Specific acid proteins of the nervous system, In: Physiological and Biochemical Aspects of Nervous Integration, F. D. Carlson (ed.), Prentice Hall, Englewood Cliffs, NJ, 1968, pp. 343-359.

2. Barondes, S. H. and Jarvik, M. E., The influence of actinomycin-D on brain RNA synthesis and on memory. J. Neurochem. 11:187-195, 1964.

3. Bondy, S. C., The ribonucleic acid metabolism of the brain. J. Neurochem. 13:955-959, 1966.

4. Glassman, E. and Wilson, J. E., The incorporation of uridine into brain RNA during short experiences. Brain Res. 21: 157-168, 1970.

5. Glassman, E., The biochemistry of learning: An evaluation of the role of RNA and protein. Ann. Rev. Biochem. 38:605-646, 1969.

6. Geel, S. E. and Timiras, P. S., Influence of growth hormone on cerebral cortical RNA metabolism in immature hypothyroid rats. Brain Res. 22:63-72, 1970.

7. Faiszt, J. and Adams, G., Role of different RNA fractions from the brain in transfer effect. Nature 220:367-368, 1968.

8. Noble, E. P. and Tewari, S., The effects of chronic ethanol ingestion on the protein synthesizing system of C57BL/6J mice. In: Biological Aspects of Alcohol Consumption, O. Forsander and K. Eriksson (eds.), The Finnish Foundation for Alcohol Studies, Helsinki, 1972, pp. 275-287.

9. Tewari, S. and Noble, E. P., Ethanol and brain protein synthesis. Brain Res. 26:469-474, 1971.

10. Fleming, E. W., Tewari, S. and Noble, E. P., Effects of chronic ethanol ingestion on brain aminoacyl-tRNA synthesis and tRNA. J. Neurochem., 1975, in press.

11. Noble, E. P. and Tewari, S., Protein and ribonucleic acid metabolism in brains of mice following chronic alcohol consumption. N. Y. Acad. Sci. 215:333-345, 1973.

12. Lee, S. Y., Mendecki, J. and Brawerman, G. A., A polynucleotide segment rich in adenylic acid in rapidly-labeled polyribosomal RNA component of mouse sarcoma 180 ascites cells. Proc. Nat. Acad. Sci., U.S., 68: 1331-1335, 1971.

13. Lim, L. and Canellakis, E. S., Adenine-rich polymer associated with rabbit reticulocyte messenger RNA. Nature 227: 710-712, 1970.

14. Darnell, J. E., Philipson, L., Wall, R. and Adesnik, M., Polyadenylic acid sequences: Role in conversion of nuclear RNA into messenger RNA. Science 174:507-510, 1971.

15. Noble, E. P., Wurtman, R. J. and Axelrod, J. A., A simple and rapid method for injecting H^3-norepinephrine into the lateral ventricle of the rat brain. Life Sci. 6: 281-291, 1967.

16. Tewari, S. and Baxter, C. F., Stimulatory effects of γ-aminobutyric acid upon amino acid incorporation into protein by a ribosomal system from immature rat brain. J. Neurochem. 16:171-180, 1969.

17. Dravid, A. R., Pete, N. and Mandel, P., An enzyme system in rat brain nuclei incorporating AMP into polyadenylate. J. Neurochem. 18:299-306, 1971.

18. Blobel, G. and Potter, V. R., Ribosomes in rat liver: An estimate of the percentage of free and membrane-bound ribosomes interacting with messenger RNA in vivo. J. Mol. Biol. 28:539-542, 1967.

19. Bloemendal, H., Littauer, U.Z., and Daniel, V., Transfer of soluble polynucleotides to microsomal RNA. Biochim. Biophys. Acta 51:66-72, 1961.

20. Lowry, O. H., Rosebrough, N. J., Farr, A. L. and Randall, R.J., Protein measurement with the folin phenol reagent. J. Biol. Chem. 193:265-275, 1951.

21. Majbaum, W., Estimation of RNA by the orcinol method of Mejbaum (1939). In: Techniques of Protein Biosynthesis. P. N. Campbell and J. R. Sargent (eds.), Vol. 1, Academic Press, NY, 1967, pp. 301-303.

22. Palade, G. E. and Siekevitz, P., Liver microsomes. An integrated morphological and biochemical study. J. Biophys. Biochem. Cytol. 2:171-201, 1956.

23. Takagi, M. and Ogata, K., Direct evidence for albumin biosynthesis by membrane bound polysomes in rat liver. Biochem. Biophys. Res. Comms. 33:55-60, 1968.

24. Hicks, S. J., Drysdale, J. W. and Munro, H. N., Preferential synthesis of ferritin and albumin by different populations of liver polysomes. Science 164:584-585, 1969.

25. Palay, S. L. and Palade, G. E., Fine structure of neurons. J. Biophys. Biochem. Cytol. 1:69-88, 1955.

26. Merits, I., Cain, J. C., Rdzok, E. J. and Minard, F. N., Distribution between free and membrane-bound ribosomes in rat brain. Experientia 25:739, 1969.

27. Kato, T. and Kurokawa, M., Studies on ribonucleic acid and homopolyribonucleotide formation in neuronal, glial and liver nuclei. Biochem. J. 116:599-609, 1970.

28. Gabrielli, F. and Baglioni, C., Translation of histone messenger RNA by homologous cell-free systems from synchronized HeLa cells. Europ. J. Biochem. 42:121-128, 1974.

29. Tewari, S., Fleming, E. W. and Noble, E. P., Alterations in brain RNA metabolism following chronic ethanol ingestion. J. Neurochem. 1975, in press.

30. Tewari, S. and Noble, E. P., Chronic ethanol ingestion by rodents: Effects on brain RNA. In: Alcohol and Abnormal Protein Biosynthesis. M. A. Rothschild, M. Oratz, and S. S. Schreiber (eds.), Pergamon Press Inc., NY, 1975.

31. Penman, S., Scherrer, K., Becker, Y. and Darnell, J., Polyribosomes in normal and poliovirus-injected HeLa cells and their relationship to messenger-RNA. Proc. Nat. Acad. Sci. 49:654-662, 1963.

32. Girard, M., Latham, H., Penman, S., and Darnell, J., Entrance of newly formed messenger RNA and ribosomes into HeLa cell cytoplasm. J. Mol. Biol. 11:187-201, 1965.

33. Campagnoni, A. T., Dutton, G. R., Mahler, H. R. and Moore, W. J., Fractionation of the RNA components of rat brain polysomes. J. Neurochem. 18:601-611, 1971.

34. Khan, A. and Wilson, J. E., Studies of turnover in mammalian subcellular particles: Brain nuclei, mitochondria and microsomes. J. Neurochem. 12:81-86, 1965.

35. Sabatini, D. D., Tashiro, Y. and Palade, G. E., On the attachment of ribosomes to microsomal membranes. J. Mol. Biol. 19:503-524, 1966.

36. Murthy, M. R. V., Free and membrane-bound ribosomes of rat cerebral cortex. J. Biol. Chem. 247:1944-1950, 1972.

37. Dutton, G. R. and Mahler, H. R., In vitro RNA synthesis by intact rat brain nuclei. J. Neurochem. 15:765-780, 1968.

38. Burr, H. and Lingrel, J. B., Poly A sequences at the 3' termini of rabbit globin mRNA's. Nature (New Biol.) 233: 41-43, 1971.

39. Adesnik, M., Salditt, M., Thomas, W. and Darnell, J. E., Evidence that all messenger RNA molecules (except histone messenger RNA) contain poly (A) sequences and that the poly (A) has a nuclear function. J. Mol. Biol. 71:21-30, 1972.

40. Mendecki, J., Lee, S. Y. and Brawerman, G., Characteristics of the polyadenylic acid segment associated with messenger ribonucleic acid in mouse sarcoma 180 ascites cells. Biochemistry 11:792-798, 1972.

41. Kolata, G. B., Control of protein synthesis (I): Poly (A) in cytoplasm. Science 185:517-518, 1974.

EFFECTS OF ETHANOL ON ELECTROLYTE METABOLISM AND NEUROTRANSMITTER RELEASE IN THE CNS

Y. Israel, F. J. Carmichael and J. A. Macdonald

Dept. Pharmacology, University of Toronto and Addiction Research Foundation, Toronto, Ontario, Canada

It has been reported that ethanol inhibits the movements of Na and K in the nerve action potential (Armstrong and Binstock, 1964; Moore et al., 1964) and also inhibits the active transport of these two ions (Israel et al, 1971). However, there is disagreement as to which of these processes is inhibited more effectively by pharmacologically relevant concentrations of ethanol (Israel et al, 1971; Wallgren, 1971; Israel et al, 1973). The active accumulation of several neurotransmitters including norepinephrine, glutamate and serotonin, has also been shown to be inhibited by ethanol, although concentrations in the lethal range are required for these effects (Israel et al, 1973; Roach et al, 1973). There is relatively little information as to the effects of ethanol on neurotransmitter release, and there is controversy in this aspect. For example, the passive (spontaneous) release of acetylcholine (ACh) has been reported to be inhibited by ethanol in brain slices (Kalant et al, 1967; Kalant and Grose, 1967) but to be increased by this drug in the neuromuscular junction (Gage, 1965; Okada, 1967; Inoue and Frank, 1967). On the other hand, it has been reported that *in vivo* ethanol inhibits the release of ACh from brain cortex and midbrain regions, (Phillis and Jhamandas, 1971; Erickson and Graham, 1973).

We have reported that 1% ethanol does not inhibit the electrically stimulated release of norepinephrine from guinea pig brain slices (Israel et al, 1973). Since tetrodotoxin, an inhibitor of the sodium current of the action potential, (Narahashi et al, 1964) completely blocked this release, this led us to conclude that the action potential was not inhibited in brain slices by this concentration of ethanol, although it is conceivable that only the adrenergic nerves were insensitive to ethanol.

In vivo studies have shown that ethanol increases the turnover of norepinephrine (NE), (Corrodi et al, 1966; Carlsson and Lindqvist, 1973; Carlsson et al, 1973; Hunt and Majchrowicz, 1974a). This could conceivably result from either an increased release or a reduced reuptake of the neurotransmitter. There is disagreement with regard to the acute *in vivo* effects of ethanol on serotonin (5HT); some groups reporting a decrease in turnover (Palaic et al. 1971; Hunt and Majchrowicz, 1974b), while others report no changes (Kuriyama et al. 1971; Frankel et al, 1974).

In vivo data showing ethanol effects on neurotransmitter release are, however, difficult to interpret since the changes observed may be secondary responses to alterations in other neuronal pathways. This is seen for example in studies with NE in which there is a good concensus that ethanol increases the turnover of this transmitter *in vivo*, while *in vitro* studies show no effects on release and only slight effects on reuptake. We have therefore undertaken a study of the effect of ethanol on neurotransmitter release in a "simpler system", such as brain slices, in which different neuronal pathways can be activated simultaneously by electrical stimulation. For this purpose, rat brain cortical slices were preincubated with radioactive neurotransmitters or precursor choline and the non-stimulated and electrically stimulated release of the transmitters in a superfusion system was then followed. The release of two transmitters was followed at the same time, for comparative purposes, by preloading brain slices with ^{3}H and ^{14}C labelled material. Experimental details have been published elsewhere (Carmichael and Israel, In Press).

Figure 1 shows the effects of various concentrations of ethanol on the non-stimulated and electrically stimulated release of ACh and NE from rat brain cortical slices. Non-stimulated release of radioactivity was allowed to proceed for 26 minutes in the presence or absence of ethanol prior to 3 minutes of electrical stimulation. During the non-stimulated period, the release of radioactivity decreased to constant levels. Following electrical stimulation there was an enhanced release of both transmitters. As can be seen, there is a dose dependent reduction in the amount of neurotransmitter released with the inhibition being greater on the cholinergic than on the adrenergic neurones. Statistically significant inhibition ($p<0.05$) was obtained with 0.25% ethanol for ACh while concentrations greater than 1% were required for the NE system.

The effect of 1% (0.22 M) ethanol on the release of ACh stimulated at different voltages was studied. As can be seen in Figure 2, there was an increase in the electrically stimulated release as the voltage was increased. There was a significant inhibition by 1% ethanol at all the voltages studied, with an

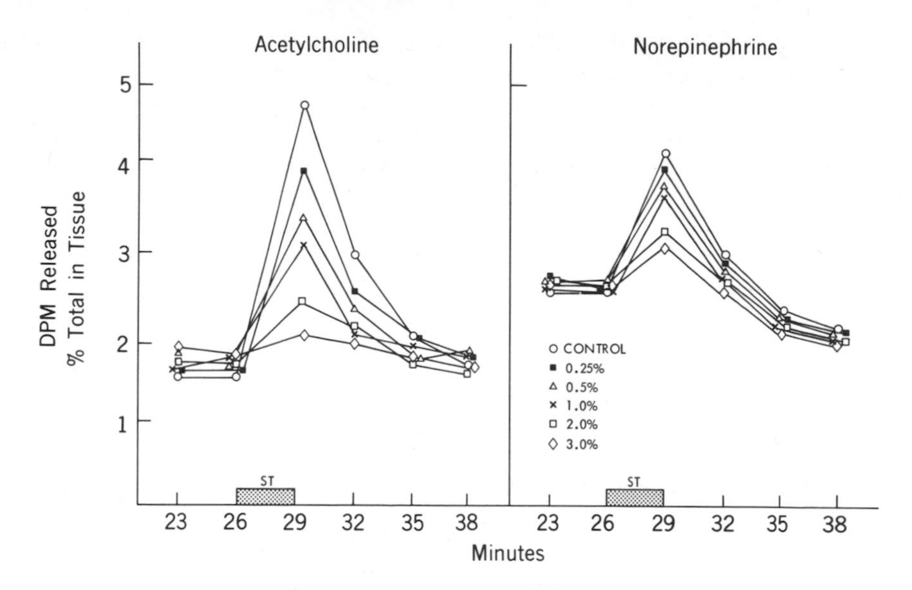

Figure 1. The effect of various concentrations of ethanol on the simultaneous efflux of radioactivity from rat cerebral cortical slices preloaded with ^{14}C-choline and ^{3}H-norepinephrine. Each point represents the mean of six experiments. The results are expressed as a percent of the total radioactivity present in the tissue at the start of the collection period (minute 23). This was determined at the end of the experiments. Ordinate: Percent of total radioactivity released per 3 min collection period. ST = electrical stimulation 25Hz; 25V.

apparently greater effect at the lower voltages; however, in the analysis of variance the voltage x treatment interaction indicated that the latter effect was not statistically significant. In this study in which ACh release and NE release were determined at the same time in the same preparation by the double label technique, there was no effect of 1% ethanol on the release of NE at any of the voltages studied.

To determine if these differences in sensitivities could be extended to other general depressants, we studied the effects of two higher alcohols and two barbiturates on the simultaneous electrically stimulated release of ACh and NE from brain cortical slices. Figure 3A shows that for all the compounds studied, the ACh release was more sensitive than that of NE. The concentrations of the alcohols required to inhibit electrically stimulated transmitter release correlated well with their lipid solubility (octanol/

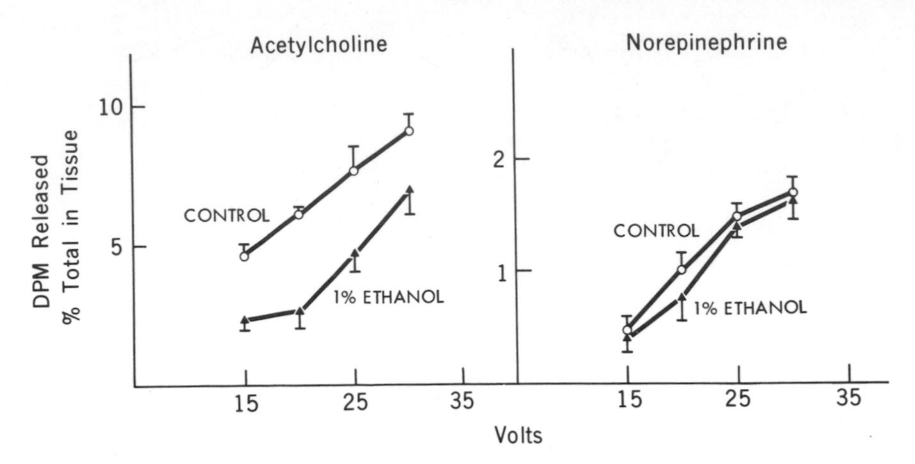

Figure 2. Effect of 1% ethanol on the radioactivity released by stimulation at various voltages from rat brain cortex tissue loaded with ^{3}H-NE (1 x 10^{-7}M) and ^{14}C-choline (5 x 10^{-6}M). Values represent the means ± S.E.M. of four experiments. Results are expressed as percent of total radioactivity initially present in the tissue released by electrical stimulation (3 min) (after subtraction of non-stimulated release).

water partition coefficient). However, the barbiturates proved to be more potent inhibitors than the alcohols when compared on the basis of lipid solubility alone. Nevertheless, if a correction for the size of the different molecules (Mullins Correction, see Seeman, 1972), is applied to arrive at the relative volume of occupation of the molecules in the lipid phase, there is a very good correction with their inhibitory potency (ACh, r = 0.9; NE, r = 0.9), (Figure 3B).

Table 1 shows the concentrations of ethanol that were required to inhibit by 50% the electrically stimulated release of ACh and NE as well as those for serotonin and glutamate. ACh release was the most sensitive while glutamate was the least sensitive to the effects of ethanol.

In order to assess if the electrically stimulated release of the different neurotransmitters was mediated by nerve action potentials rather than by transmural stimulation of the nerve endings, the effect of tetrodotoxin (TTX) was studied on the release of four neurotransmitters. TTX was a most effective inhibitor of the electrically stimulated release of all the neurotransmitters studied (Table 2), thus suggesting that the production

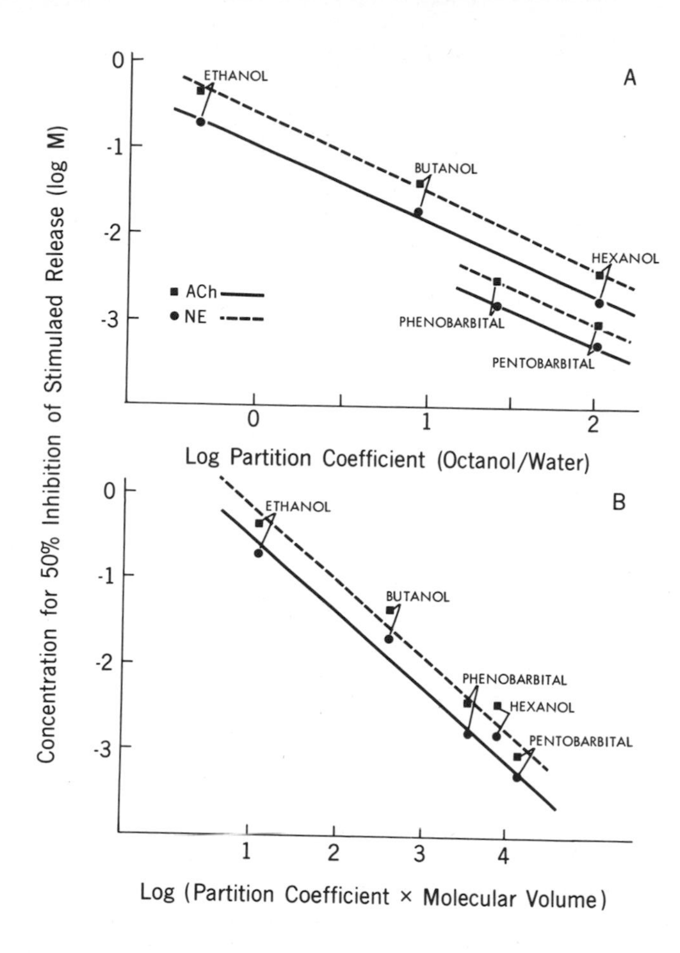

Figure 3. The relationship between lipid solubility and inhibition of the electrically stimulated release of neurotransmitters from rat brain slices preloaded with ^{14}C-choline and ^{3}H-norepinephrine. Ordinate represents the concentrations of the alcohols causing 50% inhibition of the neurotransmitter release. Abscissa: A, octanol/water partition coefficients taken from Leo et al (1971); B, partition coefficients x molecular volumes (ml drug/litre octanol, at an aqueous phase concentration of 1 mole).

of nerve action potentials precedes the liberation of the transmitters. Concentrations of the order of 10^{-9} to 10^{-8} M markedly affected the release process. It was most interesting to find that the sensitivity of the different transmitters to TTX was not

TABLE I

Concentration of Ethanol Required to Inhibit the Electrically Stimulated Release of Neurotransmitters[a] by 50%

	IC50 (M)
Acetylcholine	0.17 ± .021 (27)[b]
Serotonin	0.32 ± .044 (30)
Norepinephrine	0.42 ± .106 (26)
Glutamate	0.50 ± .228 (20)

[a]Rat brain cortical slices were preloaded with ^{14}C-choline and ^{3}H-NE, ^{14}C-5HT and ^{14}C-glutamate in a modified Krebs-bicarbonate ringer (Ca^{+2} = 0.75 mM, K^{+} = 3.60 mM, glucose 10 mM as the energy source). The effect of ethanol (0.25% to 3% w/v) on their electrically stimulated release was determined.

[b]Values presented are means ± S.E.M. The number of determinations is given in parenthesis.

constant with the IC50 values varying about 2-3 fold. This could indicate that (a) the affinity of receptors for the neurotoxin vary in the different types of neurones, or alternatively (b) that the same amount of bound neurotoxin results in different degrees of inhibition of nerve conduction thus invoking different "safety margins" for the various types of neurones. If the second alternative is correct and ethanol is postulated to act exclusively by inhibiting the sodium conductance channel, the relative degree of inhibition of the different neurotransmitters should be identical for both ethanol and TTX. This is, however, not the case. Unfortunately, it is difficult to determine if (a) or (b) are correct and thus to reach a conclusion as to the mode of action of ethanol. A separate study using saxitoxin, another neurotoxin affecting the sodium pore (Hille, 1968), instead of TTX, could help in solving this problem. If the sensitivities of the different neurotransmitter systems for saxitoxin are identical to those of TTX, the safety margin alternative would most likely be correct.

TABLE 2

Concentration of Tetrodotoxin Required to Inhibit the Electrically Stimulated Release of Neurotransmitters[a] by 50%

	IC50 (nM)
Acetylcholine	2.9 ± 0.35 (27)[b]
Serotonin	3.0 ± 0.64 (27)
Glutamate	4.7 ± 0.57 (27)
Norepinephrine	7.5 ± 2.00 (27)

[a]Rat brain cortical slices were preloaded with ^{14}C-choline and ^{3}H-NE or ^{14}C-glutamate and ^{3}H-5HT. The effect of tetrodoxin (1, 3 and 6 nM) on the electrically stimulated release of these transmitters was determined.

[b]Values presented are means ± S.E.M. The number of determinations is given in parenthesis.

Based on the findings that ethanol inhibits the intracellular changes in Na^+ and K^+ that are evoked by electrical stimulation, Nikander and Wallgren (1970) have suggested that ethanol inhibits the action potential in brain tissue. We have confirmed their results (Figure 4). Ethanol, in concentrations of 1 and 2% significantly inhibited the intracellular (non-inulin space) accumulation of Na^+ and the reduction in K^+ elicited by electrical stimulation. However, since ethanol, at these concentrations inhibits the release of some neurotransmitters, it is not possible to conclude with certainty that the effect of ethanol was exerted primarily on the action potential. The effect of ethanol on the intracellular changes in Na^+ and K^+ could also conceivably be due to a reduction in the degree of endogenous chemical stimulation of the postsynaptic nerves in the slice.

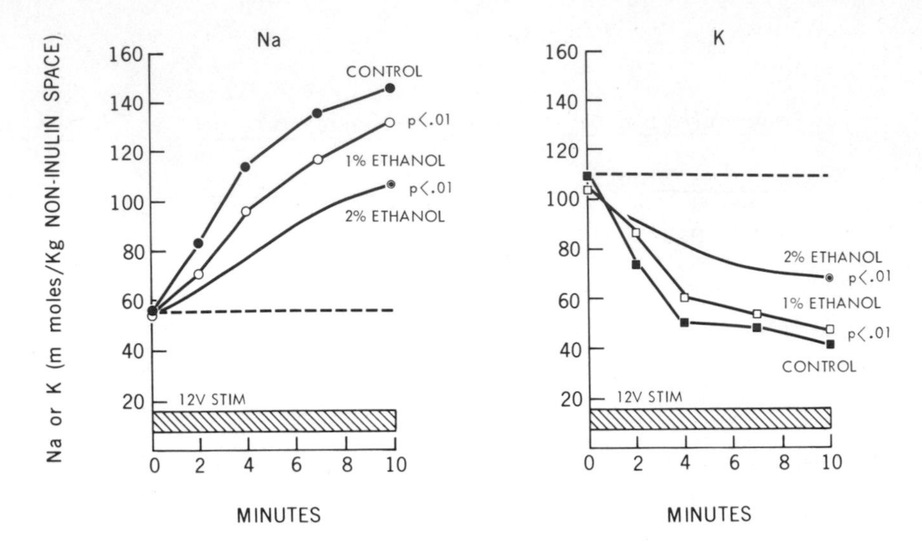

Figure 4. Effect of ethanol on the accumulation of Na^+ and loss of K^+ elicited by electrical stimulation. Rat brain cortical slices were preincubated in the absence of ethanol for 35 minutes in order to attain a steady state base line before stimulation. Stimulation (STIM) with alternating condenser pulses (12 volts, 50 Hz, time constant = 0.5 m sec) using carbon electrodes, was done according to Nikander and Wallgren (1972).

In conclusion, ethanol inhibits preferentially the release of ACh when compared to that of the other neurotransmitters studied, with significant effects being seen at pharmacologically relevant concentrations (0.25%; 0.05 M). Ethanol inhibits the electrically stimulated ion movements of Na^+ and K^+. Although the data presented here are consistent with ethanol exerting an effect on sodium conductance, it is not possible to rule out a specific effect at the presynaptic level.

References

Armstrong, C.M. and Binstock, L. (1964). J. Gen. Physiol. 48, 265.

Carlsson, A. and Lindqvist, M. (1973). J. Pharm. Pharmacol. 25, 437.

Carlsson, A., Magnusson, T., Svensson, T.H. and Waldeck, B. (1973). Psychopharmacol. 30, 27.

Corrodi, H., Fuxe, K. and Hokfelt, T. (1966). J. Pharm. Pharmacol. 18, 821.

Erickson, C.K. and Graham, D.T. (1973). J. Pharmacol. Exp. Ther. 185, 583.

Frankel, D., Khanna, J.M., Kalant, H. and LeBlanc, A.E. (1974). Psychopharmacol. 37, 91.

Gage, P.W. (1965). J. Pharmacol. 150, 236.

Hille, B. (1968). Prog. Biophys. Molec. Biol. 21, 1.

Hunt, W.A. and Majchrowicz, E. (1974). J. Neurochem. in press.

Hunt, W.A. and Majchrowicz, E. (1974). Brain Res. 72, 181.

Inoue, F. and Frank, G.B. (1967). Br. J. Pharmacol. 30, 186.

Israel, Y., Carmichael, F.J. and Macdonald, J.A. (1973). Ann. N.Y. Acad. Sci. 215, 38.

Israel, Y., Rosemann, E., Hein, S., Colombo, G. and Canessa-Fischer, M. (1971). Biological Basis of Alcoholism. Eds. Y. Israel and J. Mardones, Chapter II. Wiley, New York, N.Y.

Kalant, H. and Grose, W. (1967). J. Pharmacol. 158, 386.

Kalant, H., Israel, Y. and Mahon, M.A. (1967). Can. J. Physiol. Pharmacol. 45, 172.

Kuriyama, K., Rauscher, G.E. and Sze, P.Y. (1971). Brain Res. 26, 450.

Leo, A., Hansch, C. and Elkins, D. (1971). Chem. Rev. 71, 525.

Moore, J.W., Ulbricht, W. and Takata, M. (1964). J. Gen. Physiol. 48, 279.

Narahashi, T., Moore, J.W. and Scott, W.R. (1964). J. Gen. Physiol. 47, 965.

Nikander, P. and Wallgren, H. (1970). Acta Physiol. Scand. 80, 27A.

Nikander, P. and Wallgren, H. (1972). J. Neurochem. 19, 535.

Okada, K. (1967). Jap. J. Physiol. 17, 245.

Palaic, D.T., Desaty, J., Albert, J.M. and Panisset, J.C. (1971). Brain Res. 25, 381.

Phillis, J.W. and Jhamandas, (1971). Comp. Gen. Pharmacol. 2, 306.

Roach, M.K., Davis, D.L., Pennington, W. and Nordyke, E. (1973). Life Sci. 12, pt. I, 433.

Seeman, P. (1972). Pharmacol. Rev. 24, 583.

Wallgren, H. (1971). Handbook of Neurochem. VI, p. 509. Ed. A. Lajtha, Plenum, New York, N.Y.

DISPOSITION OF CATECHOLAMINE-DERIVED ALKALOIDS IN MAMMALIAN SYSTEMS

Virginia E. Davis, Jesse L. Cashaw and K. D. McMurtrey

Neurochemistry and Addiction Research Laboratory

Veterans Administration Hospital, Houston, Texas

The biosynthesis of alkaloids has long been considered unique to plants. Evidence evolving in recent years, however, indicates that mammalian systems may also possess this capability and the potential pharmacological consequences of biosynthesized alkaloids may be of great significance.

Condensation of catecholamines or indoleamines with carbonyl compounds yields tetrahydroisoquinoline alkaloids or tetrahydro-β-carboline alkaloids, respectively. Thus, reaction of the catecholamines, dopamine, norepinephrine, or epinephrine, with acetaldehyde the primary metabolite of ethanol, readily produces the corresponding 1-methyltetrahydroisoquinoline alkaloids. The ease with which formaldehyde or acetaldehyde condense with substituted phenethylamines to form the corresponding tetrahydroisoquinoline alkaloids suggests that the possible formation of these alkaloids as a consequence of acute or chronic ingestion of methanol or ethanol is not an unreasonable assumption. Thus, the ready production of salsolinol, the condensate of dopamine with acetaldehyde, on incubation of brain or liver homogenates with dopamine and acetaldehyde or ethanol has been demonstrated (1). Cohen and coworkers have described extensive studies of the direct condensation of catecholamines with acetaldehyde or formaldehyde in vitro and in vivo and their detailed reports of the uptake, storage and release of the simple tetrahydroisoquinoline alkaloids support the suggestion that this group of compounds can act as false neurotransmitters (2-5).

In addition to the formation of simple tetrahydroisoquinoline alkaloids derived by the direct condensation of catecholamines with the aldehyde metabolites of methanol or ethanol, the metabolism of ethanol may also precipitate a more intricate chain of events with

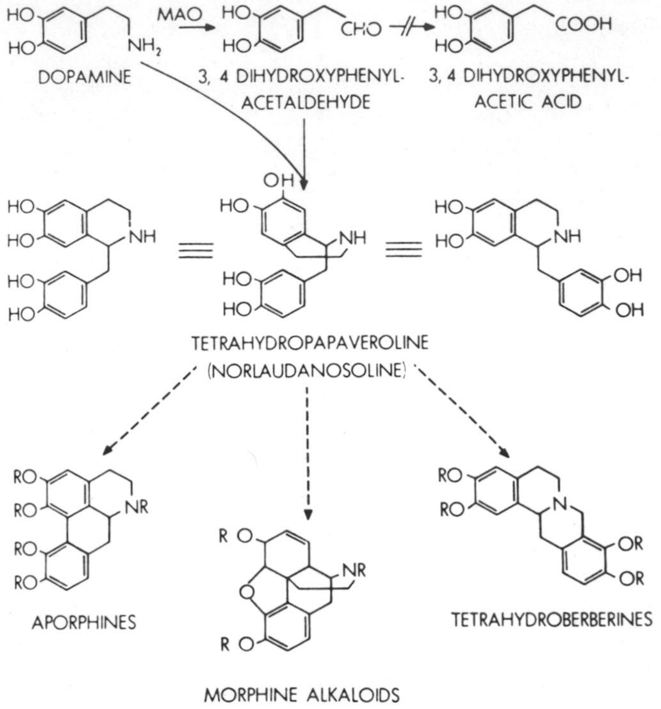

DOPAMINE

MAO

3, 4 DIHYDROXYPHENYL-ACETALDEHYDE

3, 4 DIHYDROXYPHENYL-ACETIC ACID

TETRAHYDROPAPAVEROLINE (NORLAUDANOSOLINE)

APORPHINES

MORPHINE ALKALOIDS

TETRAHYDROBERBERINES

Figure 1. Structural representation of a working hypothesis for the relationship of alcohol-evoked modifications of dopamine metabolism to alkaloid formation. Known pathways, either in vivo or in vitro, are indicated by (⟶). The site of inhibition of dopamine metabolism by acetaldehyde, the metabolite of ethanol, is shown by (⟶//⟶). The proposed possible consequence of this inhibition resulting in the formation of various complex alkaloids is illustrated by (- - - ->). The latter is not meant to be exhaustive and all inclusive. (Davis, et al., J. Pharmacol. Exp. Ther. 174: 401, 1970).

resultant formation of complex benzyltetrahydroisoquinoline alkaloids as aberrant metabolites of the catecholamines (Fig. 1). We have previously presented the postulate that drug-evoked aberrations of neuroamine metabolism may result in the endogenous formation of pharmacologically active alkaloids by mammalian systems (6-10). This concept is based on the premise that alcohol - through its primary metabolite acetaldehyde - as well as certain other sedative-hypnotic drugs such as chloral hydrate, in the case of dopamine (6-10), and barbiturates, in the case of norepinephrine (10,11), disrupt the major metabolic route for disposition of the aldehyde derivatives of the catecholamines. This disruption increases the availability of these aldehydes for participation in a Pictet-Spengler type condensation (12) with the intact neuroamines, the products of which are the corresponding pharmacologically active benzyltetrahydroisoquinoline alkaloids. For example, acetaldehyde competitively inhibits the oxidation of the dopamine-derived aldehyde to the corresponding acid with resultant increase in the steady state levels of the biogenic aldehyde and enhanced formation of tetrahydropapaveroline (THP) (6-9), a pharmacologically active (13-17) benzyltetrahydroisoquinoline alkaloid.

Benzyltetrahydroisoquinoline alkaloids are the biogenic precursors in plants of a diverse array of more complex alkaloids, e.g., the protoberberine, morphine, aporphine, as well as, papaverine, benzophenanthridine, and phthalideisoquinoline classes (18-23). The in situ formation of benzyltetrahydroisoquinoline alkaloids in man could, therefore, possibly make them available for conversions to even more complex alkaloids. Thus, certain neuropharmacological effects of alcohol and pharmacologically equivalent drugs may be mediated through various aberrant alkaloid metabolites of the neuroamines - e.g., the simple alkaloids formed by direct condensation of formaldehyde or acetaldehyde with the neuroamines, the complex benzyltetrahydroisoquinoline alkaloids, and further metabolites of the latter.

At the time of publication of the foregoing hypothesis and its initial supporting data (6-9), no tetrahydroisoquinoline alkaloids had been detected as catecholamine metabolites in intact animals or in man. However, subsequently both salsolinol and tetrahydropapaveroline were detected as urinary excretion products in patients receiving L-dopa therapy for Parkinson's disease (24). Additionally, the in vivo formation of salsolinol in dopamine-rich brain areas of rats following treatment with pyrogallol and ethanol (25) and production of tetrahydropapaveroline in the brains of rats after oral administration of L-dopa and ethanol (26) have been reported.

It should be emphasized that direct assessment of the total production of amine-derived tetrahydroisoquinoline alkaloids in mammalian systems is complicated by the probability that they are not metabolic end products but rather are extensively metabolized along diverse routes. For example, both salsolinol and tetrahydropapaveroline retain vicinal hydroxyl groups on an aromatic nucleus and it therefore seems reasonable to assume that at least one metabolic pathway of these compounds in mammals may be methylation by catechol-O-methyltransferase (COMT). In vitro studies using COMT preparations derived from either rat brain or liver demonstrated that salsolinol and tetrahydropapaveroline were excellent substrates for COMT with apparent Km values of 0.29 and 0.03 mM, respectively. The maximal velocities of salsolinol and tetrahydropapaveroline O-methylation proved to be three to five times the maximal veloci-

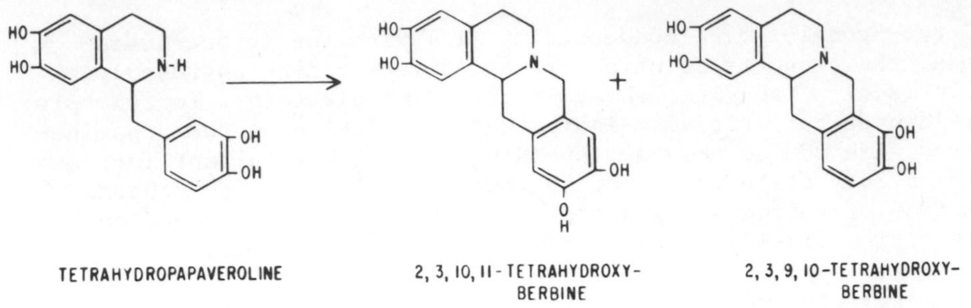

TETRAHYDROPAPAVEROLINE

2, 3, 10, 11-TETRAHYDROXY-BERBINE

2, 3, 9, 10-TETRAHYDROXY-BERBINE

Figure 2. Formation of tetrahydroprotoberberine isomers from tetrahydropapaveroline.

ties of norepinephrine and dopamine O-methylation. Not unexpectedly, both salsolinol and tetrahydropapaveroline competitively inhibited the O-methylation of catecholamines with calculated inhibitor constants (Ki values) of 0.13 and 0.02 mM, respectively (27). The complex benzyltetrahydroisoquinoline alkaloid, tetrahydropapaveroline , is of particular interest in this regard since it exceeds the simple 1-methyltetrahydroisoquinoline alkaloid, salsolinol, both in competitive inhibition of COMT and in evoking β-sympathomimetic responses. Pyrogallol-sensitive O-methylation of tetrahydroisoquinolines derived by condensation of norepinephrine with acetaldehyde or formaldehyde by rat liver and brain homogenates has also been observed (28).

Additionally, it is proposed that the benzyltetrahydroisoquinoline alkaloids may be metabolized to even more complex alkaloids by mammalian systems as they are in plants (Fig. 1). Although the enzymatic mechanisms involved in these transformations in plants are entirely undefined, it is known that these complicated conversions involve specifically directed methyl transfer and oxidative coupling reactions. We wish to focus on this aspect of the hypothesis (8) and report the formation of tetrahydroportoberberine alkaloids in experimental animals and in man (Fig. 2).

METHODS AND RESULTS

Biotransformation of Tetrahydropapaveroline in Rats

Tetrahydropapaveroline.HBr (0.4mM/Kg, 147 mg/kg) in 30% propylene glycol was injected intraperitoneally into male Sprague-Dawley rats weighing about 200 g each. Rats similarly injected with the vehicle served as controls. Urine was collected in 4 ml

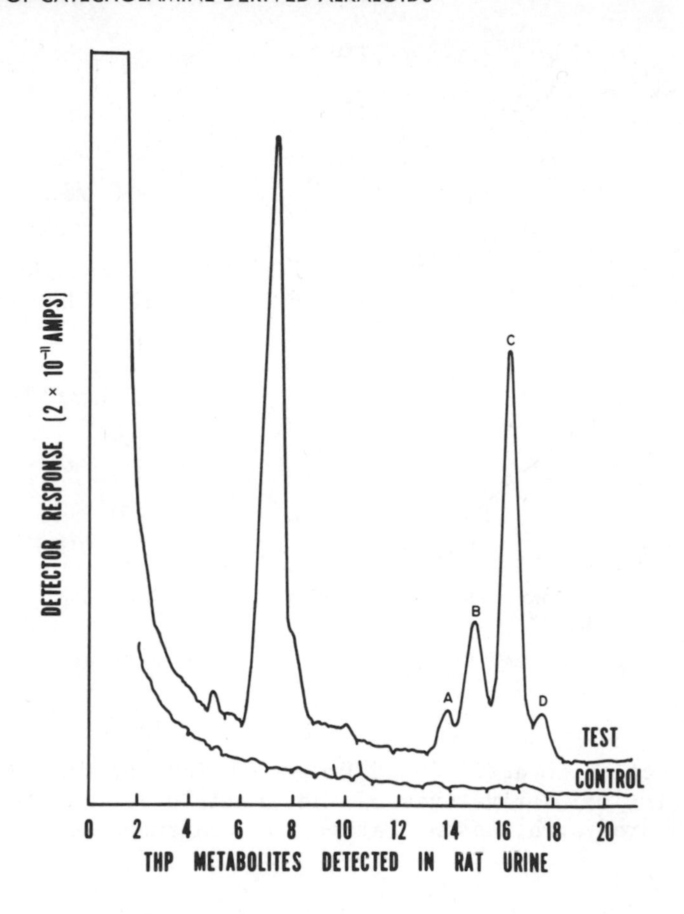

Figure 3. Gas chromatographic separation of silylated tetrahydro-protoberberine alkaloids from urine of rats pretreated with tetra-hydropapaveroline. The peaks correspond to (A) coreximine (2,11,-dihydroxy-3,10-dimethoxyberbine), (B) a 2- or 3-monomethylated derivative of tetrahydroxyberbine, (C) 2,3,10,11-tetrahydroxyber-bine, and (D) 2,3,9,10-tetrahydroxyberbine. Compound (A) is tentatively identified as coreximine on basis of retention time. Structure assignments of (B), (C), and (D) are based on gas chromatographic and mass spectral characteristics.

of 0.1 N HCl for 24 hours after the injection, diluted to 70 ml, hydrolyzed enzymatically (29), and adjusted to pH 7. Tetrahydroprotoberberine alkaloids were extracted into 4 volumes of toluene-isoamyl alcohol (3:2) and returned to one-tenth volume 0.1N HCl. Aliquots (1ml) of the HCl extracts were dried in vacuo, reacted with hexamethyl-disilazane-trimethyl-chlorosilane (9:1), 100ul, for 8 hours at 65° to prepare the trimethyl-silyl (TMS) derivatives

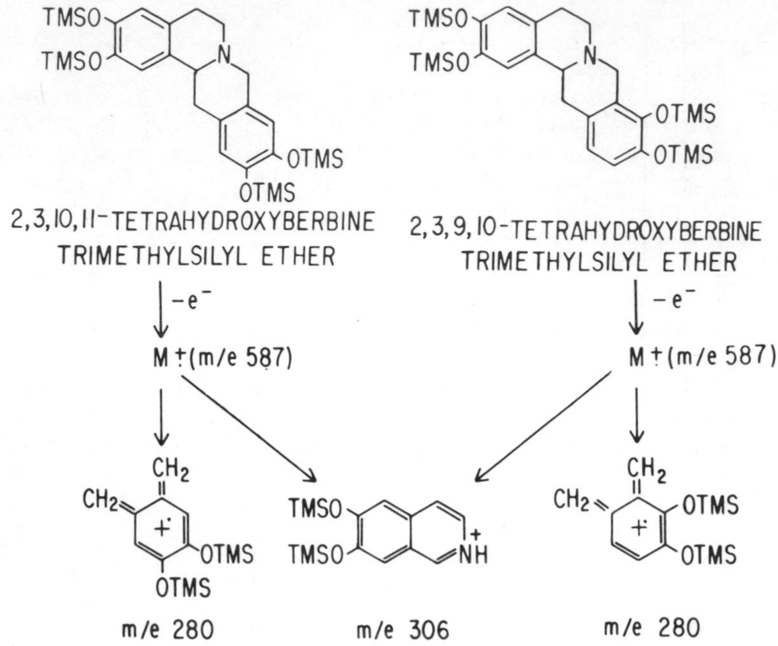

2,3,10,11-TETRAHYDROXYBERBINE TRIMETHYLSILYL ETHER

2,3,9,10-TETRAHYDROXYBERBINE TRIMETHYLSILYL ETHER

M+·(m/e 587)

M+·(m/e 587)

m/e 280

m/e 306

m/e 280

Figure 4. Electron impact induced fragmentation patterns of the trimethylsilyl (TMS) derivatives of the isomeric 2,3,10,11- and 2,3,9,10-tetrahydroxyberbines indicating the congruence of their mass spectra.

for gas chromatography and mass spectrometry. Extraction efficiency was estimated and quantification was achieved by measuring peak area, using certain authentic tetrahydroprotoberberine standards synthesized in this laboratory. Gas chromatographic conditions were as previously described for determination of O-methylated metabolites of THP (27). Mass spectra were determined with an LKB 9000 gas chromatograph-mass spectrometer equipped with a 4mm x 9 ft. 1% SE-30 column at 255°.

Evidence of THP's ready conversion to O-methylated benzyltetrahydroisoquinoline derivatives by rats in vivo is found in the significant amounts of these products excreted (30). In addition, the capability of mammalian systems to effect the synthesis of the tetracyclic ring system of protoberberine alkaloids is demonstrated by the production of four tetrahydroprotoberberine alkaloids. None of these alkaloids was detected in the urine of the control rats (Fig. 3).

In the 24-hour period following THP administration, an average of 775 μg of tetrahydroprotoberberine alkaloids was recovered from the urine of each rat. The major tetrahydroprotoberberine alkaloid metabolite excreted was (C) 2,3,10,11-tetrahydroxyberbine (498μg). Minor constituents included the isomeric (D) 2,3,9,10-tetrahydroxyberbine (44μg) and also two methylated derivatives. The latter were tentatively identified as, first, coreximine - (A) 2,11,dihydroxy-3,10-dimethoxyberbine (50μg) - with a retention time of 13.3 min and methylene unit value of 30.9, and second, a 2- or 3-monomethylated derivative (B) of 2,3,10,11-tetrahydroxyberbine (183 μg), with a retention time of 14.8 min and methylene unit value of 31.3.

The mass spectra of the silylated derivatives of two of these THP metabolites were identical and also congruent with those obtained for the authentic reference compounds - 2,3,9,10-tetrahydroxyberbine and 2,3,10,11-tetrahydroxyberbine- each having an apparent molecular ion at m/e 587 (35%), a base peak at m/e 280, and a fragment ion at m/e 306 (23%) (Fig. 4). This fragmentation pattern is characteristic of silylated tetrahydroprotoberberine alkaloids, but it does not differentiate between structural isomers. However, these isomers are resolved by gas chromatography. Comparison with authentic reference standards established that one isomer, having a retention time of 16.1 min and methylene unit value of 31.7, was 2,3,10,11-tetrahydroxyberbine; and the other, having 17.3 min retention time and a 32.0 methylene unit value, was 2,3,9,10-tetrahydroxyberbine.

Metabolism of Tetrahydropapaveroline by Rat Liver and Brain Preparations

These findings signal the existence of a previously unrecognized enzyme system in mammals that mimics the capability of plants to form the "berberine bridge" from the benzyltetrahydroisoquinoline alkaloids. Although the enzyme systems in plants effecting tetrahydroprotoberberine alkaloid biosynthesis have not been defined, it is known that the carbon atom of the "berberine bridge" can be derived from the methyl group of methionine (31). S-adenosylmethionine or methyltetrahydrofolic acid were considered likely candidates as the methyl donors for this conversion in mammalian systems.

To investigate this possibility, the portion of the soluble supernatant fraction - obtained by centrifuging homogenates of brains or livers of Sprague-Dawley rats at 46,000 x g for 60 min - precipitating at 30% to 50% saturation with ammonium sulfate - was used as the enzyme source. The enzyme preparation (15 mg protein) was incubated at 37° for 1 hour with 12.5 μmoles THP, 2.5 μmoles S-adenosyl-L-methionine (SAM), 1.25 μCi ^{14}C-SAM, 1.25 mg ascorbic acid, 1.25 μmoles $MgCl_2$ and 0.5 mmoles phosphate buffer (pH 8.0) in a final volume of 6.0 ml. The reaction was terminated by adding 6ml 0.5M phosphate buffer (pH 6.0) and immediately extracting the reaction products into 50 ml toluene-isoamyl alcohol (3:2). The metabolites were then returned to one-tenth volume 0.1N HCl and identified by gas chromatography and mass spectrometry.

Three of the four tetrahydroprotoberberine alkaloids detected as urinary THP products were also formed in vitro by both rat brain and liver preparations. Through combined gas chromatography-

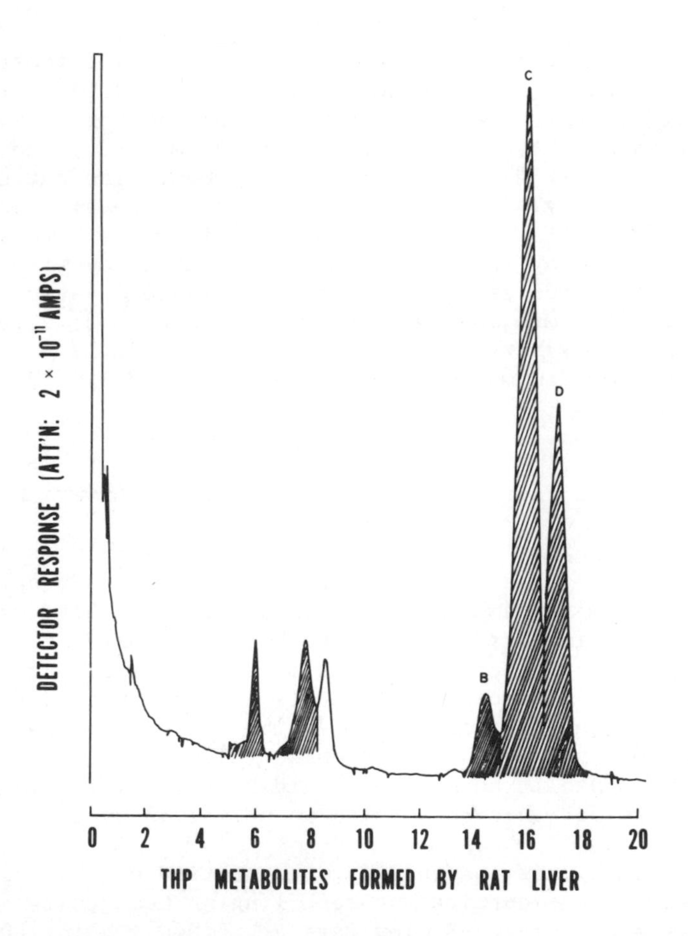

Figure 5. Gas chromatographic separation of the metabolites of tetrahydropapaveroline obtained on incubation of the soluble fraction of rat liver homogenates in the presence of ^{14}C-methyl-S-adenosylmethionine. Shaded areas indicate radioactivity. Peaks B,C and D correspond to the same compounds indicated in Figure 3.

radioassay and mass-spectral analysis, the three radio-labeled tetrahydroprotoberberine alkaloids were identified as 2,3,10,11-tetrahydroxyberbine, 2,3,9,10-tetrahydroxyberbine, and a 2- or 3-monomethylated derivative of tetrahydroxyberbine. The 2,3,10,11-tetrahydroxy isomer was the main tetrahydroprotoberberine metabolite found both in vitro and in vivo (Fig. 5).

Excretion of Tetrahydroprotoberberine Alkaloids by Man

As mentioned previously, detailed investigations of the biogenesis of plant alkaloids have clearly established that benzyltetrahydroisoquinoline alkaloids, especially tetrahydropapaveroline, are requisite intermediates in the biosynthesis of a rich variety of other complex alkaloids (18-23). Low levels of salsolinol and tetrahydropapaveroline in the urine of parkinsonian patients receiving L-dopa therapy were recently reported by Sandler and coworkers (24). Since these tetrahydroisoquinoline alkaloids can be formed by the non-enzymatic condensation of dopamine with the aldehyde, the possibility exists that these tetrahydroisoquinoline alkaloids might be formed in the bladder and not necessarily in tissues. However, identification of further metabolites of these tetrahydroisoquinoline alkaloids as urinary excretion products in man would be strongly indicative of their formation in tissues. Thus, identification of tetrahydroprotoberberine alkaloids as urinary metabolites would add support to the postulate that neuroamine-derived tetrahydroisoquinoline alkaloids not only can be formed by man (especially under appropriate pharmacological conditions) but also can be converted to even more complex alkaloids (6-10).

Urine samples were obtained from parkinsonian patients receiving 4 to 5 g of L-dopa daily and were processed as described above. Two isomeric alkaloids - 2,3,10,11-tetrahydorxyberbine and 2,3,9,10-tetrahydroxyberbine - which were found previously to be tetrahydropapaveroline metabolites both in vivo and in vitro in rats, were readily detected. Gas chromatographic retention times and mass spectral fragmentation patterns of these alkaloids were identical to the authentic reference compounds. The quantitative significance and the effects of alcohol and related drugs on the production of this class of complex alkaloids in man is presently under investigation.

DISCUSSION

The relatively small quantities of tetrahydroprotoberberine alkaloids detected as metabolites of L-dopa and tetrahydropapaveroline must be considered as semiquantitative approximations and most probably represent only a portion of the total production of these alkaloids for several reasons. The tetrahydroprotoberberine alkaloids, like their benzyltetrahydroisoquinoline precursors, are not metabolic end products in plants but are key intermediates in the biosynthesis of other complex alkaloids such as the benzophenanthridines, the protopines, and the phthalideisoquinolines (23) and the possibility exists that the tetrahydroprotoberberine alkaloids detected may be relatively short lived intermediates likewise in intact animals. Additionally, considerable technical difficulties in the quantitative determination of the tetrahydroprotoberberine alkaloids in mammalian systems have been encountered and only partially resolved to date. Recoveries were irregular due no doubt to a chemically evanescent nature. Extraction efficiencies were low and varied widely between the partially O-methylated and the tetrahydroxy compounds. The latter alkaloids were more effectively extracted. Although ample quantities of the tetrahydroxy isomers of these alkaloids have been synthesized in this laboratory, amounts of the O-methylated isomers sufficient for detailed studies of extraction effeciences and overall recovery have yet to be prepared. Notwithstanding the technical difficulties inherent in this investigational area, the ability of mammalian systems to insert the one carbon unit to form the "berberine bridge" and thus complete the skeleton of the tetrahydroprotoberberine alkaloids is established. Furthermore, the intermediacy of tetrahydropapaveroline in the biosynthesis of complex tetrahydroprotoberberine alkaloids in mammals is demonstrated. Hence, the foregoing results are in complete consonance with the ideas previously proposed (6-11).

Representatives of most classes of complex alkaloids related biosynthetically to the isoquinolines exhibit a wide range of pharmacological activity (23, 32). The specific tetrahydroprotoberberine alkaloids revealed by these studies as ultimately deriving from catecholamines in mammalian systems have not yet been examined extensively for their pharmacological effects. However, we have found that both tetrahydropapaveroline and its major tetrahydroprotoberberine metabolite, 2,3,10,11-tetrahydroxyberbine, inhibit uptake and enhance release of catecholamines in synaptosomes (33). Additionally, derivatives closely related to the tetrahydroprotoberberine alkaloids detected in this investigation have been reported to possess analgesic, sedative, and tranquillizing properties (34-36). Furthermore, a series of substituted tetrahydroprotoberberine alkaloids have been synthesized and patented as tranquilizers (37). It should be noted that both the tetrahydroisoquinoline and tetrahydroprotoberberine alkaloids possess asymmetric centers. Thus, the pharmacological action of specific isomers may be widely divergent. The stereo-dependence of the pharmacological actions of these alkaloids has received negligible attention to date. However, the reports of stereoselectivity of some substituted tetrahydroisoquinolines on lipolysis (38) and of certain tetrahydroprotoberberines on analgesic, sedative and hypnotic actions (34) suggest that the absolute configuration may be of primary importance in the pharmacological activity of these two classes of alkaloids.

SUMMARY

Tetrahydropapaveroline, the tetrahydroisoquinoline alkaloid derived from dopamine, is converted in vivo by rats and by rat liver and brain preparations to tetrahydroprotoberberine alkaloids. The latter alkaloids have also been identified for the first time in the urine of parkinsonian patients receiving L-dopa therapy. These findings suggest that man, like plants, may have the ability to elaborate several classes of alkaloids with potentially important pharmacological consequences. Thus, this newly demonstrated ability of mammalian systems to evoke the biosynthesis of benzyltetrahydroisoquinoline-derived alkaloids - a capability previously considered unique to plants - elects the tetrahydroprotoberberine alkaloids as representative of the first class of a possible constellation of complex mammalian alkaloids elaborated from the neuroamines.

ACKNOWLEDGEMENT

Supported by USPHS Grant AA00226 and the Veterans Administration. We thank Drs. E.C. and M.C. Horning (Baylor College of Medicine)for providing facilities for mass spectrometric analysis and Dr. E. Brochmann-Hanssen (University of California, San Francisco) for supplying authentic reference standards of certain tetrahydroprotoberberine alkaloids.

REFERENCES

1. Yamanaka, Y., Walsh, M.J. and Davis, V.E.: Salsolinol, an alkaloid derivative of dopamine formed in vitro during alcohol metabolism. Nature 227: 1143-1144, 1970.

2. Cohen, G. and Collins, M.: Alkaloids from catecholamines in adrenal tissue: possible role in alcoholism. Science 167: 1749-1751, 1970.

3. Greenberg, R.S. and Cohen, G.: Tetrahydroisoquinoline alkaloids: stimulated secretion from the adrenal medulla. J. Pharmacol. Exp. Therap. 184: 119-128, 1973.

4. Cohen, G.: A role for tetrahydroisoquinoline alkaloids as false adrenergic neurotransmitters in alcoholism. In M.M. Gross (ed.), Alcohol Intoxication and Withdrawal, Experimental Studies. Adv. Exp. Biol. Med. 35: 33-44, New York: Plenum Press. 1973.

5. Mytilineou, C., Cohen, G. and Barrett, R.: Tetrahydroisoquinoline alkaloids: uptake and release by adrenergic nerves in vivo. Eur. J. Pharmacol. 25: 390-401, 1974.

6. Davis, V.E.: Alcohol and aberrant metabolism of biogenic amines. In M.K. Roach, W.M. McIssac and P.J. Creaven (eds.), Biological Aspects of Alcohol, pp. 293-312. Austin: University of Texas Press. 1971.

7. Davis, V.E. and Walsh, M.J.: Alcohol, amines and alkaloids: a possible biochemical basis for alcohol addiction. Science 167: 1005-1007, 1970.

8. Davis, V.E., Walsh, M.J. and Yamanaka, Y.: Augmentation of alkaloid formation from dopamine by alcohol and acetaldehyde in vitro. J. Pharmacol. Exp. Ther. 174: 401-412, 1970.

9. Davis, V.E. and Walsh, M.J.: Effect of ethanol on neuroamine metabolism. In Y. Isreal and J. Mardones (eds.), Biological Basis of Alcoholism, pp. 73-102. New York: Wiley-Interscience. 1971.

10. Davis, V.E.: Neuroamine-derived alkaloids: a possible common denominator in alcoholism and related drug dependencies. Ann. N.Y. Acad. Sci. 215: 111-115, 1973.

11. Davis, V.E., Cashaw, J.L., McLaughlin, B.R. and Hamlin, T.A.: Alternation of norepinephrine metabolism by barbiturates. Biochem. Pharmacol. 23: 1877-1889, 1974.

12. Whaley, W.M. and Govindachari, T.R.: The Pictet-Spengler synthesis of tetrahydroisoquinolines and related compounds. In R. Adams et al. (eds.), Organic Reactions 6: 151-206. New York: Wiley. 1951.

13. Laidlaw, P.P.: The action of tetrahydropapaveroline hydrochloride. J. Physiol. (London) 40: 480-491, 1910.

14. Holtz, P., Stock, K. and Westermann, E.: Pharmakologie des Tetrahydropapaverolins und seine Entstehung aus Dopamin. Naunyn-Schmiedebergs Arch. Pharmakol. Exp. Pathol. 248: 387-405, 1964.

15. Santi, R., Bruni, A., Luciani, S., Toth, C.E., Ferrari, M.,

Fassina, G. and Contessa, A.R.: Pharmacological properties of tetrahydropapaveroline and their relation to the catecholamines. J. Pharm. Pharmacol. 16: 287-288, 1964.

16. Santi, R., Ferrari, M., Toth, C.E., Contessa, A.R., Fassina, G., Bruni, A. and Luciani, S.: Pharmacological properties of tetrahydropapaveroline. J. Pharm. Pharmacol. 19: 45-51, 1967.

17. Kukovetz, W.R. and Poch, G.: Beta-adrenerge Effekte und ihr zeitlicher Verlaug unter Tetrahydropapaverolin und Isoprenalin am Langendorff-Herzen. Naunyn-Schmeidebergs Arch. Pharmakol. Exp. Pathol. 256: 301-309, 1967.

18. Battersby, A.R.: Alkaloid Biosynthesis. Quart. Rev. 15: 259-286, 1961.

19. Bentley, K.W.: The Isoquinoline Alkaloids. New York: Pergamon Press. 1965.

20. Spenser, E.D.: Biosynthesis of the alkaloids related to norlandanosoline. Lloydia 29: 71, 1966.

21. Kirby, G.W.: Biosynthesis of the morphine alkaloids. Science 155: 170-173, 1967.

22. Robinson, T.: The Biochemistry of Alkaloids. pp. 54-71. New York: Springer Verlag. 1968.

23. Shamma, M.: The Isoquinoline Alkaloids: Chemistry and Pharmacology. In A.T. Blomquist and H. Wasserman (eds.) Organic Chemistry, 25. New York: Academic Press. 1972.

24. Sandler, M., Carter, S.B., Hunter, K.R. and Stern, G.M.: Tetrahydroisoquinoline alkaloids: in vivo metabolites of L-dopa in man. Nature 241: 439-443, 1973.

25. Collins, M.A. and Bigdeli, M.D.: Alcohol intoxication and in vivo biosynthesis of the alkaloid, salsolinol, from dopamine in rat brain. Trans. Amer. Soc. Neurochem. 5: 160, 1974.

26. Algeri, S., Baker, K.M., Frigerio, A. and Turner, A.J.: Identification and quantitation of tetrahydropapaveroline in rat brain. Proc. 21st Annual Conference Amer. Soc. Mass Spectrometry: 301-302, 1973.

27. Collins, A.C., Cashaw, J.L. and Davis, V.E.: Dopamine-derived tetrahydroisoquinoline alkaloids - inhibitors of neuroamine metabolism. Biochem. Pharmacol. 22: 2337-2348, 1973.

28. Rubenstein, J.A. and Collins, M.A.: Tetrahydroisoquinolines derived from norepinephrine-aldehyde condensations - pyrogallol-sensitive O-methylation in rat homogenates. Biochem. Pharmacol. 22: 2928-2931, 1973.

29. Davis, V.E., Brown, H., Huff, J.A., and Cashaw, J.L.: Ethanol-induced alternations of norepinephrine metabolism in man. J. Lab. Clin. Med. 69: 787-799, 1967.

30. Cashaw, J.L., McMurtrey, K.D., Davis, V.E.: Proc. 165th National Meeting Amer. Chem. Soc. Anal. Chem. Div. Abstract No. 5, 1973.

31. Gupta, R.N. and Spenser, E.D.: Biosynthetic incorporation of one carbon units into berberine and hydrastine. Can. J. Chem. 43: 133, 1965.

32. Macko, D., Douglas, B., Weisbach, J.A. and Waltz, D.T.: Studies on the pharmacology of nuciferine and related aporphines. Arch. int. Pharmacodyn. 197: 261-273, 1972.

33. Alpers, H.S., McLaughlin, B.R., Nix, W.M. and Davis, V.E.: Tetrahydroisoquinoline and tetrahydroprotoberberine alkaloids: Inhibition of catecholamine accumulation by rat brain synaptosomal preparations. Fed. Proc. 33: 511, 1974.

34. Hsu, B. and Kin, K.C.: Pharmacological study of tetrahydropalmatine and its analogs - a new type of central depressants. Arch. int. Pharmacodyn., 139: 318-327, 1962.

35. Nakanishi, H.: Pharmacological studies of xylopinine, 2,3,10, 11-tetramethoxy-5,6-13,13a-tetrahydro-8-dibenzo(a-g)quinolizine, semisynthesized from phellodendrine isolated from Phellodendron amurense Rupp. Jap. J. Pharmacol. 12: 208-222, 1962.

36. Yamamoto, H.: The central effects of xylopinine in mice. Jap. J. Pharmacol. 13: 230-239, 1963.

37. Roussel-UCLAF Patents, Substituted 3,9,10-trimethoxyberbine tranquilizers. Chem. Abstracts 69: 59475m, 1968; 1,2,3,10-tetramethoxyberbine, a tranquilizing agent. Chem. Abstracts 71: 70788c, 1969; Tranquilizing dl-1,3,10-trimethoxyberbine. Chem. Abstracts. 73: 15063d, 1970.

BIOSYNTHESIS OF TETRAHYDROISOQUINOLINE ALKALOIDS IN BRAIN AND OTHER TISSUES OF ETHANOL-INTOXICATED RATS

Michael A. Collins and Mostafa G. Bigdeli

Department of Biochemistry and Biophysics, Loyola Univ.

Stritch School of Medicine, Maywood, Illinois

INTRODUCTION

This paper summarizes some of our studies on the interactions of alcohol or amine-derived aldehydes with the catecholamines (CAs). There has been increasing research interest and speculation since 1968 in the potential formation of tetrahydroisoquinoline (TIQ) alkaloids in mammalian cells via such interactions, and in the role such TIQs may have in alcohol dependence. Reviewing, two principal TIQ theories related to alcoholism have been elaborated. (1) The initial suggestion is that ethanol (EtOH)-derived acetaldehyde (AcH) may produce "simple" (1-alkyl) TIQs from its bimolecular cyclization or condensation with CAs (Collins and Cohen, 1968; Robbins, 1968; Cohen and Collins, 1970; Yamanaka *et al*., 1970). The major products of these cyclizations, shown in Figure 1, are salsolinol derived from dopamine (DA), and the corresponding 4, 6, 7-trihydroxy-TIQ alkaloids derived from epinephrine (E) and norepinephrine (NE). For simplicity, the minor isomeric TIQs resulting from neutral aqueous cyclizations and stereochemical isomers of all products (Cohen and Collins, 1970; King *et al*., 1974) are not considered. (2) The second postulate depends upon the fact that EtOH-derived AcH, as a substrate for aldehyde dehydrogenase, may inhibit the oxidation and thereby elevate concentrations of endogenous phenylacetaldehydes which normally arise from DA and 3-methoxy-DA deaminations. Davis and Walsh (1970) suggested the EtOH-dependent augmentation of a TIQ formation route demonstrated *in vitro* a decade ago (Holtz *et al*., 1964), involving cyclization of the phenylacetaldehydes with DA to give 1-benzyl-substituted-TIQs related or identical to tetrahydropapaveroline (Figure 1). Both theories suggest the alkaloids or their metabolites have neurophysiological actions that would persist after EtOH depressant effects had dissipated. These actions may be manifested in and underlie part of the alcohol withdrawal syndrome.

Until quite recently, little direct evidence existed for either type of TIQ in mammalian cells. To date our effort has been to prove or disprove the initial alkaloid hypothesis, using gas chromatographic examination of tissues from alcohol-treated rats. The "simple" TIQs would seem to be more likely as alkaloid possibilities in the alcoholic, because AcH reacts faster than the phenylacetaldehydes, and is perhaps several orders of magnitude more concentrated than phenylacetaldehydes during alcoholism, at least in peripheral tissues. Also, our experimental methodology has favored detection of the simple TIQs relative to 1-benzyl alkaloids.

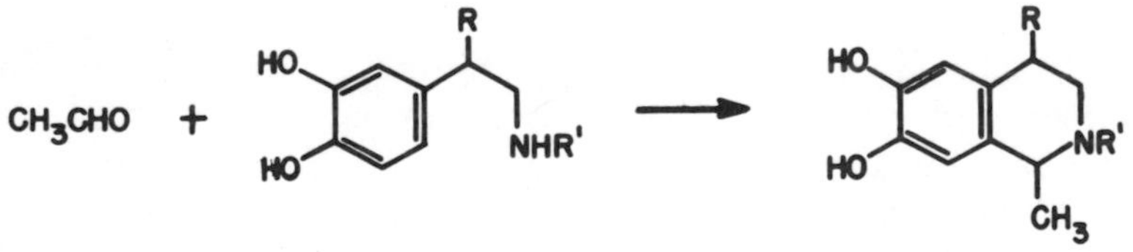

CH_3CHO + [catecholamine] → [tetrahydroisoquinoline]

AcH	Catecholamine	R	R'	Tetrahydroisoquinoline
	Dopamine (DA)	H	H	Salsolinol
	Epinephrine (E)	OH	CH_3	1,2-Dimethyl-4,6,7-trihydroxy-TIQ
	Norepinephrine (NE)	OH	H	1-Methyl-4,6,7-trihydroxy-TIQ

3,4-dihydroxyphenylacetaldehyde + DA → Tetrahydropapaveroline

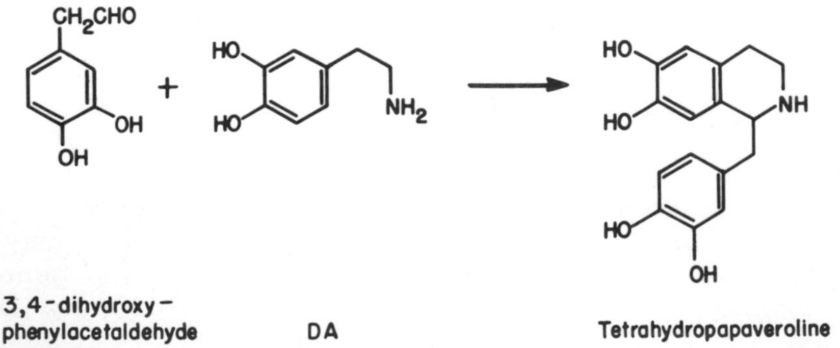

Figure 1. Reaction schemes for tetrahydroisoquinoline formation from CAs and aldehydes.

METHODS

For acute intoxication, rats were treated i. p. with 9 g/Kg of either EtOH or methanol (MeOH) or saline in three doses over 7 hrs. In most acute EtOH experiments, a single injection of the catechol-O-methyltransferase (COMT) inhibitor, pyrogallol (PG), was given i. p. an hour before EtOH, the rationale being that COMT inhibition might increase the ratio of catecholic-to-phenolic amines in the synaptic regions and thus aid the TIQ biosynthesis. As catechol TIQs themselves are substrates for COMT (Rubenstein and Collins, 1973; Collins et al., 1973) and the isolation procedure selectively extracts catechol compounds, COMT inhibition should also favor TIQ isolation. However, another result of PG pretreatment was elevation of AcH in blood (Collins et al., 1974). Non-competitive inhibition of aldehyde dehydrogenase(s) by PG is probably the principal mechanism for the observed increase in AcH levels during EtOH intoxication (Rubenstein, Collins, and Tabakoff, submitted for publication, 1974).

Analyses of tissue CAs and CA-derived TIQs were performed with sensitive electron capture gas chromatography (EC/GC). A particular method we developed involves alumina binding of catechol components subsequent to their extraction from tissues with 1N $HClO_4$, acid elution of bound catechols, volatile derivative formation using heptafluorobutryl (HFB) anhydride in acetonitrile, and EC/GC estimation on 5% GE XF 1105 or 3% OV-17 columns (Bigdeli and Collins, in press, 1974; Collins, Bigdeli and Kernozek, 1970).

The latter column, while it failed to resolve HFB-derivatives of E and NE, was applicable for brain studies because it readily separated the HFB-derivatives of salsolinol and DA, the major CA. Tissues with E and NE were analyzed on GE XF 1105, since resolution of the three HFB-CAs from one another and from two of their three respective 1-methyl TIQ products (HFB-derivatives) was obtained; HFB-DA and HFB-salsolinol were not separable on GE XF 1105, however.

Brain analyses were favored by the excellent EC responses--minimum detectable amounts 1 picogram--for the HFB-derivatives of salsolinol and DA. Less sensitive by factors of 20 - 40, but still acceptable, were the EC responses of HFB-derivatized E, NE and their respective TIQ analogs. The HFB-derivative of tetrahydropapaveroline was 300 - fold less sensitive than the analogous salsolinol derivative and therefore was too unresponsive to the EC detector to allow for its detection in tissues (Bigdeli and Collins, in press, 1974).

RESULTS

A. Acute EtOH Intoxication in PG-Pretreated Rats

Salsolinol in brain. All chromatograms of catechols extracted from the combined CA-rich brain regions (caudate nucleus, midbrain including hypothalamus, and brain stem) of PG-EtOH rats demonstrated a new component which co-chromatographed with authentic HFB-salsolinol. There was no evidence for this compound in PG-saline (control) rat brains, nor did it appear in the controls if excess AcH (100 µg/g tissue) was added at the start of the $HClO_4$-brain homogenization step (Collins and Bigdeli, 1974).

Other catechol possibilities such as epinine, N-acetyl-DA and deaminated CA and pyrogallol metabolites had shorter retention times than salsolinol and the new brain compound (Table 1). 6-Hydroxy-DA, a postulated DA metabolite *in vivo* during disease states (Stein and Wise, 1970), had a longer retention time than the new compound.

TABLE 1

Comparison of Retention Times[(a)] of Known Catechol Compounds to the Unknown Compound in the Brain of PG-EtOH Rats

Compound	Retention Time (Min.)
New substance in experimental (PG-EtOH) brain regions	6.45
Salsolinol	6.45
6-Hydroxy-DA	8.30
Epinine	3.50
N-Acetyl-DA	2.75
3, 4-Dihydroxyphenyl glycol	1.50
3, 4-Dihydroxyphenyl ethanol	1.30
3-Methoxy-catechol	1.30
Pyrogallol	1.05

(a)EC/GC of HFB-derivatives; 3% OV-17 at 160°C; N_2=30 ml/min

In the absence of mass spectrometric confirmation, the identities of unknown endogenous compounds should be substantiated by GC comparisons to authentic compounds on alternative columns (for example, see Wilk and Zimmerman, 1973). In another experiment, therefore, catechols from PG-EtOH and PG-saline brain parts were derivatized with pentafluoropropionyl (PFP) anhydride in ethyl acetate and were chromatographed on 5% SE-54, which separates PFP-DA from PFP-salsolinol. All experimental chromatograms (n=6) contained a peak absent from controls (n=5) which was identical to PFP-salsolinol.

In Table 2 are the results from one of several experiments on the effect of acute PG-EtOH treatment on brain salsolinol and CA concentrations. In this experiment, we observed a 50% depletion in brain DA and a 35% depletion in brain NE. Salsolinol concentrations in all PG-EtOH experiments ranged from 15 - 18 ng/g brain parts, or 0.7 - 0.9% of the control (PG-saline) concentrations of DA.

In further PG experiments, brain DA levels were raised by pretreatment with pargyline, a monoamine oxidase inhibitor (Table 2). An increase in the concentrations of the suspected salsolinol in PG-EtOH brain would be pharmacological proof of its TIQ identity and of the precursor role of DA. There is evidence that pargyline itself may have potentiated the AcH blood levels as well (G. Cohen, personal communication). As shown in the 2nd experiment, salsolinol was present in the combined brain regions from pargyline-PG-EtOH rats at a level of 118 ng/g, 7 - fold higher than in PG-EtOH brain (Collins and Bigdeli, Life Sciences, submitted for publication, 1974).

E-derived TIQ in adrenals. Adrenal glands from PG-EtOH and PG-saline rats were similarly extracted and analyzed for E, NE, DA 4, 6, 7-trihydroxy-TIQs. All PG-EtOH chromatograms contained a new catechol which co-chromatographed at various column temperatures on GE XF 1105 with the HFB-derivative of the major TIQ from E/AcH reactions. The absence of this compound in PG-saline controls containing AcH added during the work-up ruled against E/AcH cyclization during the extraction procedure.

In Table 3 are two experiments that resulted in significant depletion of E (47 - 60%) due to acute EtOH intoxication in PG rats. Adrenal TIQ concentrations approached 4 μg/g in PG-EtOH adrenals, or 0.7 - 1% of baseline E concentrations (Bigdeli and Collins, Life Sciences, submitted for publication, 1974).

TABLE 2

Effect of Acute EtOH on Salsolinol and CA Concentrations in Combined Brain Regions of Pyrogallol (PG) or Pargyline (PARG)-PG-Treated Rats

Experiment	n	µg/g combined brain regions ± s.d.		
		Salsolinol	DA	NE
PG-saline	(6)	undetectable	1.98±0.17	0.98±0.16
PG-EtOH	(6)	0.017±0.005	1.05±0.15*	0.63±0.08*
PARG-PG-saline	(4)	undetectable	4.44±0.42	1.26±0.20
PARG-PG-EtOH	(7)	0.118±0.02	3.27±0.28*	0.68±0.14*

PG = 250 mg/Kg 1 hr before 1st EtOH
PARG = 100 mg/Kg 24 hr and 1 hr before PG

*$p < 0.01$ Compared to saline control

TABLE 3

Effect of Acute EtOH on E-derived TIQ and CA Concentrations in Adrenals of Pyrogallol-(PG)-Treated Rats

Experiment	n	µg/g adrenals ± s.d.			
		E-Derived TIQ	E	NE	DA
1. PG-saline	(4)	undetectable	560±18	190±58	14.4±4.0
PG-EtOH	(5)	3.7±0.6	301±62*	119±48	14.0±4.9
2. PG-saline	(5)	undetectable	425±46	146±22	10.3±1.7
PG-EtOH	(7)	4.0±1.0	171±37*	125±18	15.2±2.1*

*$p < 0.01$ Compared to PG-saline control

B. Acute EtOH Intoxication Only

EC/GC examination of the brain and adrenal tissues from rats treated with EtOH only for 7 hrs failed to show detectable quantities of salsolinol or an E/AcH-derived TIQ. Blood AcH levels in experimental rats varied between 1 - 2.5 μg/ml. Furthermore, the EtOH treatment had no effect on the CA concentrations in either tissue.

C. Acute MeOH Intoxication Only

Because formaldehyde (HCHO), a MeOH metabolite, reacts so much faster in these cyclization reactions (Cohen and Collins, 1970), and earlier studies with ^{14}C-MeOH indicated the occurrence of adrenal TIQs following 3 days of MeOH metabolism (Collins and Cohen, 1970), we were interested in examining brain and adrenals from MeOH-intoxicated rats with the EC/GC method.

Adrenals. A new chromatographic component, consistently present in the adrenals of acute MeOH-treated animals and absent from those of the saline controls, co-chromatographed at various temperatures on GE XF 1105 with the major TIQ obtained routinely as a standard from E/HCHO aqueous reactions. The suspected alkaloid did not result from cyclization during preparative procedures, since it did not occur in controls which contained excess HCHO added during the $HClO_4$ extraction. As shown (Table 4), 7 hours of MeOH metabolism resulted in significant decreases in adrenal CAs (Collins, Bigdeli and Cohen, in preparation, 1974).

Brain. Contrary to the adrenal results in the same animals, the brain CAs were not affected in these acute MeOH experiments. Whole brain levels of DA and NE were unchanged after 7 hours exposure to MeOH (Table 4), and there was no GC evidence of 6, 7-dihydroxy-TIQ, the principal product from DA and HCHO, in levels above 5 μg/g brain.

DISCUSSION

These experiments show that salsolinol, a TIQ alkaloid with numerous effects on neuroenzyme systems (cf. Collins *et al*., 1973), is a detectable *in vivo* product of brain DA during EtOH metabolism. The condition apparently needed for its detection by

TABLE 4

Effect of Acute MeOH Intoxication[a)] on CA Concentrations in Adrenals and Whole Brain of Rats

		µg/g ± s.d.				
		Adrenals			Brain	
Treatment	n[b)]	E	NE	DA	DA	NE
Saline	(5)	512±59	120±4.8	13.6±0.9	0.55±0.02	0.13±0.01
MeOH	(5)	164±44*	31±4.1*	6.1±1.6*	0.48±0.07	0.12±0.01

a) 9 g/Kg MeOH in 3 i.p. doses over 7 hrs

b) No. of rats/group

*$p < 0.01$ Compared to saline

EC/GC was either sufficient AcH precursor, COMT inhibition, or both. EC/GC evidence was also obtained for the biosynthesis of an E-derived TIQ in rat adrenals in these experiments, as well as in other acute experiments with MeOH-intoxicated rats.

Two other research groups recently have shown with sensitive GC/mass spectrometry techniques ("mass fragmentography") that simply elevating endogenous DA concentrations with L-DOPA is sufficient to raise the concentrations of DA-derived TIQs within detectable limits. Turner et al. (1974) measured nanogram-quantities of tetrahydropapaveroline in brain of rats ingesting L-DOPA and the peripheral DOPA decarboxylase inhibitor, RO44602. Acute EtOH metabolism in those rats appeared to increase the concentrations of the benzyl TIQ, although a statistically significant difference could not be shown because of alkaloid instability and insufficient numbers of animals. In another study, urine from humans undergoing DOPA therapy for Parkinson's disease was shown to contain both salsolinol and tetrahydropapaveroline (Sandler et al., 1973). Experimental EtOH ingestion did result in definite increases in the levels of urinary salsolinol; levels of the urinary benzyl TIQ were unchanged. Evaluated together, these and our studies firmly establish that TIQ alkaloids can be viable, albeit quantitatively minor, in vivo metabolites of endogenous DA, E and possibly NE.

These discoveries of new CA alkaloid derivatives in vivo are potentially exciting, but many questions remain. For example, brain tetrahydropaveroline was not detectable by mass fragmentography when rats ingested EtOH only. As Turner et al. (1974) point out, this result actually casts doubt on the role of this alkaloid in EtOH dependence. Their explanation for its absence is that the biosynthesis of salsolinol from DA and AcH is possibly more favorable than tetrahydropapaveroline biosynthesis. They report no evidence for brain salsolinol, however, and in our own experiments we failed to detect salsolinol in brain regions of rats metabolizing only EtOH using an EC/GC method which is sensitive to 4 - 5 ng salsolinol/g brain. It is more likely that their EtOH-consuming rats were not sufficiently intoxicated and did not maintain significant AcH blood levels.

The source(s) of the urinary TIQs in the Sandler et al. (1973) study is also open to question, particularly because controls to determine the extent of cyclization in the urine before and during extraction apparently were not carried out. Also, since most of the administered DOPA remains outside the CNS, it is probable that a substantial amount of DA-derived TIQs not actually produced in the urine or workup were derived from peripheral, non-neuronal tissues.

Nevertheless, these studies on biosynthesis of salsolinol (and tetrahydropapaveroline) are quite relevant to human and animal alcoholic situations. Blood AcH levels in chronic "street" alcoholics during early detoxification may be as high as 8 - 10 μg/ml (Magrinat et al., 1973). Assuming these AcH measurements are valid, central and peripheral TIQ biosynthesis would be expected in these individuals. A similar possibility certainly exists in the mice-EtOH inhalation experiments of Littleton et al., (1974), in which cerebral blood AcH concentrations were reported to be surprisingly high (6 - 9 μg/g). It is interesting that Forsander has reported (this symposium) that only at similar blood levels does AcH become detectable by GC within rat brain tissue. The TIQ biosynthesis we have observed in PG-EtOH treated rats thus may be dependent upon AcH blood levels of greater than 8 - 15 μg/ml, and perhaps upon inhibition by PG of an aldehyde dehydrogenase in brain capillaries which Sippel (1974) has suggested to be part of a blood/brain barrier to AcH.

While salsolinol and E-derived TIQ concentrations are less than 1% of those of the CA precursors in our PG-EtOH experiments, their concentrations may have been greater at other times during the course of intoxication. These experiments also leave unanswered the question of site of salsolinol biosynthesis. Regional and subcellular distribution studies, similar to those performed with ^{3}H-CAs, are required to determine if the salsolinol is stored in CA vesicles, or is localized extraneuronally.

The CNS role for salsolinol, and for that matter, tetrahydropapaveroline, has yet to be clearly defined. Cohen has advanced convincing pharmacological proof for the DA-derived simple TIQs as surrogate transmitters (Mytilineou et al., 1974; Cohen, 1973). A replacement transmitter role is intriguing, but further possibilities for TIQ involvement at the molecular level need to be considered. TIQs could be envisioned as DA receptor blocking agents, for example, a possibility which might have some bearing on the clinical claims that apomorphine, a receptor stimulator, is advantageous in alcoholism treatment. The oxidation of simple TIQs to dihydro metabolites that bind to membrane and induce neuronal damage is also a speculative route of involvement which is being investigated in our laboratory. Recent studies by V. Davis and co-workers (this symposium) indicate the possibility of further cyclization of tetrahydropapaveroline to tetrahydroprotoberberine alkaloids which could be conceived as neurotoxic agents.

Finally, in regard to the biosynthesis and possible secretion of an adrenal TIQ, the interaction of this new cyclic CA-alkaloid with the peripheral β-adrenergic (adenyl cyclase) systems may be important in alcoholics, where adrenergic activation is closely associated with intoxication and withdrawal (Ogata et al., 1971).

ACKNOWLEDGMENTS

Original studies in this report were supported in part by USPHS AA00266 (formerly MH19153), Loyola University GRS funds, and Illinois MH114-11-RD. The continuing encouragement of Gerald Cohen and his research group at Mt. Sinai Medical School is gratefully acknowledged.

REFERENCES

Bigdeli, M. and Collins, M. Tissue catecholamines and potential tetrahydroisoquinoline metabolites: A gas chromatographic assay method using electron capture detection. Biochem. Medicine, in press (1974).

Cohen, G. A role for tetrahydroisoquinoline alkaloids as false adrenergic transmitters in alcoholism. In (Ed., M. M. Gross) Alcohol Intoxication and Withdrawal: Experimental Studies I, Plenum Press, N. Y., 33 (1973).

Cohen, G. and Collins, M. Alkaloids from catecholamines in adrenal tissue: Possible role in alcoholism. Science, 167: 1749 (1970).

Collins, A., Cashaw, J. and Davis, V. Dopamine-derived tetrahydroisoquinoline alkaloids--inhibitors of neuroamine metabolism. Biochem. Pharmacol., 22: 2337 (1973).

Collins, M. and Bigdeli, M. Alcohol intoxication and in vivo biosynthesis of the alkaloid salsolinol from dopamine in rat brain. Trans. Amer. Soc. Neurochem., 5: 160 (1974).

Collins, M., Bigdeli, M. and Kernozek, F. Electron capture gas chromatography studies of biogenic amine-derived isoquinoline alkaloids. Pharmacologist, 13: 309 (1971).

Collins, M. and Cohen, G. Tissue catecholamines condense with acetaldehyde to form isoquinoline alkaloids. Amer. Chem. Soc., 156th Nat. Mtg., Abst. 274 (1968).

Collins, M. and Cohen, G. Isoquinoline alkaloid biosynthesis from adrenal catecholamines during ^{14}C-methyl alcohol metabolism in rats. Fed. Proc., 29: 608 (1970).

Collins, M., Rubenstein, J., Bigdeli, M., Gordon, R. Jr. and Custod, J. Pyrogallol--a potent elevator of acetaldehyde during ethanol metabolism. In (Eds., R. Thurman et al.) "Alcohol and Aldehyde Metabolizing Systems", Academic Press, N. Y., 523 (1974).

Davis, V. and Walsh, M. Alcohol, amines and alkaloids: A possible biochemical basis for alcohol addiction. Science, 167: 1005 (1970).

Holtz, P., Stock, K. and Westermann, E. Formation of tetrahydropapaveroline from dopamine in vitro. Nature, 203: 656 (1964).

King, G., Goodwin, B. and Sandler, M. Isosalsolinol formation: A secondary reaction in the Pictet-Spengler condensation. J. Pharm. Pharmacol., 26: 476 (1974).

Littleton, J., Griffiths, P. and Ortiz, A. The induction of ethanol dependence and the ethanol withdrawal syndrome: The effects of pyrazole. J. Pharm. Pharmacol., 26: 81 (1974).

Magrinat, G., Dolan, J., Biddy, R., Miller, L. and Korol, B. Ethanol and methanol metabolites in alcohol withdrawal. Nature, 244: 234 (1973).

Mytilineou, C., Cohen, G. and Barrett, R. Tetrahydroisoquinoline alkaloids: Uptake and release by adrenergic nerves in vivo. Europ. J. Pharmacol., 25: 390 (1974).

Ogata, M., Mendelson, J., Mello, N. and Majchrowicz, E. Adrenal function and alcoholism: II. Catecholamines. In: Recent Advances in Studies on Alcoholism, U. S. Dept. Health, Education and Welfare, N. I. M. H., 140 (1971).

Robbins, J. H. Alkaloid formation by condensation of biogenic amines with acetaldehyde. Clin. Res., 16: 350 (1968).

Rubenstein, J. and Collins, M. Tetrahydroisoquinoline derived from noradrenaline-aldehyde condensations: Pyrogallol-sensitive O-methylation in rat homogenates. Biochem. Pharmacol., 22: 2928 (1973).

Sandler, M., Bonham-Carter, C., Hunter, K. and Stern, G. Tetrahydroisoquinoline alkaloids: In vivo metabolites of L-DOPA in man. Nature, 241: 439 (1973).

Sippel, H. W. The acetaldehyde content in rat brain during ethanol metabolism. J. Neurochem., 23: 451-452 (1974).

Stein, L. and Wise, C. Possible etiology of schizophorenia: Progressive damage to the noradrenergic reward system by 6-OH-DA. Science, 171: 1032 (1971).

Turner, A., Baker, K., Algeri, S., Frigerio, A. and Garattini, S. Tetrahydropapaveroline: Formation in vivo and in vitro in rat brain. Life Sciences, 14: 2247 (1974).

Wilk, S. and Zimmerberg, B. Absence of 3-methoxy-4-hydroxyphenylethanol in brain. Biochem. Pharmacol., 22: 623 (1973).

Yamanaka, Y., Davis, V. and Walsh, M. Salsolinol, an alkaloid derivative of dopamine formed in vitro during alcohol metabolism. Nature, 227: 1143 (1970).

THE EFFECTS OF ACUTE ETHANOL INTOXICATION ON CEREBRAL ENERGY METABOLISM

R.H. Nielsen, R.A. Hawkins and R.L. Veech

Laboratory of Alcohol Research, National Institute of Alcohol Abuse and Alcoholism, St. Elizabeths Hospital Washington, D.C.

It has long been known that anesthetics inhibit respiration in many tissues including brain and can, in yeast or heart extracts, lead to a reduction of cytochrome b while cytochromes c and a remain oxidized (Keilin, D. 1925 & 1929).

Barbiturate anesthesia has been found to decrease cerebral oxygen consumption in the cerebral cortex to a greater extent than that of lower brain centers (Himwich et al., 1947). In general the magnitude of the depression of cerebral oxygen consumption during anesthesia with a number of agents is about 35% indicating a very high degree of dependence of conscious mental functioning on the maintenance of maximum glucose and oxygen consumption (Himwich, 1951). The same degree of inhibition of respiration is found in brain slices by a number of anesthetic agents *in vitro* (Quastel, J.H., 1963).

Using new techniques which allow us to study a number of parameters in the whole brains of living rats, we can now investigate some of the postulated sites of ethanol action on nerve cells in a way which is somewhat less prone to difficulties of interpretation than were some previously used *in vitro* systems. In so doing perhaps we can clarify some aspects of the interaction of ethanol with nerve cells even if we cannot give a definitive answer to the question of how ethanol interacts with transmission at the synapse or with conduction along the nerve membrane. It may be that the final description of ethanol intoxication will require a degree of sophistication which is beyond the scope of current biochemical techniques. On the other hand it is also possible that major underlying biochemical events might be identified in brain. This could help our understanding of the primary site of ethanol's action from which most other events flow. To this end, we have measured several parameters in normal and ethanol intoxication rats, including the rate of brain glucose utilization, the arterio-venous differences of glucose and lactate, and a number of the energy conserving intracellular nucleotides which measure the integrity of the cellular energy producing mechanisms.

The metabolic status of brain after ethanol induced intoxication was compared to that of liver where the fundamental disorder induced by ethanol seems to be a profound decrease in the cytoplasmic free $[NAD^+]/[NADH]$ ratio. The change in this ratio results from the conversion of ethanol to acetaldehyde by the action of alcohol dehydrogenase (EC 1.1.1.1) (Lieber and Davidson, 1962; Forsander et al., 1965; Krebs, 1968; Veech et al., 1972). In this way we hoped to see what impairments if any coincided with the induction of unconsciousness resulting from the administration of ethanol.

METHODS

Rats

180-220 g. male Wistar rats were fed Purina lab chow ad libitum, housed in plastic cages and maintained in 12-hour darkness from 7 P.M. to 7 A.M. each day. When rats were fasted, they were allowed only water for 48 hours before sacrifice.

Administration of Ethanol

For the brain studies fed rats were injected intraperitoneally with 1 ml per 100 g of a 7 M/ethanol solution in 0.9% NaCl. Treated in this fashion, the rats lost their righting reflex within 3 to 3.5 minutes, and became conscious after 5 minutes. The liver studies were performed on fed rats given an I.P. dose of ethanol so as to produce a liver concentration of ethanol of 10 mM. Rats were then sacrificed 15 minutes after injection and procedures carried out as reported previously (Veech et al., 1972).

Preparation of Tissue Samples and Measurement of Metabolites

The brain tissue was removed and frozen in less than one second using a brain freeze-blowing apparatus which was previously reported (Veech et al., 1973). Samples of frozen brain were then treated at -20° in a mixture of 0.1 M HCl in absolute methanol prior to addition of perchloric acid (Veech & Hawkins, 1974) in order to prevent formation of AMP which can occur after treatment with perchloric acid alone. The enzymatic analysis of metabolites was performed as previously described (Veech et al., 1973).

Measurement of Arteriovenous differences across rat brain and the rate of cerebral glucose utilization.

The measurement of arteriovenous differences across rat brain was as described by Hawkins & Veech, (1974a). The rate of cerebral glucose utilization is measured by determining the amount of ^{14}C, derived from 2-^{14}C glucose, which is incorporated into the acid soluble metabolite pool of brain according to the equation:

$$\text{rate of glucose utilization} = \frac{^{14}\text{C accumulated by brain metabolites}}{\int_0^t \text{glucose specific activity}}$$

This method probably underestimates the rate of glucose utilization by about 10%. (See Hawkins et al., 1974b for complete details).

Results and Discussion

Table 1 shows that a single intraperitoneal dose of ethanol decreased brain glucose utilization from 0.61 to 0.46 µmoles/min/g after 60 minutes. The maximum decrease measured after 60 minutes was only 25%. While a statistically significant decrease was found 15 minutes after ethanol administration, the rate of glucose utilization measured between 5 and 10 minutes after injection did not differ significant from controls even though the ethanol treated animals lost their righting reflexes three minutes after injection. Induction of anesthesia with pentobarbital lead to a decrease of 50% from 0.62 to 0.28 µmol/min/g 37-45 minutes after injection; a considerably greater decrease than is found after ethanol administration (Hawkins et al., 1974c).

In fed adult rats glucose comprises the major metabolic fuel of cerebral respiration (Hawkins et al., 1971). In the absence of either a significant transport from the brain of incompletely combusted metabolites derived from glucose, or the intracellular accumulation of the same, one is forced to conclude that a decreased glucose utilization is indicative of a decreased brain O_2 consumption. Table 2 shows that, within the limits of errors of the method, glucose uptake by brain agreed with the rates of glucose utilization after 10 minutes and furthermore, there was no significant uptake or release of lactate into the cerebral circulation.

Taken together therefore this data obtained from rats *in vivo*

TABLE 1

THE EFFECT OF ETHANOL INTOXICATION ON GLUCOSE UTILIZATION IN RAT BRAIN IN VIVO

Values are reported as mean glucose utilization in μmoles/min/g fresh weight of brain ± S.E.M. The symbol * indicates statistical significance at the 1% level as judged by student's T Test. The number of observations is given in parentheses.

Controls (13)	7.5 min after ethanol (6)	15 min after ethanol (7)	60 min after ethanol (6)
0.61	0.55	0.49*	0.46*
±0.01	±0.04	±0.03	±0.03

TABLE 2

ARTERIO-VENOUS DIFFERENCES OF GLUCOSE AND LACTATE ACROSS RAT BRAIN 10 MINUTES AFTER INJECTION OF 60 mMOLES/Kg ETHANOL

Values are the means of 5 determination ± S.E.M.

Glucose in μMol/ml blood			Lactate in μMol/ml blood		
Artery	Vein	A-V Difference	Artery	Vein	A-V Difference
7.81	6.98	0.83	1.35	1.42	-0.07
±0.81	±0.64	±0.25	±0.16	±0.13	±0.06

confirms the earlier work obtained in humans where it was shown that during heavy alcohol intoxication, brain oxygen consumption was decreased by 25-30% (Battey et al., 1953, Fazekas et al., 1955). The rather small magnitude of the depression can also be taken to agree with the generally held view that conscious mental functioning is extremely dependent on the maintenance of maximum glucose and oxygen consumption (Himwich, 1951). The rate of onset of this inhibition of glucose utilization was slow, however, occurring about 10-15 minutes after injection, whereas unconsciousness occurred within 3 to 5 minutes. It is therefore hard to escape the conclusion that the inhibition of glucose and oxygen consumption is a secondary phenomena. In other words, the brain's oxidative metabolism seemed to decrease because the brain was "asleep" rather than the sleep being a resultant of the generalized slowdown of energy production.

Table 3 shows the results of some of the metabolite measurements made in "freeze-blown" after ethanol intoxication. Glucose, glucose 6-phosphate and citrate content of brain all increased significantly thus confirming earlier reports indicating a change in brain glucose concentration during anesthesia (Lowry et al., 1964, Mayman et al., 1964, Passonneau et al., 1971). Uncertainty as to the exact size of the extra-cellular space in brain makes it difficult to estimate the magnitude of the rise in intracellular glucose. Recent analyses done on single neurons which have been freeze-dried and then dissected out of the surrounding tissue leave little doubt that this rise in intracellular glucose during anesthesia is real (Passonneau and Lowry, 1971). At least two explanations for this rise in intracellular glucose seems possible, first, the transport process for glucose into brain may be accelerated or second the rate of glucose utilization may be decreased. No definitive answer in favor of one or the other of these hypotheses can be given at this time.

The rise in glucose-6 phosphate which accompanied the rise in glucose is unlikely to represent a simple effect on the hexokmase reaction due to the increased glucose content since the equilibrium position of the hexokmase reaction is very far to the right. Speculation as to what might be controlling the increase in glucose-6-phosphate level is beyond the scope of this paper except to say that the rise in citrate is likely also to be related to the glucose-6-phosphate elevation since both glucose-6-phosphate and citrate are related to one another through a common cytoplasmic $NADP^+$ pool (Veech et al., 1969). Factors which can control this group of metabolites in isolation from other glycolytic intermediates have so far not been identified.

In the same manner, the rather late elevation of glutamate shown here is accompanied by a decrease in aspartate levels,

TABLE 3

CONTENT OF RAT BRAIN METABOLISM AFTER ETHANOL

Values are given in μmoles/g fresh weight ±S.E.M. The number of observations is given in parentheses. The symbol * denotes statistical significance at 5% by the Mann-Whitney U Test.

	Controls (5)	7.5 min after ethanol injection (6)	15 min after ethanol injection (6)	60 min after ethanol injection (6)
Glucose	1.56 ±0.10	2.83* ±0.21	3.43* ±0.20	3.28* ±0.18
Glucose-6-P	0.147 ±0.012	0.184* ±0.008	0.180* ±0.002	0.179* ±0.008
Citrate	0.308 ±0.018	0.359* ±0.017	0.371* ±0.019	0.377* ±0.010
a-Oxoglutarate	0.217 ±0.005	0.196 ±0.013	0.205 ±0.007	0.240 ±0.012
L-Glutamate	11.8 ±0.25	12.2 ±0.33	12.1 ±0.18	13.0* ±0.19
Pyruvate	0.084 ±0.004	0.070 ±0.007	0.070 ±0.003	0.077 ±0.004
L-Lactate	1.26 ±0.05	1.20 ±0.13	1.18 ±0.06	1.10 ±0.08
Ethanol	--	88.9 ±1.1	77.9 ±3.9	47.1 ±3.6

(Veloso et al., 1972) presumably reflecting a redistribution of metabolites in accordance with the equilibrium constant of the glutamate-oxaloacetate transanimase reaction (Miller et al., 1973) as a result of the increases in brain pCO_2 which often accompany prolonged periods of anesthesia *in vivo*. We were unable to confirm earlier reports of changes in the brain content of lactate, pyruvate, malate or 2-oxoglutarate (Mayman et al., 1964, Lowry et al., 1964 and Goldberg et al., 1966) after various anesthetic agents most likely because of the different techniques for obtaining frozen brains used in this study.

From Table 4 it is obvious that intoxication with ethanol very rapidly resulted in an increase in the portion of the brain's chemical energy which is concentrated in the adenine nucleotides. These changes are very small, but statistically significant, we believe them to be real changes. It is also of interest that these changes seem to occur early in the course of ethanol intoxication. This raises the possibility that these changes may be related to a primary site of action of ethanol. In general, all of the changes in intermediary metabolite levels we have seen are very small in "brain" when compared to the events in liver following ethanol injection.

Figure 1 is a proportionate change diagram of some metabolites in liver following ethanol administration. It can be seen that four to six fold changes in metabolite concentration were common. Most of these rapid changes appeared to be due to primary decrease in the cytoplasmic $[NAD^+]$ / $[NADH]$ ratio. The decrease in $[NAD^+]$/ $[NADH]$ ratio resulted from the rapid conversion of ethanol to acetaldehyde catalyzed by alcohol dehydrogenase, and enzyme with relatively high activity in liver. The change in the $[NAD^+]$ / $[NADH]$ ratio in turn produced the large increases in [lactate] [α-glycerophosphate] and [malate] which are the reduced partners of substrate couples for very active cytoplasmic NAD-linked dehydrogenases.

From the changes in metabolite concentrations it is possible to calculate the changes in the redox states of the various pyridine nucleotide couples and in the cytoplasmic phosphorylation state (see Krebs and Veech, 1970). Table 5 gives the results of these calculations for the effects of ethanol on these ratios in brain.

From the figures given in Table 5 it is clear that there was 10% decrease in the cytoplasmic $[NAD^+]$ / $[NADH]$ ratio of brain 7.5 minutes after treatment with ethanol which disappeared by 15 minutes after injection. There was also a small decrease in $[NADP^+]$ / $[NADPH]$ ratio which was also of a transient nature.

The $[ATP]$ / $[ADP]$ X $[HPO_4^{-2}]$ ratio increased at 7.5 minutes

TABLE 4

CONTENT OF PHOSPHORYLATED NUCLEOTIDES IN RAT BRAIN AFTER ETHANOL

Values are given in μmoles/g fresh weight ±S.E.M. The number of observations is given in parentheses. The symbol * denotes statistical significance at 5% level by the Mann-Whitney U Test.

	Controls (5)	7.5 min after ethanol injection (6)	15 min after ethanol injection (8)	60 min after ethanol injection (6)
Creatine-P	3.52 ±0.11	3.90* ±0.01	3.96* ±0.04	3.91* ±0.10
ATP	2.35 ±0.01	2.50* ±0.02	2.50* ±0.05	2.53* ±0.05
ADP	0.746 ±0.042	0.661* ±0.015	0.642* ±0.010	0.645* ±0.018
AMP	0.075 ±0.008	0.060 ±0.006	0.054* ±0.004	0.055* ±0.001
P_i	2.08 ±0.06	2.01 ±0.05		

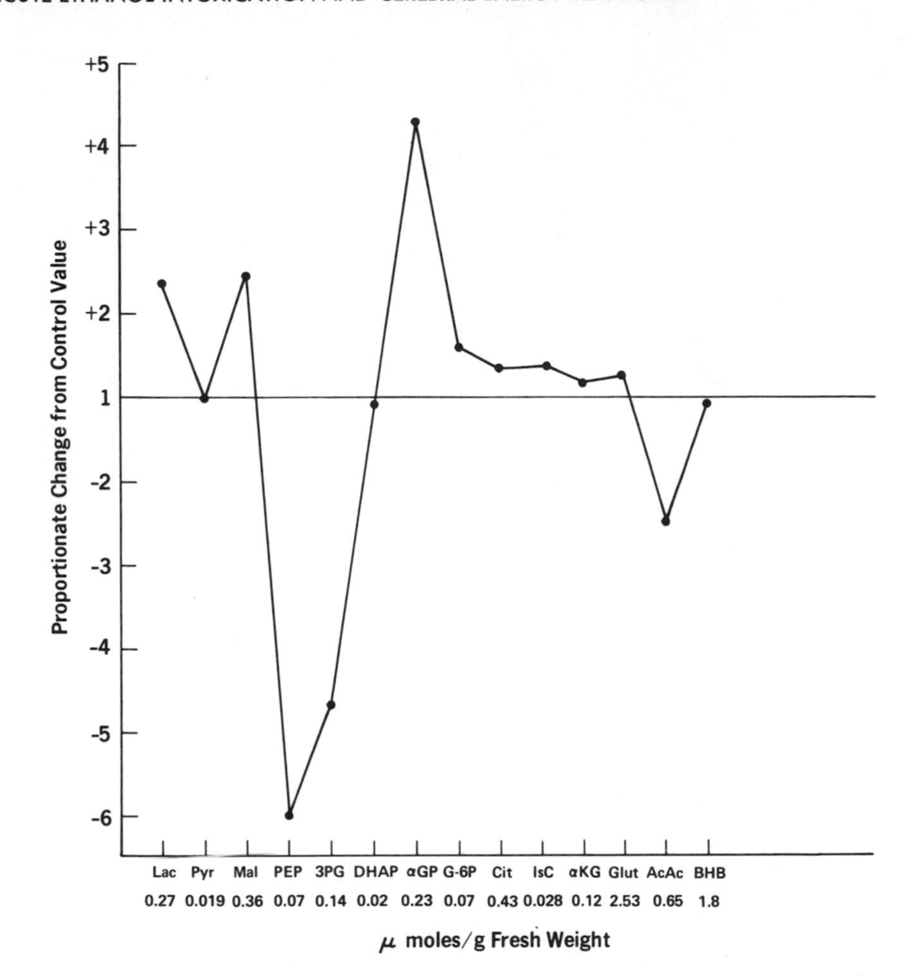

Proportionate Change Diagram of the Effects of Ethanol on Liver Metabolite Concentrations.

Control values are given on the abcissa in μmoles/g fresh weight. The proportionate change values indicate the value of liver metabolites in fed rats 15 minutes after at 10mM dose of ethanol where proportionate change = nx $\frac{\text{(experimental value)}}{\text{(control value)}}$

n = 1 when experimental value $\geq$ control value.
n = 1 when experimental value $<$ control value.

Figure 1

TABLE 5

A TIME COURSE OF THE EFFECT OF ETHANOL ON REDOX STATES IN RAT BRAINS

Values are given as means ±S.E.M. The symbol * denotes statistical significance at the 5% level as judged by the Mann-Whitney U Test. For method of calculation, see text.

	Controls (5)	7.5 min after ethanol injection (6)	15 min after ethanol injection (8)	60 min after ethanol injection (6)
Cytoplasmic LDH $\frac{[NAD^+]}{[NADH]}$	599 ± 7	533 ± 21*	544 ± 19	637 ± 25
Cytoplasmic ICDH $\frac{[NADP^+]}{[NADPH]}$	0.0160±0.0014	0.0121 ± 0.0008*	0.0125 ± 0.0006*	0.0146 ± 0.0009
$\frac{ATP}{ADP \times HPO_4^{2-}} M^{-1}$	2520	3150	3230	3235

and apparently remains elevated throughout the period of study as judged by the [ATP] and [ADP] changes. This cannot be said with certainty however because phosphate determinations in "blown brain" require lengthy extractions and cannot be performed on normal perchloric acid extracts due to the contamination of bone fragments (see Hawkins et al., 1973). Since [Pi] did not appear to change significantly at 7.5 minutes, these measurements were not performed at other times.

Figure 2 compares the very small changes in the cytoplasmic $[NAD^+]$ / [NADH] ratio found after ethanol administration in both the brain and liver of the rat. In view of the minimal changes found in redox states of brain, it seems very unlikely that ethanol exerts a significant physiological effect on brain by changing its redox states. The situation is thus very different in brain compared to liver; presumably due to the very weak alcohol dehydrogenase which exists in brain (Raskin & Sololoff, 1972).

In view of the very different responses of the redox states of liver and brain to administered ethanol, it is rather intriguing to speculate as to why the response of the [ATP] / [ADP] X $[HPO_4^{2-}]$ ratio in the two tissues should be so similar. This is particularly true in view of the linkage which is thought to exist between the ATP- and NAD- couples (Krebs & Veech, 1970). For a number of years it has been known that relatively low concentrations of ethanol (0.3%) inhibit the active transports of K^+ into cells by about 20% without changing the permiability of the cell to K^+ (Streeter and Solomon, 1954, Kalant & Israel, 1967). The Na^+ - K^+ase (Skon, 1957, Albers, 1967) is the enzyme thought to be responsible for the active transport of K^+. It is known to be inhibited in vitro by ethanol as well as by higher alcohols (Jarnefelt, 1961, Israel & Salazar, 1967, Kalant & Israel, 1967). Since the pumping of Na^+ and K^+ is probably the major energy consuming process in brain it seems likely that the following sequence of events follow the acute administration of ethanol. First ethanol inhibits the activity of the brain Na-K ATPase which leads to a decrease in the brains requirements for ATP. This decreased requirement for ATP leads to an increase in the [ATP] / [ADP] X $[HPO_4^{2-}]$ ratio which leads to a decrease in the brain oxygen consumption by increasing the redox potentials of the electron transport chain carriers around cytochrome A_3 (for a more detailed explanation see Wilson et al., 1974). The increased redox potential of the carriers around cytochrome A_3 results in a decrease in the brains O_2 consumption (see Figure 3).

While at the same time that the elevated [ATP] / [ADP] X $[HPO_4^{2-}]$ ratio is decreasing the rate of O_2 consumption in the mitochondria, it is also effecting the concentrations of a number of cytoplasmic metabolites in such a way as to decrease the rate of glycolysis.

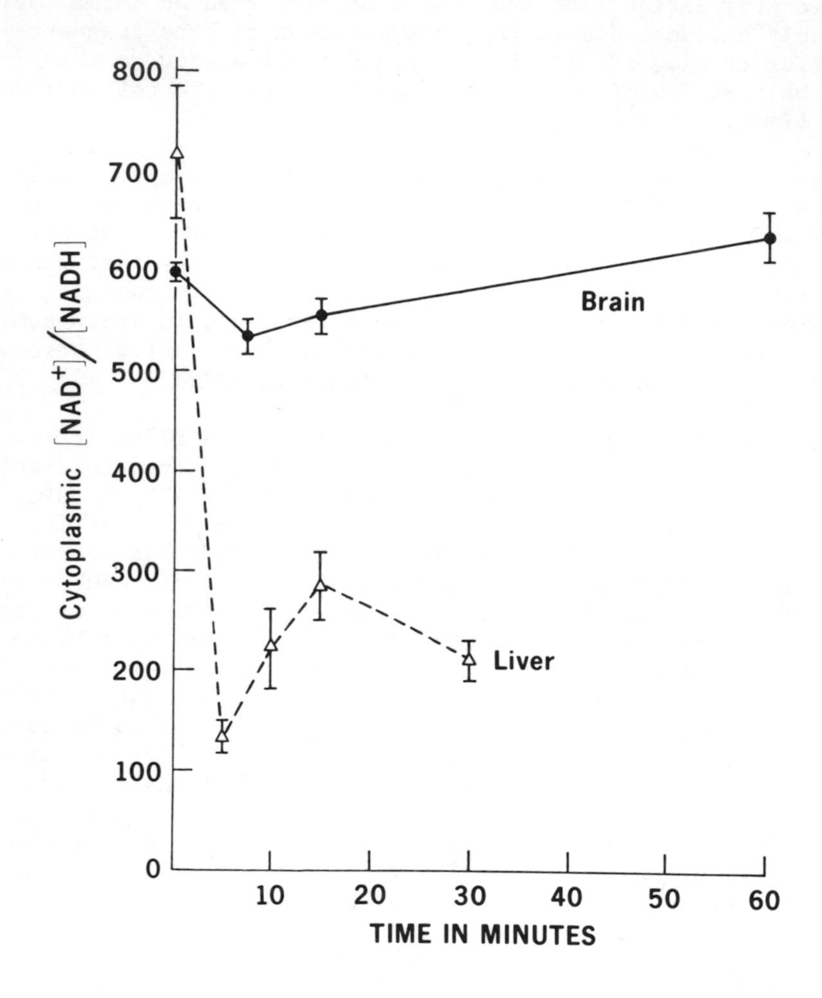

Cytoplasmic [NAD+] / [NADH] in Brain and Liver Following Ethanol Administration

Ethanol was administered at time 0; for details see text.

Figure 2

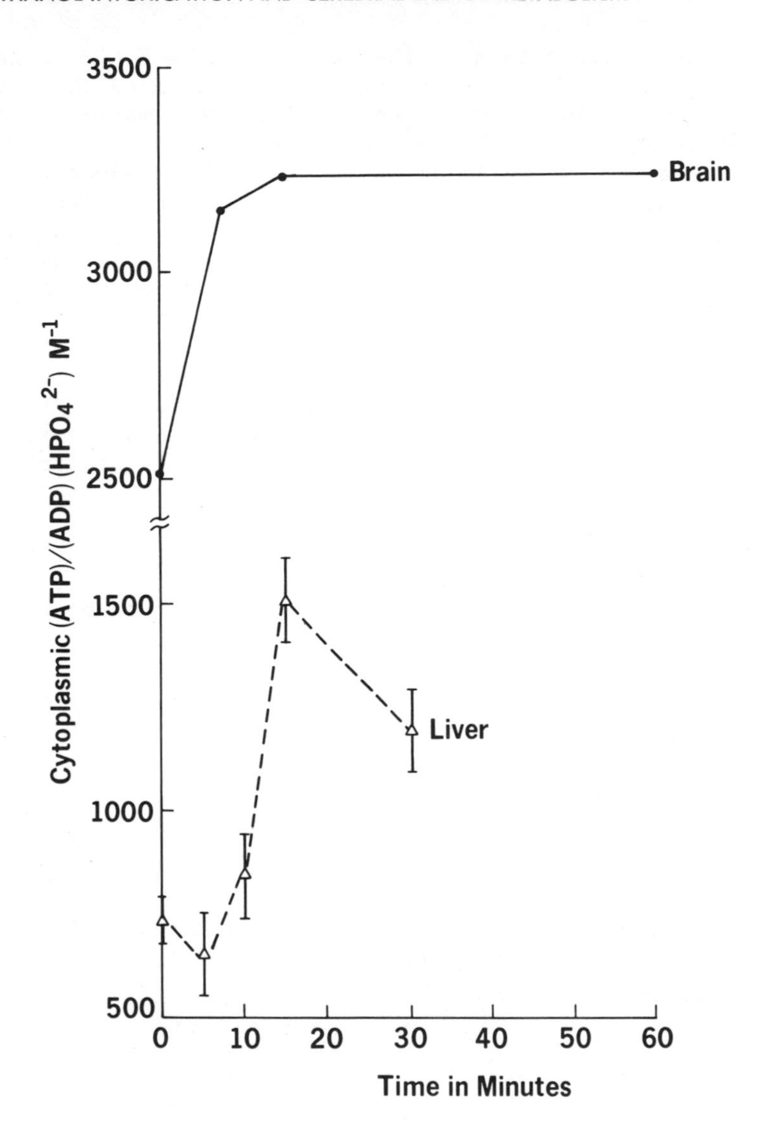

Cytoplasmic [ATP] / [ADP] X [HPO_4^{2-}] Ratio in Brain and Liver following Ethanol Administration

Ethanol was administered at time 0; for details see text.

Figure 3

While it is possible that inhibition of the Na-K ATPase in liver could account for the rather unusual elevation of the [ATP] / [ADP] X [HPO_4^{2-}] ratio found in the presence of a markedly decreased cytoplasmic [NAD^+] / [NADH] ratio, the apparent low activity of the enzyme in liver (Albers, 1967) makes this hypothesis less certain. The development of improved assay techniques for the Na-K ATPase are currently underway and should help to answer this question. At the present time however, we can only say that the change in phosphorylation potential seems to be a common response of two tissues which are most commonly subject to the adverse effects of prolonged high ethanol consumption while the redox state changes are confined to liver. Further study of the mechanisms responsible for those elevation of the phosphorylation state and a detailed study of the consequences of this elevation should help us to understand how ethanol exerts its toxic effects.

REFERENCES

Albers, R.W., 1967. Biochemical aspects of active transport. Ann. Rev. Biochem. 36, 727-756.

Battey, L.I., Heyman, A. and Patterson, J.L., 1953. Effects of ethyl alcohol on cerebral blood flow and metabolism. J. Am. Med. Ass. 152, 6-10.

Fazekas, J.F., Albert, S.N. and Alman, R.W., 1956. Combined effects of chlorpromazine and pentobarbital on cerebral hemodynamics and metabolisms. Am. J. Med. Sci. 231, 655-658.

Forsander, O.A., Raiha, N., Salaspuro, M.I. and Maenpaa, P., 1965. Influence of ethanol on the liver metabolism of fed and starved rats. Biochem. J. 94, 259-265.

Goldberg, N.D., Passonneau, J.V. and Lowry, O.H., 1966. Effects of changes in brain metabolism on the levels of citric Acid Cycle Intermediates. J. Biol. Chem. 241, 3997-4003.

Hawkins, R.A., Williamson, D.H. and Krebs, H.A., 1971. Ketone-body utilization by adult and suckling rat brain in vivo. Biochemical J. 122, 13-18.

Hawkins, R.A., Nielsen, R.C. and Veech, R.L., 1973. The measurements of the inorganic phosphate contest of brain in the presence of bone fragments. J. Neurochem. 20, 35-38.

Hawkins, R.A. and Veech, R.L., 1974a. Measurement of arteriovenous differences across rat brain. In: Research Methods in Neurochemistry, Vol. 2, (Marks, N. and Rodnight, R. eds.) pp. 161-169, Plenum Press, New York.

Hawkins, R.A., Miller, A.L., Cremer, I.E. and Veech, R.L., 1974b. Measurement of the rate of glucose utilization by rat brain in vivo. J. Neurochem. 23, 917-923.

Himwich, W.A., E. Hamburger, R. Moresca and H.E. Himwich, 1947. Brain metabolism in man: Unanesthetized and in Pentothal Narcosis. Am. J. Psychiat. 103, 689-696.

Himwich, E.E., 1951. Brain Metabolism and Cerebral Disorders. Williams & Wilkins, Baltimore.

Israel, Y. and Salazar, I., 1967. Inhibition of brain microsomal adenosine triphosphatases by general depressants. Arch. Biochem. Biophys. 122, 310-317.

Jarnefelt, J., 1961. Inhibition of the brain microsomal adenosine-triphosphatase by depolarizing agents. Biochem. Biophys. Acta 48, 111-116.

Kalant, H. and Israel, Y., 1967. Effects of ethanol on active transport of cations. In: Biochemical Factors in Alcoholism. (Maickel, R.P. ed.) pp. 25-37, Pergamon Press, Oxford.

Keilin, D., 1925. Cytochrome, a respiratory pigment, common to animals, yeast and higher plants. Proc. Roy. Soc. (London) B98, 312-339.

Keilin, D., 1929. Cytochrome and respiratory enzymes. Proc. Roy. Soc. (London) B104, 206-252.

Krebs, H.A., 1968. The effects of ethanol on the metabolic activities of the liver. Adv. Enz. Reg. 6, 467-480.

Krebs, H.A. and Veech, R.L., 1970. Regulation of the redox state of the pyridine nucleotides in rat liver. In: (Sund, H., ed.) Pyridine nucleotide-dependent dehydrogenases, pp. 413-438. Springer-Verlad, Berlin.

Lieber, C.S. and Davidson, C.E., 1962. Some metabolic effects of ethyl alcohol. Am. J. Med. 33, 319-327.

Lowry, O.H., Passonneau, J.V., Hasselberger, F.X. and Schulz, D.W., 1964. Effect of ischemia on known substrates and cofactors of glycolytic pathways in brain. J. Biol. Chem. 239, 18-42.

Mayman, C.I., Gatfield, P.D. and Brenckenridge, B.M., 1964. The glucose content of brain in anaesthesia. J. Neurochem. 11, 483-487.

Miller, A.L., Hawkins, R.A. and Veech, R.L., 1973. The mitochondrial redox state of rat brain. J. Neurochem. 20, 1393-1400.

Passonneau, J.V., Brunner, E.A., Molstad, C. and Passonneau, R., 1971. The effects of altered endocrine states and of ether anaesthesia on mouse brain. J. Neurochem. 18, 2317-2328.

Passonneau, J.V. and Lowry, O.H., 1971. Metabolic flux in single neurons during ischemia and anesthesia. Recent Adv. in Quant. Histo-and Cytochemistry, (Dubach, U.C. and Schmidt, V., eds.) pp. 197-212, Han Hoeber, Bern.

Quastel, J.H., 1963. Effects of anesthetics, depressants and tranquilizers on cerebral metabolism. In: Metabolic Inhibitors, (Hochester, R.M. & Quastel, J.H., eds.) pp. 517-538, Academic Press, New York.

Raskin, N.H. and Sokoloff, L., 1972. Enzymes catalyzing ethanol metabolism in neural and somatic tissues of the rat. J. Neurochem. 19, 273-282.

Skou, J.C., 1957. The influence of some cations on an adenosine triphosphatase from peripheral nerves. Biochem. Biophys. Acta 23, 394-401.

Streeten, D.H.P. and Solomon, A.K., 1954. The effects of ACTH and adrenal steroids on K transport in human erythrocytes. J. Gen. Physiol. 37, 643-649.

Veech, R.L., Eggleston, L.V. and Krebs, H.A., 1969. The redox state of free nicotinamide-adenine dinucleotide phosphate in the cytoplasm of rat liver. Biochem. J. 115, 609-619.

Veech, R.L., Gwynn, R. and Veloso, D., 1972. The time course of the effects of ethanol on the redox and phosphorylation states of rat liver. Biochem. J. 127, 387-397.

Veech, R.L., Veloso, D., Harris, R.L. and Veech, E.H., 1973. Freeze-blowing: A new technique for the study of brain in vivo. J. Neurochem. 20, 183-188.

Veech, R.L. and Hawkins, R.A., 1974. Brainblowing: A technique for in vivo study of brain metabolism. In: Research Methods in Neurochemistry, Vol. 2 (Marks, N. and Rodnight, R., eds.) pp. 171-182. Plenum Press, New York.

Veloso, D., Passonneau, J.V. and Veech, R.L., 1972. The effects of intoxicating doses of ethanol upon intermediary metabolism in rat brain. J. Neurochem. 19, 2679-2686.

Wilson, D.F., Stubbs, M., Veech, R.L., Erecinska, M. and Krebs, H. A., 1974. Equilibrium relations between the oxidation reduction reactions and the ATP synthesis in suspensions of isolated liver cells. Biochm. J. 140, 57-64.

THE USE OF TRITIUM AND ^{14}C LABELLED ETHANOL IN STUDIES OF ETHANOL METABOLISM AT HIGH ETHANOL CONCENTRATIONS

S. Damgaard, L. Sestoft and F. Lundquist

Department of Biochemistry A, University of Copenhagen

30, Juliane Maries Vej, 2100 Copenhagen, Denmark

Measurement of the rate of ethanol metabolism of the liver at concentrations at which the pharmacological actions of ethanol become manifest, i.e. 50-80 mM are of considerable importance. Such measurements can, however, not be performed with acceptable accuracy by determination of the decrease in ethanol concentration in preparations such as slices, isolated hepatocytes, or perfused liver, as the concentration differences are small compared to the absolute level. In the intact organism the overall metabolism of ethanol at high concentrations can be determined by measurement of the blood alcohol concentration at suitable intervals, but in this case we cannot decide in which organ the metabolism takes place. Extrahepatic metabolism may play a more important role at high than at low ethanol concentrations. The obvious solution to this problem is to measure the products formed from ethanol, not the disappearance of the substrate.

Acetate is the major metabolite formed from ethanol in liver preparations, but the proportion of ethanol recovered as acetate varies according to the nutritional state and possibly other factors as well. This substance alone is therefore not suitable for accurate measurements of ethanol metabolism.

The use of isotopically labelled ethanol would, however, appear adaptable to this purpose, as the non metabolized substrate, being volatile, is readily removed. If carbon labelled ethanol is used it is necessary to determine the radioactivity in the water and lipid soluble compounds in the medium as well as in the tissue itself and also in the CO_2 produced.

When ethanol labelled with tritium at the 1-carbon atom is employed oxidation to the level of acetate would remove all the tritium, which would end up mainly in water, but to some extent also in reduced compounds formed from the reduced nicotinamide nucleotides produced in the oxidation of ethanol to acetate, as shown by Hoberman and Carnicero (1) at low ethanol concentrations (5 mM). At the level of acetaldehyde only one half of the tritium is removed, provided the two hydrogens of the methylene group in ethanol are equally labelled, which they will be if the ethanol is produced by chemical reduction of acetaldehyde. The feasibility of using the two principles described was examined in experiments on perfused rat liver.

Materials and Methods

Female Wistar rats weighing about 150 g were used. They were either non-induced, fed normally on laboratory chow or induced by the ingestion of 10 g ethanol daily per kg body weight by stomach tube for 2 to 8 weeks and fed ad lib. Non-recirculating perfusion was used (2). The medium was Krebs-Ringer bicarbonate + 2 % bovine albumin. Washed bovine erythrocytes were added to a hematocrit of 30 %. The medium was equilibrated with atmospheric air to which 5 % carbon dioxide was added. The flow rate was about 0.7 ml. min^{-1} x g^{-1} x liver fresh wt. in all experiments.

Commercially available 1-^{3}H-ethanol was purified to a maximal contamination by water-exchangeable tritium of less than 0.1 per cent. Radioactive ethanol, whether tritium or carbon labelled, was removed from deproteinized samples of medium and tissue by chasing with absolute ethanol vapour during reflux by means of a suitable column, which retains the water. Samples as small as 2 ml may be processed in this way. Complete removal of radioactive ethanol is achieved by this technique. Details of the procedure will be published in due course.

Water samples for determination of tritium were obtained by distillation at pH 7.5 after removal of ethanol. Water soluble compounds were obtained for counting by evaporation to dryness in the presence of non-radioactive water and ethanol. Acetaldehyde in samples of perfusion medium was caught in semicarbazide by diffusion in Conway units and counted after repeated evaporation to dryness with non-radioactive ethanol. Carbon dioxide was obtained quantitatively from samples of medium,from which radioactive ethanol was removed,by diffusion into ethanolamine after addition of acid.

Oxygen uptake was calculated from the change of saturation of hemoglobin measured spectrophotometrically (Radiometer OSM 1), and the flow rate.

The standard experiments were performed as illustrated in fig. 1.

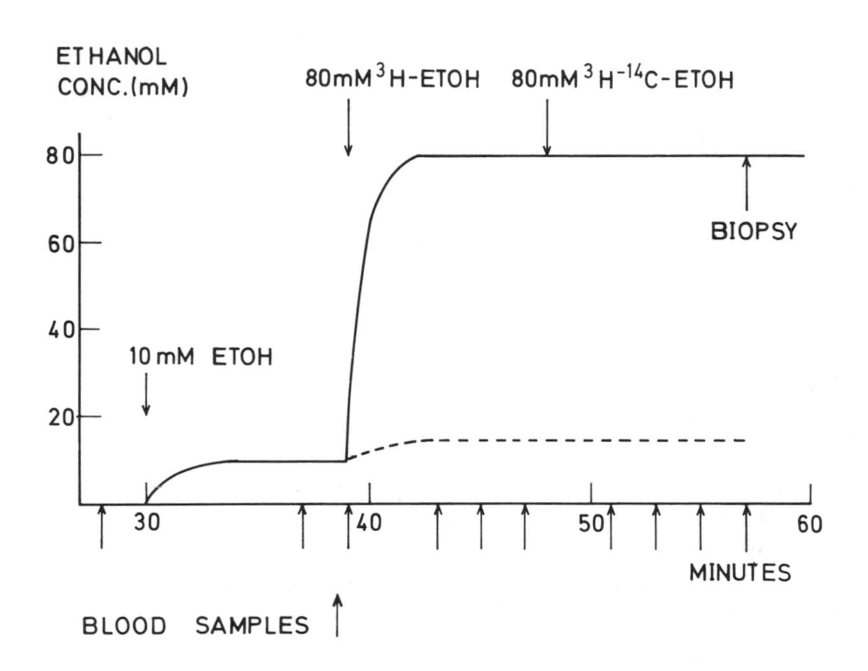

Fig.1. Schematic representation of experimental procedure. After an equilibration period of 30 min without additions to the medium ethanol was infused at a concentration of 10 mM for 9 min, in order to allow the metabolic pattern to change to one of ethanol metabolism before the experimental period proper. Tritium labelled ethanol was then given at either 15 or 80 mM for 9 min followed by a period in which both tritium and carbon labelled ethanol was infused without changing the total concentration of ethanol. Blood samples are taken as indicated and the experiment is concluded by taking a biopsy from the liver.

RESULTS

Induction of increased ethanol metabolism is believed to involve mechanisms, which are not coupled to the formation of ATP, such as oxidation by means of hydrogen peroxide or hydroxylation. One might therefore expect an increase in the total oxygen uptake in livers from induced rats compared to normal rats during ethanol metabolism. In fig. 2 experiments are shown in which the oxygen uptake and the acetate release of normal and induced rats are compared at different ethanol concentrations. The difference seems not to be significant.

The experiments in which ^{14}C ethanol was used showed the distribution of the label given in table I. It is apparent that measurement of all four groups of compounds are necessary to obtain the total rate of alcohol metabolism. Acetaldehyde contributed very little to the total radioactivity at the perfusion rate employed. The concentration in the medium calculated from the radioactivity (see Methods) was in the range around 100 μM, found in the rat also by other workers.

Table I. Distribution of ^{14}C from 1-^{14}C-ethanol

	Effluent medium		Liver	
	Carbon dioxide	Water sol. compounds	Water sol. compounds	Lipid sol. compounds
Control rats				
15 mM eth.(1)	9	42	33	16
80 mM - (2)	4	50	28	18
80 mM eth. + 4-M-pyr. (2)	10	26	34	30
Induced rats				
80 mM eth.(2)	6	50	26	18
80 mM - + 4-M-pyr. (2)	9	25	31	35

The results are given as per cent of the total radioactivity recovered.

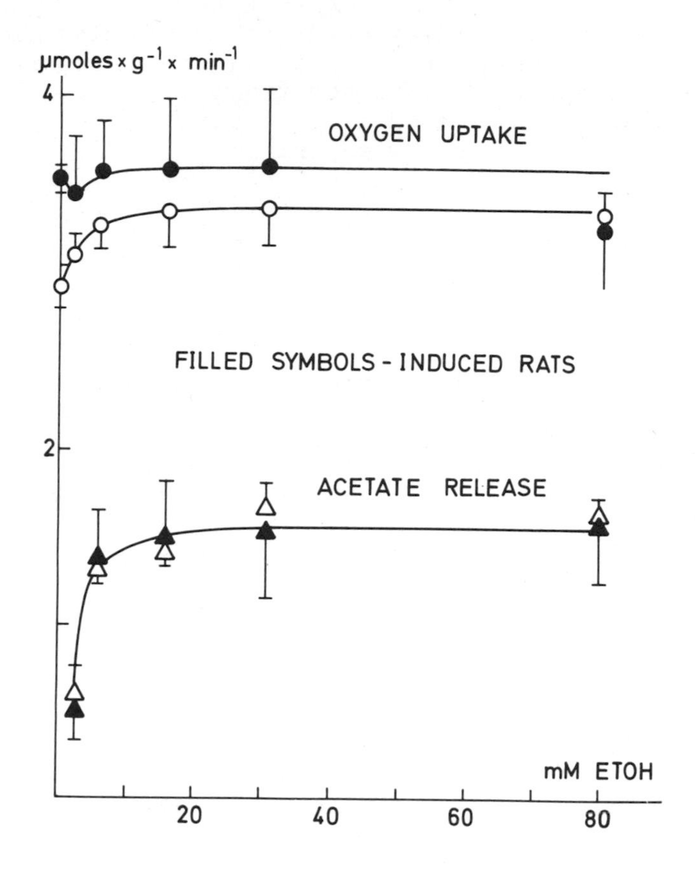

Fig. 2. Oxygen uptake and acetate release in experiments on livers from induced and control animals as a function of the ethanol concentration in the affluent medium.

In the case of tritium labelled ethanol the amount of the isotope retained in the liver was less than 5 %. It was therefore neglected in the calculations. In the perfusion medium about 90 % of the non-ethanol tritium was present in the water, except in the methyl pyrazole experiments where practically all radioactivity was found in the water, as would be expected when ADH is totally inhibited.

Table II shows the rate of tritium release to the medium at different ethanol concentrations in livers from normal and induced rats. Again the increase in both groups at high concentration compared to 15 mM ethanol was small. However, the fraction of the tritium liberation which could not be inhibited by methyl pyrazole apparently increased with increasing alcohol concentration, an observation which is in agreement with other work in this laboratory (3). Whether the non ADH fraction of ethanol metabolism is increased after induction cannot be decided on the basis of these few experiments.

Hoberman and Carnicero (1) found an isotope effect when tritium labelled ethanol was used in perfusion experiments. The tritium liberation amounted to 0.68 times the rate of ethanol metabolism at 5 mM ethanol, independently of the presence of an excess of pyruvate, which traps a considerable part of the reducing equivalents as lactate. In two experiments we found the rate of tritium release to be 0.78 times the rate of ethanol metabolism calculated from the production of ^{14}C labelled metabolites at an ethanol concentration of 15 mM. Table III gives the results of a few experiments performed at 80 mM ethanol on liver from induced and control rats. Even though the material is limited the results suggest that the sum of the non-ADH pathways has a larger isotope effect than the ADH-pathway. This would be in agreement with the observations of Gang et al. (4).

DISCUSSION

In some experiments the ethanol metabolism calculated from ^{14}C was not higher in livers from induced rats than in those from control animals, and as seen in fig. 2 the oxygen uptake was not increased significantly. The reason for this may be that the intake of ethanol has not been sufficient to ensure adequate induction or that the access to oxygen has been insufficient. In Liebers experiments, in which ethanol

Table II. Rate of tritium release from 1-^{3}H ethanol

Ethanol conc.	15 mM	50 mM	80 mM
	(μmoles ethanol x min^{-1} x g^{-1} ± S.E.M.)		
Control rats	0.63±0.09(4)	0.52±0.16(3)	0.68±0.06(4)
- + 4-M-Pyr.	-	0.032	0.20
Induced rats	0.64-0.56(2)	0.60±0.16(3)	0.74±0.09(3)
- + 4-M-Pyr.	-	0.095	0.28

The figures represent the amount of tritium released to the perfusion medium. The concentration of 4-methyl pyrazole (4-M-Pyr.), when used, was 1-5 mM, sufficient to ensure complete inhibition of ADH at all ethanol concentrations used.

Table III. The isotope effect in ^{3}H-ethanol metabolism

	Metabolism calc.from ^{14}C (A)	Liberation of tritium (B)	A/B
	(µmoles x min^{-1} x g^{-1} liver)		
Control rats			
80 mM ethanol (2)	1.08	0.75	1.44
80 mM eth.+ 4-M.Pyr. (2)	0.86	0.205	4.2
Induced rats			
80 mM ethanol (2)	1.78	0.79	2.25
80 mM eth.+4-M-pyr. (2)	1.04	0.22	4.7

Column A is the sum of radioactive carbon compounds produced. Column B is the non-ethanol tritium in the effluent perfusion medium. The rate of ethanol metabolism in these experiments is somewhat lower than that arrived at from fig. 2 and table I.

was incorporated into a liquid diet, about 14 g of ethanol was consumed per kg and day. It cannot be excluded that the increase from 10 to 14 g, and the more continous administration in Liebers experiments is of decisive importance for the induction process. The other possibility is connected with the fact that at the flow rate used any major increase in oxygen consumption is not possible. Though the oxygen saturation of the effluent medium did not fall below 15 % in any experiment, it is possible that due to hydrodynamic factors part of the liver cells still suffer from lack of oxygen as suggested by Krebs (5). This possibility is under investigation.

The isotope effect of tritium in ethanol metabolism is difficult to interpret in a complicated system like the liver. If splitting of the C-H bond were rate limiting a very large effect should be observed. This has indeed been reported with alcohol dehydrogenase under special circumstances using deuterium labelled ethanol (6). What is observed in the liver is the overall result of a series of elementary reactions, which may depend on a number of conditions, such as the pathways followed and the degree of saturation of the enzymes involved.

The experiments shown in Table III suggest that the magnitude of the 'isotope effect' depends on the contribution of the non-ADH pathways to alcohol metabolism. It is therefore not possible simply by measuring the liberation of tritium in an *in vitro* or *in vivo* system to evaluate the rate of ethanol metabolism. In suitable cases it should, however, by combination of tritium and carbon labelling as outlined here, be possible to estimate quantitatively the contribution of pathways other than ADH, provided that the 'isotope effect' for all participating reactions can be shown to be reasonably constant.

REFERENCES

1. Hoberman, H.D. and Carnicero, H.H. in Alcohol and Aldehyde Metabolizing Systems (Ed. R.G. Thurman *et al.*), Acad. Press, New York 1974, p. 395-407.
2. Sestoft, L. Biochem. Biophys. Acta 343, 1-16 (1974).
3. Grunnet, N., Quistorff, B. and Thieden, H.I.D. Eur. J. Biochem. 40, 275-282 (1973).
4. Gang, H., Cederbaum, A.I. and Rubin, E. Biochem. Biophys. Res. Comm. 54, 264-269 (1973).
5. Krebs, H.A. In Regulation of Hepatic Metabolism (Ed. F. Lundquist and N. Tygstrup). Munksgaard, Copenhagen 1974, p. 806.
6. Gershman, H. and Abeles, R.H. Arch. Biochem. Biophys. 154, 659-674 (1973).

ENZYMES OF BIOGENIC ALDEHYDE METABOLISM

Jean P. von Wartburg, Denis Berger, Margret M. Ris and Boris Tabakoff*

Medizinisch-chemisches Institut, University of Berne, Berne, Switzerland

The terms biogenic aldehydes, biogenic alcohols and biogenic acids have been coined (1, 2) to describe the deaminated metabolites of the neurotransmitter amines (i.e. biogenic amines). Much recent evidence indicates that these products have a physiologic role in the CNS. For example, deaminated metabolites of both serotonin (3) and norepinephrine (4) have been theorized to be involved in transitions between REM and slow wave sleep. Serotonin metabolites have also been postulated to control body temperature (5, 6) and sleep-wakefulness (7) in animals. Biogenic aldehydes have been reported to inhibit both Na^{+} - K^{+} and Mg^{++} activated ATP'ases in brain synaptosomes (8). However, the characteristics of the enzymes responsible for the metabolism of biogenic aldehydes, and the factors which may regulate this metabolism in vivo have only recently come under scrutiny (2, 9).

Scheme 1 illustrates some of the major pathways of metabolism for biogenic amines such as serotonin, norepinephrine and dopamine. Previous work has demonstrated that blockade of the oxidation or reduction of biogenic aldehydes by pharmacologic agents such as

* Current address: Dept. of Biochemistry, The Chicago Medical School, Chicago, 60612 Ill., USA

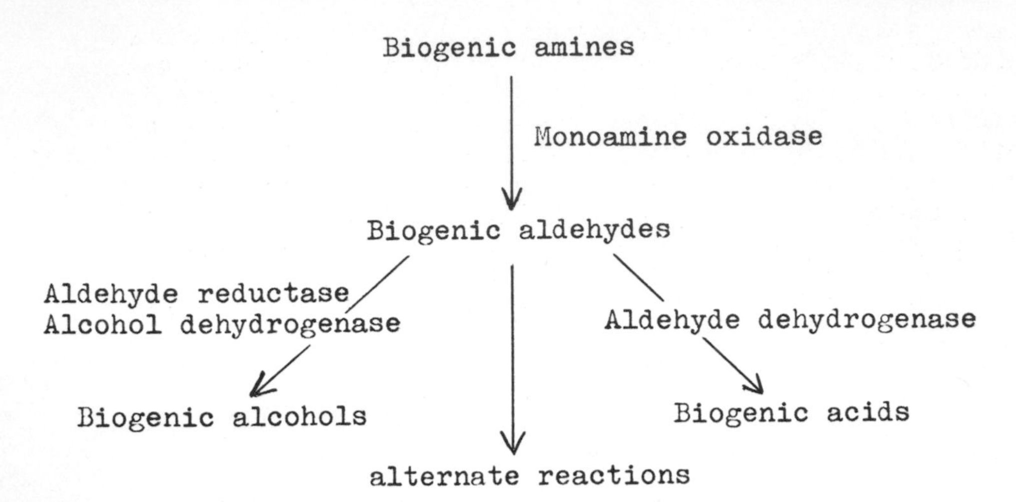

Scheme 1: Metabolism of biogenic amines

ethanol or its metabolites and sedative hypnotics can shunt these aldehydes into alternate reactions. Two such reactions are the formation of tetrahydroisoquinoline alkaloids (10) or the binding of biogenic aldehydes to neuronal components (11). These pathways may prove to be of primary importance for certain symptoms associated with long term ingestion of ethanol or other CNS depressants.

While the study of aldehyde metabolism in the brain is in a developmental stage, ethanol as well as aldehyde metabolism in the liver has been widely studied (12, 13). The knowledge gained from the study of liver enzymes can serve as a model for studying the enzyme heterogeneity, enzyme mechanisms and genetic factors controlling aldehyde metabolism in the CNS. By now the existence of a biochemical individuality with respect to the liver enzymes of ethanol and aldehyde metabolism is well established. In particular the genetic control of the multiple molecular forms of human liver alcohol dehydrogenase has been investigated recently (14, 15). These dimeric isoenzymes are formed by combination of four different subunits (16). In addition to this heterogeneity, a polymorphism also exists for human alcohol dehydrogenase and the term "atypical" variant was introduced (17). The variant enzyme is mainly characterized by a higher specific activity for alcohol oxidation and a shift of the pH optimum to a lower value. Although it has been shown that in individuals carrying this genetic trait ethanol disappearance from blood is not faster than in normal individuals (18, 19), an initially higher rate of ethanol metabolism may be expected for a short period ending when the reoxidation of the reduced coenzyme becomes rate limiting. Hence, these individuals may attain a higher level of circulating acetaldehyde and be affected by direct pharmacological actions of this ethanol metabolite. It is of interest to note that individual and racial differences have been observed with respect to flushing after ethanol ingestion (20). Further differences may be expected due to interference with the redox state of the coenzymes, interaction with enzymes by acetaldehyde as a competitive substrate and condensation with biogenic amines to form isoquinoline alkaloids (21). Finally, a faster rate of production of biogenic alcohols in the liver may be predicted in "atypical" individuals on the basis of the enzyme mechanism of liver alcohol dehydrogenase. Although many of these phenomena are primarily expected to take place in liver and other peripheral tissues, secondary effects on the CNS have to be considered.

Progress has been made recently in our laboratory with respect to the elucidation of the prevalence and the structure of "atypical" human liver alcohol dehydrogenase (22). According to the genetic model proposed by Smith et al. (14, 15) one would expect three genotypes due to this polymorphism. However, only a bimodal distribution, i.e. normal and "atypical", of alcohol dehydrogenase activities have been observed using several screening tests (18). This fact may be the result of a dominance of one allelic gene, broad variation in isoenzyme patterns, lability of the "atypical" enzyme in homozygous individuals, low gene frequency, or inadequacy of the screening methods.

With the use of CM-cellulose chromatography and hybridization techniques we found that the isoenzyme BB from livers of "normal" individuals was homogeneous, while the one from "atypical" individuals was heterogeneous. In accord with this, the partial sequence analysis of isoenzyme BB from "normal" livers

Table 1: Partial amino acid sequence of "normal" and "atypical" human liver alcohol dehydrogenase

Residue No[a)]				
Horse[b)]	... Asp·Lys	·Phe·Ala·Lys	·Ala·Lys	·Glu ...
Human B_1	... Asp·Lys	·Phe·Ala·Lys	·Ala·Lys	·Glx ...
Human B_2	... Asp·Lys	·Phe·Pro·Lys	·Ala·Lys	·Glx ...
Peptide No[c)]	/ 2	/ 24a,b	/ 11	/ 3

a) Number of amino acid residue in the sequence of horse liver alcohol dehydrogenase

b) According to Jörnvall (23)

c) According to Berger et al. (22)

(see Table 1) disclosed only one tripeptide (number 24a) bearing the sequence phenylalanine-alanine-lysine. However, the isoenzyme BB from "atypical" livers contained an additional tripeptide (number 24b) not found in samples from "normal" isoenzyme BB. This tripeptide had the sequence phenylalanine-proline-lysine. Thus, it was concluded that the alanine containing tripeptide found in the "atypical" samples originated from the "normal" subunit B_1 of the enzyme, while the proline containing tripeptide originated from the "atypical" subunit B_2.

This substitution of alanine-230 in subunit B_1 to proline in subunit B_2 is of interest since it is in the area corresponding to the α-helix C of the coenzyme binding structure of the horse enzyme (24). It is known that alanine is a strong former of α-helices while proline breaks such helical structures. This substitution, therefore, would be expected to result in changes in the secondary and tertiary structures of the subunit B_2 which are probably responsible for the witnessed increase in specific activity and the shift in pH optimum.

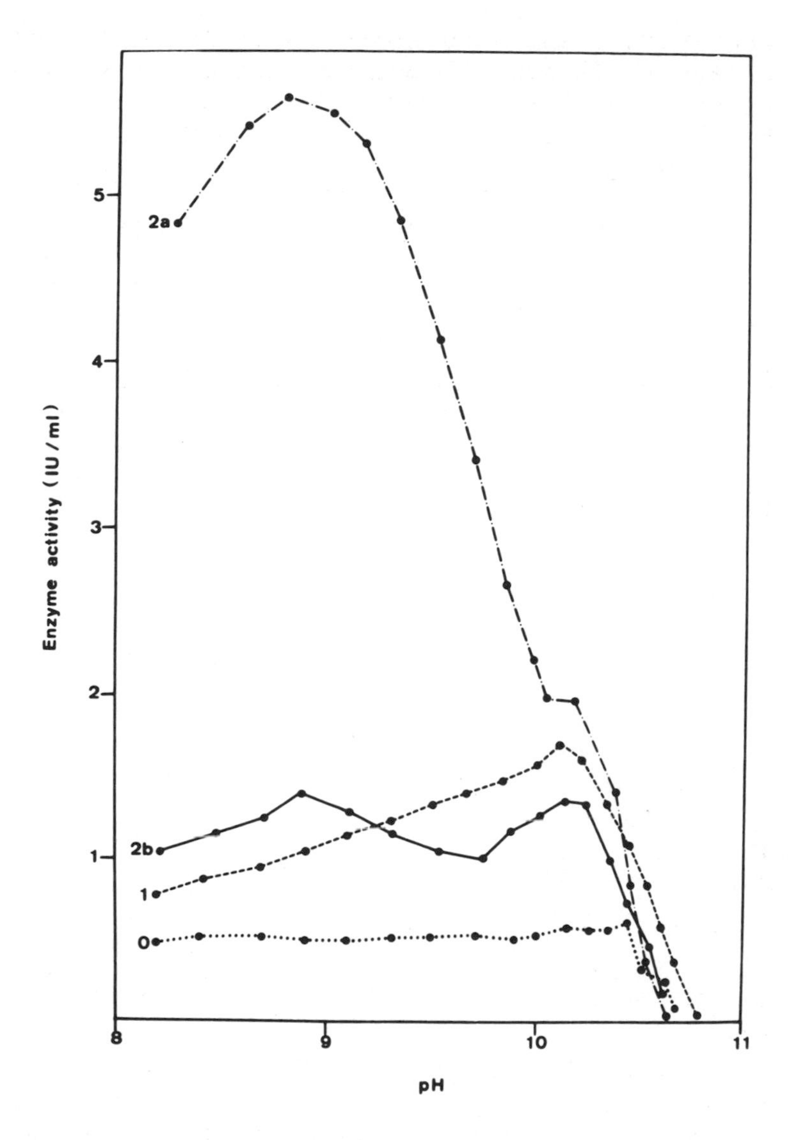

Figure 1: pH rate profiles of ethanol oxidation by alcohol dehydrogenase in homogenates of human livers.

Conditions: 66 mM glycine-NaOH buffer, 16 mM ethanol, 1.6 mM NAD. (0) non classifiable samples without distinct pH optimum and low activity. (1) "normal" livers with one pH optimum. (2) "atypical" livers with two distinct pH optima; (a) and (b) represent extreme curves of this type.

Livers of the "normal" phenotype with a pH rate profile of type 1 in Fig. 1 would correspond to the homozygous genotype b_1b_1. Livers of the "atypical" phenotype with a pH rate profile of type 2 would represent the heterozygous genotype b_1b_2. Since the estimated frequency of the allele b_2 is extremely low (14) few homozygous individuals with the homozygous "atypical" genotype b_2b_2 can be expected in the European populations which have been studied. In contrast to these findings a frequency of over 90 % for "atypical" individuals has been reported from Japan (25). However, we feel that the frequency values calculated from data using ratios of the enzyme activity at only two pH values in autopsy material have to be considered with caution; a large number of livers, when tested for alcohol dehydrogenase activity over a wide range of pH values were found to be nonclassifiable (type 0 in Fig. 1). Therefore, attempts at determining frequencies for the "atypical" variant by the use of pH activity ratios should utilize activity determinations at intermediary pH values, as well as the pH optima of the two phenotypes. In addition, the term "atypical" should be deleted from use and this enzyme polymorphism should be considered as a manifestation of the multiple molecular forms of group 3 in the nomenclature recommended by IUB (26).

Although many questions remain to be answered regarding various aspects of the heterogeneous array of liver alcohol dehydrogenases, the presence of alcohol dehydrogenase in brain and its involvement in the pharmacologic and biochemical effects of ethanol is even less clear (for reviews see 12 and 13). Recently, using an assay system in which ethanol metabolism is coupled to the production of propanediol from lactaldehyde (27), Raskin and Sokoloff (28) demonstrated the presence of a small alcohol dehydrogenase activity in brain. This observation was followed by a matching report of changes in the redox state of metabolite couples in mouse brain during ethanol intoxication (29). However, further work by Veech et al. (30) showed that only little change in the ratios of NAD to NADH and NADP to NADPH in rat brain was evident after ethanol administration, and this small effect could be explained by an alteration in pCO_2 in brain due to ethanol induced anesthesia.

Forsander (31) has demonstrated that acetaldehyde is undetectable ($< 1\ \mu M$) in brain until blood levels of acetaldehyde exceed 200 mg/100 ml. This may be explained by the high aldehyde dehydrogenase activity in rat brain which could effectively metabolize lower amounts of acetaldehyde entering the brain and/or produced in the brain. Although small amounts of alcohol dehydrogenase activity may not contribute a great deal to the maintenance of an acetaldehyde level, brain alcohol dehydrogenase may function normally to metabolize some endogenous constituents and the presence of ethanol may interfere with this metabolism. The possibility that brain alcohol dehydrogenase may catalyze the conversion of succinic semialdehyde to γ-hydroxybutyric acid has been proposed recently (32). Since our laboratories have in the last years been concerned with enzymes that metabolize biogenic aldehydes and since alcohol dehydrogenase would be capable of metabolizing such compounds (33) we became interested in establishing the presence and character of brain alcohol dehydrogenase.

The experimental results reported in Table 2 were obtained by application of the coupled redox reaction with ethanol and lactaldehyde as substrates described by Gupta and Robinson (27) for horse liver alcohol dehydrogenase and applied to rat brain alcohol dehydrogenase by Raskin and Sokoloff (28). Compounds were added to incubation mixtures containing rat brain cytosol (0.3 ml supernatant of 20 % homogenates), sodium phosphate (50 mM, pH 7.4) and D- or L-lactaldehyde in a total volume of 1.2 ml. Other additions are listed in Table 2. Reactions were incubated at 37°C for 30 min and propanediol concentrations were determined as described by Jones and Riddick (34).

Although high reagent blank values were obtained, the addition of the enzymes in rat brain cytosol produced a substantial increase in propanediol formation from D-lactaldehyde in the presence of either NAD or NADH (Table 2). The fact that this enzymatic production of propanediol was not inhibited by pyrazole suggests that it is mediated by an aldehyde reductase (see further description of these enzymes below) distinct from alcohol dehydrogenase. That aldehyde reductase can participate in the production of propanediol is further substantiated by the finding that the highest rates were observed with NADPH as a coenzyme, and that

Table 2: Activity of brain alcohol dehydrogenase and other aldehyde metabolizing enzymes. For details see text.

Enzyme	EtOH 200mM	Pyrazole 1 mM	Coenzyme	Lactaldehyde 8 mM	Activity nMol $min^{-1}g^{-1}$	n
+	-	-	-	D	6.0	2
-	+	-	NAD (0.5mM)	"	8.6 ± 4.0	5
+	-	-	"	"	18.2 ± 5.6	5
+	+	-	"	"	21.4 ± 7.0	5
+	+	+	"	"	18.8 ± 7.8	5
-	+	-	NADH (0.16mM)	"	9.7 ± 5.0	5
+	-	-	"	"	20.0 ± 5.9	5
+	+	-	"	"	28.0 ± 8.7	5
+	+	+	"	"	22.6 ± 7.6	5
+	-	+	"	"	19.4	2
-	+	-	NADP (0.5mM)	"	4.8	2
+	-	-	"	"	24.4	2
+	+	-	"	"	22.1	2
+	+	+	"	"	25.5	2
-	+	-	NADPH (0.16mM)	"	3.2	2
+	-	-	"	"	49.6	2
+	+	-	"	"	50.7	2
+	+	+	"	"	50.5	2
-	+	-	NAD (0.5mM)	L	12.5 ± 5.7	4
+	-	-	"	"	14.6 ± 4.8	4
+	+	-	"	"	24.7 ± 11.4	4
+	+	+	"	"	16.7 ± 1.1	4
+	-	+	"	"	14.1 ± 4.8	3
-	+	-	NADH (0.16mM)	"	13.4 ± 1.8	3
+	-	-	"	"	18.2 ± 2.6	3
+	+	-	"	"	24.6 ± 6.2	3
+	+	+	"	"	16.4 ± 2.7	3
+	-	+	"	"	17.8 ± 5.1	3

also this activity was not affected by either pyrazole or ethanol. It was of interest to note that the addition of oxidized cofactors resulted in enhanced formation of propanediol. Since both NAD- and NADP-linked aldehyde dehydrogenase occur in rat brain cytosol (35) this finding can be explained by the following way: D-lactaldehyde and the oxidized pyridine nucleotides would serve as substrates for aldehyde dehydrogenase resulting in the production of NADH and NADPH, which could then be utilized by aldehyde reductase to produce propanediol from D-lactaldehyde. Additions of ethanol and pyrazole had no effect in the presence of NADP. In contrast, when NAD was the cofactor, ethanol produced a small increase in activity which was inhibited by pyrazole. Although the variation between different experiments was large, this increase and inhibition were consistent within each experiment. Analogous effects of ethanol and pyrazole were found with NADH as a cofactor. These results indicate that there is a contribution of a NAD-dependent, pyrazole sensitive alcohol dehydrogenase to propanediol formation from D-lactaldehyde in addition to the aldehyde reductase mediated production. Such alcohol dehydrogenase is better demonstrated with the analogous experiments using L-lactaldehyde (Table 2).

Gupta and Robinson (27) reported nearly a hundredfold stimulation of acetaldehyde production when lactaldehyde was added to incubation mixtures containing horse liver alcohol dehydrogenase, ethanol and NAD. It is probable that rat brain alcohol dehydrogenase conforms to the same type of mechanism as liver alcohol dehydrogenase. Hence, the rate limiting step (i.e. the dissociation of the enzyme-NADH-complex) would be bypassed in the coupled assay system. Thus, a significant increase in ethanol oxidation may be expected when the coupled system is used to assay enzymatic activity. Therefore, direct assays of alcohol dehydrogenase activity would be more appropriate. Such assays can be accomplished by either measuring the production of NADH due to oxidation of an alcohol or measuring the oxidation of NADH due to reduction of an aldehyde. The determination of an alcohol oxidation in brain cytosol is inherently difficult and less sensitive than aldehyde reduction. However, our results in Table 2 and those published previously (2, 36) indicate that aldehyde reductases may interfere with an assay of alcohol dehydrogenase relying on determination of aldehyde reduction in presence of NADH. Therefore, a partial purification of alcohol dehydrogenase was attempted.

Taking into account the known differences in molecular weight of alcohol dehydrogenase (approximately 80'000) and aldehyde reductase (approximately 40'000) a separation was accomplished by gel chromatography on Sephadex G-100. Two peaks of aldehyde metabolizing activity were noted when L-lactaldehyde and NADH were used as substrates. From the elution volumes of the two peaks approximate molecular weights of 80'000 for peak I and of 44'000 for peak II were estimated. Enzyme activity in peak II had the characteristics of aldehyde reductase (see below) and no alcohol dehydrogenase activity was found using the coupled assay system. Only little NADPH-linked activity was found in peak I, but alcohol dehydrogenase was present and could be assayed directly by monitoring NADH production fluorimetrically and spectrophotometrically with ethanol as substrate. This brain alcohol dehydrogenase activity was inhibited by pyrazole and its pH optimum was found to be approximately at pH 9.8. The lower pH optimum (pH 7.6) reported by Raskin and Sokoloff (28) for this enzyme is probably due to the use of the coupled assay system and would reflect the maximizing of aldehyde reduction rather than ethanol oxidation per se. The specific activity of the brain alcohol dehydrogenase in peak I was approximately 0.06 nMol/min/mg of protein when assayed fluorimetrically at pH 7.4 and 5.1 uMol/min/mg of protein when determined in the coupled assay system. The increase in specific activity measured in the coupled assay is similar to that reported for the horse liver alcohol dehydrogenase (27). The total activity per gram rat brain was estimated to be 0.4 nMol/min at physiologic pH based on spectrophotometric assays of activity in peak I.

On the other hand, another class of oxidoreductases, i.e. aldehyde reductases, is present in brain, and has been shown to be quite active in metabolizing endogenously produced aldehydes, particularly the biogenic aldehydes. We have isolated several forms of aldehyde reductase from bovine (37), rat and human brain (2). Reports from other laboratories indicate that similar enzymes are present in pig (38, 39) and monkey brain (40) and reductase activity has also been shown to be present in mouse brain (41). In general the aldehyde reductases (alcohol : NADP oxidoreductase E.C.1.1.1.2) differ from alcohol dehydrogenase (alcohol : NAD oxidoreductase E.C.1.1.1.1) by their insensitivity to pyrazole inhibition, their preference for NADPH as cofactor and their low molecular weight ($\sim$44'000).

Four multiple molecular forms of aldehyde reductase have been isolated from human brain (2), while rat, pig and bovine brain contain at least two forms (2, 39, 42). These isoenzymes were numbered H 4.1, 4.2, 4.3 and 4.4 for human brain and R 4.1 and 4.2 for rat brain (2). All forms studied so far can utilize NADPH as a cofactor. However, only one form isolated to date (H 4.2 and R 4.2) can utilize NADH as well. Similar findings have been reported for a reductase from pig brain (39) and bovine brain (42). In general these enzymes have a broad substrate specificity for various aldehydes, but significant differences can be demonstrated between each of the multiple forms. Substrates for the enzyme fractions 4.2 of both species include aliphatic aldehydes, such as acetaldehyde and propionaldehyde.

Further differences between the multiple forms can be noted in their sensitivities toward inhibition by various pharmacological agents (2, 43 and Table 3). Not only is the activity of aldehyde reductases sensitive to inhibition by synthetic hypnotics, sedatives and anticonvulsants, but certain compounds endogenous to brain inhibit aldehyde reductase. Of particular interest is the fact that biogenic acids are good inhibitors mainly of rat enzyme 4.1 and human enzyme 4.3.

With this in mind, it is of interest to note that both a single dose and "chronic" administration (44) of ethanol produces an elevation in brain 5-hydroxyindoleacetic acid (5-HIAA) levels in mice (41). There were no significant changes in serotonin levels in brain at any time after ethanol treatments in our experiments. The rise in brain 5-HIAA after a single dose of ethanol lasts only for 2-3 hours, while chronic administration of ethanol results in a prolonged increase in brain 5-HIAA levels for approximately 30 hours (see Table 4). These changes in brain 5-HIAA mirror the time course for withdrawal symptoms seen after chronic ethanol administration to mice.

Table 3: Inhibition of human and rat brain aldehyde reductases by biogenic acids and drugs

Compound	Conc.	Human				Rat	
	µM	4.1	4.2	4.3	4.4	4.1	4.2
5-Hydroxyindoleacetic acid	500	22	27	28	21	52	20
4-Hydroxy-3-methoxy-mandelic acid	500	47	22	23	28	45	15
3,4-Dihydroxyphenyl-acetic acid	500	8	55	53	38	69	25
4-Hydroxyphenyl-acetic acid	500	18	45	52	32	70	25
Phenobarbital	1000	6	55	88	33	95	34
Diphenylhydantoin	500	8	49	89	16	98	74
2-Phenyl-2-methyl-succinimide	250	6	55	95	13	77	31
2-Phenyl-2-ethyl-glutarimide	500	0	8	52	8	62	39

Results represent % inhibition. Enzyme activity was determined spectrophotometrically under the following assay conditions: 500 µM p-nitrobenzaldehyde, 160 µM NADPH, enzyme protein, in 100 mM sodium phosphate buffer pH 7.0

The increase in CNS 5-HIAA levels after ethanol administration have also been noted in cerebrospinal fluid of cats (45) and in human alcoholics during ethanol ingestion (46). We have recently performed an analysis of possible mechanisms by which ethanol may alter 5-HIAA levels in the CNS. Several possibilities were examined: 1. Since brain serotonin turnover has been shown to be dependent on availability of tryptophan, the effects of ethanol of brain tryptophan levels and tryptophan uptake from plasma were examined. 2. Serotonin turnover in CNS was examined after ethanol administration by the methods described by Neff and Tozer (47). This methodology would reflect both, a functional increase in the activity of serotonergic neurons and an increase in the activity of enzymes synthesizing serotonin from tryptophan (i.e. tryptophan hydroxylase and aromatic amino acid decarboxylase). One of course, has to keep in mind that tryptophan hydroxylase is the rate limiting enzyme in this pathway and thus changes in decarboxylase activity would not be readily evident by use of this methodology. 3. The activity of MAO, and mitochondrial and cytosolic aldehyde dehydrogenase, as well as NADPH dependent aldehyde reductase, were measured after ethanol administration, since these enzymes are responsible for catabolism of serotonin. 4. The ability of CNS to remove 5-HIAA in the presence of ethanol was examined. No differences were found in plasma or brain tryptophan levels or in the rate of uptake of peripherally injected ^{14}C-tryptophan. These determinations were done during times that brain 5-HIAA was signifioantly elevated in ethanol treated animals. The turnover of serotonin was significantly depressed (rather than increased) in brains of mice treated with ethanol. A similar reduction in turnover of serotonin after a single dose of ethanol has also been reported in rat brain by Hunt and Majchrowicz (48) and mouse brain by Kuriyama et al. (49). No significant increases in MAO, aldehyde dehydrogenase or aldehyde reductase activities were noted in brain after either a single dose or chronic ethanol administration.

However, the removal of 5-HIAA from the CNS of animals treated with ethanol was significantly inhibited (see Table 5). The transport of 5-HIAA from brain is mediated by an energy dependent, probenecid sensitive process (50). The egress of 5-HIAA from brain may be directed to blood or 5-HIAA may exit via the cerebro-

Table 4: Increase in brain 5-HIAA levels after a single dose or chronic[1] ethanol administration

Treatment	Time[4] (hrs)	5-HIAA (ng/gm brain)	p[5]
saline	2	225 ± 26	
ethanol[2]	2	327 ± 24	0.01
saline	10	225 ± 44	n.s.
ethanol[2]	10	207 ± 17	
control[3]	0.16	245 ± 36	
ethanol[3]	0.16	396 ± 42	< 0.01
control[3]	20	266 ± 19	
ethanol[3]	20	354 ± 14	< 0.01
control[3]	36	263 ± 28	
ethanol[3]	36	304 ± 25	n.s.

1) Chronic ethanol administration to C57Bl/6J mice was by inhalation as described by Goldstein (44)

2) Ethanol was administered as a single dose 3 gm/kg i.p.

3) Mice were kept in chambers in which the atmosphere either contained or did not contain ethanol (ca. 8 mg/liter air) for three days before determinations of brain 5-HIAA after withdrawal. For further experimental details see Tabakoff et al. (41).

4) Time after last exposure to ethanol

5) P values of 0.05 or lower were taken to indicate significant differences between control and ethanol treated mice

Table 5: Inhibition of removal of 5-HIAA from mouse brain by ethanol

Experiment I	dpm x 10^{-3}/gm brain	
Time (min)	saline	ethanol
0	92.4 ± 10.6	87.0 ± 9.7
30	38.4 ± 12.7	47.8 ± 12.4
60	16.8 ± 3.1	27.2 ± 5.8
Experiment II		
60	14.0 ± 1.9	12.0 ± 2.3

Mice were injected intraventricularly with ^{14}C-5-HIAA (0.1 μCi/10 μl) thirty minutes (experiment I) or five hours (experiment II) after i.p. injection of either ethanol (3gm/kg) or saline. Mice were decapitated either immediately after injection of ^{14}C-5-HIAA or at various times thereafter and the radioactivity remaining in brain was determined at these times. For further experimental detail see Tabakoff et al. (53).

spinal fluid (51). The tissue concerned with clearance of 5-HIAA from the CSF is the choroid plexus of the lateral and fourth ventricles and these transport systems are similar to neuronal 5-HIAA transport systems (52). Thus, not only was ethanol shown to inhibit the exit of intraventricularly injected ^{14}C-5-HIAA (Table 5) from CNS but was also found to be an inhibitor of 5-HIAA uptake by isolated choroid plexus. Significant inhibition of 5-HIAA uptake by choroid plexus was noted at concentrations of ethanol that are found in vivo ($< 0.05M$) after ethanol administration. This inhibition was shown to be reversible and the kinetic parameters are now under investigation.

Since the distribution of 5-HIAA, as serotonin, has been shown to be localized in particular areas of brain (54), one may postulate that considerable amounts of this organic acid may become concentrated in certain brain areas after ethanol administration. In addition, the transport of other biogenic acids may also be inhibited by ethanol. Thus, other neuronal systems may also be subject to an increased concentration of organic acids. These acids may become concentrated in the neurons where they are produced or diffuse to neighbouring neurons. The increased concentration of organic acids may affect directly the neuronal membrane excitability pattern (55).

On the other hand, our previous studies mentioned above indicate that biogenic acids would also inhibit the metabolism of aldehydes by aldehyde reductase and may lead to an accumulation of these reactive compounds in neurons. The possible consequences of an increase in biogenic aldehydes in the CNS has been discussed previously (1, 43) in the light of alterations in the action potential by binding aldehydes to neuronal membranes (56) and formation of physiologically active tetrahydroisoquinoline.

Our present work indicates that ethanol may influence biogenic aldehyde metabolism via the biogenic acids as well as by way of acetaldehyde production. These studies would support the conclusion that ethanol administration has a profound effect on biogenic amine metabolism at several levels, and would indicate that certain changes in physiologic functions which result from chronic ethanol ingestion are mediated by the metabolites of the biogenic amines.

Acknowledgments: The excellent technical assistance from Ms. M. Kaufmann and B. Callahan is greatfully acknowledged.
Original work was supported in part by grants from the Swiss National Science Foundation No. 3.8340.72, NIMH No. AA00233 and Deutsche Forschungsgemeinschaft Wa 302 to J.P. v.W.; and grants from NIAAA, NIH, St. of Illinois, Dept. of Mental Health, U.S. Brewers Association, Hoffman-La Roche Foundation and Swiss National Science Foundation to B.T., B.T. is a Schweppe Foundation Fellow.

REFERENCES

1. Tabakoff, B., Ungar, F, and Alivisatos, S.G.A. Adv. in Exper. Med. and Biol. 35, 45 (1973)

2. Ris, M.M. and von Wartburg, J.P. Eur. J. Biochem. 37, 69 (1973)

3. Jouvet, M. Science 163, 32 (1969)

4. Jones, B.E. Brain Res. 39, 121 (1972)

5. Barofski, I. and Feldstein, A. Experientia 26, 99o (197o)

6. Gessner, P.K. and Soble, A.G. J.Pharmacol. Exp. Ther. 186, 276 (1973)

7. Sabelli, H.C., Giardina, W.J., Alivisatos, S.G.A., Seth, P.K. and Ungar, F. Nature 223, 73 (1969)

8. Tabakoff, B. Res. Comm. in Clin. Path. and Pharmacol. 7, 621 (1974)

9. Tabakoff, B., Groskopf, W., Anderson, R. and Alivisatos, S.G.A. Biochem. Pharmacol. 23, 17o7 (1974)

1o. Davis V.E., Cashaw, J.L., Mc Laughlin, B.R. and Hamlin, T.A. Biochem. Pharmacol. 23, 1877 (1974)

11. Tabakoff, B., Ungar, F. and Alivisatos, S.G.A. Nature, New Biology 238, 126 (1972)

12. Kissin, B. and Begleiter, H. "The Biology of Alcoholism", Vol. 1 - 3, Plenum Press, New York-London (1971 - 1974)

13. Wallgren, H. and Barry, H. "Actions of Alcohol" Vol. 1 and 2, Elsevier Publishing Company, Amsterdam-London-New York (197o)

14. Smith, M., Hopkinson, D.A. and Harris, H. Ann. Hum. Genet., London 35, 243 (1972)

15. Smith, M., Hopkinson, D.A. and Harris, H. Ann. Hum. Genet., London 34, 251 (1971)

16. von Wartburg, J.P., Berger, D., Bühlmann, C., Dubied, A. and Ris, M.M. in "Alcohol and Aldehyde Metabolizing Systems", Eds. Thurman, R.G. et al. Academic Press, Inc., New York-London, p. 33 (1974)

17. von Wartburg, J.P., Papenberg, J. and Aebi, H. Canad. J. Biochem. 43, 889 (1965)

18. von Wartburg, J.P. and Schürch, P.M. Ann. N.Y. Acad. Sci. 151, 936 (1968)

19. Edwards, J.A. and Evans, P.D.A. Clin. Pharmacol. Ther. 8, 824 (1967)

2o. Ewing, J.A., Rouse, B.A. and Pellizzari, E.D. Am. J. Psychiatry 131, 2o6 (1974)

21. Cohen, G. and Collins, M.A. Science 167, 1749 (197o)

22. Berger, D., Berger, M. and von Wartburg, J.P. Eur.J. Biochem. (1974); in press

23. Jörnvall, H. Eur. J. Biochem. 16, 25 (197o)

24. Brändén, C.I., Eklund, H., Nordström, B., Boiwe, T., Söderlund, G., Zeppezauer, E., Ohlson, I. and Akeson, A. Proc. Nat. Acad. Sci. U.S.A. 7o, 2439 (1973)

25. Ogata, S. and Mizohata, M. Japan. J. Stud. Alcohol 8, 33 (1973)

26. IUPAC-IUB Commission on Biochemical Nomenclature (CBN) Eur. J. Biochem. 24, 1 (1971)

27. Gupta, N.K. and Robinson, W.G. J. Biol. Chem. 235, 16o9 (196o)

28. Raskin, N.H. and Sokoloff, L. J. Neurochem. 22, 427 (1974)

29. Rawat, A.K., Kuriyama, K, and Mose, J. J. Neurochem. 2o, 23 (1973)

3o. Veech, R.L. in "Alcohol and Aldehyde Metabolizing Systems", Eds. Thurman, R.G. et al., Academic Press, Inc., New York-London, p. 383 (1974)

31. Forsander, O., "The Acetaldehyde Level of the Body During Ethanol Oxidation" " (this volume)

32. Taberner, P.V. Biochem. Pharmacol. 23, 1219 (1974)

33. von Wartburg, J.P. in "Biology of Alcoholism", Eds. Kissin, B. and Begleiter, H. Plenum Press, New York-London, Vol. 1, p. 63 (1971)

34. Jones, L.R. and Riddick, J. A. Analyt. Chem. 29, 1214 (1957)

35. Ris, M.M., Deitrich, R.A. and von Wartburg, J.P. J. Neurochem. (1974), in preparation

36. Tabakoff, B., Vugrincic, C., Anderson, R. and Alivisatos, S.G.A. Biochem. Pharmacol. 21, 1457 (1972)

37. Tabakoff, B. and Erwin, V.G. J. Biol. Chem. 245, 3262 (1970)

38. Turner, A.J. and Tipton, K.F. Eur. J. Biochem. 30, 361 (1972)

39. Turner, A.J. and Tipton, K.F. Biochem. J. 130, 765 (1972)

40. Bronaugh, R.L. and Erwin, V.G. J. Neurochem. 21, 809 (1973)

41. Tabakoff, B. and Boggan, W.O. J.Neurochem. 22, 759 (1974)

42. Erwin, V.G., Heston, W.D. and Tabakoff, B. J.Neurochem. 19, 2269 (1972)

43. von Wartburg, J.P., Ris, M.M. and White, T.G. Proc. 8th Congress CINP, Copenhagen, North-Holland Publishing Company, Amsterdam-London and Avicenum, Czechoslovak Medical Press, Prague (1973)

44. Goldstein, D. J. Pharmacol. Exp. Ther. 180, 203 (1972)

45. Tabakoff, B., Ritzmann, R.F., Boggan, W.O. and Radulovacki, M. Transactions of Meeting American Soc. for Neurochem. Vol. 5, p. 159 (1974)

46. Zarcone, V. "Effects of Alcohol on CSF Biogenic Amines and Sleep in Man" (this volume)

47. Neff, N.H. and Tozer, T.N. Adv. in Pharmacol. 6A, 97 (1968)

48. Hunt, W.A. and Majchrowicz, E. Brain Res. 72, 181 (1974)

49. Kuriyama, K., Rauscher, G.E. and Sze, P.Y. Brain Res. 26, 450 (1971)

50. Neff, N.H., Tozer, T.N. and Brodie, B.B. J. Pharmacol. Exp. Ther. 158, 214 (1967)

51. Guldberg, H.C. in "Metabolism of Amines in Brain" Ed. Hopper, G., MacMillan (London) p. 55 (1969)

52. Guldberg, H.C., Ashcroft, G.W. and Crawford, T.B.B. Life Sci. 5, 1571 (1966)

53. Tabakoff, B., Ritzmann, R.F. and Boggan, W.O. J. Neurochem. (1974), submitted for publication

54. Eccleston, D., Ashcroft, G.W., Moir, A.T.B., Parker-Rhodes, A., Lutz, W. and O'Mahoney, D.P. J. Neurochem. 15, 947 (1968)

55. Tower, D.B. in "Basic Mechanisms of the Epilepsies" Eds. Jaspers, H.H. et al., Little Brown and Co., Boston, Mass.,p. 611 (1969)

56. Shrager, P.G., Strickholm, A. and Macey, R.I. J. Cell Physiol. 74, 91 (1969)

VARIATIONS IN THE ACETALDEHYDE LEVEL OF SOME TISSUES IN HYPER- AND HYPOTHYROID RATS

Olof A. Forsander

Research Laboratories of the State Alcohol Monopoly (Alko), Box 350, SF-00101 Helsinki 10, Finland

Since small amounts of ethanol are always present in the blood (Krebs and Perkins 1970), minute concentrations of acetaldehyde might also be expected to be present. Large amounts, however, can only be found after alcohol consumption. Ethanol is oxidized almost exclusively in the liver, consequently the acetaldehyde in the blood must originate from this tissue.

Different authors have reported very different values for the acetaldehyde content of the blood (see Truitt and Welsh 1971 for a review). These differences have resulted partly from the lack of specificity in the methods used for the aldehyde determination and probably also from a nonenzymatic formation of acetaldehyde from ethanol during the analytical procedures (Sippel 1973). There also are, however, large individual differences in the rate of aldehyde metabolism (Eriksson 1973). Furthermore, the metabolism can also be greatly affected by different physiological conditions (Forsander *et al.* 1969).

In the present experiments, the physiological state of rats were altered by treatment with triiodothyronine or propylthiouracil, a procedure which has previously been shown to influence the ethanol metabolism (Rawat and Lundquist 1968, Ylikahri *et al.* 1968). The acetaldehyde levels in various tissues were then determined after alcohol administration.

MATERIALS AND METHODS

Male albino rats of Wistar origin, and given ordinary laboratory food and tap water *ad libitum*, were used for the experiments. Hypothyroidism was induced by daily administration of 5 mg per 100 g body weight of a 0.5 % solution of propylthiouracil (obtained from Eli Lilly and Co.) by stomach tube for three weeks. Hyperthyroidism was induced by daily treatment with 20 µg per 100 g body weight of 3.3'5-triiodo-L-thyronine (Sigma Chemical Co.), given intraperitoneally in saline, also for three weeks.

Fasting was effected by food depletion for twenty hours before testing.

Oxygen consumption of the experimental animals in basal conditions was recorded at the end of the third week by means of a Beckman Oxygen Analyser, Model E 2; the method used was that described by Depocas and Hart (1957). The basal oxygen consumption of the propylthiouracil treated group was about 30 per cent lower than that of the control group, and that of the triiodothyronine treated group about 30 per cent higher than that of the control group; this confirms that the animals did develope pronounced hypo- and hyperthyroidism.

Alcohol administration. Ethanol, 1.5 g per kg of body weight, was given to all animals intraperitoneally as a 10 % (w/v) solution in saline. The animals were anesthetized with 40 mg of pentobarbital (Mebunalum NFN) per kg of body weight 15 minutes before sampling.

Sampling. Sixty minutes after the ethanol injection, blood samples were drawn from the tip of the tail, the abdomen was opened, a heparinized aspiration needle was inserted into the hepatic vein, and 2 ml of blood was drawn into a heparinized plastic catheter. The liver was quickly frozed *in situ* with aluminium tongs precooled in liquid nitrogen. The frozen tissue was crushed in a morter with the addition of liquid nitrogen and suspended in three parts of ice could 0.6 M perchloric acid. The precipitate was centrifuged down and the supernatant was neutralized with 2 M-K_2HPO_3 to pH 6.5-6.9. For estimation of the acetaldehyde in the brain, the roof of the skull of the anesthetized animal was quickly removed and 0.4-0.8 g of the brain tissue (excluding the pons-medulla, cerebellum, olfactory bulbs and the lateral parts of the cerebral hemispheres) was obtained by the freeze-stop technique as with the liver. Mixed arterio-venous blood was obtained from the cranial cavity immediately after removal of the brain sample.

Ethanol and acetaldehyde were measured on a Perkin-Elmer F 40 head-space gas liquid chromatograph, as reported previously (Sippel 1972).

Lactate and pyruvate were assayed enzymatically as described by Hohorst et al. (1959). Enzymes and coenzymes were obtained from C.F. Boehringer.

RESULTS

The acetaldehyde content of the peripheral blood, the cerebral blood, the liver and the brain is shown in Table 1. It was highest in the liver where it was formed, much lower in the cerebral blood and still lower in the peripheral blood. When the aldehyde present in the cerebral blood was subtracted from the value found for the freeze-clamped brain it could be concluded that the brain tissue contained only small traces of acetaldehyde. This has also been reported earlier by Sippel (1974).

The differences in the acetaldehyde content of the various tissues must depend on differences in extrahepatic acetaldehyde oxidation. A rough calculation based on the values given in the literature for the liver blood flow, the previously measured alcohol oxidation rate, and the liver content of acetaldehyde given in Table 1, showed that 95-99 per cent of the aldehyde formed from

Table 1. Acetaldehyde content of peripheral and cerebral blood, liver and brain after alcohol administration to normal fed rats[a]

Blood		Liver	Brain
Peripheral	Cerebral		
21	63	190	about 0

[a] Expressed as nmoles per ml blood or nmoles per g wet tissue. Means of experiments with 4 animals are given. Ethanol, 1.5 g per kg body weight was given intraperitoneally. Acetaldehyde was measured as described in the Methods section. (C.J.P. Eriksson and H. Sippel, personal communication).

ethanol in the liver is also oxidized there. Only 0.1-0.5 per cent is lost through the lungs while the rest is oxidized in various tissues (see Deitrich 1966). Fig. 1.

The blood acetaldehyde content rises with the blood alcohol level (Hald _et al._ 1949, Lindros _et al._ 1972). This same trend can be seen in Freund and O'Hollaren's (1965) work with man when lower alcohol doses were used. The same content of ethanol can,

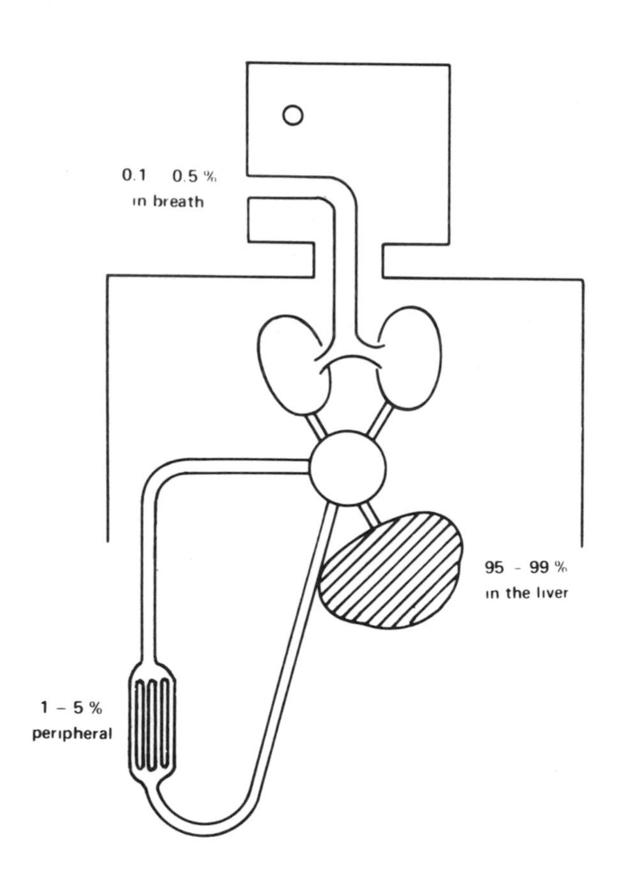

Fig. 1. Ethanol is oxidized mainly in the liver where 95-99 per cent of the acetaldehyde formed is immediately oxidized. 0.1-0.5 per cent of the aldehyde transported out from the liver is expired by the breath. The rest, 1-5 per cent is oxidized extrahepatically (K. Lindros and C.J.P. Eriksson, personal communication).

however, produce different aldehyde levels. Sheppard et al. (1970) found that the C57BL/6J and the DBA/2J mice strains had significantly different acetaldehyde levels in the blood after ingestion of the same amount of alcohol. Eriksson (1973) has found that this is true also for two rat strains which were developed on the basis of their voluntary alcohol consumption. Different physiological states can also affect the acetaldehyde level of the body: e.g., fasting strongly depresses it. As can be observed from Table 2 not only fasting but also hypo- or hyperthyroidism can influence the aldehyde content of the liver and the blood. The level was much lower in animals treated with triiodothyronine than in normal animals and still lower in propylthiouracile treated animals.

The acetaldehyde concentration in the liver must be established as a result of the kinetic properties of the enzyme system or systems forming acetaldehyde and the enzymes oxidizing it. In circulating liver perfusion studies it has been shown that the acetaldehyde concetration is controlled by the redox state, since

Table 2. Acetaldehyde content of the liver after ethanol administration to normal, triiodothyronine-treated and propylthiouracil-treated rats, fed and fasted[a]

	liver nmoles per g wet wt.
Normal	
fed	135 + 22
fasted	22 + 19
Triiodothyronine treated	
fed	83 + 31
fasted	58 + 29
Propylthiouracil treated	
fed	65 + 15
fasted	21 + 2

[a] Means + S.D. of 6-9 experiments are given. Ethanol, 1.5 g per kg body weight, was given intraperitoneally. Acetaldehyde was measured from the freeze clamped liver as described in the Methods section. Triiodothyronine and propylthiouracil treatments are also described there. (From Forsander et al. 1969).

the ethanol-acetaldehyde redox pair is in redox equilibrium with the free NAD-NADH couple of the liver cytosol (Scholz 1968, Krebs 1969, Fellenius 1973). This has been found not to be true for a noncirculating liver perfusion (Lindros et al. 1972).

Table 3 shows the measured ethanol/acetaldehyde ratios of the intact livers from normal, triiodothyronine and propylthiouracile trated rats and the same ratios calculated from the lactate/pyruvate ratios of the livers. In these calculations it was assumed that the lactate/pyruvate ratio is 1.7 times higher than the ethanol/acetaldehyde ratio when the two systems are in equilibrium (Lindros et al. 1972). It can be seen that in none of the physiological states is the ethanol-acetaldehyde couple in redox equilibrium with the free NAD-NADH redox system of the liver cytosol.

DISCUSSION

The toxicity and the pharmacological effects of acetaldehyde are much stronger than those of ethanol. In experiments with isolated cow adrenal gland it has been shown that an aldehyde concentration of as low as 1.5×10^{-5}M induces a release of catecholamines (Schneider 1971). Lahti and Majchrowicz (1967) have found

Table 3. The measured and calculated ethanol/acetaldehyde ratios of the livers of normal, triiodothyronine-treated and propylthiouracile-treated fed rats[a]

	ethanol/acetaldehyde	
	measured	calculated
normal	218	22
triiodothyronine	355	19
propylthiouracile	453	46

[a] The measured acetaldehyde values are taken from Table 2. In these experiments the mean ethanol content of the blood was 31.5 nmoles per g wet liver. The ethanol/acetaldehyde ratio is calculated from the lactate/pyruvate ration given by Forsander et al. (1969). It is assumed that the lactate/pyruvate ratio is 1.7 times higher than the ethanol/acetaldehyde ratio.

a very marked effect of aldehyde on serotonin metabolism when the concentration was 1.0 M. Even lower concentration would probably also have an influence on enzymatic reactions. It can also be assumed that the acetaldehyde formed from ethanol during alcohol oxidation will have pharmacological effects on the intact body.

Some time after alcohol consumption ethanol is spread evenly throughout the water phase of the body. This is not the case with acetaldehyde. Since the aldehyde can be oxidized in most tissues no general aldehyde pool is formed, and an analysis of the aldehyde content of the blood does not give any informations on the content of other parts of the body. As can be seen from Table 1 no acetaldehyde could be measured in the brain even though considerable amounts of aldehyde were present in the peripheral blood. Kesäniemi (1974a) found that the content of ethanol in the blood and in the milk of lactating women after alcohol consumption was rather similar. In contrast to this, no aldehyde was found in the milk although the level of the peripheral blood was rather high. Similarly, Kesäniemi (1974b) could find no acetaldehyde in the rat fetus, although the maternal aortic blood contained rather high levels. The content of ethanol, however, was approximately the same in the maternal blood and in the fetal tissue.

Acetaldehyde is formed from ethanol by alcohol dehydrogenase but can also be produced by the catalase and the MEOS systems. The first mentioned enzyme is, however, accepted to be the most important for ethanol oxidation. The acetaldehyde formed can be oxidized by a variety of aldehyde dehydrogenases located in the cytosol, the microsomal fraction and the mitochondria of the liver cell. These enzymes have widely different K_m values for acetaldehyde and differ also in other respects (see Tottmar 1974). It would, therefore, be difficult to estimate the kinetics for acetaldehyde oxidation in the intact liver under the same conditions that exists when ethanol is oxidized. Consequently, a calculation of the theoretical acetaldehyde level during ethanol oxidation in the way proposed by Lundquist and Wolthers (1958) may not be feasible.

Theorell and Bonnichsen (1951), using from the velocity constants of the alcohol dehydrogenase reactions, calculated that at equilibrium there would be 60 times more ethanol than acetaldehyde present in the system. An ethanol/acetaldehyde ratio of 22 can be calculated from the equilibrium constants for the alcohol dehydrogenase (Bäcklin 1958) and lactate dehydrogenase (Williamson <u>et al.</u> 1967) if the lactate/pyruvate ratios were 37. The ratio so calculated by Theorell and Bonnichsen and the value calculated from the equilibrium constants are astonishly similar when taking into account the approximations made in the calculations.

The values obtained for the intact livers shown in Table 3, are, however, much higher than either of the calculated ratios, thus demonstrating that the ethanol-acetaldehyde pair are not in equilibrium with the redox level of the free NAD-NADH system of the liver cytosol. An equilibrium is probably reached in a liver perfusion with recirculating medium, where the acetaldehyde formed during one passage through the liver is carried back with the medium at the next passage and aldehyde accumulates until an equilibrium is established. This is not the case in the intact animal where the aldehyde formed in the liver is oxidized extrahepatically and probably only a small amount is carried back to the liver with the portal blood.

REFERENCES

Bäcklin, K.-J., 1958. The equilibrium constant of the system ethanol, aldehyde, DPN^+, DPNH and H^+, Acta Chem. Scand. 12: 1279 - 1285.

Deitrich, R.A., 1966. Tissue and subcellular distribution of mammalian aldehyde oxidizing capacity, Biochem. Pharmacol. 15: 1911 - 1922.

Depocas, F. and Hart, J.S., 1957. Use of the Pauling oxygen analyzer for measurement of oxygen consumption of animals in open-circuit systems and in short-lag, closed circuit apparatus, J. Appl. Physiol. 10: 388 - 392.

Eriksson, C.J.P., 1973. Ethanol and acetaldehyde metabolism in rat strains genetically selected for their ethanol preference, Biochem. Pharmacol. 22: 2283 - 2292.

Fellenius, E., 1973. The interaction between ethanol and lipid metabolism in the rat liver. Thesis, Uppsala University.

Forsander, O.A., Hillbom, M.E. and Lindros, K.O., 1969. Influence of thyroid function on the acetaldehyde level of blood and liver of intact rats during ethanol metabolism. Acta Pharmacol. Toxicol. 27: 410 - 416.

Freund, G. and O'Hollaren, P., 1965. Acetaldehyde concentrations in alveolar air following a standard dose of ethanol in man, J. Lipid Res. 6: 471 - 477.

Hald, J., Jacobsen, E., and Larsen, V., 1949. Formation of acetaldehyde in the organism in relation to dosage of Antabus and to alcohol-concentration in blood, Acta Pharmacol. 5: 179 - 188.

Hohorst, H.J., Kreutz, F.H.,and Bücher, Th., 1959. Über Metabolitgehälte und Metabolit-Konzentrationen in der Leber der Ratte, Biochem. Z. 332: 18 - 46.

Kesäniemi, Y.A., 1974a. Ethanol and acetaldehyde in the milk and peripheral blood of lactating women after ethanol administration, Brit. Common. 81: 84 - 86.

Kesäniemi, Y.A., 1974b. In vivo contents of ethanol and acetaldehyde in the rat fetus during maternal ethanol oxidation, to be published.

Kesäniemi, Y.A., 1974c. Metabolism of ethanol and acetaldehyde in intact rats during pregnancy, Biochem. Pharmacol. 23: 1157 - 1162.

Krebs, H.A., 1969. Role of equilibria in the regulation of metabolism, Curr. Top. Cell. Regul. 1: 45 - 55.

Krebs, H.A. and Perkins, J.R., 1970. The physiological role of liver alcohol dehydrogenase. Biochem. J. 118: 635 - 644.

Lahti, R.A. and Majchrowicz, E., 1967. The effects of acetaldehyde on serotonin metabolism, Life Sci. 6: 1399 - 1406.

Lindros, K.O., Vihma, R., and Forsander, O.A., 1972. Utilization and metabolic effects of acetaldehyde and ethanol in the perfused rat liver, Biochem. J. 126: 945 - 952.

Lundquist, F. and Wolthers, H., 1958. The kinetics of alcohol elimination in man, Acta Pharmacol. Toxicol. 14: 265 - 289.

Rawat, A.K. and Lundquist, F., 1968. Influence of thyroxine on the metabolism of ethanol and glycerol in rat liver slices, Eur. J. Biochem. 5: 13 - 17.

Scholz, R., 1968. Untersuchungen zur Redoxkompartmentierung bei der hämoglobinfrei perfundierten Rattenleber, in Stoffwechsel der isoliert perfundierten Leber. (W. Steib und R. Scholz, eds.) pp. 25-47, Springer Verlag, Berlin.

Schneider, F.H., 1971. Acetaldehyde-induced catecholamine secretion from the cow adrenal medulla, J. Pharmacol. Exp. Ther. 177: 109 - 118.

Sheppard, J.R., Albersheim, P., and McClearn, G., 1970. Aldehyde dehydrogenase and ethanol preference in mice, J. Biol. Chem. 245: 2876 - 2882.

Sippel, H.W., 1972. Thiourea an effective inhibitor of the non-enzymatic ethanol oxidation in biological extracts, Acta Chem. Scand. 26: 3398 - 3400.

Sippel, H.W., 1973. Non-enzymatic ethanol oxidation in biological extracts, Acta Chem. Scand. 27: 541 - 550.

Sippel, H.W., 1974. The acetaldehyde content in rat brain ethanol metabolism, Biochem. Pharmacol., in press.

Theorell, H. and Bonnichsen, R., 1951. Studies on liver alcohol dehydrogenase, Acta Chem. Scand. 5: 1105 - 1126.

Tottmar, O., 1974. Aldehyde dehydrogenase in rat liver. Thesis, Uppsala University.

Truitt, E.B. and Walsh, M.J., 1971. The role of acetaldehyde in the actions of ethanol, in The Biology of Alcoholism, Vol. 1 (B. Kissin and H. Begleiter, eds.) pp. 161 - 195, Plenum Press, New York.

Williamson, D.H., Lund, P., and Krebs, H.A., 1967. The redox state of the freenicotinamide-adenine dinucleotide in the cytoplasm and mitochondria of rat liver, Biochem. J. 103: 514 - 527.

Ylikahri, R.H., Mäenpää, P.H., and Hassinen, J.E., 1968. Ethanol-induced changes of cytoplasmic redox state as modified by thyroxine treatment, Ann. Med. Exp. Biol. Fenn. 46: 137 - 142.

FACTORS CONTROLLING THE RATE OF ALCOHOL DISPOSAL BY THE LIVER

Hans A. Krebs and Marion Stubbs

Metabolic Research Laboratory, Nuffield
Department of Clinical Medicine
Radcliffe Infirmary, Oxford, England

INTRODUCTION

The acute effects of ethanol on the nervous system depend on how rapidly ethanol is removed from the blood circulation after the intake. Thus it is of interest to study the factors responsible for the removal of ethanol in the liver, the main site of ethanol disposal. We therefore examined systematically the rate of alcohol removal by the liver, using isolated hepatocytes as experimental material. This is convenient material because many parallel experiments can be carried out on one homogeneous suspension. Tests in many laboratories so far have not revealed any major differences in the behaviour of the intact perfused liver and the isolated hepatocytes. When suitable conditions are chosen quantitative differences disappear, or are minimal (Krebs, Cornell, Lund and Hems, 1974).

The main reaction by which alcohol is removed is that catalyzed by alcohol dehydrogenase, involving interaction with NAD^+ according to the equation

$$\text{Ethanol} + NAD^+ \longrightarrow \text{acetaldehyde} + NADH + H^+$$

followed by the dehydrogenation of acetaldehyde to acetic acid according to the equation

$$\text{Acetaldehyde} + NAD^+ \longrightarrow \text{acetate} + NADH + H^+$$

It should be mentioned that part of the dehydrogenation of acetaldehyde occurs in the cytoplasm and part in the mitochondria (Marjanen, 1972) whilst the first step of dehydrogenation of ethanol to acetaldehyde is limited to the cytosol.

The question, then, which we have studied is which factors determine the rate of the alcohol dehydrogenation reaction. One of these is the capacity of alcohol dehydrogenase, but under some conditions the dehydrogenation of ethanol ceases either because equilibrium is reached in the alcohol dehydrogenase system, or because free (i.e. available) NAD^+ has fallen to a low level having been converted to NADH. Hence it would be expected that the removal of alcohol is accelerated by processes which reconvert NADH to NAD^+. This can be achieved by a variety of different processes, the main process being the transfer to the mitochondria and oxidation by the respiratory chain. This oxidation is obligatorily coupled with the synthesis of ATP and it is therefore to be expected that the rate of ethanol removal is increased when the rate of ATP synthesis is increased. There is, however, another factor determining the rate of oxidation by the respiratory chain of NADH produced in the cytoplasm. This is the rate of transfer of NADH from the cytoplasm to the mitochondria which involves a special mechanism because NADH does not readily penetrate the mitochondrial membrane.

METHODS

Liver cells were prepared from female Wistar rats, starved for 48 h, by the method of Berry and Friend (1969) with the modifications described by Cornell, Lund, Hems and Krebs (1973) and Krebs et al. (1974). Between 20 and 40 mg dry wt. of liver cells per flask were used. Experiments were either carried out in manometer cups as described by Krebs et al. (1974) when oxygen uptake was to be measured or, when ethanol removal was to be measured, in specially designed flasks (see Fig. 1). The flasks minimised the loss of added ethanol through evaporation by limiting the gas space to about 4 ml, still adequate for the oxygen requirements of the cells. The flasks were sealed with Suba seals (Suba-Seal Works, Barnsley, England) after all reagents including cells but excepting ethanol had been introduced.

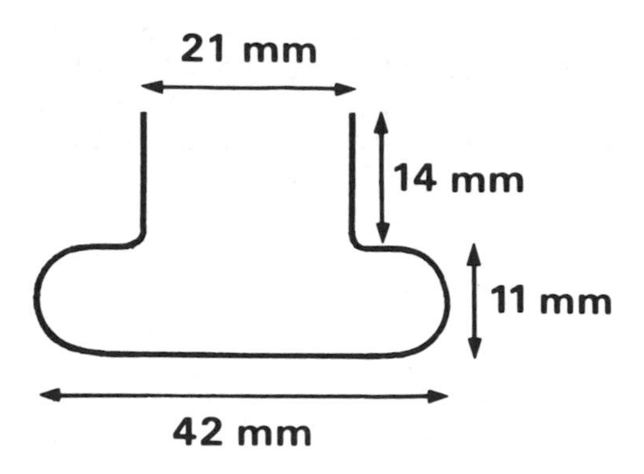

Figure 1

Flask for incubation of cell suspension (4 ml) for experiments with ethanol. The dimensions refer to external values. Total volume is 14 ml. The gas space is about 10 ml.

Gassing was carried out by inserting 2 hypodermic needles through the seal, through one of which the gas entered. The ethanol was then injected through the seal with the aid of a 100 μl Hamilton syringe. For accurate determination of the amount of ethanol added, two extra flasks were set up in each experiment to which no cells were added, but which in every other respect were treated identically with the test flasks. Ethanol was measured in a perchloric acid extract (final concentration 2%) of the cell suspension by the principle of Bonnichsen (1963) but instead of pyrophosphate-semicarbazide buffer Tris recrystallised from methanol was used.

Acetate was measured by diffusion of the volatile acid into distilled water in a Conway unit (Conway, 1958). Anhydrous sodium sulphate (10 g) was placed in the outer ring and 1 ml of deionised distilled water was in the centre. Of the perchloric extracts 1 ml was spread onto the sodium sulphate. The units, covered with glass slides, were left overnight. The estimation of acetate was performed by the following manometric technique based on the release of stoicheiometric amounts of CO_2 from a bicarbonate buffer by the acetic acid. The contents of the centre of the Conway unit were transferred quantitatively into a 5 ml graduated tube by washing with about 1 ml distilled water. The volume was made up to 2.2 ml and 2 ml were placed in a small (10 ml) Warburg manometer cup which contained in the side arm 0.4 ml of 0.1 $NaHCO_3$ previously gassed with O_2:CO_2 (95:5). After a period of equilibration at 25°C the contents of the side arm were tipped into the main compartment and the positive pressure changes were recorded. Calculations were made assuming 22.24 μl equals 1 μmol CO_2 which had been released by 1 μmol acetic acid. A correction was made for the carbonate content of the $NaHCO_3$ solution, measured in a control. This control contained 2 ml deionised distilled water in the main compartment. On adding the bicarbonate solution from the side arm a negative pressure change occurred owing to a change in the pH of the bicarbonate solution which caused CO_3^{2-} to react with CO_2.

RESULTS

Acetate is known to be the main product of ethanol metabolism in the liver in relatively short periods of incubation (Williamson et al., 1969). On prolonged incubation, when the acetate concentration in the medium has substantially risen, the rate of oxidation of acetate becomes significant and less acetate is found than alcohol removed. Another reason for some discrepancy between ethanol removal and acetate formation is the fact that some ethanol is oxidised to the stage of acetaldehyde only. In the present experiments the yield of acetate was of the order of 75% of the ethanol removed (Table 1). Acetate formation was mainly used as an indicator of ethanol removal because loss of ethanol by evaporation introduces an irregular error unless incubations are performed in the special flasks, described in the Methods section, but these flasks could not be used when the O_2 uptake was to be measured. Table 1 also confirms for the isolated liver cells earlier observations on other liver preparations that oxygen uptake is not affected by the presence of ethanol.

The experiments recorded in Table 2 show the effects of various additions which increase the O_2 consumption and the formation of acetate from ethanol. Fructose is known to consume much ATP on account of its rapid phosphorylation, and to raise the rate of O_2 consumption in connection with the increased turnover of ATP (Mäenpää, Raivio and Kekomäki, 1968; Woods, Eggleston and Krebs,

Table 1. Oxygen Uptake, Acetate Formation and Ethanol Removal by Isolated Rat Liver Cells in the Presence of 5, 10 and 20 mM Ethanol

Conditions were as described in text. The values are expressed as means $\pm$ S.D. when there are three or more experiments with the number of observations in parentheses.

Ethanol added	μmol/g wet wt./min			$\frac{\text{Acetate}}{\text{Ethanol}} \times 100$
	Oxygen uptake	Acetate formation	Ethanol removal	
None	2.48 $\pm$ 0.12 (3)	$<$0.1	-	-
5 mM	2.35 $\pm$ 0.15 (6)	0.61 $\pm$ 0.12 (4)	-0.74 $\pm$ 0.43 (3)	82.5
10 mM	2.51 $\pm$ 0.49 (7)	0.75 $\pm$ 0.13 (11)	-0.98 $\pm$ 0.34 (4)	76.5
20 mM	2.22 $\pm$ 0.50 (3)	0.88 (1)	-1.18 (2)	74.4

Table 2. The Effect of Various Substrates on Oxygen Uptake and Acetate Formation by Isolated Rat Liver Cells

Conditions were as described in text. The values are µmol/g wet wt./min with means ± S.D. when there are three or more experiments with the number of observations in parentheses. 10 mM Ethanol was present in all tests.

Additions	Oxygen uptake	Acetate formation
None	-2.51 ± 0.49 (7)	0.75 ± 0.13 (11)
20 mM Fructose	-4.17 (1)	1.70 (2)
10 mM Dihydroxyacetone	-4.47 (1)	2.16 (2)
10 mM Pyruvate, 10 mM NH_4Cl	-6.04 (1)	2.31 ± 0.08 (3)
10 mM Lactate, 2 mM Lysine, 10 mM NH_4Cl, 1 mM Ornithine	-7.02 (2)	1.91 ± 0.50 (3)
1 mM Asparagine	-3.93 (1)	1.56 ± 0.24 (4)
0.025 mM Dinitrophenol	-4.63 (2)	1.21 ± 0.18 (3)

1970). Dihydroxyacetone is known to act in a similar way. Pyruvate and ammonium chloride cause urea to be synthesized and this involves extra ATP synthesis. Pyruvate also causes a cytosolic conversion of NADH through the lactate dehydrogenase reaction (Berry and Werner, 1973; Smith and Newman, 1959). The addition of lactate, lysine, ornithine and ammonium chloride causes maximal rates of gluconeogenesis and urea synthesis (both ATP consuming reactions) and accordingly they increase ethanol removal and acetate formation. The common feature of the additions so far discussed is the acceleration of the rate of NADH oxidation in the mitochondria on account of an increased ATP requirement.

A different mechanism of accelerated oxidation of NADH is brought about by the addition of the uncoupler DNP which results in a stimulation of ethanol removal, as already reported by Videla and Israel (1970) and Israel et al. (1970).

The increase in ethanol removal by 1 mM asparagine is of a special nature. Asparagine, after hydrolysis to aspartate, promotes the transfer of NADH from cytosol to mitochondria through the malate-aspartate shuttle.

The involvement of the shuttle in ethanol metabolism stems from the fact that the NADH formed by the alcohol dehydrogenase reaction cannot be transferred directly to the mitochondria for oxidation by the respiratory chain, because, as already stated, of permeability barriers. The transfer of the reducing equivalents of NADH is achieved by the reaction

$$\text{NADH} + \text{oxaloacetate} \longrightarrow \text{malate} + \text{NAD}^+ \qquad (1)$$

occurring in the cytosol. The malate enters the mitochondria readily where the above reaction occurs in reverse

$$\text{malate} + \text{NAD}^+ \longrightarrow \text{oxaloacetate} + \text{NADH} \qquad (2)$$

By this process the cytoplasmic NADH becomes mitochondrial NADH. It follows from (1) that oxaloacetate is required in the cytoplasm and this arises by the reaction

$$\text{aspartate} + \alpha\text{-oxoglutarate} \rightarrow \text{oxaloacetate} + \text{glutamate} \qquad (3)$$

Under some conditions the amount of aspartate available limits the rate of this reaction. Hence an addition of asparagine, which undergoes hydrolysis to aspartate and ammonia in the cytosol, promotes reaction (3). As aspartate does not readily enter the liver cell while asparagine does, the addition of asparagine is more effective than the addition of aspartate.

Relatively small amounts of aspartate are sufficient because it is regenerated by reaction (4)

$$\text{oxaloacetate} + \text{glutamate} \rightarrow \text{aspartate} + \alpha\text{-oxoglutarate} \quad (4)$$

This reaction participates in the transfer of oxaloacetate from the mitochondria to the cytoplasm, a direct transfer being impracticable because of the permeability characteristcs of the inner mitochondiral membrane. The products of reaction (4) are transferred to the cytosol where reaction (3) takes place, thus making oxaolacetate available for reaction (1). The glutamate produced in reaction (3) enters the mitochondria and becomes available for reaction (4). It will be noted that in this shuttle between mitochondria and cytosol the carriers (malate, aspartate, glutamate, oxoglutarate) are not used up; small quantities are therefore sufficient (for details see Cornell, Lund and Krebs, 1974).

One of the outstanding questions of alcohol metabolism is that of the role of enzymes other than alcohol dehydrogenase. Alcohol dehydrogenase is known to be inhibited by pyrazole (Theorell and Yonetani, 1963) and if alcohol dehydrogenase were the only enzyme responsible for alcohol removal under our test conditions then 1 mM pyrazole should inhibit alcohol removal completely. According to Table 3 the inhibition is only 44% when ethanol only is added. However, pyrazole completely inhibits the extra ethanol oxidation which occurs on addition of lactate or pyruvate plus ammonium chloride. The results suggest but does not prove that some ethanol oxidation is independent of alcohol dehydrogenase.

A way of testing the question of whether the malate-aspartate shuttle is an essential component of ethanol oxidation is the addition of a specific inhibitor of transamination, such as aminooxyacetate. This is known to inhibit processes involving transamination (Hopper

Table 3. The Effect of Pyrazole on Acetate Formation by Isolated Rat Liver Cells

Conditions were as described in text. The values are expressed as means ± S.D. when there are three or more experiments with number of observations in parentheses. 10 mM Ethanol was present in all tests.

Additions	Pyrazole (mM)	Acetate formation µmol/g wet wt./min
None	0	0.75 ± 0.13 (11)
	1	0.42 ± 0.13 (5)
10 mM Lactate	0	1.40 ± 0.23 (4)
	1	0.37 (2)
10 mM Pyruvate 10 mM NH_4Cl	0	2.31 ± 0.08 (3)
	1	0.41 (2)

and Segal, 1972), such as gluconeogenesis from lactate (Rognstad and Katz, 1970). As shown in Table 4 inhibition of ethanol oxidation is only 50% with 0.1 mM aminooxyacetate although 0.01 mM aminooxyacetate causes a 70% inhibition of gluconeogenesis from lactate, and at very high concentrations of aminooxyacetate (10 mM) ethanol oxidation is even increased. No satisfactory explanation can be offered for this stimulating effect of aminooxyacetate.

The incomplete inhibition of ethanol oxidation by aminooxyacetate suggests that the malate-aspartate shuttle is not essential and other authors have also come to the conclusion that the transfer of reducing equivalents of NADH may be partially effected by the α-glycerolphosphate shuttle (Berry and Kun, 1972; Rognstad and Clark, 1974).

Table 4. The Effect of Aminooxyacetate on Acetate Formation in Isolated Rat Liver Cells in the Presence of 10 mM Ethanol

Conditions were as described in text. The values are expressed as means ± S.D. when there are three or more experiments with the number of observations in parentheses.

Additions		
Lactate (mM)	Aminooxyacetate (mM)	Acetate formed μmol/g wet wt./min
0	0	0.75 ± 0.13 (11)
0	0.1	0.38 (2)
0	1.0	0.70 ± 0.14 (7)
0	10.0	1.34 ± 0.16 (4)
10	0	1.40 ± 0.23 (4)
10	0.1	0.64 (1)
10	1.0	0.92 ± 0.03 (3)
10	10.0	1.37 ± 0.25 (3)

DISCUSSION AND SUMMARY

The present experiments show that the rate of ethanol removal by isolated liver cells can be increased by the addition of substances which promote the oxidation of NADH by the mitochondrial electron transport chain. Fructose, or lactate, or ammonium salts, or dinitrophenol act by causing an increased turnover of the adenine nucleotides. These substances accelerate NADH oxidation because of the obligatory coupling of electron transport with phosphorylation. Asparagine acts mainly by promoting the transfer of NADH from the cytosol to mitochondria.

The fact that the rate of alcohol removal can be accelerated by the addition of the above mentioned substances means that the capacity of alcohol dehydrogenase itself is not rate limiting under many conditions; rather it is the reconversion of NADH into NAD^+ which limits the rate (see Videla and Israel, 1970; Israel et al., 1970).

The findings bear on the question of whether it is practicable to accelerate the rate of ethanol disposal *in vivo*. The use of fructose in humans for this purpose has been discussed for some twenty years but the results obtained appear to be inconsistent. The mechanism of action of fructose is now understood, and it is known that exceptionally high concentrations of fructose are required for significant effects. Inconsistent results reported in the literature on humans can be probably resolved on the basis of variations in the dose of fructose and in the blood levels reached. In any case there is a risk of depleting the liver of adenine nucleotides by large doses of fructose and for this reason the use of fructose is not advisable.

Stimulation of ATP synthesis by increased rates of hepatic gluconeogenesis and urea synthesis is more likely to be a practical proposition. A meal rich in protein may be expected to accelerate the disposal of ethanol.

An extra supply of asparagine which promotes the shuttle under experimental conditions is not likely to be effective *in vivo* because the tissue is probably saturated with the components of the shuttle.

ACKNOWLEDGEMENTS

This work was supported by grants from the Medical Research Council and the National Institute on Alcohol Abuse and Alcoholism (no. 1 R01 AA00381).

REFERENCES

Krebs, H.A., Cornell, N.W., Lund, P. and Hems, R. Isolated liver cells as experimental material. In (Eds.) F. Lundquist and N. Tygstrup, Regulation of Hepatic Metabolism. Alfred Benzon Symposium, Vol. 6, pp. 718-743, Munksgaard, Copenhagen, 1974.

Marjanen, L. Intracellular localisation of aldehyde dehydrogenase in rat liver. Biochem. J., Vol. 127, 633-639, 1972.

Berry, M.N. and Friend, D.S. High yield preparation of isolated rat liver parenchymal cells. J. Cell Biol., Vol. 43, 506-520, 1969.

Cornell, N.W., Lund, P., Hems, R. and Krebs, H.A. Acceleration of gluconeogenesis from lactate by lysine. Biochem. J., Vol. 134, 671-672, 1973.

Bonnichsen, R. Ethanol: determination with alcohol dehydrogenase and DPN. In (Ed.) H.U. Bergmeyer, Methods in Enzymatic Analysis, pp. 285-286, Academic Press, New York and London, 1963.

Conway, E.J. Microdiffusion Analysis and Volumetric Error. Lockwood. London, 1958.

Williamson, J.R., Scholz, R., Browning, E.T., Thurman, R.G. and Fukami, M.H. Metabolic effects of ethanol in perfused rat liver. J. Biol. Chem., Vol. 244, 5044-5054, 1969.

Mäenpää, P.H., Raivio, K.O. and Kekomäki, M.P. Liver adenine nucleotides: Fructose induced depletion and its effect on protein synthesis. Science, Vol. 161, 1253-1254, 1968.

Woods, H.F., Eggleston, L.V. and Krebs, H.A. The cause of hepatic accumulation of fructose-1-phosphate on fructose loading. Biochem. J., Vol. 119, 501-510, 1970.

Berry, M.N. and Werner, H.V. Depression of the stimulation of ethanol oxidation by fructose or pyruvate in liver cells from hyperthyroid animals. Biochemical Society Transactions, Vol. 1. 190-193, 1973.

Smith, M.E. and Newman, H.W. The rate of ethanol metabolism in fed and fasting animals. J. Biol. Chem., Vol. 234, 1544-1549, 1959.

Videla, L. and Israel, Y. Factors that modify the metabolism of ethanol in rat liver and adaptive changes produced by its chronic administration. Biochem. J., Vol. 118, 275-281, 1970.

Israel, Y., Khanna, J.M. and Lin, R. Effect of 2,4 dinitrophenol on the rate of ethanol elimination in the rat *in vivo*. Biochem. J., Vol. 120, 447-448, 1970.

Cornell, N.W., Lund, P. and Krebs, H.A. The effect of lysine on gluconeogenesis from lactate in rat hepatocytes. Biochem. J., (in press).

Theorell, H. and Yonetani, T. Liver alcohol dehydrogenase - DPN - Pyrazole complex: a model of a ternary intermediate in the enzyme reaction. Biochem. Z., Vol. 338, 537-553, 1963.

Hopper, S. and Segal, H.L. Kinetic studies of rat liver glutamic alanine transaminase. J. Biol. Chem., Vol. 237, 3189-3195, 1962.

Rognstad, R. and Katz, J. Gluconeogenesis in the kidney cortex: effects of D-malate and aminooxyacetate. Biochem. J., Vol. 116, 483-491, 1970.

Berry, M.N. and Kun, E. Rate limiting steps of gluconeogenesis in liver cells as determined with the aid of fluor-dicarboxylic acids. Eur. J. Biochem., Vol. 27, 395-400, 1972.

Rognstad, R. and Clark, D.G. Effects of aminooxyacetate on the metabolism of isolated liver cells. Arch. Biochem. Biophys., Vol. 161, 638-651, 1974

THE ROLE OF HYDROGEN PEROXIDE PRODUCTION AND CATALASE IN HEPATIC ETHANOL METABOLISM

R. G. Thurman, N. Oshino and B. Chance

Johnson Research Foundation

University of Pennsylvania, Philadelphia, Pa.

INTRODUCTION

Catalase is present in high concentrations in liver (Sumner & Dounce, 1937), and the existence of its complex with hydrogen peroxide has been previously demonstrated in bacterial systems (Chance, 1952) as well as in liver (Sies and Chance, 1970).

Catalase decomposes hydrogen peroxide in two different ways: the "catalatic" and "peroxidatic" reactions.

$$\text{Catalase} + H_2O_2 \longrightarrow \text{Catalase-}H_2O_2 \quad (1)$$

$$\text{Catalase-}H_2O_2 + H_2O_2 \longrightarrow \text{Catalase} + 2H_2O_2 + O_2 \quad (2)$$

$$\text{Catalase-}H_2O_2 + {>}C{<}^{OH}_{H} \longrightarrow \text{Catalase} + 2H_2O + {>}C{=}O \quad (3)$$

The "catalatic" reaction (Eq. 1 and 2) occurs when the catalase-H_2O_2 complex (compound I) reacts with a molecule of H_2O_2 to generate H_2O and oxygen. The "peroxidatic" reaction (Eq. 1 and 3) occurs when compound I reacts with a variety of hydrogen donors such as methanol (Chance, 1974), ethanol (Jacobstein, 1952), nitrite (Heppel & Porterfield, 1949) and formate (Abei et al. 1957) to generate the corresponding keto compound and probably represents a detoxification mechanism for these small molecules.

More recently, Chance and his co-workers have utilized the direct read-out of catalase-H_2O_2 from perfused organs and *in vivo* as a means of measuring the rates of H_2O_2 production under various metabolic conditions (Sies & Chance, 1970; Oshino et al. 1975A). The "peroxidatic" reaction, first described by Keilin and Hartree nearly thirty years ago (Keilin & Hartree, 1945), is a function of the ethanol concentration and the turnover number of the catalase reaction (Chance, 1974). For all practical purposes, however, the turnover number of catalase, and the subsequent oxidation of ethanol to acetaldehyde via catalase-H_2O_2 is dependent upon the rate of generation of hydrogen peroxide by the hepatocyte.

Recent evidence has also shown that the Keilin-Hartree peroxidation is operative in microsomal ethanol oxidation (Thurman et al. 1972; Roach et al. 1969; Thurman & Scholz, 1973; Isselbacher & Carter, 1974). Whether or not another unique enzyme exists in microsomes in addition to catalase is still a matter of controversy. An excellent review of the debate of the role of catalase in microsomal ethanol oxidation is now available (Estabrook, 1974).

The quantitative role of catalase in ethanol oxidation has been consistently discounted, primarily because ethanol oxidation was affected little, if at all, *in vivo* by a fairly specific inhibitor of catalase, aminotriazole (Laser, 1955) in older studies. However, recent studies by Thurman et al. (Thurman et al. 1975) indicate that alcohol dehydrogenase-independent ethanol oxidation by perfused liver is sensitive to aminotriazole at high ethanol concentrations and that it can be a considerable fraction of ethanol oxidation at high ethanol concentrations (50 to 80 mM). Moreover, stimulation of hydrogen peroxide production with substrates for peroxisomal flavoproteins such as glycolate, urate and D-amino acids markedly activates ethanol oxidation by perfused kidney and liver (Thurman and McKenna, 1974; McKenna & Thurman, 1974).

The purpose of this paper is to review the recent developments in technology which have lead to a better understanding of the rates of H_2O_2 production by the cell and the role of catalase-H_2O_2 in alcohol metabolism.

CATALATIC AND PEROXIDATIC REACTIONS OF CATALASE

It is well established that the "peroxidatic" reaction (Eq. 1 and 3) of catalase with hydrogen donors is predominant at low concentrations of H_2O_2. However, at high H_2O_2 concentrations, as the probability of catalase-H_2O_2 reacting with H_2O_2 is increased, the "catalatic" or oxygen producing reaction is favored (Eq. 1 and 2).

The relationship between the proportion of peroxidatic activity, the ethanol concentration, the catalase concentration and the rate of hydrogen peroxide production is demonstrated in Fig. 1. First, at any fixed ratio of H_2O_2 generation rate to catalase heme concentration over the range between 0.33 to 100 mM ethanol [Fig. 1], the % peroxidatic activity increases with the ethanol concentration. Generally, however, as the ratio $[H_2O_2]$/[Catalase Heme] per minute is increased, which really means increasing the rate of H_2O_2 generation, the % "peroxidatic" reaction decreases. Basically, therefore, the rate of catalase-dependent acetaldehyde generation from ethanol in the liver is a function of the ethanol concentration and the rate of H_2O_2 generation (Fig. 1).

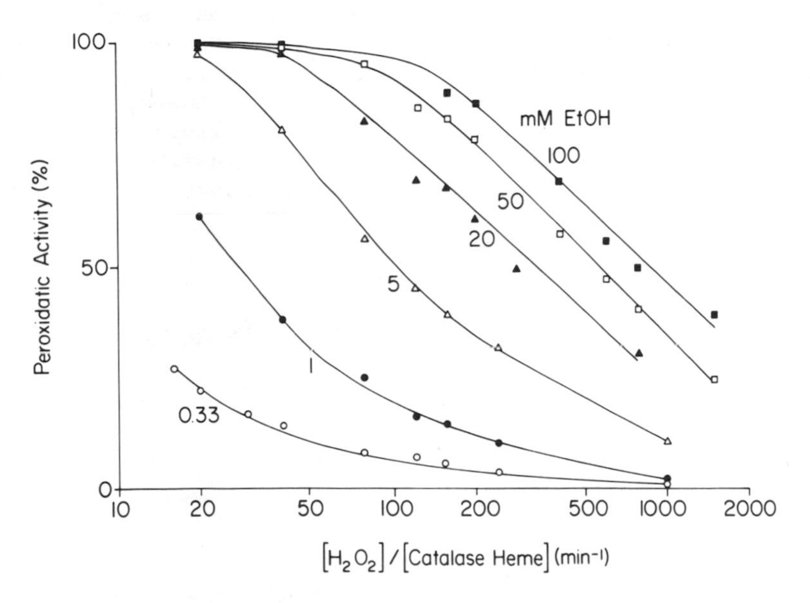

FIGURE 1. RELATIONSHIP BETWEEN PEROXIDATIC ACTIVITY, CATALASE, H_2O_2 GENERATION RATE, CATALASE AND ETHANOL CONCENTRATIONS. Assay conditions and procedures were reported previously (Oshino et al., 1974). Experiments were performed with isolated, purified rat liver catalase.

RATES OF CELLULAR HYDROGEN PEROXIDE PRODUCTION

Recent technical developments from this laboratory (Sies & Chance, 1970; Oshino et al., 1975 B) have made it possible to measure the rate of generation of H_2O_2 by subcellular fractions and perfused organs.

1. Subcellular Fractions

In isolated subfractions two methods have been employed to detect H_2O_2 generation: the scopoletin fluorescence method (Thurman et al. 1972; Roach et al., 1969) and the cytochrome c peroxidase method. Scopoletin, a fluorescence dye, reacts with H_2O_2 to yield a non-fluorescent product in the presence of horseradish peroxide. Cytochrome c peroxidase also reacts with H_2O_2 to give a spectrally distinct compound and is a sensitive, specific, and accurate indicator for H_2O_2 based on the spectral properties of the enzyme substrate-complex of yeast cytochrome c peroxidase (Yonetani & Ray, 1965).

All subcellular fractions examined demonstrated the ability to generate H_2O_2 endogenously, albeit at very low rates. However, the process was stimulated by the addition of suitable hydrogen donors (e.g. glycolate addition to peroxisomes, etc.). For example, Boveris et al. (1972) employing the cytochrome c peroxidase method, have measured the rates of H_2O_2 production in mitochondria, microsomes, cytosol, and a fraction rich in peroxisomes in the absence of added hydrogen donors. Rates of H_2O_2 production have been estimated as follows in nmoles/min/g liver: mitochondria, 12; microsomes, 42; peroxisomes, 30; and supernatant, 4. Thus, while the specific activity of peroxisomes is the greatest of any subcellular fraction, the fact that liver contains large amounts of endoplasmic reticulum per unit weight makes it the greatest overall contributor to H_2O_2 production.

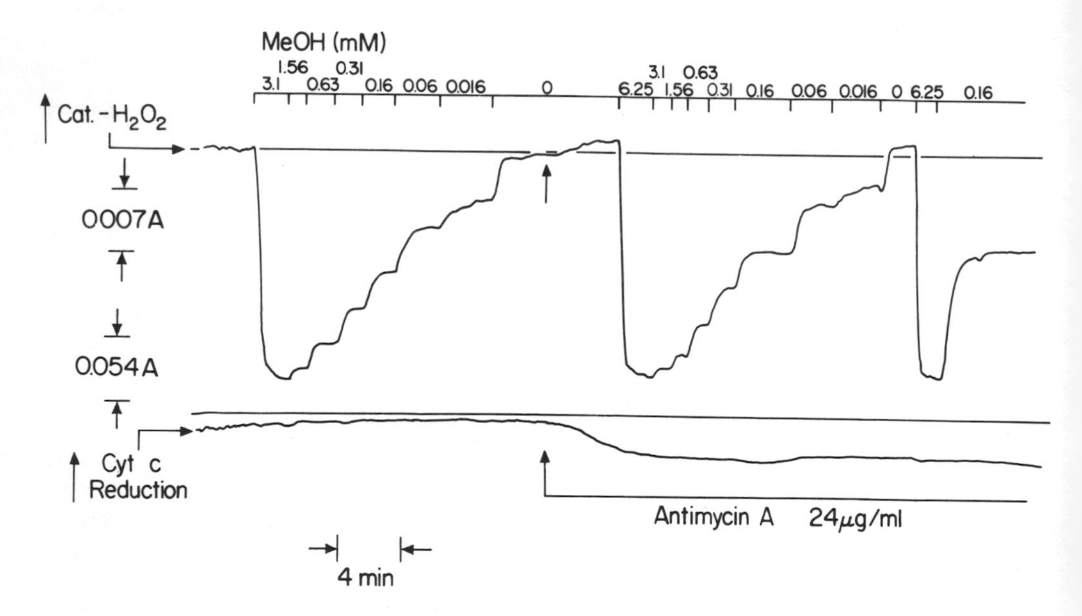

FIGURE 2. EFFECT OF METHANOL ON CATALASE-H_2O_2 AND CYTOCHROME c IN PERFUSED RAT LIVER. Simultaneous measurement of catalase compound I (640-660 nm) and cytochrome c (550-540 nm) was performed as described previously (Oshino et al., 1973). Step-wise addition of methanol in presence and absence of antimycin A as described in Figure.

2. Perfused Organs: Methodology

Following the fundamental developments of Sies and Chance (1970) who demonstrated catalase-H_2O_2 spectrally in perfused liver, Chance and his co-workers have extended this technique to the determination of average rates of H_2O_2 production in perfused organs, primarily in hemoglobin-free perfused rat livers. A typical experiment is shown in

Fig. 2. The steady-state concentration of catalase-H_2O_2 is detected in the red with the wave-length pair 660-640 (Fig. 2, upper line), employing a special time-sharing spectrophotometer developed in the Johnson Research Foundation. When any peroxidatic substrate for catalase-H_2O_2 is infused, catalase-H_2O_2 is rapidly decreased (e.g. 3.1 mM methanol). By varying the methanol concentration it is possible to titrate the steady-state concentration of catalase-H_2O_2 in the liver. This experiment also demonstrates that such titrations are not influenced by inhibition of the mitochondrial respiratory chain (antimycin A), even though H_2O_2 production is stimulated slightly (Fig. 2). With such data, it is possible to calculate the concentration of alcohol necessary to half-titrate the catalase-H_2O_2 complex. The mathematical background for the calculation of rates of H_2O_2 production is beyond the intention of this review. However, interested readers are referred elsewhere (Chance et.al.,1974). Simply put, equations have been developed based upon the stoichiometry of the catalase reaction with H_2O_2 (Eq. 1). They allow for the calculation of rates of H_2O_2 production in intact organs using the destruction of catalase-H_2O_2 by peroxidatic substrates (ethanol, methanol, etc.,). Specifically, the concentration of methanol necessary to destroy half of the catalase-H_2O_2 complex is utilized to calculate rates of H_2O_2 production.

3. Perfused Organs: Rates of Hydrogen Peroxide Production

Figure 3 shows the rate of hepatic H_2O_2 production as a function of oxygen pressure from 1 to 5 atmospheres (for experimental details see Oshino et.al., 1975 A). Endogenous rates of H_2O_2 production were about 50 nmoles/min^{-1}.g^{-1}. This rate was only slightly effected by glucose, but could be increased to nearly 200 nmoles/min^{-1}.g^{-1} by the addition of fatty acids proving that the latter can act as generators of H_2O_2. This experiment suggests that catalase-dependent ethanol oxidation would be about doubled by the addition of fatty acids. Until now, such experiments have not been performed. However, total hepatic ethanol metabolism is depressed by octanoate (Thurman & Scholz, 1975), most likely due to competition between ethanol and fatty acyl CoA compounds for intramitochondrial NAD+, since both processes elevate the pyridine nucleotide redox state. On the other hand, substrates for the peroxisomal H_2O_2-generating flavins such as glycolate and urate strongly activate H_2O_2 production (Fig. 3). When a cocktail containing glucose, glycolate, urate and octanate was added, presumably maximal rates of H_2O_2 production were obtained. Only under these latter conditions did oxygen pressure elevate rates of H_2O_2 production to values around 1400 nmoles/min/g (or 84 µmoles/g/hour, a value equal to or greater than the rate of ethanol metabolism in perfused livers). Thus, this spectral technology has provided for the first time accurate accounting of the rates of intracellular H_2O_2 production and its activation by H_2O_2-generating substrates. The significance of these findings with regard to ethanol metabolism will be discussed in more detail below.

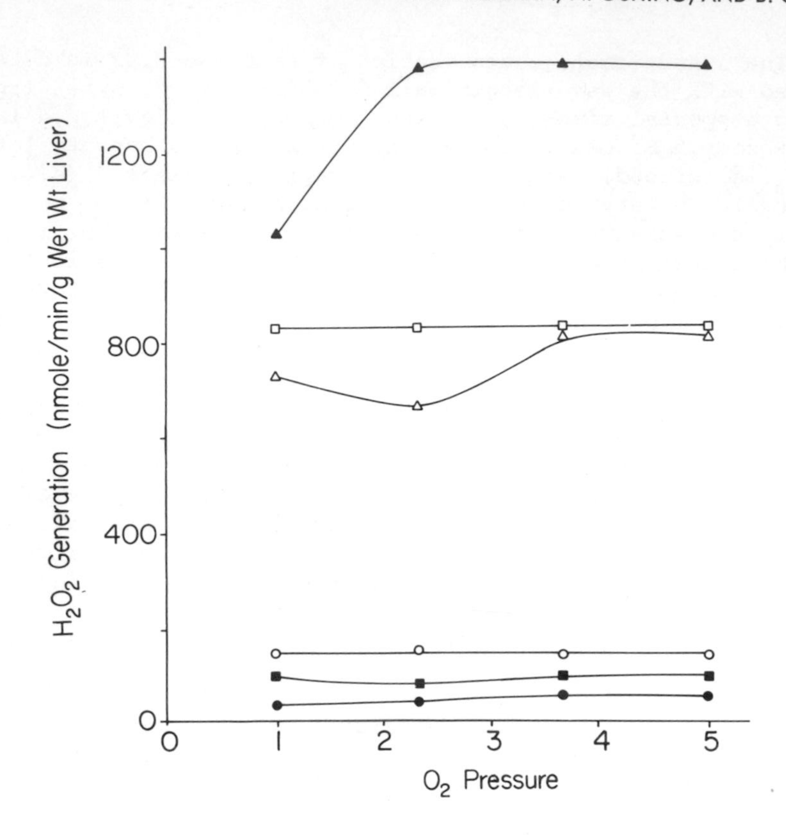

FIGURE 3. EFFECT OF OXYGEN PRESSURE ON RATE OF H_2O_2 PRODUCTION WITH VARIOUS SUBSTRATES. The rate of H_2O_2 production was estimated from the result of the methanol titration (as in Fig. 2). Each point is the mean of 2-4 experiments: □ Urate 0.75 mM; Δ glycolate 1 mM; 0 octanoate 1 mM; ■ lactate 1 mM plus pyruvate 0.15 mM; ● glucose 5 mM; ▲ mixture of all of above.

ROLE OF CATALASE IN MICROSOMAL ETHANOL OXIDATION

It is now ten years since Orme-Johnson and Ziegler (1965) first described the production of acetaldehyde from ethanol when microsomes were incubated in the presence of NADPH and oxygen. Unfortunately, in spite of a great number of publications which have followed, in an attempt to explain this phenomenon, the precise mechanism of microsomal ethanol oxidation remains controversial. Shortly after the initial observations, Mary Roach and her co-workers (Roach et al., 1969) showed that inhibitors of catalase markedly diminished acetaldehyde production from ethanol by microsomes and concluded that a large proportion if not all of microsomal ethanol oxidation was due to catalase. However, Lieber and De Carli (1970) attempted to differentiate catalase-H_2O_2 from NADPH-dependent ethanol oxidation and concluded that the microsomal system should be accounted for by a unique interaction between ethanol and cytochrome P-450. This experiment however, was later criticized on theoretical grounds by Chance and Oshino (Oshino et al., 1974). Subsequently, Thurman et al. (1972) demonstrated that microsomes could generate H_2O_2 when incubated with NADPH and oxygen, and that the rate of generation was sufficient to account for rates of acetaldehyde production from ethanol. They also confirmed Mary Roach's initial finding that inhibitors of catalase (e.g. azide) markedly diminished microsomal ethanol oxidation and observed that microsomal ethanol oxidation and microsomal H_2O_2 productions had identical K_m's for oxygen and were equally sensitive to inhibition by carbon monoxide. Furthermore, microsomal ethanol oxidation was shown to be sensitive to inhibition by substrates for catalase (formate) and H_2O_2-utilizing systems. They concluded that NADPH interacts with the endoplasmic reticulum to generate H_2O_2, which in turn reacts with catalase to form catalase-H_2O_2, allowing the peroxication of ethanol to acetaldehyde (Thurman et al., 1972).

Lieber and De Carli (1970) demonstrated that induction of the endoplasmic reticulum followed pretreatment of animals with a diet containing ethanol. This experiment has been used as an argument for a unique microsomal ethanol oxidizing system as well as an explanation for the adaptive increase in ethanol metabolism following chronic pretreatment with ethanol. On the other hand, since the induction of endoplasmic reticulum following chronic pretreatment with ethanol is accompanied by an increase in NADPH-dependent H_2O_2 production by microsomes (Thurman, 1973), this experiment sheds doubt upon the distinctness of the microsomal ethanol oxidizing system from catalase. Subsequently, a large body of literature has appeared clearly indicating that the adaptive increase in ethanol metabolism following chronic pretreatment with ethanol involves alcohol dehydrogenase associated with the induction of an ATPase activity, most likely the sodium pump (Israel et al., 1974; McCaffrey & Thurman, 1974).

Later experiments attempted to resolve the controversy between those that favored a unique catalase system and those that advocated a unique cytochrome P-450 system by separating catalase from cytochrome P-450 in detergent solubilized microsomal preparations (Lieber et al., 1974; Thurman, et al., 1974). Unfortunately, the results of these experiments have been equivocal. Several laboratories (Thurman et al., 1974; Vermilion, et al., 1975) have not been able to detect acetaldehyde productions from ethanol employing two different methods by the reconstituted drug metabolizing systems of Lu and Coon (Lu & Coon, 1968; Lu et al., 1969), (see Table I). On the other hand, Lieber and his co-workers have isolated an active

TABLE I

COMPARISON OF SOLUBILIZED MICROSOMAL PREPARATIONS

Literature	Animal Pretreatment	Acetaldehyde Formation From Ethanol nmoles x min^{-1} x mg^{-1}	Catalase Activity	Azide Sensitivity	Additional Comments
Mezey et al.	Phenobarbital	6.9	"low"	0.1 mM - No 1.0 mM - yes	Equal activity with H_2O_2 generating system
Thurman & Scholz	Phenobarbital	not detectable	not detectable	--	--
Teschke et al.	Ethanol	0.25	not detectable	--	Inhibited by Carbon Monoxide
Thurman et al.	Ethanol	not detectable	not detectable	--	----
Lieber et al.	Ethanol and control	10.0 to 25.0	not detectable	0.1 mM - No 1.5 mM - Yes	--
Vermillion et al.	Phenobarbital	not detectable	"low"	--	--
Teschke visiting Thurman	Ethanol	< 2.0	"low"	1.0 mM - Yes	Inhibited by H_2O_2 utilizing system
Thurman visiting Teschke	Ethanol	21.0 to 35.0	No ? Yes	not measured	catalase not detected by H_2O_2 disappearance, but present by molecular weight determination

ethanol oxidase which they claim to be free of catalase (Teschke et al., 1974A). This latter statement should, however, be regarded with caution, as samples provided to Dr. Oshino by Dr. Teschke of Dr. Lieber's laboratory contained catalase activity and we have detected a protein in Dr. Lieber's preparation on SDS polyacrylamide gel electrophoresis which had an rf value identical to rat liver catalase. More recently, claims and counterclaims from various laboratories concerning the mechanism of this phenomenon continue to be published. Thurman et al. showed that the ratio of acetaldehyde to propionaldehyde produced by ethanol and propanol respectively was identical irrespective of whether one incubated microsomes with NADPH or with an H_2O_2 generating system and concluded that catalase is responsible for a predominant portion (if not all) of microsomal ethanol oxidation. Conversely, Teschke et al. report the ability to differentiate the unique microsomal ethanol oxidizing system from catalase based upon the different selectivity of the two systems to ethanol, proponol and butanol (Teschke et al., 1974 B). This approach, however, represents a certain redundancy in science, since longer chain alcohols were shown to be non-reactors with the microsomal system over ten years ago (Orme-Johnson & Ziegler, 1965). Lastly, Vatsis and Schulman (1974 and 1973) reported that microsomal preparations from acatalasemic mice possessing a thermo-labile hepatic catalase can oxidize drugs, but not ethanol. They conclude, therefore, that microsomal ethanol oxidation has a unique requirement for catalase (Vatsis & Schulman, 1974; Vatsis and Schulman, 1973). This viewpoint, however, has been challenged by Lieber and his co-workers (Lieber and DeCarli, 1974). Thus, this controversy continues. It is our viewpoint that most if not all of microsomal ethanol oxidation can be accounted for by catalase (see Fig. 4). A summary of the debate following the presentation of five papers on this subject will serve as further background for the interested student (Estabrook, 1974).

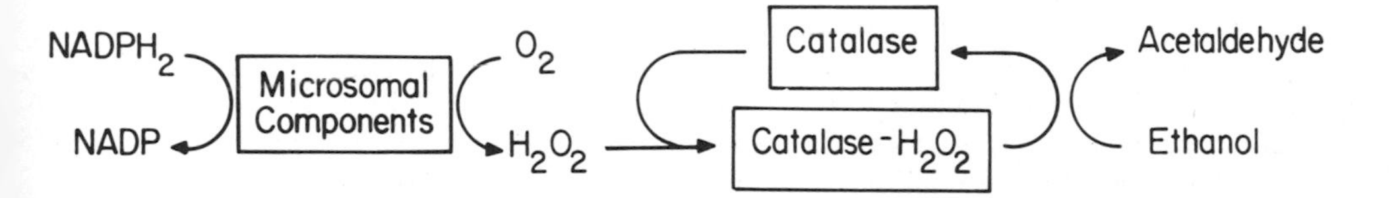

FIGURE 4. SCHEME DEPICTING ROLE OF CATALASE AND H_2O_2 IN MICROSOMAL ETHANOL OXIDATION.

PHYSIOLOGICAL ROLE OF MICROSOMAL ETHANOL OXIDATION

The question arises if the peroxidatic conversion of ethanol to acetaldehyde by catalase employing H_2O_2 generation by NADPH and oxygen by enzymes of the endoplasmic reticulum occurs *in vivo*. Thurman and Scholz (1973) demonstrated that menadione stimulated both NADPH-dependent H_2O_2 formation and ethanol oxidation in microsomes. If NADPH-dependent H_2O_2 generation contributed to ethanol oxidation in the intact liver one would expect menadione to stimulate ethanol oxidation by the perfused liver. This was not the case, allowing for the conclusion that production of H_2O_2 by the flavoproteins and cytochromes of the endoplasmic reticulum is inconsequential in hepatic ethanol metabolism (Thurman and Scholz, 1973). Furthermore, a sublethal dose of carbon tetrachloride administered to rats strongly suppressed ethanol utilization in microsomes, but had no effect on ethanol utilization by liver slices or whole animals adding further support to the conclusion that the endoplasmic reticulum does not participate in ethanol metabolism *in vivo* (Khana et.al., 1971). Finally, quantitative titration of ethanol metabolism in the perfused liver in combination with a qualitative read-out of intermediates of alcohol dehydrogenase, catalase, and drug metabolizing systems demonstrated that alcohol dehydrogenase and catalase totally account for hepatic ethanol metabolism (Thurman et al., 1975), see below.

The inescapable conclusion from these three experiments is that whatever the mechanism of microsomal ethanol oxidation is *in vitro*, its contribution to overall hepatic ethanol metabolism is nil.

QUANTITATIVE TITRATION OF HEPATIC ETHANOL METABOLISM

In the past several years, Theorell and his co-workers have developed a variety of pyrazole derivatives (e.g. 4-methylpyrazole, 4-propylpyrazole, etc.,). These inhibitors have been shown to be very potent inhibitors of alcohol dehydrogenase *in vitro* and they markedly diminish ethanol metabolism by perfused organs (McCaffrey & Thurman, 1974; Lindros et al., 1975). In addition, pretreatment of animals with aminotriazole, an inhibitor of catalase, has been shown to abolish methanol oxidation in several species (Kinard et al., 1956; Nelson et al. 1957). We have evaluated ethanol metabolism by the perfused rat liver with a combination of biophysical techniques (pyridine nucleotide fluorescence, absorption of catalase-H_2O_2 and cytochrome P-450)in conjunction with inhibitors of alcohol dehydrogenase and catalase.

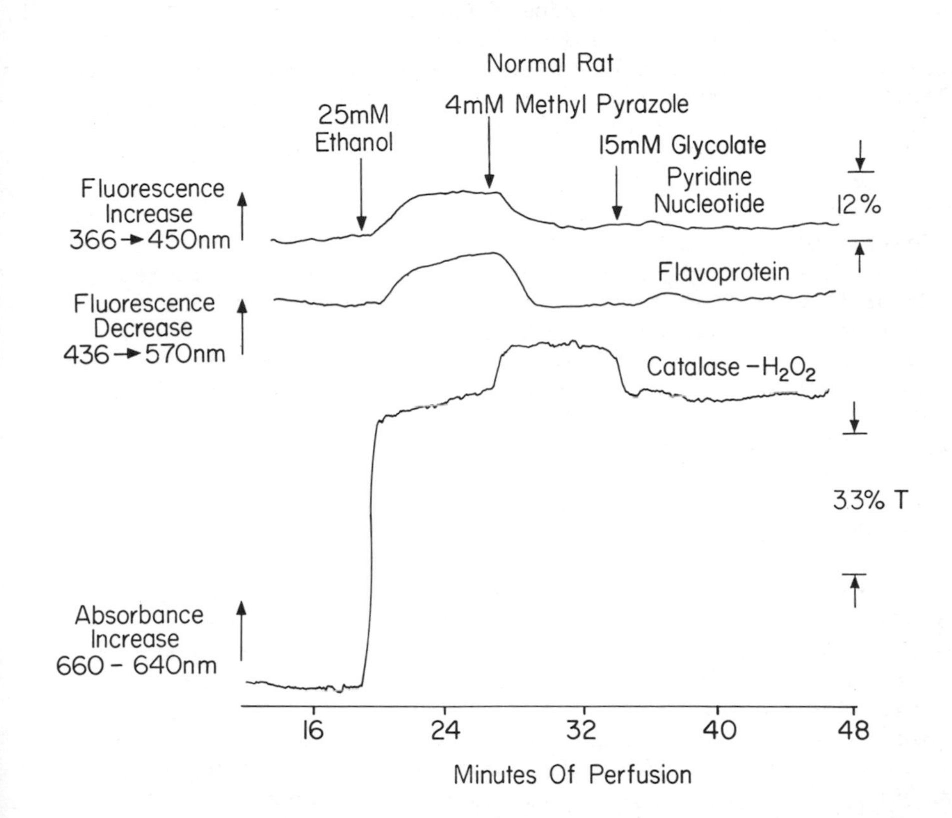

FIGURE 5. EFFECT OF ETHANOL, 4-METHYLPYRAZOLE, AND GLYCOLATE ON PYRIDINE NUCLEOTIDE AND FLAVOPROTEIN FLUORESCENCE AND ABSORPTION OF CATALASE-H_2O_2 IN PERFUSED RAT LIVER. Pigments were monitored as described previously (Sies & Chance, 1970; Oshino et al., 1973; Thurman et al., 1975). Conditions similar to Fig. 2.

In general, no inhibitor is totally specific in any biological system. However, selectivity at least with regard to a few systems can be monitored. Thurman et.al. checked the selectivity of 4-methylpyrazole and aminotriazole upon alcohol dehydrogenase, catalase and cytochrome P-450 systems in perfused rat liver (Thurman, et al., 1975). This was done employing techniques developed by Chance and his co-workers to monitor the surface fluorescence of pyridine nucleotides, catalase-H_2O_2, and cytochrome P-450 in the perfused rat liver (Sies & Chance, 1970; Oshino et al., 1973; Thurman et al., 1975). These pigments measure qualitatively the activity of alcohol dehydrogenase, catalase and cytochrome P-450, respectively. An example of the first two of these read-out techniques is presented in Figure 5. Following the addition of 25 mM ethanol, a decrease in the steady-state concentration of catalase-H_2O_2 is observed. In addition, the fluorescence of flavoproteins reflects the transfer of hydrogen into the mitochondrial space. Following the addition of the inhibitor of alcohol dehydrogenase, 4-methylpyrazole, pyridine nucleotide and flavoprotein fluorescence both returned to their original baselines indicating that the metabolism of ethanol via the alcohol dehydrogenase pathway was abolished. On the other hand, catalase-H_2O_2 was still functional under these conditions. In fact, the steady-state concentration of catalase-H_2O_2 was actually increased by the addition of an H_2O_2 generating substance, sodium glycolate (Fig. 5).

Following pretreatment with aminotriazole (Thurman et al., 1975), the pyridine nucleotide fluorescence responded to the addition of ethanol, but the catalase-H_2O_2 signal did not, indicating that catalase was inactive in the presence of this inhibitor. These studies have, therefore, established that 4-methylpyrazole and aminotriazole act specifically on alcohol dehydrogenase and catalase respectively under these conditions. Thus, the combination of 4-methylpyrazole and aminotriazole abolish both alcohol dehydrogenase and catalase-mediated ethanol metabolism by the perfused rat liver.

Titration of ethanol metabolism between the concentration ranges of 20-80 mM (approximately 100-400 mg %) showed a slight dependency upon ethanol concentration (Thurman et al., 1975). When the titration was repeated in the presence of 4-methylpyrazole (Fig. 6), the rate of ethanol metabolism was also increased as a function of ethanol concentrations. In an attempt to identify the nature of the 4-methylpyrazole-insensitive ethanol metabolism we have taken the working hypothesis that either alcohol dehydrogenase, catalase, or the so-called microsomal system can account for this phenomenon (Thurman, 1975 The subsequent addition of antimycin A, which inhibits the reoxidation of NADH and should theoretically suppress the activity shown in Fig. 6, if it were due to alcohol dehydrogenase, was without effect. Similarly, inhibition of cytochrome P-450 mediated reactions by the addition of metyrapone was also without effect. This experiment confirms the conclusion reached above that cytochrome P-450 is not in-

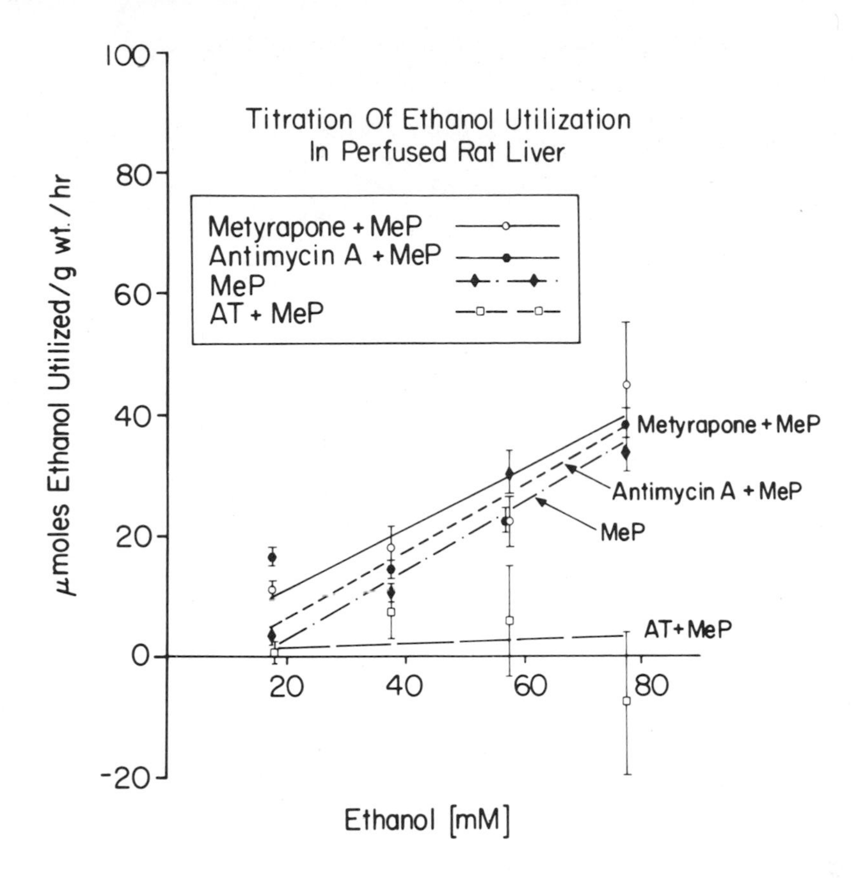

FIGURE 6. EFFECT OF 4-METHYLPYRAZOLE, METYRAPONE, ANTIMYCIN A, AND AMINOTRIAZOLE ON ETHANOL UTILIZATION BY PERFUSED RAT LIVER AS A FUNCTION OF ETHANOL CONCENTRATION. Mep, 4-methylpyrazole 4 mM; metyrapone, 1 mM; antimycin A, 25 mg/ml; aminotriazole, 1 gm/kg i.p. 1 hour before perfusion experiment. See inset.

volved in ethanol oxidation in the whole organ. On the other hand, when 4-methylpyrazole was added to livers from aminotriazole pretreated animals, ethanol metabolism could not be distinguished from vaporization controls (Thurman et al., 1975), Figure 6. These results indicate that at low concentrations of ethanol, ethanol metabolism is predominantly, if not totally, due to alcoholdehydrogenase. As the ethanol concentration is increased, an increasing contribution of the catalase-H_2O_2 system is observed so that at 80 mM ethanol, ethanol metabolism is approximately 50% due to alcohol dehydrogenase and 50% catalase dependent. These whole organ studies fail to supply any evidence for a unique microsomal ethanol oxidation system.

ACTIVATION OF CATALASE-DEPENDENT ETHANOL METABOLISM BY H_2O_2 GENERATING SUBSTRATES

The experiments described above demonstrate a) that the catalase reaction in the whole organ is rate-limited by the supply of H_2O_2, and b) that this supply can be markedly increased by the addition of substrates of peroxisomal flavoproteins such as glycolate, urate and D-amino acids.

Thurman and McKenna have utilized this information in their study of catalase-dependent ethanol metabolism by perfused rat livers from normal and ethanol pretreated animals (Thurman & McKenna, 1974). The addition of glycolate activated the control rate of ethanol metabolism four or five-fold in their studies without causing any concomitant changes in the pyridine nucleotide linked redox couples lactate/pyruvate, or β-hydroxybutyrate/acetoacetate (Fig. 7). It was observed that H_2O_2-generating substrates such as urate were poor activators of ethanol metabolism at low ethanol concentrations, but they did produce a marked stimulation if the ethanol concentration was between 20-40 mM (Thurman & McKenna, 1974). However, glycolate is highly toxic and urate is normally deposited in synovial spaces in several disease syndromes. Thus, the practical implications of employing either of these agents in elevating ethanol metabolism in man must be discounted. McKenna and Thurman (1974) took advantage of the fact that the peroxisomal space also contains an H_2O_2-generating flavoprotein in their extension of these studies. They showed that D-alanine which is less toxic than glycolate or urate, although not without its toxic manifestations (nephrotoxicity in high doses; interference with protein synthesis) activated ethanol metabolism by over 50%. While these results are less dramatic than those produced by glycolate and urate, a full study of a wide variety of D-aminoacids is presently underway.

Several years ago, von Wartburg and Eppenberger (1961) demonstrated that D- and L-aminoacids activated ethanol metabolism in kidney slices. They suggested that both alcohol dehydrogenase and catalase are responsible for ethanol metabolism by the kidney. We have repeated their studies in the isolated perfused rat kidney in conjunction with Dr. Holger Franke of the University of Kiel. In contrast to von Wartburg and Eppenberger, we were unable to detect ethanol metabolism by the perfused rat kidney (Table 2). However, the addition of D-alanine produced a very high rate of ethanol metabolism as well as a marked stimulation of the basal rate of oxygen uptake in this preparation. The addition of 4-methylpyrazole produced only about a 20% inhibition of this stimulated rate of ethanol metabolism. This indicates that D-alanine acts to generate H_2O_2 for catalase under these conditions. Thus, H_2O_2-generating substrates activate catalase-dependent ethanol metabolism in both liver and kidney. If a non-toxic H_2O_2-generating substrate could be found, it could theoretically be useful as a "sobriety" pill at moderate to high ethanol concentrations.

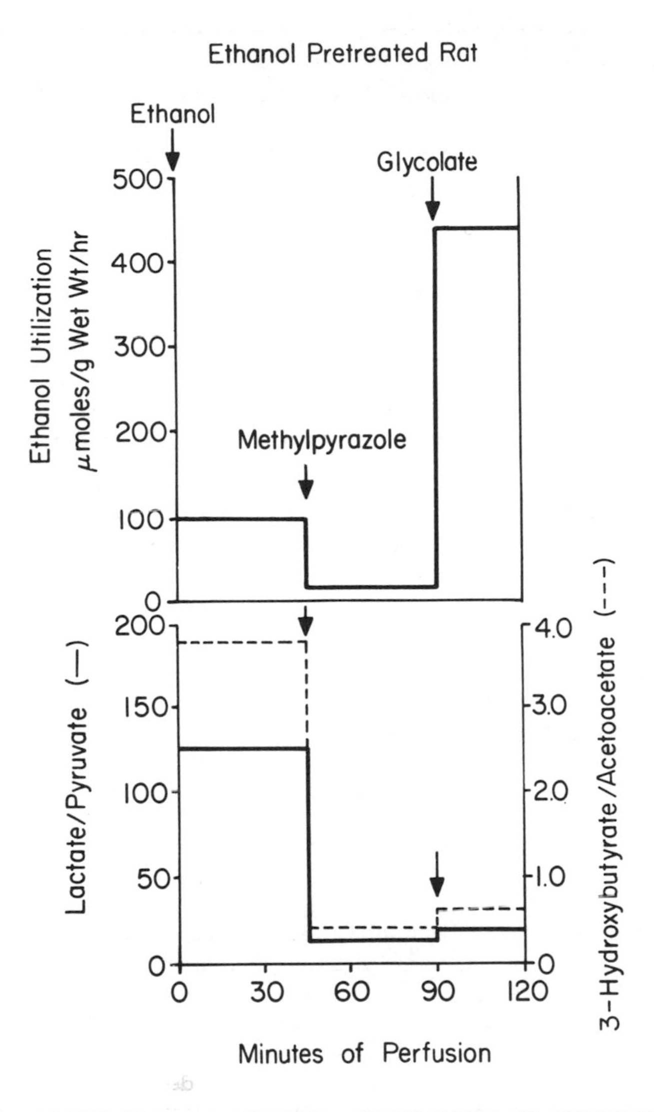

FIGURE 7. ACTIVATION OF ETHANOL METABOLISM IN PERFUSED LIVERS FROM ETHANOL PRETREATED RATS BY SODIUM GLYCOLATE. Rats were pretreated 3-5 weeks with a diet containing ethanol (Porta et al., 1968). Ethanol, lactate, pyruvate, 3-hydroxybutyrate and acetoacetate were determined enzymatically on deproteinized samples (Thurman & McKenna, 1974). Methylpyrazole and glycolate were added as indicated by arrows. Mean of three experiments.

TABLE II. EFFECT OF D-ALANINE ON ETHANOL UTILIZATION AND OXYGEN UPTAKE BY PERFUSED RAT KIDNEY (*)

	Ethanol µmoles/g wet/hr	Oxygen µmoles/g wet/hr
Control	-0.7 ± 8.1	343.0 ± 2.0
+ D-Alanine (30 mM)	158.1 ± 24.6	410.2 ± 7.3
+ D-Alanine and 4-Methylpyrazole (4 mM)	126.6 ± 16.1	—

(*) Rat kidneys were perfused as described previously (Franke et al., 1972). Otherwise, conditions were identical to Fig. 6.

CONCLUSIONS

Recent technical developments from the Johnson Research Foundation make it possible to determine rates of hydrogen peroxide production in a wide variety of biological materials.

Substrates for flavoproteins located within the peroxisomal space such as glycolate, urate, and D-amino acids strongly activate H_2O_2 production in both liver and kidney.

Catalase has been identified as a component in microsomal ethanol oxidation.

Quantitative evaluation of ethanol metabolism as a function of ethanol concentration showed that nearly all ethanol oxidation at low ethanol concentrations (less than 20 mM) was dependent upon alcohol dehydrogenase. However, at higher concentrations (80 mM), catalase participates in up to 50% of the rate.

Glycolate, urate and D-amino acids strongly activate catalase-dependent ethanol metabolism in liver and kidney.

ACKNOWLEDGEMENTS

Supported in part by grants from the National Institute of Alcohol Abuse and Alcoholism (AA-00288; AA-00292) and a Career Development Award from the National Institute of Mental Health (K 2 MH 70, 155;RGT). The authors thank Dr. Holger Franke, University of Kiel, for his assistance with the kidney perfusion experiments.

REFERENCES

ABEI, J., H. KOBLET, J.P. VON WARTBURG, 1957, Uber den Mechanismus der biologischen Methanoloxydation, Helv. Physiol. Acta, 15, 384.

BOVERIS, A., N. OSHINO and B. CHANCE, 1972, The cellular production of hydrogen peroxide, Biochem. J. 128, 617.

CHANCE, B., 1947, An intermediate compound in the catalase-hydrogen peroxide reaction, Acta Chem. Scand. I, 236.

CHANCE, B., 1952, The state of catalase in the respiring bacterial cell, Science, 116, 202.

CHANCE, B., N. OSHINO, T. SUGANO and D. JAMIESON, 1974, Role of catalase in ethanol metabolism, in: Alcohol and Aldehyde Metabolizing Systems, eds. R.G. Thurman, T. Yonetani, J.R. Williamson, and B. Chance (Academic Press, New York), p. 169.

ESTABROOK, R.W., 1974, The role of catalase in microsomal ethanol oxidation, in: Alcohol and Aldehyde Metabolizing Systems, eds. R.G. Thurman, T. Yonetani, J.R. Williamson, and B. Chance (Academic Press, New York)p. 559.

FRANKE, H., H. HULAND, K. UNSICKER and C. WEISS, 1972, Studies on the substrate uptake of rat kidney, in: Das Wissenschaftliche Taschenbuch, ed. M. Hohenegger, (W. Goldmann Verlag, Munich).

HEPPEL, L.A. and V.T. PORTERFIELD, 1949, Metabolism of Inorganic Nitrite and Nitrate Esters. I. The coupled oxidation of nitrite by peroxide forming systems and catalase, J. Biol. Chem. 178, 549.

ISRAEL, Y., J. BERNSTEIN and L. VIDELA, 1974, On the mechanism of the changes in liver oxidative capacity produced by chronic alcohol ingestion; in: Alcohol and Aldehyde Metabolizing Systems, eds. R.G. Thurman, T. Yonetani, J.R. Williamson, and B. Chance, (Academic Press, New York) p. 493.

ISSELBACHER, K.J. and E.A. CARTER, 1974, Peroxidatic Oxidation of ethanol by microsomes, in: Alcohol and Aldehyde Metabolizing Systems, eds. R.G. Thurman, T. Yonetani, J.R. Williamson and B. Chance, (Academic Press, New York), p. 271.

JACOBSEIN, E.,1952, The metabolism of ethyl alcohol, Nature (London) 169, 645.

KEILIN, D. and E.F. HARTREE, 1945, Properties of catalase. Catalysis of coupled oxidation of alcohols, Biochem. J. 39, 293.

KHANNA, J.M., H. KALANT, E. LIN and G.O. BUSTOS, 1971, Effect of carbon tetrachloride treatment on ethanol metabolism, Biochem. Pharmacol. 20, 3269.

KINARD, F.W., G.H. NELSON and M.G. HAY, 1956, Catalase activity and ethanol metabolism in the rat. Proc. Soc. Exptl. Biol. Med. 92, 772.

LASER, H.,1955, Peroxidatic Activity of catalase, Biochem. J. 61, 122.

LEMBERG, R. and E.O. FOULKES, 1946, Reaction between catalase and hydrogen peroxide, Nature (London), 161, 131.

LIEBER, C.S. and L.M. DECARLI, 1970, Hepatic microsomal ethanol oxidizing system, J. Biol. Chem. 245, 2505.

LIEBER, C.S. and L. M. DECARLI, 1974, Oxidation of ethanol by hepatic microsomes of acatalasemic mice, Biochem. Biophys. Res. Comm. 60, 1187.

LIEBER, C.S., R. TESCHKE, Y. HASUMURA, and L.M. DECARLI, 1974, Interaction of ethanol with liver microsomes, in: Alcohol and Aldehyde Metabolizing Systems, eds. R.G. Thurman, T. Yonetani, J. R. Williamson, and B. Chance, (Academic Press, New York)p. 243.

LINDROS, K.O., N. OSHINO, R. PARILLA, AND J.R. WILLIAMSON, 1975, Characteristics of ethanol and acetaldehyde oxidation on flavin and pyridine nucleotide fluorescence changes in perfused rat liver, J. Biol. Chem. 249, 7956.

LU, A.Y., and M.J. COON, 1968, Role of hemoprotein P-450 in fatty acid ω-hydroxylation in a soluble enzyme system from liver microsomes, J. Biol. Chem. 243, 1331.

LU, A.Y., K.W. JUNK and M.J. COON, 1969, Resolution of the cytochrome P-450-containing ω-hydroxylation system of liver microsomes into three components, J. Biol. Chem. 244, 3714.

MCCAFFREY, T.B. and R.G. THURMAN, 1974, Mechanism of the adaptive increase in ethanol utilization due to chronic prior treatment with alcohol, in: Alcohol and Aldehyde Metabolizing Systems, eds. R.G. Thurman, T. Yonetani, J.R. Williamson and B. Chance, (Academic Press New York) p. 483.

MCKENNA, W.R. and R.G. THURMAN, 1974, Activation of ethanol utilization in perfused livers from normal and ethanol pretreated rats, Fed. Proc. 33, 554.

NELSON, G.H., F.W. KINARD, G.C. HULL and M.G. HAY, 1957, The effect of aminotriazole an alcohol metabolism and hepatic enzyme activities in several species. Quart. J. Stud. Alc., 18, 343.

ORME-JOHNSON,W.H. and D.M. ZIEGLER, 1965, Alcohol mixed function oxidase activity of mammalian liver microsomes, Biochem. Biophys. Res. Comm. 21, 78.

OSHINO, N., B. CHANCE, H. SIES, and TH. BUCHER, 1973, The role of hydrogen peroxide generation in perfused rat liver and the reaction of catalase compound I and hydrogen donors, Arch. Biochem. Biophys. 154, 117.

OSHINO, N., D. JAMIESON and B. CHANCE, 1975, The properties of H_2O_2 production under hyperoxic and hypoxic conditions of the perfused rat liver, Biochem. J. in press.

OSHINO, N., D. JAMIESON, I. SUGANO and B. CHANCE, 1975, Optical measurement of the catalase-H_2O_2 intermediate (compound I) in the liver of anesthetized rats and its implication to H_2O_2 production in situ, Biochem. J., in press.

OSHINO, N., R. OSHINO and B. CHANCE, 1974, The properties of catalase "peroxidatic" reaction and its relationship to microsomal methanol oxidation, in: Alcohol and Aldehyde Metabolizing Systems, eds. R. G. Thurman, T. Yonetani, J.R. Williamson, B. Chance (Academic Press, New York) p. 231.

PORTA, E.A.,L.A. CESAR and C.A. GOMEZ-DUMM, 1968, A new experimental approach in the study of chronic alcoholism. I. Effects of high alcohol intake in rats fed a commercial laboratory diet. Lab. Invest. 18, 352.

ROACH, M.K., W.N. REESE, and P.J. CREAVEN, 1969, Ethanol oxidation in the microsomal fraction of rat liver. Biochem. Biophys. Res. Comm. 36, 596.

SIES, H. and B. CHANCE, 1970, The steady-state level of catalase compound I in isolated hemoglobin-free perfused rat liver. Febs. Letters, 11, 172.

SUMNER, J.B. and A.L. DOUNCE, 1937, Crystalline Catalase, J. Biol. Chem. 121, 417.

TESCHKE, R., Y. HASUMURA, and C.S. LIEBER, 1974, Hepatic microsomal ethanol-oxidizing systems: solubilization, isolation and characterization, Arch. Biochem. Biophys. 163, 404.

TESCHKE, R., Y. HASUMURA, and C.S. LIEBER, 1974, NADPH-dependent oxidation of methanol, ethanol, propanol and butanol by hepatic microsomes, BBRC 60, 861 pp. 851-857.

TESCHKE, R., Y. HASUMURA, and C.S. LIEBER, 1974, NADPH-dependent oxidation of methanol, ethanol, propanol and butanol by hepatic microsomes, Clin. Res. 22, 324. (b).

THURMAN, R.G. 1973, Induction of hepatic microsomal NADPH-dependent hydrogen peroxide production by chronic prior treatment with ethanol, Mol. Pharmacol. 9, 670.

THURMAN, R.G., 1975, in press, Quantitative roles of alcohol dehydrogenase and catalase in hepatic alcohol metabolism, Proc. of NIAAA Meeting, Washington.

THURMAN, R.G., S. HESSE and R. SCHOLZ, 1974, The role of NADPH-dependent hydrogen peroxide formation and catalase in hepatic microsomal ethanol oxidation, in: Alcohol and Aldehyde Metabolizing Systems, eds. R.G. Thurman, T. Yonetani, J.R. Williamson, B. Chance (Academic Press, New York) p. 257.

THURMAN, R.G. , H.G. LEY and R. SCHOLZ, 1972, Hepatic microsomal ethanol oxidation, Eur. J. Biochem. 25, 420.

THURMAN, R.G. and W.R. MCKENNA, 1974,Activation of ethanol utilization in perfused livers from normal and ethanol pretreated rats, Hoppe-Seyler's Zeitschrift f. Physiolog. Chemie, 355, 336.

THURMAN, R.G., W.R. MCKENNA, H.J. BRENTZEL and S. HESSE, 1975, in press, Significant pathways of hepatic ethanol metabolism, Fed. Proc.

THURMAN, R.G. and R. SCHOLZ, 1973, The role of hydrogen peroxide and catalase in hepatic microsomal ethanol oxidation, in Drug Metabolism and Disposition, I, 441.

THURMAN, R.G. and R. SCHOLZ, 1975, in press, The effect of octonoate on ethanol metabolism, Fed. Proc.

VATSIS, K.P. and M.P. SCHULMAN, 1973, Absence of ethanol metabolism in "acatalatic" hepatic microsomes, Biochem. Biophys. Res. Comm. 52, 588.

VATSIS, K.P. and M.P. SCHULMAN, 1974, "Acatalatic" hepatic microsomes metabolize drugs but not ethanol, in: Alcohol and Aldehyde Metabolizing Systems, eds. R.G. Thurman, T. Yonetani, J.R. Williamson, B. Chance, (Academic Press, New York), p. 287.

VERMILION, J., R. KASCHNITZ and M.J. COON, 1975, Ethanol oxidation and drug hydroxylation in rat liver microsomes and reconstituted enzyme preparations containing cytochrome P-450. Cited by R.G. Thurman and W. McKenna in "Pharmacology of Alcohol", ed. E. Majchrowicz (Plenum Press, New York), in press.

VON WARTBURG, J.P. and H.M. EPPENBERGER, 1961, Vergleichende Untersuchungen uber den oxidativen Abbau von 1-C^{14}-Athanol und 1-C^{14} Azetat in Leber und Niere. Helv. Phys. Acta, 19, 303.

YONETANI, T., and G. RAY, 1965, Studies on cytochrome c peroxidase. I. Purification and some properties, J. Biol. Chem. 240, 4503.

EFFECT OF CHRONIC ALCOHOL CONSUMPTION ON ETHANOL AND ACETALDEHYDE METABOLISM*

C.S. Lieber, L.M. DeCarli, L. Feinman, Y. Hasumura,

M. Korsten, S. Matsuzaki and R. Teschke

Section and Laboratory of Liver Disease and Nutrition

Bronx Veterans Administration Hospital and Department

of Medicine, Mount Sinai School of Medicine of the City

University of New York, New York, New York

ABSTRACT

Hepatic metabolism of ethanol to acetaldehyde by the alcohol dehydrogenase (ADH) pathway is associated with the generation of reducing equivalents as NADH. Conversely, reducing equivalents are consumed when ethanol oxidation is catalyzed by the NADPH dependent microsomal ethanol oxidizing system (MEOS). Since the major fraction of ethanol metabolism proceeds via ADH and since the oxidation of acetaldehyde also generates NADH, an excess of reducing equivalents is produced. This explains a variety of effects following acute ethanol administration, including hyperlactacidemia, hyperuricemia, enhanced lipogenesis and depressed lipid oxidation. To the extent that ethanol is oxidized by the alternate MEOS pathway, it slows the metabolism of other microsomal substrates. Following chronic ethanol consumption, adaptive microsomal changes prevail, which include enhanced ethanol and drug metabolism, and increased lipoprotein production. Eventually, injury develops with alterations of the rough endoplasmic reticulum and structural and functional abnormalities of the mitochondria.

* Most of the original studies were supported, in part, by grants from the National Institute of Alcohol Abuse and Alcoholism and the National Institute of Arthritis, Metabolism and Digestive Diseases and projects from the Veterans Administration.

Ethanol exerts different effects on hepatic cellular metabolism, depending mainly on the duration of its intake. In the presence of ethanol following an acute load, a number of hepatic functions are inhibited including lipid oxidation and microsomal drug metabolism. In its early stages, chronic ethanol consumption produces adaptive metabolic changes in the endoplasmic reticulum which result in increased metabolism of ethanol and drugs and accelerated lipoprotein production. Prolongation of ethanol intake damages cell organelles and may result in hepatic lesions such as alcoholic hepatitis and cirrhosis. The purpose of this review is to describe some of the mechanisms involved in each of these stages and, when possible, to relate these changes to the metabolism of ethanol and acetaldehyde.

A. PATHWAYS OF HEPATIC ETHANOL OXIDATION

Ethanol can be synthesized endogenously in trace amounts, but it is primarily an exogenous compound that is readily absorbed from the gastrointestinal tract. Only 2 to 10 per cent of that absorbed is eliminated through the kidneys and lungs; the rest must be oxidized in the body, principally in the liver (Lieber, 1970).

I. Alcohol Dehydrogenase (ADH)

The main hepatic pathway for ethanol disposition involves alcohol dehydrogenase (ADH), an enzyme of the cell sap (cytosol) that catalyzes the conversion of ethanol to acetaldehyde. Hydrogen is transferred from ethanol to the cofactor nicotinamide adenine dinucleotide (NAD), which is converted to its reduced form (NADH) (Figure 1A). As a net result, ethanol oxidation generates an excess of reducing equivalents in the liver, primarily as NADH. The latter is responsible for a variety of metabolic abnormalities (Figure 2).

Although alcohol dehydrogenase may account for the bulk of ethanol oxidation in the normal state, ethanol metabolism was found to persist even in the presence of pyrazole (a potent ADH inhibitor) both in vivo (Lieber and DeCarli, 1972) and in vitro in isolated perfused liver (Papenberg et al., 1970), liver slices (Lieber and DeCarli, 1970b) and isolated liver cells (Thieden, 1971; Grunnet et al., 1973). Furthermore, in the presence of pyrazole, glucose labeling from (1R-^{3}H) ethanol was nearly abolished, while $H^{3}HO$ production was inhibited less than 50%. In view of the stereospecificity of ADH for (1R-^{3}H) ethanol, these findings again suggest "the presence of a significant pathway not mediated by cytosolic ADH" (Rognstad and Clark, 1974). The rate of this nonADH mediated oxidation varied depending on the concentrations of ethanol used, from 20-25% (Lieber and DeCarli, 1970b, 1972; Papenberg et al., 1970)

A. $$CH_3CH_2OH + NAD^+ \xrightarrow{ADH} CH_3CHO + NADH + H^+$$

B. $$CH_3CH_2OH + NADPH + H^+ + O_2 \xrightarrow{MEOS} CH_3CHO + NADP^+ + 2H_2O$$

C. $$NADPH + H^+ + O_2 \xrightarrow{\text{NADPH Oxidase}} NADP^+ + H_2O_2$$
+
$$H_2O_2 + CH_3CH_2OH \xrightarrow{\text{Catalase}} 2H_2O + CH_3CHO$$

D. $$\text{HYPOXANTHINE} + H_2O + O_2 \xrightarrow{\text{Xanthine Oxidase}} \text{XANTHINE} + H_2O_2$$
+
$$H_2O_2 + CH_3CH_2OH \xrightarrow{\text{Catalase}} 2H_2O + CH_3CHO$$

Figure 1.
Ethanol oxidation in the liver by A, alcohol dehydrogenase (ADH); nicotinamide adenine dinucleotide (NAD), nicotinamide adenine dinucleotide, reduced form (NADH). B, the microsomal ethanol-oxidizing system (MEOS); nicotinamide adenine dinucleotide phosphate, reduced form (NADPH), nicotinamide adenine dinucleotide phosphate (NADP). C, a combination of NADPH oxidase and catalase; or D, xanthine oxidase and catalase.

to half or more (Thieden, 1971; Grunnet et al., 1973) of the total ethanol metabolism. Additional evidence that this pyrazole insensitive residual ethanol metabolism is not ADH mediated was derived from the fact that the cytosolic redox state was unaffected (Grunnet and Thieden, 1972). These findings raise the question of the nature of the ADH independent pathway. Theoretically, two enzyme systems could account for the latter, namely the microsomal ethanol oxidizing system or catalase.

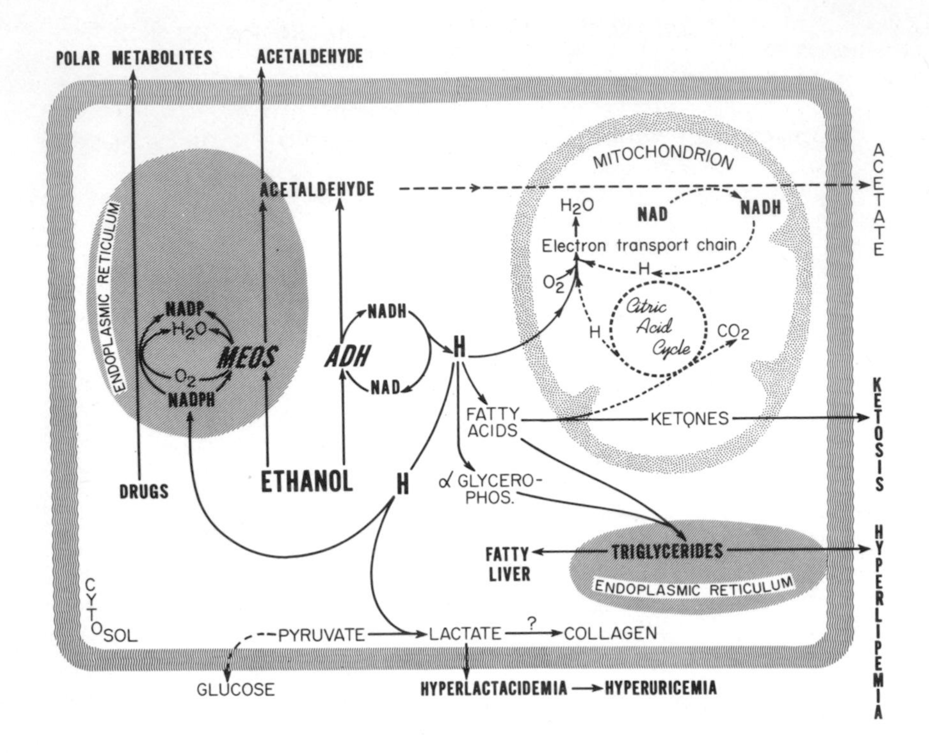

Figure 2:
Metabolism of ethanol in the hepatocyte and schematic representation of its link to fatty liver, hyperlipemia, hyperuricemia, hyperlactacidemia and ketosis. Pathways which are decreased by ethanol are represented by dashed lines. ADH, alcohol dehydrogenase; MEOS, microsomal ethanol-oxidizing system; NAD, nicotinamide adenine dinucleotide; NADH, nicotinamide adenine dinucleotide, reduced form; NADP, nicotinamide adenine dinucleotide phosphate; NADPH, nicotinamide adenine dinucleotide phosphate, reduced form.

II. Microsomal Ethanol Oxidizing System (MEOS)

The first indication of an interaction of ethanol with the microsomal fraction of the hepatocyte was provided by the morphologic observation that in rats, ethanol feeding results in a proliferation of the smooth endoplasmic reticulum (SER) (Iseri et al., 1964, 1966). This increase in SER resembles that seen after the administration of a wide variety of xenobiotic compounds including known hepato-toxins (Meldolesi, 1967), numerous therapeutic agents (Conney, 1967) and food additives (Lane and Lieber, 1967). Most of these substances which induce a proliferation of the SER are metabolized, at least in part, in the microsomal fraction of the hepatocyte which comprises the SER. The observation that ethanol produces proliferation of the SER raised the possibility that, in addition to its oxidation by ADH in the cytosol, ethanol may also be metabolized by the microsomes. A microsomal system capable of methanol oxidation was described (Orme-Johnson and Ziegler, 1965) but its capacity for ethanol oxidation was extremely low. Subsequently, a microsomal ethanol oxidizing system with a rate of ethanol oxidation 10 times higher than reported by Orme-Johnson and Ziegler (1965) was described (Lieber and DeCarli, 1968, 1970b). The striking increase in the nonADH fraction of ethanol metabolism with increasing ethanol concentrations (Thieden, 1971; Grunnet et al., 1973) is consistent with the known K_m for ADH and MEOS: whereas the former has a K_m varying from 0.5 to 2 mM (Reynier, 1969; Makar and Mannering, 1970), the latter has a value of 8-9 mM (Lieber and DeCarli, 1970b). The in vitro K_m of MEOS agrees well with the corresponding value of the pyrazole insensitive pathway of 8.8 mM in vivo (Lieber and DeCarli, 1972), suggesting that MEOS may play a significant role in ethanol metabolism.

Differentiation of MEOS from alcohol dehydrogenase was achieved by subcellular localization, pH optimum in vitro, cofactor requirements (Figure 1B) and effects of inhibitors such as pyrazole (Lieber et al., 1970; Lieber and DeCarli, 1973). Studies with inhibitors have also indicated that a major fraction of the ethanol oxidizing activity in microsomes is independent from catalase (Lieber and DeCarli, 1968, 1970b, 1973; Lieber et al., 1970) an enzyme which is considered to be a contaminant of the microsomal fraction rather than a component of the membrane of the endoplasmic reticulum itself (Redman et al., 1972). The concept that hepatic microsomes contain a catalase independent pathway for ethanol oxidation has been supported by various studies (Hildebrandt and Speck, 1973; Hildebrandt et al., 1974; Mezey et al., 1973). Similarly, a clear dissociation of the NADPH-dependent from a H_2O_2-mediated ethanol oxidation in microsomes has been demonstrated by the use of aminotriazole in vitro (Khanna et al., 1970). Aminotriazole is an inhibitor of catalase (Papenberg et al., 1970) and also of microsomal enzymes (Kato, 1967): it completely abolished the H_2O_2 dependent peroxidation of ethanol by inactivation of catalase, whereas the NADPH mediated microsomal ethanol oxidation was only slightly reduced (Khanna et al., 1970). Thus, under experimental conditions with complete abolition of the peroxidatic activity of catalase, the NADPH dependent ethanol oxidation still proceeded at a significant rate; this again dissociates the NADPH dependent MEOS activity from a process involving catalase-H_2O_2.

Some other groups, however, have attributed all of the ethanol oxidizing activity in microsomes to NADPH oxidase dependent H_2O_2 generation (Figure 1C) combined either with peroxidatic activity of catalase alone (Thurman et al., 1972) or with catalase and another unidentified enzyme (Roach et al., 1969) or even with catalase

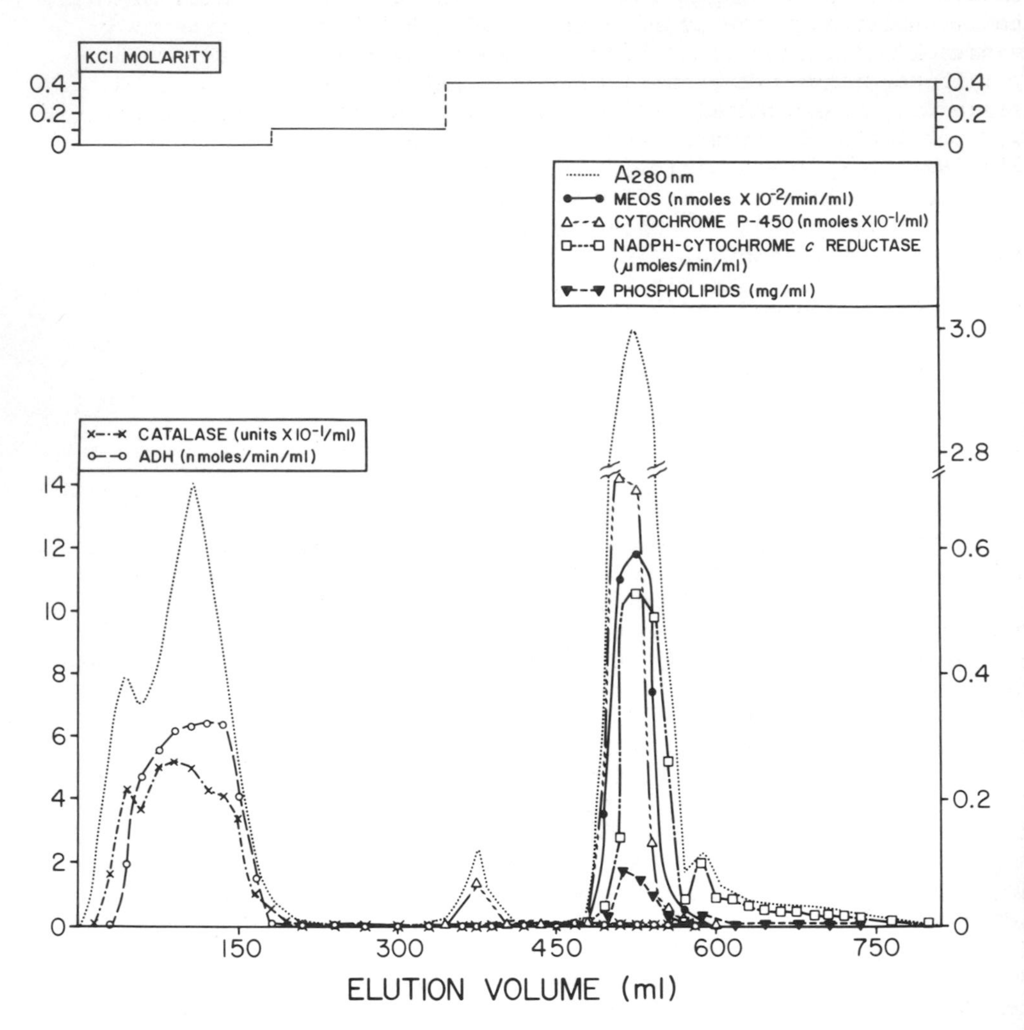

Figure 3.
Separation of MEOS from ADH and catalase activities by ion exchange column chromatography on DEAE-cellulose. Sonicated microsomes from rats fed laboratory chow were further solubilized by treatment with sodium desoxycholate and put onto a DEAE-cellulose column (2.5 x 45 cm). The separation of the enzyme activities was achieved by a stepwise increase of the salt gradient. Data from Teschke et al. (1974a).

and alcohol dehydrogenase (Isselbacher and Carter, 1970). It had also been reported that microsomes from acatalasemic mice fail to oxidize ethanol (Vatsis and Schulman, 1973), but this claim has now been retracted (Vatsis and Schulman, 1974). Indeed, hepatic microsomes of acatalsemic mice subjected to heat inactivation displayed decreased catalatic activity but NADPH dependent MEOS remained active and unaffected (Table 1). Even without heat inactivation, in the acatalasemic strain, the NADPH dependent metabolism was much more active than the H_2O_2 mediated one, whereas microsomes of control mice displayed equal rates of H_2O_2 and NADPH dependent ethanol oxidation (Table 2). These results therefore support the conclusion that hepatic microsomes of normal and acatalasemic mice contain a NADPH mediated ethanol oxidizing system which is catalase independent (Lieber and DeCarli, 1974B).

More recently, MEOS was solubilized and isolated from alcohol dehydrogenase and catalase activities (Figure 3) by DEAE-cellulose column chromatography (Teschke et al., 1972, 1974a; Mezey et al., 1973). Differentiation of MEOS from ADH was then shown by the failure of NAD^+ to promote ethanol oxidation at pH 9.6, by cofactor requirements, by the apparent K_m for ethanol (7.2 mM), by the insensitivity of the microsomal ethanol oxidizing system to the ADH inhibitor pyrazole and by the failure of added ADH to increase the ethanol oxidation. MEOS was also distinguished from a process involving catalase-H_2O_2 by the lack of catalatic activity, by the apparent K_m for oxygen (8.3 µM), by the insensitivity to the catalase inhibitors azide and cyanide, and by the lack of a H_2O_2 generating system (glucose-glucose oxidase) to sustain ethanol oxidation in the isolated column fraction (Teschke et al., 1974a).

Table 1

EFFECT OF HEAT (37°C) PRETREATMENT ON THE ACTIVITY OF THE NADPH DEPENDENT MICROSOMAL ETHANOL OXIDATION (MEOS) AND CATALASE IN CONTROL (Cs^a) AND ACATALASEMIC (Cs^b) MICE.

Washed microsomes (from 250 mg liver tissue) were incubated in phosphate buffer (80 mM) at pH 7.4 with 5 mM magnesium chloride, 0.3 mM NADP, 8 mM sodium isocitrate, 2 mg/ml isocitric dehydrogenase and 50 mM ethanol. Ethanol oxidation is expressed in nmoles acetaldehyde produced/min/mg protein and catalatic activity in Lück units/mg protein (average of 2 experiments). Data from Lieber and De Carli (1974).

Duration of heat pretreatment (minutes)	Cs^a mice		Cs^b mice	
	MEOS	Catalase	MEOS	Catalase
0	20.8	8.0	15.6	4.7
60	15.6	7.2	13.4	0.91

Table 2

HEPATIC MICROSOMAL ETHANOL OXIDATION IN CONTROL (Cs^a) AND ACATALASEMIC (Cs^b) MICE

Washed microsomes (from 250 mg tissue) were incubated in phosphate buffer (80 mM) at pH 7.4 with 5 mM magnesium chloride and either a NADPH generating system, consisting of 0.3 mM NADP, 8 mM sodium isocitrate and 2 mg/ml isocitric dehydrogenase (crude-type I, Sigma Chemical Co.) or an H_2O_2 producing system, namely 3.3 mM hypoxanthine and 0.02 units/ml xanthine oxidase. The final ethanol concentration was 50 mM. Ethanol oxidation is expressed in nmoles acetaldehyde produced/min/mg protein. Data from Lieber and DeCarli (1974).

	NADPH dependent	H_2O_2 dependent	P
Cs^a n = 6	17.1 ± 3.6	16.0 ± 3.1	NS
Cs^b n = 6	11.8 ± 4.8	4.8 ± 2.6	<0.01

Thurman et al. (1974a) apparently failed to obtain substantial ethanol oxidation following column chromatography of microsomes. A careful examination of their technique by one of us (Teschke, R.) during a visit to Philadelphia revealed however that they had deviated from our published procedure in many significant steps. In addition, it was claimed by Thurman et al. (1974a) that catalatic activity was present in one of Dr. Teschke's fractions. Actually, this sample had no such activity when freshly prepared. The small catalatic activity which developed after five weeks of storage can be fully accounted for by the conversion of cytochrome P-450 to P-420 and the well-known catalatic activity of cytochrome P-420 (Thurman et al., 1974a). Our fresh column fractions however, which exhibited striking MEOS activity, contained either no or only trace amounts of cytochrome P-420. Furthermore, a decrease of cytochrome P-450 content which was achieved under a variety of experimental conditions and resulted in partial conversion to P-420, diminished the ethanol oxidizing activity to a similar degree. The activity was restored, however, when fractions rich in cytochrome P-450 were added to the incubation medium. Thus, it appears that cytochrome P-450 is involved in MEOS activity (Teschke et al., 1974a) rather than P-420 as claimed by others (Thurman et al., 1974a).

In a more recent collaborative study carried out during a visit by Drs. Thurman and Hesse to New York (R. Teschke, R.G. Thurman, Y. Hasumura, S. Hesse and C.S. Lieber: Unpublished observation)

striking ethanol oxidation was observed in column chromatographic fractions incubated with NADPH itself or a NADPH generating system (Table 3). The rates of acetaldehyde produced were similar to those obtained in a previous study (Teschke et al., 1972). Various freshly prepared MEOS containing fractions were tested by Drs. Thurman and Teschke under a variety of experimental conditions, but no catalatic activity could be detected, confirming thereby our previous findings (Teschke et al., 1972, 1974a). Similarly, with an active H_2O_2 generating system (glucose-glucose oxidase) no acetaldehyde was produced in the MEOS fractions, clearly indicating that the column fractions were devoid also of peroxidatic activity of catalase (Table 3). Ethanol oxidation with the H_2O_2 generating system could only be demonstrated when exogenous catalase was added (Table 3). Thus, using specific and sensitive methods, MEOS activity could be clearly differentiated from an enzymatic process involving peroxidatic activity of catalase, microsomal H_2O_2 generation or both. Our results as well as other reports (Hildebrandt et al., 1973, 1974; Mezey et al., 1973) therefore fail to support the concept of the rate limiting role of H_2O_2 generation in microsomal ethanol oxidation (Thurman et al., 1972). Similarly, the low rate of microsomal H_2O_2 generation of 1.5-1.7 nmoles/min/protein (Boveris et al., 1972; Thurman et al., 1972) as measured with the specific cytochrome c peroxidase method can hardly account for the rate of 8 nmoles ethanol oxidized/min/mg protein (Thurman et al., 1972). In addition, MEOS activity could be dissociated from microsomal NADPH oxidase activity (Hasumura et al., 1974b), which generates the H_2O_2 in microsomes (Gillette et al., 1957).

Of particular interest regarding the nature of MEOS are studies with different alcohols as substrates. Previously, a NADPH dependent oxidation of methanol and ethanol, but not of propanol or butanol was reported in the microsomal fraction (Orme-Johnson and Ziegler, 1965). This was considered as evidence for an obligatory role of catalase in microsomal alcohol oxidation (Thurman et al., 1972) since catalase reacts peroxidatically primarily with methanol and ethanol, but not with alcohols with longer aliphatic chains (Chance, 1947). More recently, however, a NADPH dependent microsomal alcohol oxidizing system (MAOS) was described capable of metabolizing methanol, ethanol, propanol and butanol to their respective aldehydes in whole hepatic microsomes as well as in column fractions which contained the microsomal components cytochrome P-450, NADPH-cytochrome c reductase and phospholipids, but no ADH or catalase activity (Teschke et al., 1974b, 1974c). In whole hepatic microsomes, the oxidation rate of ethanol is approximately twice that of butanol in the presence of NADPH (Table 4). With a H_2O_2 generating system, rates similar to the NADPH dependent oxidation were achieved with ethanol, whereas butanol is a substrate only for the NADPH dependent microsomal system but, unlike ethanol is not a substrate for catalase-H_2O_2. The latter finding is in excellent agreement with previous reports regarding the substrate specificity

Table 3

ISOLATION OF A MEOS FRACTION OBTAINED FROM HEPATIC MICROSOMES OF FEMALE RATS FED AN ALCOHOL CONTAINING LIQUID DIET FOR 5 WEEKS (Collaborative study of Drs. Teschke, Thurman, Hasumura, Hesse and Lieber in New York, December, 1973).

The purified MEOS fraction was prepared as described (Teschke et al., 1974a). MEOS activity was determined with ethanol (50 mM), NADPH (0.4 mM) or a NADPH generating system consisting of 0.4 mM $NADP^+$, 8 mM sodium isocitrate and 2 mg per ml of isocitrate dehydrogenase. Peroxidatic activity of catalase was measured with a H_2O_2 generating system consisting of 10 mM glucose and 1.0 µg glucose oxidase per ml. Bovine liver catalase (Sigma Chemical Company, St. Louis, Mo.) (300 units/flask) was added to some of the flasks. All incubation media (final volume 3.0 ml) contained 1 mM EDTA and 5 mM $MgCl_2$. The protein concentration per flask was 3.0 and 2.0 mg for pool 1 and 2, respectively. Absence of catalatic activity in the MEOS fraction was verified by measurement of the H_2O_2 disappearance at 240 nm (Lück, 1963).

Incubation medium	Purified MEOS fraction Pool 1	Purified MEOS fraction Pool 2
	nmoles acetaldehyde/ min/flask	
NADPH	104.2	44.5
NADPH generating system	105.0	40.5
H_2O_2 generating system	0	0
H_2O_2 generating system and catalase	52.7	48.7

of catalase (Chance, 1947; Chance and Oshino, 1971; Keilin and Hartree, 1945). Chronic ethanol feeding resulted in a striking increase of MAOS activity with all four alcohols (methanol, ethanol, propanol and butanol) as substrates (Table 5). These experiments therefore show that the adaptive response observed after chronic ethanol consumption can be ascribed predominantly to a catalase-H_2O_2 independent mechanism since it was also demonstrated with propanol and butanol as substrates (Table 5) which virtually fail to react peroxidatically with catalase-H_2O_2 (Table 4).

It has been suggested that MEOS could account for the alcohol dehydrogenase independent pathway of alcohol metabolism in vivo (Lieber and DeCarli, 1972) and in vitro in liver slices (Lieber and DeCarli, 1970b), isolated parenchymal liver cells (Thieden,

Table 4

SUBSTRATE SPECIFICITY OF THE MICROSOMAL ALCOHOL OXIDIZING SYSTEM (MAOS)

The incubations were carried out with systems generating either NADPH (0.4 mM $NADP^+$, 8 mM sodium isocitrate and 2 mg per ml of isocitrate dehydrogenase) or H_2O_2 (10 mM glucose and 0.7 µg per ml of glucose oxidase). The values represent the average results of three experiments. Data from Teschke et al. (1974b).

Substrate	NADPH	H_2O_2
	nmoles aldehyde/min/mg protein	
Methanol	7.7	7.8
Ethanol	9.5	9.2
Propanol	4.6	0.2
Butanol	3.9	0

1971; Grunnet et al., 1973) and perfused liver (Papenberg et al., 1970). Evidence for a microsomal component in ethanol oxidation in perfused liver was also provided by others (Scholz et al., 1971). It has been shown that the rate of ethanol utilization was much less in the fasted than in the fed state, which is consistent with a lack of NADPH generation due to depletion of glycogen, resulting in insufficient supply of this particular cofactor required for the NADPH dependent microsomal ethanol oxidation (Lieber and DeCarli, 1970b). This discrepancy in metabolic rates of ethanol oxidation between the fasted and fed state persisted even in the presence of mitochondrial inhibitors (Scholz et al., 1971) supporting the view that an extramitochondrial system was involved. Similar results were obtained with aminopyrine, a typical substrate for the NADPH dependent microsomal drug detoxifying enzyme system. It is noteworthy that the rates of ethanol oxidation are strikingly enhanced in patients with glycogenosis type I (Papenberg, 1971). In this disorder, the content of hepatic glucose-6-phosphate is increased possibly resulting in increased rate of generation of NADPH, the cofactor of MEOS.

Indirect evidence that MEOS activity may play a role in vivo can also be derived from the fact that other drugs (such as barbiturates) which increase total hepatic MEOS activity were found to enhance rates of blood ethanol clearance (Fischer, 1962; Lieber and DeCarli, 1972; Mezey and Robles, 1974). Some other studies failed to verify this effect (Klaassen, 1969; Tephly et al., 1969). In the latter investigation, however, long-acting barbiturates

Table 5

EFFECT OF CHRONIC ETHANOL CONSUMPTION (6-8 WEEKS) ON THE ACTIVITY OF MAOS

Female rats were pair-fed nutritionally adequate liquid diets containing either ethanol or dextrin as controls. The values represent means ($\pm$SEM) of six pairs. Data from Teschke et al. (1974b).

Substrate	Control rat	Ethanol fed rat	P	Control rat	Ethanol fed rat	P
	nmoles aldehyde/min/ mg micros. protein			nmoles aldehyde/min/ 100 g BW		
Methanol	7.3$\pm$0.6	12.7$\pm$1.0	<0.01	983$\pm$45	1650$\pm$83	<0.001
Ethanol	9.9$\pm$0.4	14.9$\pm$0.8	<0.01	1364$\pm$57	1948$\pm$83	<0.01
Propanol	5.8$\pm$0.6	11.1$\pm$0.9	<0.01	808$\pm$102	1461$\pm$150	<0.02
Butanol	4.4$\pm$0.3	7.4$\pm$0.7	<0.01	575$\pm$49	966$\pm$95	<0.01

were used and ethanol clearance was tested in close association with barbiturate administration, at a time when blood barbiturate levels were probably elevated. Under these conditions, it was found that barbiturates interfere with blood ethanol clearance (Lieber and DeCarli, 1972).

III. Catalase

It has been shown that catalase is capable of oxidizing ethanol in vitro in the presence of a H_2O_2 generating system (Keilin and Hartree, 1945) (Figure 1C and 1D). However, a significant role of catalase in ethanol metabolism has been denied by many (Bartlett, 1952; Lester and Benson, 1970; Papenberg et al., 1970: Feytmans and Leighton, 1973). It is generally accepted that the H_2O_2 mediated ethanol peroxidation by catalase is limited by the rate of H_2O_2 generated rather than the amount of catalase itself. Thus, an indirect answer to the question whether catalase plays a role in ethanol metabolism can be derived from the rate of H_2O_2 generated in the liver. The physiological rate of H_2O_2 production has been estimated to be 3.6 µmole/h/g of liver (Boveris et al., 1972) which represents 2% of the in vivo rate of ethanol oxidation of 178 µmole/h/g of liver (Lieber and DeCarli, 1972) assuming 3.5 g of liver per 100 g of BW. Actually, the rate of 2% of ethanol being oxidized possibly by a catalase-H_2O_2 mediated mechanism is probably an overestimate, since not all of the H_2O_2 generated in the liver can be utilized by the peroxidatic reaction of catalase (Oshino et al., 1973b). Furthermore, the H_2O_2 produced by the microsomes (which comprise the endoplasmic reticulum upon subcellular fractionation) does not contribute to ethanol oxidation in perfused liver (Thurman and Scholz, 1973) although this fraction furnishes almost half of the total hepatic H_2O_2 generation (Boveris et al., 1972). Similarly, comparing the rate of H_2O_2 generation of 3.6 µmole/h/g of liver (Boveris et al., 1972) with the pyrazole insensitive ethanol oxidation of 53 µmole/h/g of liver in vivo (Lieber and DeCarli, 1972) reveals that catalase-H_2O_2 can account for not more than 5% of the nonADH mediated pathway. Studies with perfused livers have also shown a similar striking discrepancy between the rate of H_2O_2 generation with that of ethanol oxidation. The rate of H_2O_2 production in perfused liver of 3 µmole/h/g of liver (Oshino et al., 1973a) accounts for less than 5% of the rate of ethanol oxidation of 150 µmoles/h/g of liver and for less than 10% of the ADH independent pathway estimated to be 40-50 µmole ethanol oxidized/h/g of perfused liver (McKenna and Thurman, 1974). It is noteworthy that the breakdown of the catalase-H_2O_2 intermediate (Compound I) by ethanol has a K_m of less than 0.6 mM for ethanol (Chance and Oshino, 1971; Theorell et al., 1974) which is more than one order of magnitude lower than the apparent K_m of the ADH-independent pathway of 8.8 mM for ethanol in vivo (Lieber and DeCarli, 1972). Finally, the hepatic oxygen concentration which was estimated to be less than 50 µM may be far too low for significant rates of peroxisomal H_2O_2 generation which has an apparent K_m of 100 μM for oxygen (Boveris et al., 1972). To evaluate a possible role of catalase in ethanol metabolism, aminotriazole (a catalase inhibitor) has been widely used. However, no significant change of the metabolic rate of ethanol oxidation was observed, both in vivo (Smith, 1961; Roach et al., 1972; Feytmans and Leighton, 1973) and in vitro in liver slices (Smith, 1961) and perfused liver (Papenberg et al., 1970). Similarly, in all these experiments pretreatment with aminotriazole failed to inhibit the pyrazole insensitive pathway for ethanol metabolism in vivo and in vitro even under experimental conditions in which the perfusion medium was supplemented with additional aminotriazole. Only one single report recently described the complete inhibition of the pyrazole insensitive pathway for ethanol oxidation by pretreatment with aminotriazole, a compound which was claimed to completely block catalase (Thurman et al., 1974b). However, in our hands pretreatment with aminotriazole under identical experimental conditions decreased, but failed to completely block catalase activity, as demonstrated by its catalatic and by its peroxidatic reaction (Table 6); these results agree with several previous reports (Kinard et al., 1956; Lieber and DeCarli, 1970; Feytmans and Leighton, 1973). Furthermore, aminotriazole is not a specific inhibitor for catalase; indeed, this compound also inhibits the activity of microsomal enzymes, namely the metabolism of type I (aminopyrine) and of type II (aniline) binding substrates (Table 6). Though the inhibitory effect of aminotriazole on microsomal functions is less striking than that on catalase (Table 6), this compound could theoretically depress the microsomal ethanol oxidizing system under the conditions employed by Thurman et al. (1974b). Similarly, an impairment of other microsomal functions following aminotriazole has been previously reported (Kato, 1967). Finally, aminotriazole inhibits alcohol dehydrogenase activity (Feytmans and Leighton, 1973), again indicating its lack of specificity.

In other experiments, sodium azide was used to assess the potential role of catalase in ethanol metabolism by liver slices. Azide in a concentration of 1.0 mM has previously been shown not to interfere with the activity of the isolated hepatic microsomal ethanol oxidizing system (Teschke et al., 1974a). Pyrazole, an ADH inhibitor (Lieber and DeCarli, 1972) significantly ($P<0.001$) decreased ethanol metabolism, whereas both compounds sodium azide and pyrazole together failed to further decrease the ethanol oxidation already diminished by pyrazole alone (Table 7). To test whether sodium azide entered the cells and effectively blocked catalase, dl-alanine was used to stimulate ethanol metabolism by activating peroxisomal H_2O_2 generation. As expected, dl-alanine increased ethanol metabolism, but this enhancement was completely abolished by sodium azide in a concentration which left the pyrazole insensitive ethanol oxidation unchanged (Table 7). Thus, these results show that catalase may participate under certain conditions in ethanol

Table 6

EFFECT OF AMINOTRIAZOLE ON CATALASE AND MICROSOMAL ENZYME ACTIVITIES

Male chow fed rats (BW 360-420 g) were injected with 3-amino-1,2,4 triazole (1 g/kg BW i.p.) in saline one hour before sacrifice. Control rats received saline alone. Washed microsomes were prepared as described (Lieber and DeCarli, 1972), and enzyme assays were performed and activities expressed as previously reported (Teschke et al., 1974a). Peroxidatic activity of catalase was measured with 50 mM ethanol, 10 mM glucose and 0.7 μg per ml of glucose oxidase. Values represent means (±SEM) from 8 rats in each group and are expressed per mg of microsomal protein. Data from R. Teschke, Y. Hasumura and C.S. Lieber (unpublished observation).

Enzyme	Control	Aminotriazole	P
Catalatic activity (units/mg)	1.59±0.10	0.27±0.05	<0.001
Peroxidatic activity (nmoles acetaldehyde/min/mg)	6.89±0.28	0.57±0.15	<0.001
Aminopyrine Demethylase (nmoles formaldehyde/min/mg)	7.98±0.45	6.90±0.43	<0.05
Aniline Hydroxylase (nmoles p-aminophenol/min/mg)	1.06±0.07	0.71±0.04	<0.001

Table 7

EFFECT OF PYRAZOLE AND SODIUM AZIDE ON ETHANOL METABOLISM BY LIVER SLICES

Liver slices (500 mg weight) of male chow fed rats (BW 380-440 g) were added to a medium (5.0 ml) containing Krebs-Ringer buffer (pH 7.4), ethanol (50 mM) and, when indicated, the following compounds: pyrazole (2 mM), sodium azide (1 mM), and dl-alanine (40 mM). The incubations were carried out under O_2 (95%) and CO_2 (5%) which was flushed for 5 min into closed 50 ml Erlenmeyer flasks. The incubation mixture was allowed to equilibrate for a half hour at 37°C. Incubations were then carried out for 0,30,60 and 90 min at which intervals aliquots (100 µl) of the incubation medium were harvested, and ethanol disappearance was measured by gas-liquid chromatography. Control incubations were performed with boiled liver slices in the presence of ethanol. Values represent means (±SEM) of five experiments. Unpublished data from R. Teschke, Y. Hasumura and C.S. Lieber.

Incubation medium	Ethanol oxidation
	µmoles ethanol consumed/h/g of liver
Control	55.3±7.1
+ Pyrazole	22.3±3.8
+Pyrazole + Sodium Azide	26.8±5.6
+ Pyrazole + dl-Alanine	58.3±9.3
+ Pyrazole + dl-Alanine + Sodium Azide	23.8±4.4

metabolism, provided the peroxisomal H_2O_2 production is stimulated. However, under physiological conditions, catalase appears to play little if any role and can not account quantitatively for the ADH independent pathway of ethanol metabolism. Conversely, the pyrazole and azide insensitive pathway is most probably due to the activity of the microsomal ethanol oxidizing system.

B. DIRECT EFFECTS OF ETHANOL OXIDATION ON HEPATIC AND INTERMEDIARY METABOLISM

I. Role of Excessive Hepatic NADH Generation by the ADH Pathway

As shown in figure 2, the oxidation of ethanol results in the transfer of hydrogen to NAD. The resulting enhanced NADH/NAD ratio, in turn, produces a change in the ratio of those metabolites that are dependent for reduction on the NADH-NAD couple. It was therefore proposed that the altered NADH/NAD ratio is responsible for a number of metabolic abnormalitites associated with alcohol abuse (Lieber and Davidson, 1962).

a. Hyperlactacidemia, hyperuricemia, acidosis. The enhanced NADH/NAD ratio reflects itself in an increased lactate/pyruvate ratio that results in hyperlactacidemia (Lieber et al., 1962a; 1962b) because of both decreased utilization and enhanced production of lactate by the liver. The hyperlactacidemia contributes to acidosis and also reduces the capacity of the kidney to excrete uric acid, leading to secondary hyperuricemia (Lieber et al. 1962a). Alcohol induced ketosis may also promote the hyperuricemia. The latter may be related to the common clinical observation that excessive consumption of alcoholic beverages frequently aggravates or precipitates gouty attacks (Newcombe, 1972). Alcoholic hyperuricemia can be readily distinguished from the primary variety by its reversibility upon discontinuation of ethanol abuse (Lieber et al. 1962a). A fascinating but as yet hypothetical consequence of the increased availability of lactate may be the stimulation of collagen production and increased hepatic collagen proline hydroxylase activity which conceivably play a role in collagen accumulation (Feinman and Lieber, 1972).

b. Enhanced lipogenesis and depressed lipid oxidation. The increased NADH/NAD ratio also raises the concentration of α-glycerophosphate (Nikkila and Ojala, 1963) that favors hepatic triglyceride accumulation by trapping fatty acids. In addition, excess NADH promotes fatty acid synthesis (Gordon, 1972a; Lieber and Schmid, 1961) possibly by the elongation pathway or transhydrogenation to nicotinamide adenine dinucleotide phosphate (NADP). Theoretically, enhanced lipogenesis can be considered a means for disposing of the excess hydrogen. Some hydrogen equivalents can be transferred into the mitochondria by various "shuttle" mechanisms. Since the activity of the citric acid cycle is depressed (Forsander et al., 1965; Lieber et al., 1967), partly because of a slowing of the reactions of the cycle that require NAD, the mitochondria will use the hydrogen equivalents originating from ethanol, rather than from the oxidation through the citric acid cycle of two carbon fragments derived from fatty acids. Thus, fatty acids that normally serve as the main energy source of the liver are supplanted by ethanol. Decreased fatty acid oxidation by ethanol has been demonstrated in liver slices

(Lieber and Schmid, 1961; Blomstrand et al., 1973), perfused liver (Lieber et al., 1967), isolated hepatocytes (Ontko, 1973) and in vivo (Blomstrand and Kager, 1973). This results in the deposition in the liver of dietary fat, when available, or fatty acids derived from endogenous synthesis in the absence of dietary fat (Lieber and Spritz, 1966; Lieber et al., 1966, 1969; Mendenhall, 1972).

II. Interaction of Ethanol with Microsomal Functions

Interactions of the effects of ethanol and various drugs have been widely recognized (Forney and Hughes, 1968). Intoxicated individuals are more susceptible to several medications (Soehring and Schüppel, 1966). These various effects are usually attributed to additive or synergistic effects of alcohol and various drugs on the central nervous system. However, the existence of an at least partially common microsomal system for ethanol and drug metabolism sheds new light on the interaction of ethanol and drug metabolism. The increased susceptibility of the inebriated individual could be explained, at least in part, by the effect of ethanol on microsomal drug-detoxifying enzymes. It has indeed been found that ethanol inhibits the metabolism of a variety of drugs in vitro (Ariyoshi et al., 1970; Rubin et al., 1970a; Rubin and Lieber, 1968a; Schüppel, 1971). With some systems, such as aniline hydroxylase, this inhibition is of a competitive nature (Rubin et al., 1970a; Cohen and Mannering, 1973). These in vitro effects may explain the observation that in vivo, simultaneous administration of ethanol and drugs slows the rate of drug metabolism (Rubin et al., 1970a). Drugs also inhibit ethanol oxidation by microsomes in vitro in a way which has been considered as strong evidence for a catalase independent fraction of ethanol metabolism in hepatic microsomes (Hildebrandt et al., 1974). In addition, some drugs also inhibit alcohol dehydrogenase (Sutherland et al., 1960).

C. ADAPTIVE METABOLIC CHANGES FOLLOWING CHRONIC ETHANOL INTAKE

It is common knowledge that chronic alcohol consumption produces increased tolerance to ethanol. This is generally attributed to central nervous system adaptation. In addition, recent studies have shown the development of metabolic adaptation, that is an accelerated clearance of alcohol from the blood. Furthermore, there is an associated increased capacity to metabolize other drugs as well. Moreover, the liver acquires an enhanced capacity to rid itself of lipids through lipoprotein secretion into the blood stream. It is noteworthy that these functions which adaptively increase after chronic ethanol feeding involve to a large extent the activity of the hepatic smooth endoplasmic reticulum, which undergoes significant change after chronic alcohol consumption. It was indeed observed ten years ago that ethanol feeding results in a proliferation of the smooth membranes of the hepatic endoplasmic reticulum (Iseri et al., 1964, 1966). This ultramicroscopic finding was subsequently confirmed (Lieber and Rubin , 1968; Rubin et al., 1968; Carulli et al., 1971) and established on a biochemical basis by the demonstration of an increase in both phospholipids and total protein content of the smooth membranes (Ishii et al., 1973). Its functional counterparts include accelerated metabolism of drugs (including ethanol) and lipoprotein production.

I. Accelerated Ethanol Metabolism After Chronic Ethanol Consumption

Regular drinkers tolerate large amounts of alcoholic beverages, mainly because of central nervous system adaptation. In addition, alcoholics develop increased rates of blood ethanol clearance, so-called metabolic tolerance (Kater et al., 1969a; Ugarte et al., 1972). Experimental ethanol administration also results in an increased rate of ethanol metabolism (Lieber and DeCarli, 1970b; Misra et al., 1971; Tobon and Mezey, 1971). The mechanism of this acceleration is the subject of debate.

a. Increase of ethanol metabolism related to the ADH pathway. There is a controversy over whether ethanol consumption affects activities of hepatic ADH, with most investigators reporting no change or even decreases (Lieber, 1973). In alcoholics, liver ADH was found to be lowered even in the absence of liver damage (Ugarte et al., 1967). Extrahepatic ADH, particularly the gastric one, has been reported to increase after alcohol feeding (Mistilis and Garske, 1969) but this has not been confirmed either after acute or chronic ethanol administration (deSaint-Blanquat et al., 1972).

Actually, the question of whether there is a moderate change in hepatic ADH activity may not have direct bearing on the problem of rates of alcohol metabolism since it is generally recognized that ADH activity is usually not the rate-limiting factor in that pathway. There are numerous examples, discussed elsewhere (Lieber, 1973), of the lack of correlation between rates of ethanol oxidation and hepatic ADH activity.

One mechanism proposed as contributing to the acceleration of ADH dependent-ethanol metabolism after ethanol consumption is based on increased NADH reoxidation, for instance because of enhanced ATPase activity (Bernstein et al., 1973). Mitochondrial mechanisms which have been postulated include enhanced shuttling of the H equivalent from the cytosol to the mitochondria after chronic ethanol feeding. We failed to find evidence in favor of this possibility (Cederbaum et al., 1973). In general, it must be pointed out that if following chronic ethanol consumption, changes affecting the ADH pathway (such as ATPase activity) were responsible exclusively for the acceleration of ethanol metabolism, the latter should be fully abolished by pyrazole treatment, but this was not the case (Lieber and DeCarli, 1970b, 1972). This raises the possibility of the involvement of nonADH pathways.

b. Non-ADH related acceleration of ethanol metabolism. Following chronic ethanol consumption, MEOS significantly increases in activity (Lieber and DeCarli, 1968, 1970b). Calculations show that when corrected for microsomal losses during the preparative procedure, the rise in MEOS activity can account for one-half to two-thirds of the increase in blood ethanol clearance (Lieber and DeCarli, 1972). The unaccounted for difference may actually result from a secondary increase in oxidation via ADH, a pathway limited by the rate of NADH reoxidation. This could be accelerated by an increase in MEOS activity, since the latter is associated with NADPH utilization, and the NADPH-NADP and NADH-NAD systems are linked (Veech et al., 1969). Indeed, acute administration of ethanol led to a smaller shift of the reduced state in livers from animals adapted to ethanol than in those from control rats (Domschke et al., 1974) both in cytosolic and mitochondrial redox levels (Table 8). Moreover, evidence is accumulating that NADH may serve as partial electron donor for microsomal-drug-detoxifying systems (Cohen and Estabrook, 1971).

There is also some debate over whether in rats, ethanol feeding enhances catalase activity. Both an increase (Carter and Isselbacher, 1971) and no change (von Wartburg and Rothlisberger, 1961; Hawkins et al., 1966; Lieber and DeCarli, 1970b) have been reported. In man there was no increase (Ugarte et al., 1972). This question, however, may not be fully relevant to the rate of ethanol metabolism since, as discussed before, peroxidative metabolism of ethanol in the liver is probably limited by the rate of hydrogen peroxide formation rather than by the amount of available catalase (Boveris et al., 1972). Ethanol consumption does, however, enhance the activity of hepatic NADPH oxidase (Lieber and DeCarli, 1970a; Carter and Isselbacher, 1971), which, as illustrated in figure 1C, generates H_2O_2. Theoretically, this mechanism could contribute to ethanol metabolism in vivo (and to its increase after chronic ethanol consumption) by furnishing the H_2O_2 needed for peroxidative oxidation of ethanol. As discussed before, however, the amount of H_2O_2 generated by the liver is small (Boveris et al., 1972; Oshino et al., 1973a) and even when increased by ethanol consumption could not account for the rate of ethanol clearance observed (Israel et al., 1973; Lieber, 1973). Similarly, under conditions where microsomal H_2O_2 generation was stimulated by addition of menadione, ethanol metabolism was not altered in perfused liver, indicating that H_2O_2 generated by the microsomal NADPH oxidase does not participate in the overall process of ethanol metabolism (Thurman and Scholz, 1973). Moreover, ethanol was found recently to inhibit NADPH oxidase activity (Teschke et al., 1974a).

Table 8

EFFECT OF AN ACUTE DOSE OF ETHANOL ON THE REDOX LEVEL IN ETHANOL PRETREATED RATS AND THEIR CONTROLS

The values compiled are means (±SD) of the ratios calculated for each individual animal (7 pairs). The P value indicates the significance of the difference between the ethanol treated rats and the corresponding controls. Data from Domschke et al. (1974).

	Ethanol Pretreated Rats	Control Rats
Cytoplasm		
$(NAD)/(NADH_2)$	485*± 96	292± 70
$(NADP)/(NADPH_2)$	.00198*±.00036	.00138±.00031
Mitochondria		
$(NAD)/(NADH_2)$	2.25*±0.79	0.82±0.27
$(NADP)/(NADPH_2)$	2.56*±0.54	1.49±0.53

*P<0.01

II. Stimulation of the Microsomal Drug Metabolizing Enzymes

a. Enhanced drug metabolism (drug tolerance). The proliferation of hepatic SER induced by ethanol has a functional counterpart: an increased activity of a variety of microsomal drug-detoxifying enzymes (Rubin and Lieber, 1968; Ariyoshi et al., 1970; Carulli et al., 1971; Joly et al., 1973a; Misra et al., 1971). Ethanol also increases the content of microsomal cytochrome P-450 and the activity of NADPH-cytochrome P-450 reductase (Rubin et al., 1968; Joly et al., 1973a). These increases occur in the smooth membranes (Ishii et al., 1973; Joly et al., 1973a). Furthermore, ethanol feeding raises the hepatic phospholipid content (Lieber et al., 1965) including that of the smooth microsomal membranes (Ishii et al., 1973). Moreover, it has been shown that microsomal cytochrome P-450, a reductase, and phospholipids play a key role in the microsomal hydroxylation of various drugs (Lu et al., 1969). Therefore, the increase in the activity of hepatic microsomal drug-detoxifying enzymes and in the content of cytochrome P-450 induced by ethanol ingestion offers a likely explanation for the recent observation that ethanol consumption enhances the rate of drug clearance in vivo. The tolerance of the alcoholic to various drugs has been generally attributed to central nervous system adaptation (Kalant et al., 1970). However, there is sometimes a dissociation in the time course of the decreased drug sensitivity of the animal and the occurrence of central nervous system tolerance: the decreased drug sensitivity was found to precede the central nervous system tolerance (Ratcliffe, 1969). Thus, in addition to central nervous system adaptation, metabolic adaptation must be considered. Indeed, it has been shown recently that the rate of drug clearance from the blood is enhanced in alcoholics (Kater et al., 1969b). Of course, this could be due to a variety of factors other than ethanol, such as the congeners and the use of other drugs so commonly associated with alcoholism. Our studies showed, however, that administration of pure ethanol with non-deficient diets under metabolic ward conditions resulted in a striking increase in the rate of blood clearance of meprobamate and pentobarbital (Misra et al., 1971). Similarly, an increase in the metabolism of aminopyrine (Vesell et al., 1971) and tolbutamide (Carulli et al., 1971) was found. Furthermore, the capacity of liver slices from animals fed ethanol to metabolize meprobamate was also increased (Misra et al., 1971) which clearly shows that ethanol consumption affects drug metabolism in the liver itself, independent of drug excretion or distribution.

b. Increased CCl_4 toxicity in alcoholics. The stimulation of microsomal enzyme activities also applies to those which convert exogenous substrates to toxic compounds. For instance, CCl_4 exerts its toxicity only after conversion in the microsomes. Alcohol pretreatment remarkably stimulates the toxicity of CCl_4 (Figure 4). These experiments were carried out at a time when the ethanol had disappeared from the blood to rule out the increase of the toxicity of CCl_4 due to the presence of ethanol (Traiger and Plaa, 1972). The potentiation of the CCl_4 toxicity by ethanol pretreatment may be accounted for by the increased production of toxic compounds of CCl_4 since the conversion of $^{14}CCl_4$ to $^{14}CO_2$ was significantly accelerated in microsomes of ethanol pretreated rats (Figure 5). Similarly, pretreatment of rats with phenobarbital, a well known inducer of the hepatic microsomal drug metabolizing system, increased CCl_4 hepatotoxicity (Garner and McLean, 1969) concomitant with an enhanced production of toxic metabolites of CCl_4 (Diaz-Gomez et al., 1973). Thus, the clinical observation of the enhanced susceptibility of alcoholics to the hepatotoxic effect of CCl_4 (Moon, 1950) may be, at least in part, due to an increased activation and biotransformation of CCl_4. It is likely that a larger number of other toxic agents will be found to display a selective injurious action in the alcoholic. This side effect is possibly an undesirable consequence of the "adaptive" response to chronic ethanol consumption.

III. Increase in Microsomal Functions Related to Lipid Metabolism

a. Cholesterol metabolism. The various functions of the endoplasmic reticulum include cholesterol synthesis. Increased cholesterol synthesis after ethanol (Lefevre et al., 1972) may have a microsomal basis akin to that after barbiturate and may explain, in

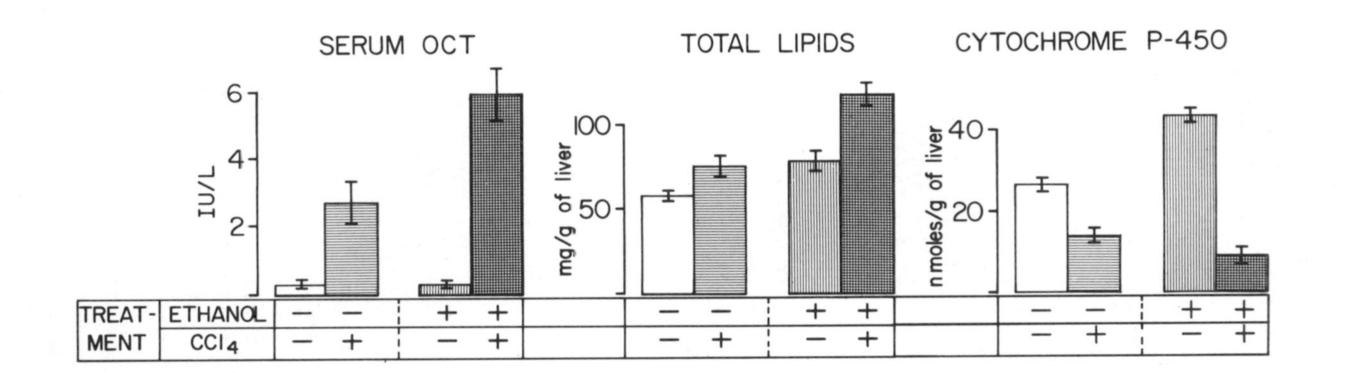

Figure 4.
Increased CCl_4 hepatotoxicity after chronic ethanol consumption in rats. Rats were treated in pairs for 4-5 weeks with a nutritionally adequate liquid diet containing either ethanol (36% of total calories) or isocaloric carbohydrate. CCl_4 (0.5 ml per kg) was given intragastrically 24 hours before sacrifice. Data from Hasumura et al. (1974a).

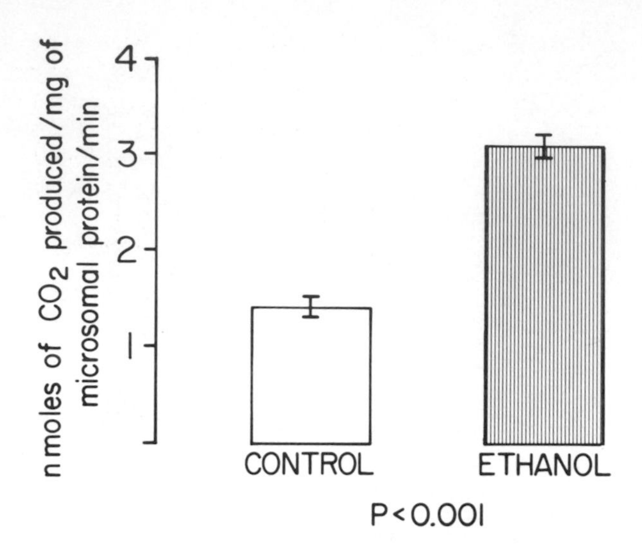

Figure 5.
Effect of chronic ethanol consumption on hepatic microsomal biotransformation of $^{14}CCl_4$ to $^{14}CO_2$ in vitro. $^{14}CCl_4$ was incubated with liver microsomes obtained from rats fed ethanol or carbohydrate (control) in the presence of a NADPH generating system. Data from Hasumura et al. (1974a).

part, the accumulation of cholesterol ester observed in the liver after feeding alcohol (Lieber et al., 1963; Lefevre et al., 1972) especially with a cholesterol-free diet. When ethanol is given with cholesterol-containing diets, decreased cholesterol catabolism, evidenced by a reduction in bile-acid production and turnover after alcohol feeding, plays a major part.

b. Alcoholic hyperlipemia. In both man (Losowsky et al., 1963) and the rat (Baraona and Lieber, 1969) ethanol administration produces mild hyperlipemia, involving especially the very low density lipoproteins. Incorporation into lipoprotein of intragastrically administered ^{3}H-palmitate and intravenously injected ^{14}C-lysine is increased (Baraona and Lieber, 1969) suggesting enhanced lipoprotein production. Fatty acids are esterified, and lipoproteins are formed in the endoplasmic reticulum. Furthermore, chronic feeding of ethanol increases hepatic lipoprotein production, even when ethanol is not present at the time of testing, which reveals an increased capacity for lipoprotein synthesis (Baraona et al., 1973). Moreover, ethanol consumption enhances the activity of hepatic microsomal L-α-glycerophosphate acyltransferase (Joly et al., 1973b). The mechanism of the alterations of these microsomal functions produced by ethanol has not been clarified. It could be linked directly to the fact that ethanol can be oxidized at this key metabolic site. Ethanol could also induce hepatic production of lipoproteins indirectly by enhancing the availability of fatty acids either by decreasing their oxidation or by enhancing synthesis, as alluded to before. Increased glycerolipid production has indeed been found after ethanol consumption (Mendenhall et al., 1969). Ethanol feeding was observed to enhance the activity of glycosyl-transferase in the Golgi apparatus (Gang et al., 1973) and to increase the synthesis of the protein moiety of lipoproteins (Baraona and Lieber, 1969). In some individuals, the response is markedly exaggerated because of a fatty diet (Brewster et al., 1966; Verdy and Gattereau, 1967; Barboriak and Meade, 1968; Wilson et al., 1970; Kudzma and Schonfeld, 1971) or because of some underlying abnormality of lipid or carbohydrate metabolism, such as a forme fruste of essential hyperlipemia (Losowsky et al., 1963; Mendelson and Mello, 1973; Ginsberg et al., 1974), pancreatitis (Kessler et al., 1967), diabetes or prediabetes (Losowsky et al., 1963; Chait et al., 1972) or an increased susceptibility to ethanol itself (Kudzma and Schonfeld, 1971).

Contrasting with the hyperlipemia, which is commonly associated with the administration of moderate to large amounts of ethanol, an extremely high dose has been reported to decrease serum triglyceride (Dajani and Kouyoumjian, 1967), very low density lipoproteins (Madsen, 1969), high density lipoproteins (Koga and Hirayama, 1968) and the incorporation of glucosamine into the carbohydrate moiety of serum lipoproteins (Mookerjea and Chow, 1969) in the rat. These inhibitory effects can be related to the observation that addition of ethanol to isolated perfused livers depresses lipid secretion (Schapiro et al., 1964). This may be due to the high ethanol concentration used. Indeed, more recently, when livers were perfused with ethanol in concentrations more in keeping with in vivo conditions, no inhibition of lipoprotein secretion was found (Gordon, 1972a). The depressive effects of high concentrations of ethanol compared to the adaptive response to lower ones may be a reflection of the hepatotoxic effect of ethanol.

D. INJURIOUS EFFECTS ASSOCIATED WITH ALCOHOLISM

In addition to the adaptive changes described before, chronic alcohol consumption is associated with functional and structural changes which can be interpreted as injurious, particularly in the rough endoplasmic reticulum and in the mitochondria.

I. Alteration of the Rough Endoplasmic Reticulum

After chronic ethanol consumption the membranes of the rough endoplasmic reticulum (RER) appear decreased on electron microscopy (Iseri et al., 1966; Lane and Lieber, 1966; Rubin and Lieber, 1967; Lieber and Rubin, 1968), and this reduction has now been substantiated by chemical fractionation (Ishii et al., 1973). One of the main functions of the rough endoplasmic reticulum is protein synthesis. This subject has not been extensively studied after chronic ethanol consumption, but acute ethanol administration or addition of ethanol to perfused isolated livers was found to depress protein synthesis (Rothschild et al., 1971; Jeejeebhoy et al., 1972). It is also of special interest that the ultrastructural changes found in the early stages of the alcoholic fatty liver are identical to those seen in more severe stages of alcoholic liver disease namely alcoholic hepatitis (Svoboda and Manning, 1964) which raises the question of the possible role of the fatty liver as a precursor to the hepatitis to be discussed in another paper (Lieber and DeCarli, this symposium).

II. Development of Mitochondrial Injury

Alcoholics are known to have profound mitochondrial changes in their liver (Svoboda and Manning, 1964) which are associated with increased serum activity of the intramitochondrial enzyme glutamate dehydrogenase (Konttinen et al., 1970). From these clinical observations, however, it was impossible to assess whether the mitochondrial changes were a direct result of chronic ethanol intake or were secondary to other factors such as dietary deficiencies. Recent studies have incriminated alcohol itself as the responsible agent and have clarified some functional counterparts of the ultrastructural lesions.

a. Ultrastructural changes of mitochondria. Chronic alcohol consumption results in striking mitochondrial alterations which include swelling and disfiguration of mitochondria, disorientation of the cristae, and intramitochondrial crystalline inclusions (Svoboda and Manning, 1964; Kiessling and Pilstrom, 1971). Similarly, in the rat, isocaloric substitution of ethanol for carbohydrate in otherwise adequate diets leads to enlargement and alterations of the configurations of the mitochondria (Iseri et al., 1966), indicating that ethanol itself or one of its metabolites causes the alterations rather than dietary deficiencies. Mitochondrial changes similar to those seen in chronic alcoholics were also produced by isocaloric substitution of ethanol for carbohydrate in baboons (Lieber et al., 1972) and in man, both in alcoholics (Lane and Lieber, 1966; Lieber and Rubin, 1968) and in nonalcoholics (Rubin and Lieber, 1968b). Mitochondrial alterations occurred under a variety of conditions, which included high protein, low fat, and choline-supplemented diets (Lieber and Rubin, 1968; Rubin and Lieber, 1968b). Degenerated mitochondria were conspicuous and the debris of these degraded organelles was also found within autophagic vacuoles and residual vacuolated bodies (Rubin and Lieber, 1967). The striking structural changes of the mitocondria are associated with corresponding functional abnormalities.

b. Alterations of mitochondrial functions. These injured mitochondria have a reduction in cytochrome a and b content (Rubin et al., 1970) and in succinic dehydrogenase activity (Rubin et al., 1970b; Oudea et al., 1970) although in one study (Videla and Israel, 1970) succinic dehydrogenase activity measured in total liver homogenates was reported to be increased in ethanol-fed rats. The respiratory capacity of the mitochondria was found to be depressed (Kiessling and Pilstrom, 1968; Rubin et al., 1972) using pyruvate and succinate as substrates. Other substrates were also found to have reduced oxidation by the mitochondria of ethanol-fed rats, except for α-glycerophosphate, the oxidation of which was reported by some to be increased (Kiessling, 1968) or unchanged (Pilstrom and Kiessling, 1972) whereas others found it to be decreased (Rubin et al., 1972). Oxidative phosphorylation was found to be selectively altered at site I (Cederbaum et al., 1974a). Since the structural changes of the mitochondria persist, the question arose as to whether these in turn could be responsible for some alterations in lipid metabolism beyond those which were attributed to the altered redox change (Gordon, 1972b). The first indication that ethanol consumption may result in more persistent metabolic changes arose from the observation that alcohol ingestion is associated with a progressive increase in ketonemia and ketonuria, which was most pronounced in the fasting state (Lieber et al., 1970). The ketonemia may aggravate the acidosis of the hyperlactacidemia (Lieber et al., 1963) and, on occasion, may lead to severe alcoholic ketoacidosis (Jenkins et al., 1971). The capacity for ethanol to produce ketonemia was found to be greater than that of fat itself, provided however that fat was present in the diet. Thus, fat seems to play a permissive role (Lefevre et al., 1970). Preliminary observations indicated that mitochondria obtained from ethanol-fed rats, when incubated in vitro, even in the absence of ethanol, display decreased capacity to oxidize fatty acid, but enhanced β-oxidation which is possibly responsible for the increased ketogenesis (Toth et al., 1973). Decreased fatty acid oxidation, whether as a function of the reduced citric acid cycle activity (secondary to the altered redox potential) or whether as a consequence of permanent changes in mitochondrial structure, offers the most likely explanation for the deposition of fat in the liver after chronic alcohol ingestion, especially fat derived from the diet (Lieber and Spritz, 1966; Lieber et al., 1966; Mendenhall, 1972). It is noteworthy that high concentrations of acetaldehyde, the product of ethanol metabolism, mimick the defects produced by chronic ethanol consumption on oxidative phosphorylation at site I (Cederbaum et al., 1974b). One may wonder to what extent chronic exposure to acetaldehyde is the cause for the defect observed after chronic ethanol consumption.

III. Effects of Metabolities of Ethanol: Acetaldehyde and Acetate

Acetaldehyde is the first major "specific" oxidation product of ethanol, whether the latter is oxidized by the classic alcohol dehydrogenase of the cytosol or by the more recently described microsomal system.

The exact fate of acetaldehyde is still the subject of debate. That acetyl-CoA is formed from ethanol is indicated by the observation that ethanol-C^{14} can be traced to a variety of metabolites of which acetyl-CoA is a precursor, such as fatty acids and cholesterol, as reviewed elsewhere (Lieber, 1967). It is noteworthy that a large fraction of the carbon skeleton of ethanol is incorporated in hepatic lipids after ethanol administration (Brunengraber et al., 1974; Scheig, 1971). The acetaldehyde which results from the oxidation of ethanol could be converted to acetyl-CoA via acetate. The reverse possibility, namely that ethanol is converted directly to acetyl-CoA which in turn could be either incorporated into various metabolites or yield acetate, has not been ruled out. In any event, acetate has been found to markedly increase in the blood after ethanol administration (Lundquist et al., 1962; Crouse et al., 1968).

Although in vitro, the liver can readily utilize acetate, in vivo most of the acetate is metabolized in peripheral tissues (Katz and Chaikoff, 1955). The effects of a rise of circulating acetate on intermediary metabolism in various tissues have not been defined, except for adipose tissue where it was found to be responsible, at least in part, for the decreased release of free fatty acids (FFA) and the fall of circulating FFA (Crouse et al., 1968).

The metabolic effects of acetaldehyde have long remained unclear, even though it had been speculated that this compound may contribute to the complications of alcoholism (Truitt and Duritz, 1966). Lack of information resulted from the difficulties of accurately measuring and handling acetaldehyde because of its very low concentration (except after Antabuse administration), rapid metabolism, high chemical reactivity and low boiling point (21°C). In recent years Truitt (1971; Duritz and Truitt, 1964) studied acetaldehyde levels after oral administration of ethanol to alcoholics and non alcoholics using their improved gas-chromatographic methods but no statistically significant differences were found. Freund and O'Hollaren (1965) found a plateau of acetaldehyde concentration in human alveolar air after ethanol administration, but the relation to ethanol metabolism was not clarified. Majchrowicz and Mendelson (1970) also found high acetaldehyde levels in alcoholics, but they concluded that the amount of acetaldehyde contained as a congener in alcoholic beverage was related to its concentration in the blood, because of the lack of relationship between blood ethanol and acetaldehyde concentration in their clinical study.

More recently, Korsten et al. (1974) found a relationship between blood acetaldehyde and ethanol levels in humans after intravenous administration of ethanol yielding concentrations high enough to saturate intra-hepatic ethanol oxidizing system, including ADH, MEOS and Catalase. This study revealed that the acetaldehyde level remained relatively constant despite wide variations in blood ethanol above 150 mg/100 ml, but the acetaldehyde plateau abruptly terminated when ethanol concentration reached a mean level of 110 mg/100 ml, a concentration which corresponds to MEOS desaturation (Figure 6). Moreover, the plateau level of acetaldehyde was significantly higher in alcoholics (42. 7 $\pm$ 1.2 nmole/ml than in non-alcoholics (26.5 $\pm$ 1.5) (Figure 7).

The finding that at high blood ethanol concentration blood acetaldehyde is maintained at a relatively constant level implies that, under these conditions, release of acetaldehyde from the liver and disappearance of acetaldehyde from the blood must be equal. Since the concentrations of acetaldehyde observed during this plateau were far above those needed to saturate the major pathway for acetaldehyde dehydrogenation (Marjanen, 1972; Tottmar et al., 1973), acetaldehyde metabolism must be constant, and it follows that acetaldehyde production must also be constant. If the latter is true, it implies that the enzyme system involved in ethanol oxidation is saturated and maintains a constant reaction velocity independent of substrate levels. However, this study revealed that at a mean ethanol level of 110 mg/100 ml, acetaldehyde levels abruptly decreased. Acetaldehyde disappearance should still be constant, since the acetaldehyde concentrations were still far above those needed to saturate acetaldehyde dehydrogenation. Since the acetaldehyde levels drop at this point, it follows that production of acetaldehyde decreased. The main pathway for hepatic ethanol oxidation to acetaldehyde proceeds via alcohol dehydrogenase (ADH). However, given the K_m of alcohol dehydrogenase (.5-2 mM) (Makar and Mannering, 1970; Reynier, 1969), it is obvious that at an ethanol concentration of 110 mg/100 ml (24 mM), ADH is still fully saturated. Another hepatic oxidizing system has been described (the microsomal ethanol oxidizing system or MEOS), which has a K_m of about 8 mM for ethanol in microsomes

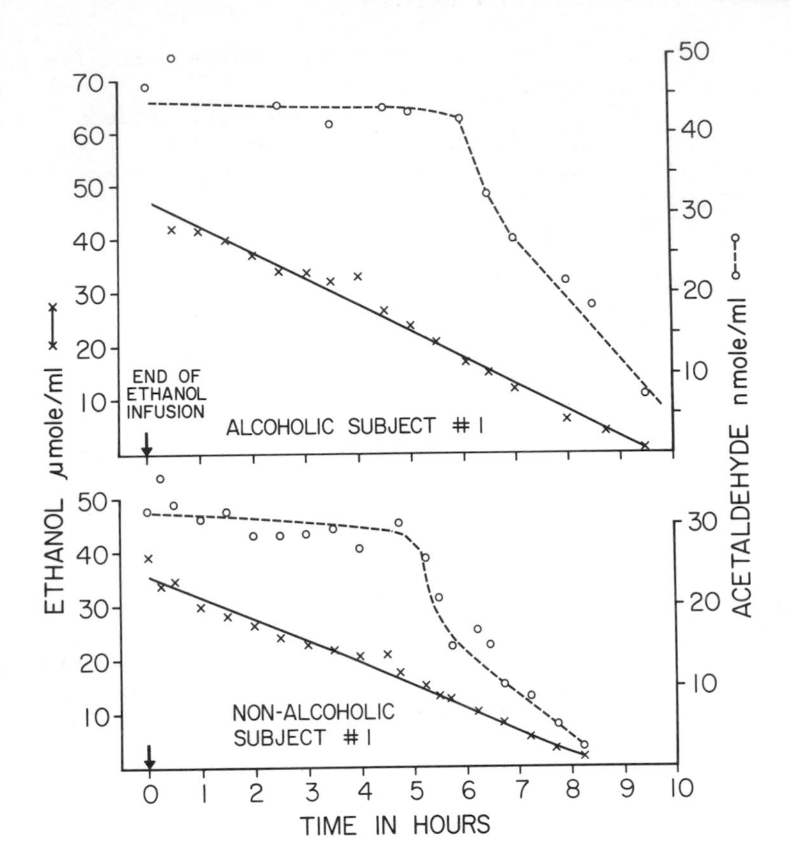

Figure 6.
Blood acetaldehyde and ethanol levels following intravenous alcohol infusion in an alcoholic and a non-alcoholic subject. The plateau acetaldehyde level of the alcoholic fluctuated around a higher mean than that of the non-alcoholic. In both, however, acetaldehyde sharply declined at an ethanol concentration of 18-20 μmole/ml. Data from Korsten et al. (1974).

and 8.8 mM in vivo (Lieber and DeCarli, 1970b, 1972). The activity of such a system can be expected to decrease at blood ethanol levels associated with the drop in blood acetaldehyde. Catalase represents a second possible non-ADH pathway for ethanol oxidation because of its similar K_m, but its capacity is limited as discussed before.

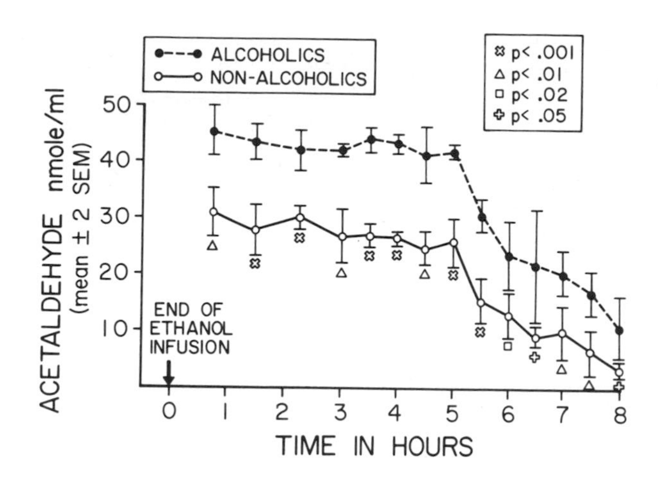

Figure 7.
Comparison of blood acetaldehyde levels of alcoholic and non-alcoholic subjects following i.v. alcohol infusion. The significance level of the difference of the means is noted. Data from Korsten et al. (1974).

Thus, the observed abrupt fall in blood acetaldehyde concentration at blood ethanol levels which are fully saturating for ADH suggests the in vivo operation of a non-ADH pathway of alcohol metabolism. This is substantiated by the kinetics of the ethanol disappearance curve seen in rats after pyrazole treatment (Lieber and DeCarli, 1972).

The microsomal ethanol oxidizing system has been shown to increase in activity after chronic alcohol feeding. This induction may explain, at least in part, the higher plateau levels in alcoholics. On the other hand, alcohol related mitochondrial damage in the liver described before could also explain this higher level of acetaldehyde in alcoholics through decreased disposition. Whether ethanol consumption alters the activity of acetaldehyde dehydrogenase is the subject of debate, with both increases (Dajani et al., 1963; Horton, 1971) and no changes (Raskin and Sokoloff, 1972; Redmond and Cohen, 1971) reported. However, in our laboratory, decreased rates of acetaldehyde dehydrogenation were observed using intact mitochondria of chronically alcohol fed rats, in spite of the increase of enzymatic activity after the destruction of the mitochondrial membrane (Hasumura, Teschke and Lieber, unpublished observation).

It follows that acetaldehyde dehydrogenation which is predominantly intramitochondrial (Grunnet, 1973; Marjanen, 1972), and the reoxidation of the co-factor NADH to NAD (which is rate-limiting in alcohol dehydrogenase catalyzed ethanol oxidation (Theorell and Bonnichsen, 1951) and is also, in part, dependent on mitochondrial integrity) might both be impaired by chronic alcohol consumption. Thus, on the one hand, defective acetaldehyde dehydrogenation may retard acetaldehyde clearance. The higher acetaldehyde concentrations that result may in turn enhance the functional disturbance of the mitochondria (by reducing the activity of various shuttles involved in the disposition of reducing equivalents and by inhibiting oxidative phosphorylation (Cederbaum et al. 1974b). On the other hand, the rate of ADH catalyzed ethanol oxidation might be inhibited by failure to reoxidize NADH by the usual intramitochondrial routes. In this respect induction of non-ADH pathways would be considered compensatory. Indeed, such induction may explain why in other human studies reported (Kater et al., 1969a; Mezey and Tobon, 1971), the rate of ethanol clearance of alcoholics is either equal to or faster than that of non-alcoholics. This "induction" may be the source of greater production of the toxic metabolite (acetaldehyde) resulting in a "vicious cycle" (Figure 8): acetaldehyde causes mitochondrial dysfunction which in turn promotes higher acetaldehyde levels; this circular process could result in progressive liver damage.

These findings increase our understanding not only of ethanol metabolism but possibly of the pathogenesis of some of its complications. Numerous toxic effects have been attributed to acetaldehyde (Walsh, 1971). In addition to the release of catecholamines (Eade, 1959), acetaldehyde has been shown to participate in and to favor the condensation reactions of biogenic amines (Davis and Walsh, 1970; Cohen and Collins, 1970). The products of these interactions could have addictive properties if sufficient amounts were generated in vivo. Finally, myocardial protein synthesis was impaired by acetaldehyde (Schreiber et al., 1972, 1974) at concentrations comparable to those found in our study; this effect might contribute to the development of alcoholic cardiomyopathy. Thus, the high levels of blood acetaldehyde found at high alcohol concentrations may have fundamental pathogenetic consequences in the alcoholic since he has significantly higher blood acetaldehyde levels than the non-alcoholic. It is conceivable that the decreased capacity of injured mitochondria to dispose of the toxic acetaldehyde may initiate a self sustaining mechanism of liver injury by promoting the accumulation of acetaldehyde which aggravates the mitochondrial dysfunction. As discussed elsewhere in this symposium (Lieber and DeCarli), the organelle abnormalities of the fatty liver are followed by a stage of cell necrosis (alcoholic hepatitis) and is, conceivably, its precursor. The hepatitis, in turn, may trigger the development of cirrhosis.

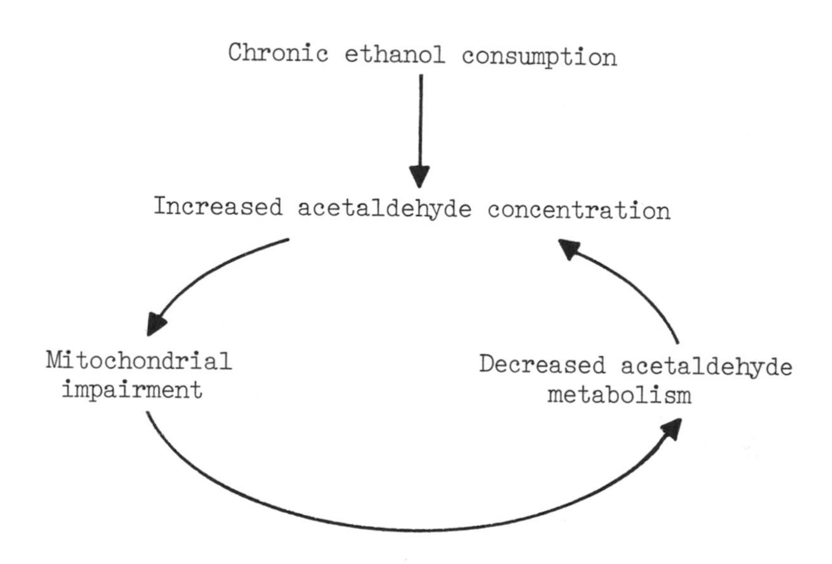

Figure 8.
Possible relationship between ethanol metabolism, altered hepatic acetaldehyde levels and mitochondrial impairment.

REFERENCE

Ariyoshi, T., Takabatake, E. and Remmer, H. 1970, Life Sci. 9:361.

Baraona, E. and Lieber, C.S. 1969, J. Clin. Invest. 49:769.

Baraona, E., Pirola, R.C. and Lieber, C.S. 1973, J. Clin. Invest. 52:296.

Barboriak, J.J. and Meade, R.C. 1968, Amer. J. Med. Sci. 255:245.

Bartlett, G.R. 1952, Quart. J. Stud. Alc. 13:583.

Bernstein, J., Videla, L. and Israel, Y. 1973, Biochem. J. 134:515.

Blomstrand, R. and Kager, L. 1973, Life Sci. 13:113.

Blomstrand, R., Kager, L. and Lantto, O. 1973, Life Sci. 13:1131.

Boveris, A., Oshino, N. and Chance, B. 1972, Biochem. J. 128:617.

Brewster, A.C., Lankford, H.G., Schwartz, M.G. and Sullivan, J.F. 1966, Amer. J. Clin. 19:255.

Brunengraber, H., Boutry, M., Lowenstein, L. and Lowenstein, J.M. Alcohol and Aldehyde Metabolizing Systems. edited by R.C. Thurman, T. Yonetani, J.R. Williamson and B. Chance. New York, Academic Press, 1974, p. 329.

Carter, E.A. and Isselbacher, K.J. 1971, Ann. N.Y. Acad. Sci. 179: 282.

Carulli, N., Manenti, F., Gallo, M. and Salvioli, G.F. 1971, Eur. J. Clin. Invest. 1:421.

Cederbaum, A.I., Lieber, C.S., Toth, A., Beattie, D.S. and Rubin, E. 1973, J. Biol. Chem. 248:4977.

Cederbaum, A.I., Lieber, C.S. and Rubin, E. 1974a, Arch. Biochem. Biophys. in press.

Cederbaum, A.I., Lieber, C.S. and Rubin, E. 1974b, Arch. Biochem. Biophys. 161:26.

Chait, A., February, A.W., Mancini, M. and Lewis, B.L. 1972, Lancet ii:62.

Chance, B. 1947, Acta Chem. Scand. 1:236.

Chance, B. and Oshino, N. 1971, Biochem. J. 122:225.

Cohen, B.S. and Estabrook, R.W. 1971, Arch. Biochem. Biophys. 143: 54.

Cohen, G. and Collins, M. 1970, Science 167:1749.

Cohen, G.M. and Mannering, G.J. 1973, Molec. Pharmacol. 9:383.

Conney, A.H. 1967, Pharmacol. Rev. 19:317.

Crouse, J.R., Gerson, C.D., DeCarli, L.M. and Lieber, C.S. 1968, J. Lipid Res. 9:509.

Dajani, R.M. and Kouyoumjian, C.S. 1967, J. Nutr. 91:535

Dajani, R.M., Danielski, J. and Orten, J.M. 1963, J. Nutr. 80:196.

Davis, V.E. and Walsh, M.J. 1970, Science 167:1749.

deSaint-Blanquat, G., Fritsch, P. and Derache, R. 1972, Path. Biol. 20:249.

Diaz Gomez, M.I., Castro, J.A., deFerreyra, E.C., D'Actosa, N. and deCastro, C.R. 1973, Toxicol. Appl. Pharmacol. 25:534.

Domschke, S., Domschke, W. and Lieber, C.S. 1974, Life Sci. in press.

Duritz, G. and Truitt, E.B. 1964, Quart. J. Stud. Alc. 25:498.

Eade, N.R. 1959, J. Pharmacol. Exp. Ther. 127:29.

Feinman, L. and Lieber, C.S. 1972, Science 176:795.

Feytmans, E. and Leighton, F. 1973, Biochem. Pharmacol. 22:349.

Fischer, H.-D. 1962, Biochem. Pharmacol. 11:307.

Forney, R.B. and Hughes, F.W. 1968, Combined Effects of Alcohol and Other Drugs. Springfield, Illinois, C.C. Thomas.

Forsander, O.A., Maenpaa, P.H. and Salaspuro, M.P. 1965, Acta Chem. Scand. 19:1770.

Freund, G. and O'Hallaren, P. 1965, J. Lipid Res. 6:471.

Gang, H., Lieber, C.S. and Rubin, E. 1973, Nature (New Biology) 243:123.

Garner, R.C. and McLean, A.E.M. 1969, Biochem. Pharmacol. 18:645.

Gillette, J.R., Brodie, B.B. and La Du, B.N. 1957, J. Pharmacol.

Exp. Ther. 119:532.

Ginsberg, H., Olefsky, J., Farquhar, J.W. and Reaven, G.M. 1974, Ann. Intern. Med. 80:143.

Gordon, E.R. 1972a, Biochem. Pharmcol. 21:2991.

Gordon, E.R. 1972b, Canad. J. Biochem. 50:949.

Grunnet, N. 1973, Eur. J. Biochem. 35:236.

Grunnet, N. and Thieden, H.I.D. 1972, Life Sci. (Part II) 11:983.

Grunnet, N., Quistorf, B. and Thieden, H.I.D. 1973, Eur. J. Biochem. 40:275.

Hasumura, Y., Teschke, R. and Lieber, C.S. 1974a, Gastroenterology 66:415.

Hasumura, Y., Teschke, R. and Lieber, C.S. 1974b, Clin. Res. 22:360.

Hawkins, R.D., Kalant, H. and Khanna, J.M. 1966, Canad. J. Physiol. Pharmacol. 44:241.

Hildebrandt, A.B. and Speck, M. 1973, Arch. Pharmacol. 277:165.

Hildebrandt, A.G., Speck, M. and Roots, I. 1974, Naunyn-Schmiedeberg's Arch. Pharmacol. 281:371.

Horton, A.A. 1971, Biochim. Biophys. Acta 253:514.

Iseri, O.A., Gottlieb, L.S. and Lieber, C.S. 1964, Fed. Proc. 23:579.

Iseri, O.A., Lieber, C.S. and Gottlieb, L.S. 1966, Amer. J. Path. 48:535.

Ishii, H., Joly, J.-G. and Lieber, C.S. 1973, Biochim. Biophys. Acta 291:411.

Israel, Y., Videla, L., MacDonald, A. and Bernstein, J. 1973, Biochem. J. 134:532.

Isselbacher, K.J. and Carter, E.A. 1970, Biochem. Biophys. Res. Commun. 39:530.

Jeejeebhoy, K.N., Phillips, M.J., Bruce-Robertson. A., Ho, J. and Sodtke, U. 1972, Biochem. J. 126:1111.

Jenkins, D.W., Eckel, R.W. and Craig, J.W. 1971, J A M A 217:177.

Joly, J.-G., Ishii, H., Teschke, R., Hasumura, Y. and Lieber, C.S. 1973a, Biochem. Pharmacol. 22:1532.

Joly, J.-G., Feinman, L., Ishii, H. and Lieber, C.S. 1973b, J. Lipid Res. 14:337.

Kalant, H., Khanna, J.M. and Marshman, J. 1970, J. Pharmacol. Exp. Ther. 175:318.

Kater, R.M.H., Carulli, N. and Iber, F.L. 1969a, Amer. J. Clin. Nutr. 22:1608.

Kater, R.M.H., Roggin, G., Tobon, F., Zieve, P. and Iber, F.L. 1969b, Amer. J. Med. Sci. 258:35.

Kato, R. 1967, Jap. J. Pharmacol. 17:56.

Katz, J. and Chaikoff, I.L. 1955, Biochim. Biophys. Acta 18:87.

Keilin, D. and Hartree, E.F. 1945, Biochem. J. 39:293.

Kessler, J.I., Miller, M., Barza, D. and Mishkin, S. 1967, Amer. J. Med. 42:968.

Khanna, J.M., Kalant, H. and Lin, G. 1970, Biochem. Pharmacol. 19:2493.

Kiessling, K.H. 1968, Acta Pharmacol. et. Toxicol. 26:245.

Kiessling, K.H. and Pilstrom, L. 1968, Quart. J. Stud. Alcohol 29:819.

Kiessling, K.H. and Pilstrom, L. 1971, Cytobiologie 4:339.

Kinard, F.W., Nelson, G.H. and Hay, M.G. 1956, Proc. Soc. Exp. Biol. Med. 92:772.

Klaassen, C.D. 1969, Proc. Soc. Exp. Biol. Med. 132:1099.

Koga, S. and Hirayama, C. 1968, Experientia 24:438.

Konttinen, A., Hartel, G. and Louhija, A. 1970, Acta Med. Scand. 188:257.

Korsten, M., Matsuzaki, S., Feinman, L. and Lieber, C.S. 1974, New Eng. J. Med. in press.

Kudzma, D.J. and Schonfeld, G. 1971, J. Lab. Clin. Med. 77:384.

Lane, B.P. and Lieber, C.S. 1966, Amer. J. Path. 49:593.

Lane, B.P. and Lieber, C.S. 1967, Lab. Invest. 16:342.

Lefevre, A.F., Adler, H. and Lieber, C.S. 1970, J. Clin. Invest. 49:1775.

Lefevre, A.F., DeCarli, L.M. and Lieber, C.S. 1972, J. Lipid Res. 13:48.

Lester, D. and Benson, G.D. 1970, Science 169:282.

Lieber, C.S. 1967, Ann. Rev. Med. 18:35.

Lieber, C.S. 1970, Gastroenterology 59:930.

Lieber, C.S. 1973, Gastroenterology 65:821.

Lieber, C.S. and Davidson, C.S. 1962, Amer. J. Med. 33:319.

Lieber, C.S. and DeCarli, L.M. 1968, Science 162:917.

Lieber, C.S. and DeCarli, L.M. 1970a, Science 170:78.

Lieber, C.S. and DeCarli, L.M. 1970b, J. Biol. Chem. 245:2505.

Lieber, C.S. and DeCarli, L.M. 1972, J. Pharmacol. Exp. Ther. 181:279.

Lieber, C.S. and DeCarli, L.M. 1973, Drug Metabolism and Disposition 1:428.

Lieber, C.S. and DeCarli, L.M. 1974, Biochem. Biophys. Res. Commun. in press.

Lieber, C.S. and Rubin, E. 1968, Amer. J. Med. 44:200.

Lieber, C.S. and Schmid, R. 1961, J. Clin. Invest. 40:394.

Lieber, C.S. and Spritz, N. 1966, J. Clin. Invest. 45:1400.

Lieber, C.S., Jones, D.P., Losowsky, M.S. and Davidson, C.S. 1962a, J. Clin.Invest. 41:1863.

Lieber, C.S., Leevy, C.M., Stein, S.W., George, W.S., Cherrick, G.R., Abelmann, W.H. and Davidson, C.S. 1962b, J. Lab. Clin. Med. 59:826.

Lieber, C.S., Jones, D.P., Mendelson, J. and DeCarli, L.M. 1963, Trans. Ass. Amer. Physicians 76:289.

Lieber, C.S., Jones, D.P. and DeCarli, L.M. 1965, J. Clin. Invest. 44:1009.

Lieber, C.S., Spritz, N. and DeCarli, L.M. 1966, J. Clin. Invest. 45:51.

Lieber, C.S., Lefevre, A.F., Spritz, N., Feinman, L. and DeCarli, L.M. 1967, J. Clin. Invest. 46:1451.

Lieber, C.S., Spritz, N. and DeCarli, L.M. 1969, J. Lipid Res. 10:283.

Lieber, C.S., Rubin, E. and DeCarli, L.M. 1970, Biochem. Biophys. Res. Commun. 40:858.

Lieber, C.S., DeCarli, L.M., Gang, H., Walker, G. and Rubin, E. Medical Primatology-1972. edited by E.I. Goldsmith and J. Moor-Jankowski, Basel, S. Karger, Part 3, 1972, p. 270.

Losowsky, M.S., Jones, D.P., Davidson, C.S. and Lieber, C.S. 1963, Amer. J. Med. 35:794.

Lu, A.Y.H., Strobel, H.W. and Coon, M.J. 1969, Biochem. Biophys. Res. Commun. 36:545.

Lück, H. Methods of Enzymatic Analysis. edited by H.U. Bergmeyer, New York, Academic Press, 1963, p. 885.

Lundquist, F., Tygstrup, N., Winkler, K., Mellemgaard, K. and Munck-Petersen, S. 1962, J. Clin. Invest. 41:955.

Madsen, N.P. 1969, Biochem. Pharmacol. 18:261.

Majchrowicz, E. and Mendelson, J.H. 1970, Science 168:1100.

Makar, A.B. and Mannering, G.J. 1970, Biochem. Pharmacol. 19:2017.

Marjanen, L. 1972, Biochem. J. 127:633.

McKenna, W.R. and Thurman, R.G. 1974, Fed. Proc. 33:554.

Meldolesi, J. 1967, Biochem. Pharmacol. 16:125.

Mendelson, J.H. and Mello, N.K. 1973, Science 180:1372.

Mendenhall, C.L. 1972, J. Lipid Res. 13:177.

Mendenhall, C.L., Bradford, R.H. and Furman, R. 1969, Biochim. Biophys. Acta 187:501.

Mezey, E. and Robles, E.A. 1974, Gastroenterology 66:248.

Mezey, E. and Tobon, F. 1971, Gastroenterology 61:707.

Mezey, E., Potter, J.J. and Reed, W.D. 1973, J. Biol. Chem. 248:1183.

Misra, P.S., Lefevre, A., Ishii, H., Rubin, E. and Lieber, C.S. 1971, Amer. J. Med. 51:346.

Mistilis, S.P. and Garske, A. 1969, Aust. Ann. Med. 18:227.

Mookerjea, S. and Chow, A. 1969, Biochim. Biophys. Acta 184:83.

Moon, H.D. 1950, Amer. J. Path. 26:1041.

Newcombe, D.S. 1972, Metabolism 21:1193.

Nikkila, E.A. and Ojala, K. 1963, Proc. Soc. Exp. Biol. Med. 113:814.

Ontko, J.A. 1973, J. Lipid Res. 14:78.

Orme-Johnson, W.H. and Ziegler, D.M. 1965, Biochem. Biophys. Res. Commun. 21:78.

Oshino, N., Chance, B., Sies, H. and Bucher, T. 1973a, Arch. Biochem. Biophys. 154:117.

Oshino, N., Oshino, R. and Chance, B. 1973b, Biochem. J. 131:555.

Oudea, M.C., Launay, A.N., Oueneherne, S. and Oudea, P. 1970, Rev. Europ. Etudes Clin. et. Biol. XV:48.

Papenberg, J. 1971, In: Alcohol and the Liver. W. Gerok, K. Sickinger and H. Hennekeuser: Schattauer Verlag, New York, p. 45.

Papenberg, J., von Wartburg, J.P. and Aebi, H. 1970, Enzym. Biol. Clin. 11:237.

Pilstrom, L. and Kiessling, K.H. 1972, Histochemie 32:329.

Raskin, N.H. and Sokoloff, L. 1972, Nature 236:138.

Ratcliffe, F. 1969, Life Sci. 8:1051.

Redman, C.M., Grab, D.J. and Irukulla, R. 1972, Arch. Biochem. Biophys. 152:496.

Redmond, G. and Cohen, G. 1971, Science 171:387.

Reynier, M. 1969, Acta Chem. Scand. 23:1119.

Roach, M.K., Reese, W.N. and Creaven, P.J. 1969, Biochem. Biophys. Res. Commun. 36:596.

Roach, M.K., Khan, M., Knapp, M. and Reese, W.N. 1972, Quart. J. Stud. Alcohol 33:751.

Rognstad, R. and Clark, D.G. 1974, Eur. J. Biochem. 42:51.

Rothschild, M., Oratz, M., Mongelli, J. and Schreiber, S.S. 1971, J. Clin. Invest. 50:1812.

Rubin, E. and Lieber, C.S. 1967, Gastroenterology 52:1.

Rubin, E. and Lieber, C.S. 1968a, Science 162:690.

Rubin, E. and Lieber, C.S. 1968b, New Eng. J. Med. 278:869.

Rubin, E., Hutterer, F. and Lieber, C.S. 1968, Science 159:1469.

Rubin, E., Gang, H., Misra, P.S. and Lieber, C.S. 1970a, Amer. J. Med. 49:801.

Rubin, E., Beattie, D.S. and Lieber, C.S. 1970b, Lab. Invest. 23:620.

Rubin, E., Beattie, D.S., Toth, A. and Lieber, C.S. 1972, Fed. Proc. 31:131.

Schapiro, R.H., Drummey, G.D., Shimizu, Y. and Isselbacher, K.J. 1964, J. Clin. Invest. 43:1338.

Scheig, R. 1971, Gastroenterology 60:751.

Scholz, R., Hansen, W. and Thurman, R.G. 1971, In: Metabolic changes induced by Alcohol. edited by G.A. Martini and Ch. Bode. Springer Verlag, New York, p. 101.

Schreiber, S.S., Briden, K., Oratz, M. and Rothschild, M.A. 1972, J. Clin. Invest. 51:2820.

Schreiber, S.S., Oratz, M., Rothschild, M.A., Reff, F. and Evans, C. 1974, J. Molec. Cell. Cardiol. 6:207.

Schüppel, R. Alcohol and the Liver. edited by W. Gerok, K. Sickinger and H.H. Hennekeuser. New York, Schattauer Verlag, 1971, p. 227.

Smith, M.E. 1961, J. Pharmacol. 134:233.

Soehring, K. and Schüppel, R. 1966, Deutsch Med. Wschr. 91:1892.

Sutherland, V.C., Burbridge, T.N., Adams, J.E. and Simon, A. 1960, J. Appl. Physiol. 15:189.

Svoboda, D.J. and Manning, R.T. 1964, Amer. J. Pathol. 44:645.

Tephly, T.R., Finelli, F. and Watkins, W.D. 1969, Science 166:627.

Teschke, R., Hasumura, Y., Joly, J.-G., Ishii, H. and Lieber, C.S. 1972, Biochem. Biophys. Res. Commun. 49:1187.

Teschke, R., Hasumura, Y. and Lieber, C.S. 1974a, Arch. Biochem. Biophys. 163:404.

Teschke, R., Hasumura, Y. and Lieber, C.S. 1974b, Biochem. Biophys. Res. Commun. 60:851.

Teschke, R., Hasumura, Y. and Lieber, C.S. 1974c, Clin. Res. 22:324.

Theorell, H. and Bonnichsen, R. 1951, Acta Chem. Scand. 5:1105.

Theorell, H., Chance, B., Yonetani, T. and Oshino, N. 1974, Arch. Biochem. Biophys. 151:434.

Thieden, H.I.D. 1971, Acta Chem. Scand. 25:3421.

Thurman, R.G. and Scholz, R. 1973, Drug Metabolism and 1:441.

Thurman, R.G., Ley, H.G. and Scholz, R. 1972, Eur. J. Biochem. 25:420.

Thurman, R.G., Hesse, S. and Scholz, R. Alcohol and aldehyde metabolizing systems. edited by R.G. Thurman, T. Yonetani, J.R. Williamson and B. Chance. New York, Academic Press, 1974 a, p. 257.

Thurman, R.G., McKenna, W.R., Brentzel, H.J. and Hesse, S. 1974 b, Fed. Proc. in press.

Tobon, F. and Mezey, E. 1971, J. Lab. Clin. Med. 77:110.

Toth, A., Lieber, C.S., Cederbaum, A.I., Beattie, D.S. and Rubin, E. 1973, Gastroenterology 64:198.

Tottmar, S.O.C., Pettersson, H. and Kiessling, K.H. 1973, Biochem. J. 135:577.

Traiger, G.J. and Plaa, G.L. 1972, J. Pharmacol. Exp. Ther. 183:481.

Truitt, E.B. Biological Aspects of Alcohol. edited by M.K. Roach, W.M. McIsaac and P.J. Creaven. Austin and London, University of Texas Press, 1971 p. 212.

Truitt, E.B. and Durtiz, G. Biochemical Factors in Alcoholism.

edited by R.P. Maickel, Elmsford, New York, Pergamon Press, 1966, p. 61.

Ugarte, G., Pereda, T., Pino, M.E. and Iturriaga, H. 1972, Quart. J. Stud. Alcohol 33:698.

Ugarte, G., Pino, M.E. and Insunza, I. 1967, Amer. J. Dig. Dis. 12:589.

Vatsis, K.P. and Schulman, M.P. 1973, Biochem. Biophys. Res. Commun. 52:588.

Vatsis, K.P. and Schulman, M.P. 1974, Fed. Proc. 33:554.

Veech, R.L., Eggleston, L.V. and Krebs, H.A. 1969, Biochem. J. 115:609.

Verdy, M. and Gattereau, A. 1967, Amer. J. Clin. Nutr. 20:997.

Vesell, E.S., Page, J.G. and Passananti, G.T. 1971, Clin. Pharmacol. Ther. 12:192.

Videla, L. and Israel, Y. 1970, Biochem. J. 118:275.

von Wartburg, J.P. and Rothlisberger, M. 1961, Helv. Physiol. Acta 19:30.

Walsh, M.J. Biological Aspects of Alcohol. edited by M.K. Roach, W.M. McIsaac and P.J. Creaven. Austin and London, University of Texas Press, 1971, p. 233.

Wilson, D.E., Schreibman, P.H., Brewster, A.C. and Arky, R.A. 1970, J. Lab. Clin. Med. 75:264.

THE EFFECTS OF ETHANOL ON TRYPTOPHAN PYRROLASE ACTIVITY AND THEIR COMPARISON WITH THOSE OF PHENOBARBITONE AND MORPHINE

A. A.-B. Badawy and M. Evans

University Hospital of Wales, Addiction Unit Research Laboratory, Whitchurch Hospital Cardiff CF4 7XB, U.K.

Tryptophan pyrrolase (L-tryptophan-O_2 oxidoreductase EC 1.13.11.11) is the first and rate-limiting haem-dependent enzyme of the hepatic kynurenine-nicotinic acid pathway of tryptophan degradation. The importance of this pathway to general body metabolism is evident from the fact that its end-products are the important redox cofactors NAD^+ and $NADP^+$. It is also important to the brain because, being the quantitatively most important of all the tryptophan metabolic pathways, its activity can determine the extent of the brain uptake of the amino acid precursor of 5-hydroxytryptamine (5-HT or serotonin), and there is evidence of an inverse relation between liver tryptophan pyrrolase activity and brain 5-HT concentration (Curzon, 1969). The role of 5-HT in the regulation of mood (Curzon, 1969; Lapin & Oxenkrug, 1969), and its implication in the actions of alcohol have been previously discussed (Badawy & Evans, 1973b).

We have previously reported the activation of rat liver tryptophan pyrrolase by acute (single dose) ethanol administration, and its inhibition followed by induction respectively after chronic treatment and subsequent withdrawal of this drug (Badawy & Evans, 1973a,b). Similar effects to those of ethanol were also reported (Badawy & Evans, 1973d) for chronic phenobarbitone administration and subsequent withdrawal. In the present paper, the possible mechanisms for all these changes are examined further, new experiments with morphine are reported for the first time and possible common mechanisms of action of these three drugs of dependence are suggested.

MATERIALS AND METHODS

Chemicals

The sources of various chemicals and the administration of some of them have been described previously (Badawy & Evans, 1973 a,b,c,d,e).

Animals

Male Wistar rats (150-250g) and male golden hamsters (84-129g) were fed cube diet M.R.C. no. 41B and water ad libitum whereas male Dunkin-Hartley albino guinea-pigs (326-842g) were maintained on diet TR2 and hay. Rats used for chronic studies weighed 150g each at the beginning of the experiment. All animals were killed between 12.00h and 15.00h by stunning and cervical fracture.

Chronic Administration and Withdrawal of Drugs

All three drugs of dependence were freely given in drinking water in increasing concentrations to prevent tolerance. Withdrawal was started by replacing each drug solution with drinking water. Ethanol was administered in increasing doses (48h apart) of 5%, 7.5% and finally 10% (v/v). Rats were then given this latter concentration for the remainder of the experimental period. Phenobarbitone sodium was given in increasing concentrations (48h apart) of 1, 1.5, 2, 2.5 and finally 3mg/ml of drinking water, and the animals were maintained thereafter on the latter concentration. Morphine sulphate was administered in amounts (48h apart) of 0.1, 0.15, 0.2, 0.3 and 0.4mg/ml. Rats were then given the last concentration till the end of the study period.

Chemical and Enzymic Determinations

Tryptophan pyrrolase activity was determined in liver homogenates by measuring the formation of kynurenine from L-tryptophan (Feigelson & Greengard, 1961) in either the absence (holoenzyme activity) or the presence (total enzyme activity) of added haematin as detailed previously (Badawy & Evans, 1973e,1974). The apoenzyme activity, calculated by difference, was then used to measure the ratio of holoenzyme/apoenzyme activities. This ratio is indicative of the degree of haem saturation of the apoenzyme. Serum free fatty acids were determined by the method of Mikac-Dević et al.(1973) and serum and liver tryptophan by that of Denkla & Dewey (1967). The procedure described by the latter authors was used to prepare the serum total (acid-soluble) tryptophan, while liver total tryptophan was prepared by the method of Badawy & Smith (1972) who also described the preparation, by ultrafiltration, of serum free (non-protein-bound) tryptophan. Direct determination of NAD^+, NADH, $NADP^+$ and NADPH concentrations was performed on pieces of liver as described by Slater & Sawyer (1962) and Slater et al. (1964).

RESULTS

1. Acute Effects of Drugs on Tryptophan Pyrrolase Activity

Ethanol: The time-course of the effects of a single intraperitoneal injection of ethanol (5ml/kg) on rat liver tryptophan pyrrolase activity is shown in Fig.1. Three changes can be observed: an initial (5min) doubling of both holoenzyme and total pyrrolase activities, a remarkable degree of haem saturation of the apoenzyme (at 3h) and finally an overall increase in both enzyme activities with a peak at 5h after the injection of ethanol.

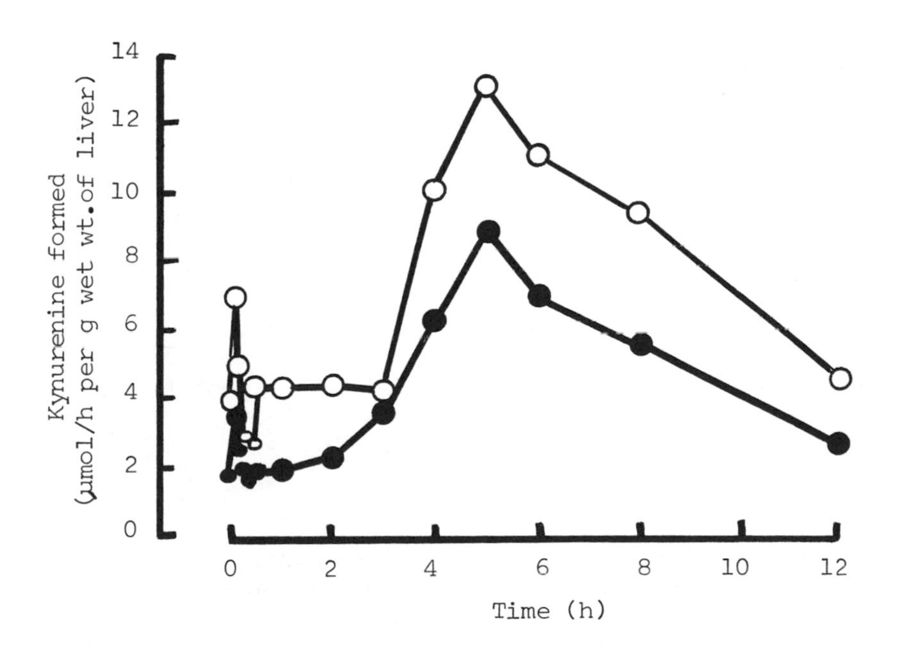

Fig.1. Effects of a single ethanol injection on rat liver tryptophan pyrrolase activity

Ethanol (5ml/kg) was injected intraperitoneally at zero-time. Each point represents the mean value for four rats except that the zero-time points are means for 20 animals. O, Total enzyme activity; ●, holoenzyme activity (reproduced from Badawy & Evans, 1973a)

Table 1. Effects of administration of ethanol, acetaldehyde, norepinephrine, epinephrine or palmitic acid on rat liver tryptophan pyrrolase activity.

Each rat received, 4h before being killed, an intraperitoneal injection of either 0.9% NaCl, dimethylformamide (0.1ml/100g), ethanol (5ml/kg as a 25% v/v solution in 0.9%NaCl), acetaldehyde (156mg/kg in 0.9% NaCl), norepinephrine or epinephrine (2mg base each/kg, in 0.9% NaCl) or palmitic acid (100mg/kg body wt. in dimethylformamide; 0.1ml/100g). The pyrrolase was determined in either the absence (holoenzyme activity) or in the presence (total enzyme activity) of added haematin (2μM). Values are means±S.E.M. of each group of four rats.

	Kynurenine formed (μmol/h per g wet wt.of liver)		
Treatment	Holoenzyme activity	Total enzyme activity	Holoenzyme/apo-enzyme ratio
0.9% NaCl	1.7±0.2	4.6±0.4	0.59
Dimethylformamide	1.7±0.2	4.7±0.4	0.57
Ethanol	5.1±0.5	8.3±0.5	1.59
Acetaldehyde	6.2±0.7	9.8±0.5	1.72
Norepinephrine	4.8±0.6	5.8±0.7	4.80
Epinephrine	4.4±0.6	6.0±0.7	2.75
Palmitic acid	6.1±0.4	9.2±1.0	1.97

Since ethanol has previously been found to have no effect on the pyrrolase activities in vitro, a number of substances which are formed from, or released by, ethanol have been tested as possible candidates for mediation of the ethanol effect (Table 1). Acetaldehyde, norepinephrine, epinephrine and palmitic acid all activated the enzyme by an ethanol-like substrate- or cofactor-type mechanism involving an increased haem saturation of the apo-enzyme.

The possible role of these substances in mediating the ethanol effect was examined by pretreatment of rats with various drugs (Table 2). Pyrazole, an alcohol dehydrogenase inhibitor, did not prevent the ethanol activation of tryptophan pyrrolase whereas the aldehyde dehydrogenase inhibitor disulfiram, which had an effect of its own, caused a moderate (33%) potentiation of the ethanol effect. Depletion of catecholamines by reserpine completely prevented the effect of ethanol. Reserpine increased the holoenzyme activity by approximately threefold by a hormonal mechanism, which is consistent with reserpine's known activating effect on the pituitary-adrenal axis. The expected additive effect of ethanol as a substrate-type activator did not occur, presumably because reserpine depleted catecholamines.

Table 2. Effects of pretreatment of rats with various drugs on the activation of rat liver tryptophan pyrrolase by ethanol.

Each rat received, 4h before being killed, an intraperitoneal injection of either 0.9% NaCl or ethanol (5ml/kg body wt.) Pyrazole (272mg/kg) and ergotamine (2.5mg/kg) were given 20 and 30 min respectively before ethanol or 0.9% NaCl. Disulfiram (100mg/kg), dibenamine (50mg/kg), dibenzyline (10mg/kg) and theophylline (100mg/kg) were injected 1h before ethanol or 0.9% NaCl. Propranolol was given at -2, 0 and +2h (20mg/kg each dose) whereas reserpine (20mg/kg) was given 24h before ethanol or 0.9% NaCl. For simplicity, only the holoenzyme activity is shown. Values are means±S.E.M. of each group of four rats.

	Kynurenine formed (μmol/h per g.wet wt. of liver)		Inhibition of the ethanol activation (%)
Pretreatment	0.9% NaCl	Ethanol	
Nil	1.7±0.2	5.1±0.5	-
Pyrazole	1.4±0.1	6.0±0.9	0
Disulfiram	3.4±0.2	7.9±0.9	0
Reserpine	4.6±0.3	4.6±0.2	100
Propranolol	1.9±0.3	5.2±0.3	3
Dibenamine	4.9±0.5	5.7±0.7	76
Dibenzyline	4.3±0.6	4.6±0.2	91
Ergotamine	2.8±0.3	2.6±0.4	100
Theophylline	1.6±0.1	12.3±0.5	-215

The adrenergic beta-blocker propranolol did not prevent the ethanol activation of the enzyme, whereas the alpha-blockers dibenamine, dibenzyline and ergotamine, which increased the enzyme activity (by a hormonal and not a substrate-type mechanism), strongly (76, 91 and 100% respectively) prevented the ethanol effect. Theophylline, which prevents the destruction of cyclic AMP by inhibiting the phosphodiesterase system, caused a 214% potentiation of the ethanol activation of tryptophan pyrrolase.

The administration of ethanol, or palmitic acid caused an increase in both serum free fatty acids and serum free (ultrafiltrable) tryptophan (Table 3). The total serum tryptophan was decreased by ethanol but not by palmitic acid. Liver total tryptophan in (μg/g wet wt. of liver ±S.E.M. of each group of 4 rats) was significantly (P=0.001) increased from a control value of 12.79±0.31 to 14.53±0.65, 15.39±0.23, 19.29±0.11, 18.24±0.12, 18.19±0.70, 19.15±0.46 and 19.80±0.43 at 0.5, 1, 1.5, 2, 2.5, 3 and 4h respectively after an injection of ethanol (5ml/kg). Liver tryptophan was likewise increased (P=0.001) at 2h after intraperitoneal injection of palmitic acid (100 mg/kg) or norepinephrine (2mg base/kg) to 18.02±0.27 and 18.66±0.48 respectively.

Table 3. Effects of ethanol and palmitic acid administration on rat serum free fatty acids and serum free (ultrafiltrable) and total (acid-soluble) tryptophan

Each rat received, 2h before being killed, an intraperitoneal injection of either 0.9% NaCl, ethanol (5ml/kg) or palmitic acid (100mg/kg body wt.) Values are means±S.E.M. of each group of four rats.

Injection	Serum free fatty acids (mmole/L)	Serum tryptophan (μg/ml) free	Total
0.9% NaCl	0.31±0.04	1.37±0.15	13.85±1.03
Ethanol	1.10±0.09	3.17±0.20	7.65±0.70
Palmitic acid	1.66±0.14	3.09±0.45	11.74±0.46

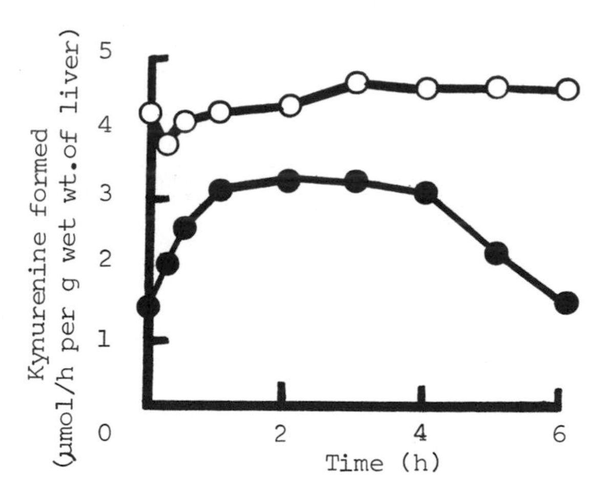

Fig.2. Effects of a single injection of morphine on rat liver tryptophan pyrrolase activity

Morphine sulphate (50mg/kg) was injected intraperitoneally at zero-time. Each point represents the mean value for four rats. O, Total enzyme activity; ●, holoenzyme activity.

Phenobarbitone: A single intraperitoneal injection of phenobarbitone (100mg/kg body wt.) did not alter the holoenzyme nor the total pyrrolase activity at 4h or 24h. Only at 24h after administration of phenobarbitone in drinking water (1mg/ml) did some change occur. The holoenzyme activity rose by 75% whereas the total activity remained unaltered.

Morphine: The time-course of the effects of an acute intraperitoneal injection of morphine sulphate (50mg/kg body wt.) on rat liver tryptophan pyrrolase activity is shown in Fig.2. No significant change in the total enzyme activity occurred, but that of the holoenzyme was **significantly** increased (P=0.05-0.001) as early as 0.25h after the injection, reached a maximum (2.3-fold increase) at 2-3h and finally fell to normal after 6h. Morphine sulphate added in vitro, even at concentrations as high as 2 mM had no effect on tryptophan pyrrolase activities of normal rat liver homogenates

2. Chronic and Withdrawal Effects of Drugs on Tryptophan Pyrrolase Activity

Ethanol: Five replicate series of experiments were performed to examine the effects of chronic ethanol administration and subsequent withdrawal on the activity of rat liver tryptophan pyrrolase. The results of a typical experiment are shown in Fig. 3. Chronic ethanol consumption did not significantly alter the holoenzyme activity. On the other hand, the total activity gradually decreased, reaching the holoenzyme values by 14-15 days of ethanol treatment. The apoenzyme activity, calculated by difference, was then zero, and remained completely inhibited for the remainder of the experimental period. In other experiments, not reported here, this inhibition persisted for as long as 80 days of chronic ethanol consumption.

When ethanol was withdrawn after 20 days of chronic consumption, the total pyrrolase activity returned to normal after 5 days, then both holoenzyme and total activities rose by 2.5-fold (P=0.02) 3 days later and finally fell to normal 11 days after withdrawal.

To see if this inhibition is specific for the apoenzyme activity, ethanol was administered chronically by the same procedure as in rats, to two other species known to lack any detectable apoenzyme activity, namely the guinea-pig (Hvitfelt & Santti,1972; Badawy & Evans, 1974) and the golden hamster (unpublished observation). The results of this comparative study are shown in Table 4. Whereas the rat liver enzyme showed the usual inhibition specific to the apoenzyme activity, the only form of the pyrrolase (holoenzyme) in the guinea-pig and the golden hamster was not inhibited but was increased for periods of up to 39 days of chronic ethanol treatment. Guinea pig liver (Badawy & Evans, 1974) and hamster liver (unpublished work) tryptophan pyrrolase exhibit different properties from those of the rat enzyme including the response to activators and inhibitors. Further work is required to see whether the difference in response to chronic ethanol treatment of these two species is due to these different properties or to any other factor(s).

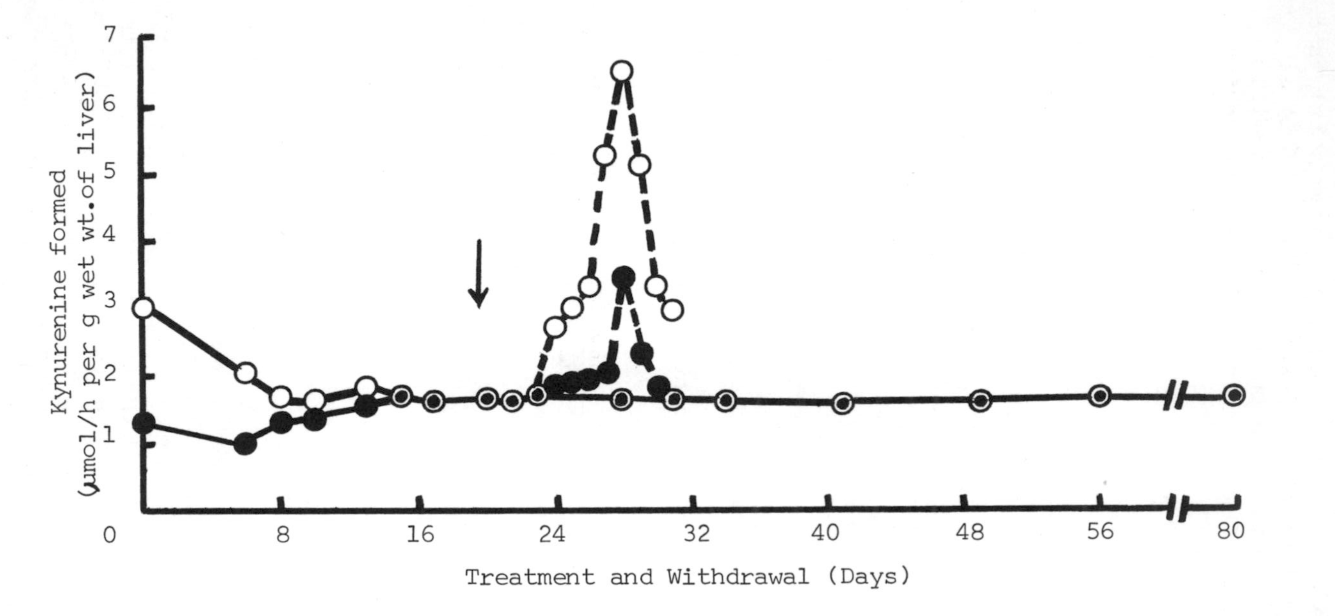

Fig.3. Effects of chronic ethanol administration and subsequent withdrawal on the activity of rat liver tryptophanopyrrolase

The methods of administration and withdrawal are described in the Materials and Methods section. Each point represents the mean value for four rats except that the zero-time points are means for 20 animals. The arrow indicates that withdrawal started after 20 days of ethanol treatment. O, Total enzyme activity; ●, holoenzyme activity; ——, activities during treatment; ---, activities during withdrawal. Reproduced (from Badawy & Evans, 1973b) by kind permission of the editor, Professor M. M. Gross.

Table 4. Effects of chronic ethanol administration on liver tryptophan pyrrolase activity of rats, guinea-pigs and golden hamsters.

Ethanol was administered as the only source of drinking water in increasing (v/v) concentrations of 5% for 2 days, 7.5% for 2 days and finally 10% for the remainder of the experimental period. The pyrrolase was determined in liver homogenates either in the absence (holoenzyme activity) or in the presence (total enzyme activity) of added haematin (2μM).

		Days of ethanol drinking			
		0	23	32	39
Rat	Holoenzyme	1.3 ± 0.1	1.9 ± 0.1	-	1.8 ± 0.3
	Total enzyme	4.5 ± 0.4	2.2 ± 0.2	-	2.0 ± 0.2
Guinea pig	Holoenzyme	1.9 ± 0.1	2.1 ± 0.4	3.1 ± 0.3	2.6 ± 0.3
	Total enzyme	1.9 ± 0.2	1.7 ± 0.3	2.5 ± 0.1	2.7 ± 0.1
Hamster	Holoenzyme	2.2 ± 0.1	4.4 ± 0.4	6.5 ± 0.4	5.4 ± 0.8
	Total enzyme	2.5 ± 0.3	-	6.8 ± 0.7	4.1 ± 0.7

The mechanism of inhibition of the apoenzyme activity by chronic ethanol administration was then examined. The addition to the incubation mixture of excessive amounts (4–8μM) of haematin did not reverse the inhibition. The results in Table 5 show that cortisol was equally effective in chronic ethanol-treated, as in normal, rats in inducing the pyrrolase. On the other hand, the injection of tryptophan, salicylate or ethanol failed to activate the enzyme in alcohol-treated rats. The administration of fructose, pyruvate, or phenazine methosulphate, and the addition in vitro of NAD^+ or $NADP^+$ had no effect on normal rat liver pyrrolase activities, but in chronic alcohol-treated rats the inhibition of the total activity was completely reversed by all these treatments. Chronic ethanol administration into rats caused a significant increase in the concentration of liver NADH and NADPH, and, accordingly, a change in the ratio of concentrations of NAD^+/NADH and NADPH/$NADP^+$ (Table 6).

The effect of tryptophol, which is produced in excess by ethanol administration, was also examined. At 0.1mM, tryptophol did not affect the holoenzyme activity (control value was 1.63 ± 0.11 and tryptophol value was 1.50 ± 0.11). The total pyrrolase activity, however, was inhibited from 3.85 ± 0.36 to 1.72 ± 0.14 ($P=0.005$). The addition of excess haematin reversed the inhibition of the total pyrrolase activity by tryptophol. The control holoenzyme and total enzyme activities were 1.60±0.11 and 4.01±0.41. The total activity in the presence of 0.1mM-tryptophol was decreased to 1.70±0.11 but when haematin was added in excess (4μM) the activity was corrected to 3.82±0.53.

Table 5. Comparison of normal and chronic ethanol-treated rat liver tryptophan pyrrolases after various treatments in vivo and in vitro

Chronic ethanol treatment was performed as described in the Materials and Methods section and also in Table 4. Rats were tested between 20-30 days of ethanol treatment. Both normal and chronic ethanol-treated rats received intraperitoneal injections of either 0.9% NaCl, cortisol acetate (20mg/kg), tryptophan (200mg/kg), sodium salicylate (400mg/kg), ethanol (5ml/kg), fructose (400mg/kg) or sodium pyruvate (200mg/kg) 4h before being killed. Phenazine methosulphate (P.M.S.) (10mg/kg) and 0.9% NaCl were given at 1h before death. In the experiments in vitro, NAD^+ and $NADP^+$ were added at final concentrations of 2mM. Values are means±S.E.M. of each group of four rats except that the untreated (nil) groups had means of 30 and 20 rats for the normal and chronic alcohol rats respectively.

	Kynurenine formed (μmol/h per g wet wt. of liver)			
	Normal rats		Alcohol-treated rats	
Treatment	Holoenzyme activity	Total enzyme activity	Holoenzyme activity	Total enzyme activity
Nil	1.7±0.1	3.8±0.2	1.6±0.04	1.6±0.03
0.9% NaCl	1.7±0.2	4.6±0.4	1.6±0.1	2.6±0.2
Cortisol	6.4±0.7	16.4±1.7	5.9±0.3	15.6±1.9
Tryptophan	6.2±0.4	9.3±0.8	1.7±0.1	3.2±0.1
Salicylate	5.4±0.4	8.8±0.9	2.3±0.3	4.8±0.3
Ethanol	6.3±0.5	10.1±0.9	2.0±0.1	4.6±0.7
Fructose	2.3±0.1	4.7±0.5	1.4±0.1	4.8±0.2
Pyruvate	2.4±0.2	5.4±0.4	2.5±0.3	6.3±0.5
0.9% NaCl	1.7±0.1	4.5±0.3	1.4±0.1	2.0±0.3
P.M.S.	1.7±0.1	3.3±0.2	1.9±0.04	5.7±0.3
Control	1.4±0.1	3.0±0.4	1.6±0.1	1.9±0.2
NAD^+	1.6±0.1	3.7±0.4	1.9±0.1	6.2±1.2
Control	1.8±0.2	4.4±0.2	1.6±0.1	1.9±0.2
$NADP^+$	1.6±0.2	4.7±0.6	2.0±0.4	5.0±0.9

Table 6. Comparison of liver nucleotides in normal rats with those in rats chronically-treated with ethanol, phenobarbitone and morphine.

Chronic administration of ethanol, phenobarbitone and morphine was performed as described in the Materials and Methods section and the rats were tested after complete inhibition of apotryptophan pyrrolase activity was established. Normal rats were of the same age and body wt. Values (expressed as µg/g wet wt. of liver) represent means±S.E.M. of each group of 4-6 rats, **=P=0.001 and *=P=0.05

Chronic Treatment	Nil	Ethanol	Phenobarbitone	Morphine
NAD^+	437±12	427±14	**292±6	380±28
NADH	148±4	**205±3	*123±8	149±4
NAD^++NADH	585	632	415	529
NAD^+/NADH	2.95	2.08	2.37	2.55
NADPH	301±5	**423±18	**393±7	**369±11
$NADP^+$	67±3	74±10	75±3	66±3
NADPH+$NADP^+$	367	497	468	435
NADPH/$NADP^+$	4.64	5.72	5.24	5.59

Phenobarbitone: The time-course of the effects of chronic phenobarbitone administration and subsequent withdrawal on the activity of rat liver tryptophan pyrrolase is shown in Fig.4. After 1 day of treatment, the holoenzyme activity rose by 75% ($P<0.05$) whereas the total activity was unchanged. The holoenzyme activity then returned to the basal value and remained unaltered ($P>0.1$) until the end of the experiment. The total pyrrolase activity was significantly decreased after 3 days of phenobarbitone treatment and remained inhibited (P=0.05-0.005) for the rest of the study period. The apoenzyme activity (obtained by difference) was strongly (80-100%) inhibited for the same length of time. Withdrawal of phenobarbitone caused an increase in both holoenzyme and total pyrrolase activities of 91-106% after 1 day and of 145-175% (i.e. approx.2.5-fold) after 2 days. The activities then fell to normal 6 days after withdrawal.

Haematin added in vitro did not reverse pyrrolase inhibition by chronic phenobarbitone treatment. The results in Table 7 show that **induction by cortisol occurred and that activation of pyrrolase by tryptophan, salicylate or ethanol was either abolished or greatly de**creased in phenobarbitone-treated rats. The administration into these rats of phenazine methosulphate but not fructose or pyruvate effectively reversed the inhibition. NAD^+ and $NADP^+$ added in vitro, also reversed the inhibition in liver homogenates of chronic phenobarbitone-treated rats. Chronic phenobarbitone treatment significantly increased the liver NADPH concentration and decreased NAD^+ and NADH concentrations (Table 6).

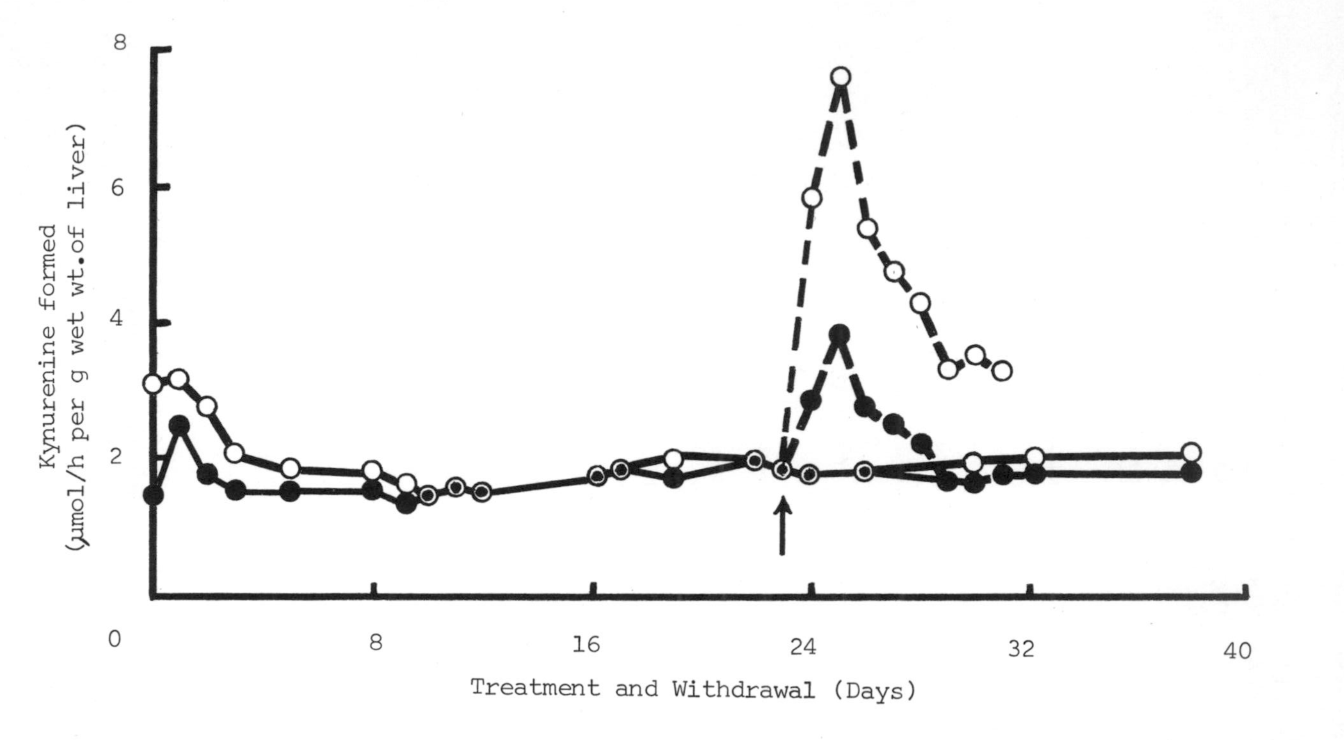

Fig.4. Effects of chronic phenobarbitone administration and subsequent withdrawal on rat liver tryptophan pyrrolase activity.

The methods of administration and withdrawal are described in the Materials and Methods section. Each point represents the mean value for four rats. The arrow indicates that withdrawal started after 23 days of phenobarbitone treatment. O, Total enzyme activity; ●, holoenzyme activity; ——, activities during treatment; ---, activities during withdrawal. (Reproduced from Badawy & Evans, 1973d).

Table 7. Tryptophan pyrrolase activity in liver homogenates from rats chronically-treated with phenobarbitone or morphine after various treatments in vivo and in vitro.

Chronic administration of phenobarbitone or morphine was performed as described in the Materials and Methods section, and the rats were tested between 15-30 days of treatment. All injections and additions in vitro are as described in Table 5. Values are means±S.E.M. of each group of four rats except that the untreated (nil) groups are means of 20 rats each. The basal pyrrolase activities for the phenobarbitone series (i.e. before starting the experiment were 1.5±0.04 and 5.4±0.5 and the corresponding ones for the morphine series were 1.3±0.1 and 3.4±0.1). Some of the results with rats chronically-treated with phenobarbitone are reproduced here from a previous publication (Badawy & Evans, 1973d).

Kynurenine formed (μmol/h per g wet wt.of liver)

Treatment	Phenobarbitone-treated rats: Holoenzyme activity	Phenobarbitone-treated rats: Total enzyme activity	Morphine-treated rats: Holoenzyme activity	Morphine-treated rats: Total enzyme activity
Nil	1.6±0.06	1.7±0.07	1.4±0.04	1.6±0.05
0.9% NaCl	1.7±0.1	2.6±0.2	1.2±0.04	2.5±0.3
Cortisol	5.5±0.3	12.8±0.9	3.7±0.2	9.0±0.6
Tryptophan	2.0±0.2	3.8±0.4	1.2±0.1	3.1±0.4
Salicylate	3.2±0.6	5.3±0.7	1.9±0.2	5.7±0.5
Ethanol	1.8±0.1	3.6±0.5	2.9±0.6	4.6±0.7
Fructose	2.6±0.2	3.2±0.2	1.4±0.3	2.4±0.2
Pyruvate	2.1±0.2	2.3±0.1	1.5±0.0	2.8±0.3
0.9% NaCl	2.3±0.2	2.9±0.3	1.6±0.1	2.2±0.2
P.M.S.	2.5±0.3	4.6±0.5	2.5±0.2	4.6±0.6
Control	2.1±0.1	2.6±0.2	1.3±0.1	1.3±0.2
NAD^+	2.5±0.3	5.9±0.2	1.4±0.2	3.7±0.5
Control	1.8±0.1	2.2±0.2	1.4±0.1	1.5±0.2
$NADP^+$	2.3±0.3	5.8±0.7	1.6±0.1	6.8±0.8

Morphine: The time-course of the effects of chronic morphine administration and subsequent withdrawal on rat liver tryptophan pyrrolase activity is shown in Fig.5. Chronic morphine administration did not significantly alter the holoenzyme activity but the total activity was significantly decreased (34%, $P<0.05$) after 2 days and remained inhibited (41-60%, $P=0.005-0.001$) for the remainder of the experimental period. The apoenzyme activity (obtained by difference) was therefore strongly inhibited (80-100%) for the same length of time. Withdrawal of morphine brought about an approx. 2.5-fold increase ($P=0.02-0.01$) in both holoenzyme and total pyrrolase

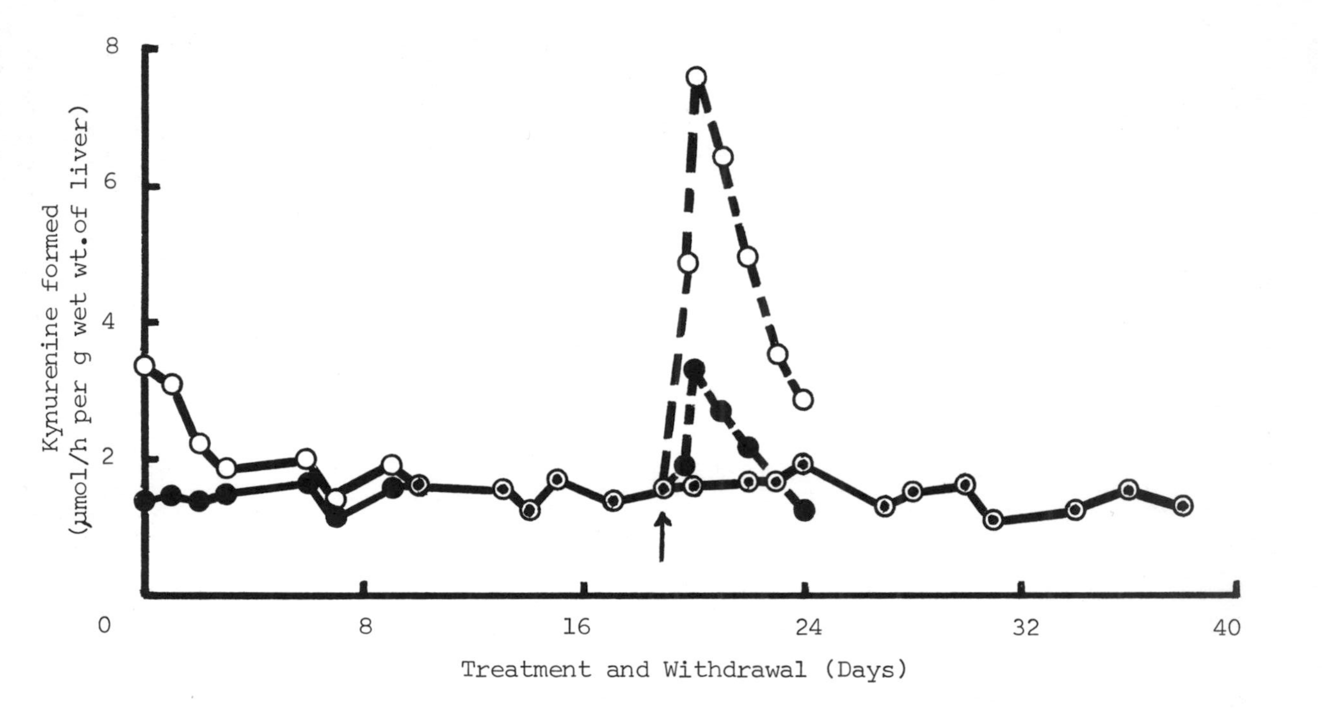

Fig.5. Effects of chronic morphine administration and subsequent withdrawal on rat liver tryptophan pyrrolase activity.

The methods of administration and withdrawal are described in the Materials and Methods section. Each point represents the mean value for four rats except that the zero-time points are means for 20 animals. The arrow indicates that withdrawal started after 19 days of morphine treatment. O, Total enzyme activity; ●, holoenzyme activity; ——, activities during treatment; ---, activities during withdrawal.

activities after 1 day, and normal levels were reached 4-5 days after withdrawal.

The inhibition of the total pyrrolase activity by chronic morphine administration was not reversed by the addition of excess haematin to the assay mixture in vitro. A reversal of the inhibition was obtained by the addition in vitro of NAD^+ and $NADP^+$ or by the administration in vivo, of phenazine methosulphate but not fructose or pyruvate (Table 7). It can also be seen in this Table that the administration of cortisol but not tryptophan, salicylate or ethanol, increased the pyrrolase activities in chronic morphine-treated rats.

The effects of chronic morphine administration on liver nucleotide concentrations is shown in Table 6. Morphine increased the liver NADPH concentration and altered the ratio of the concentrations of $NADPH/NADP^+$.

DISCUSSION

1. Acute Effects of Drugs on Tryptophan Pyrrolase Activity

Ethanol: Brodie et al.,(1961) were the first to report an approximately threefold increase in rat liver tryptophan pyrrolase activity at 6h after an oral dose of ethanol (4.8g/kg), which was prevented by pretreatment with the adrenergic alpha-blocker dibenamine. The authors considered this activation to be a manifestation of the stimulation, by ethanol, of the pituitary-adrenal axis. Sardesai & Provido (1972) reported increases (not exceeding twofold) of both holoenzyme and total pyrrolase activities at 2-6h after an intraperitoneal injection of ethanol (5ml/kg), and suggested that this effect may be mediated by either corticosteroids or the cofactor haematin. The activation of the rat liver enzyme, previously reported by us (Badawy & Evans, 1972; 1973a,b) shows three distinct changes: an early stimulation (at 5min), an increased haem saturation of the apoenzyme (at 3h) and an overall enhancement (at 4h and thereafter).

The initial increase in both holoenzyme and total pyrrolase activities was previously suggested to be associated with ethanol's own metabolism by alcohol dehydrogenase. This is because pyrazole, an alcohol dehydrogenase inhibitor (Lester et al.,1968), prevents this stimulation, NADH or ethanol plus NAD^+ cause a similar stimulation in vitro, and preincubation with pyrazole prevents the stimulation by ethanol plus NAD^+ but not that by NADH (Badawy & Evans, 1973a,b,c).

The almost complete saturation of the apoenzyme with its haem activator observed at 3h after ethanol (Fig.1) resembles that caused by the substrate tryptophan (Greengard & Feigelson, 1961) as well as by the haem (cofactor) precursor 5-aminolaevulinate (Druyan & Kelly,

1972; Badawy & Evans, 1973e), and precedes, as is also the case after tryptophan, an overall increase in both holoenzyme and total pyrrolase activities. This increase (at 4h after ethanol) is not prevented by actinomycin D, and it is additive with that of cortisol but not tryptophan, and it is associated with a holoenzyme/apoenzyme ratio similar to the tryptophan-activated enzyme (above 1) but different from that of the basal or cortisol-induced enzyme (less than 1) (Badawy & Evans, 1973a,b). These results suggest that ethanol activates tryptophan pyrrolase by a substrate- or cofactor-type mechanism and does not act by a hormonal-type mechanism. Although ethanol enhances 5-aminolaevulinate synthetase activity (Shanley et al., 1968), current considerations regarding haem synthesis, and other evidence (Tyrrell & Marks, 1972) make the possibility of ethanol activating the pyrrolase by direct action on the cofactor less likely, although this cannot be confirmed without further examination of haem synthesis after ethanol.

A tryptophan-mediated activation of the pyrrolase by ethanol is more likely for a number of reasons. Ethanol releases catecholamines which have been shown (Curzon & Knott, 1974) to release tryptophan from its binding sites on plasma proteins. This displacement may be mediated by free fatty acids whose release into the circulation follows activation of adipose tissue lipase by cyclic AMP in response to the released catecholamines (see review by Butcher et al., 1968). Free fatty acids have been shown (Curzon et al., 1973) to displace plasma protein-bound tryptophan in vitro, and agents that increase plasma free fatty acids in vivo cause the release of protein-bound tryptophan (Curzon & Knott, 1974). Finally, salicylate which displaces albumin-bound tryptophan in vitro (McArthur & Dawkins, 1969) and in vivo in human serum (Smith & Lakatos, 1971) and in rat serum and liver (Badawy & Smith, 1972) also activates tryptophan pyrrolase by a substrate mechanism associated with increased liver tryptophan concentrations (Badawy & Smith, 1971, 1972).

The above considerations suggest that ethanol activates tryptophan pyrrolase by a substrate mechanism and this is supported by the following findings. Ethanol activates adipose tissue lipase leading to an increase in serum free fatty acid concentration (Table 3). Furthermore, both catecholamines and palmitic acid activate tryptophan pyrrolase (Table 1) and both ethanol and palmitic acid (Table 3), as well as catecholamines (Curzon & Knott, 1974) release serum or plasma protein-bound tryptophan. The ethanol activation of tryptophan pyrrolase is prevented by reserpine (which depletes catecholamines) and by the adrenergic alpha-blockers dibenamine, dibenzyline and ergotamine (Table 2). The ergotamine effect is interesting because although this drug blocks alpha receptors, it is also known to specifically block hepatic beta-receptor activity (Ellis et al., 1953) with which cyclic AMP is associated (Robison et al., 1967). The effect of theophylline which prevents the breakdown of cyclic AMP was therefore examined. As can be seen in Table 2 theophylline produced a marked potentiation of the ethanol activation of tryptophan pyrrolase. Finally, the administration of ethanol, palmitic acid or norepinephrine causes an increase in the concentration of tryptophan in the liver (see text). The proposed mechanism for the substrate-type activation of rat liver tryptophan pyrrolase by acute ethanol administration is summarized in Fig.6.

Although acetaldehyde activates tryptophan pyrrolase (Table 1), this is not required for the ethanol effect as suggested by the lack of any significant effect of pyrazole or disulfiram on the ethanol activation of the enzyme (Table 2).

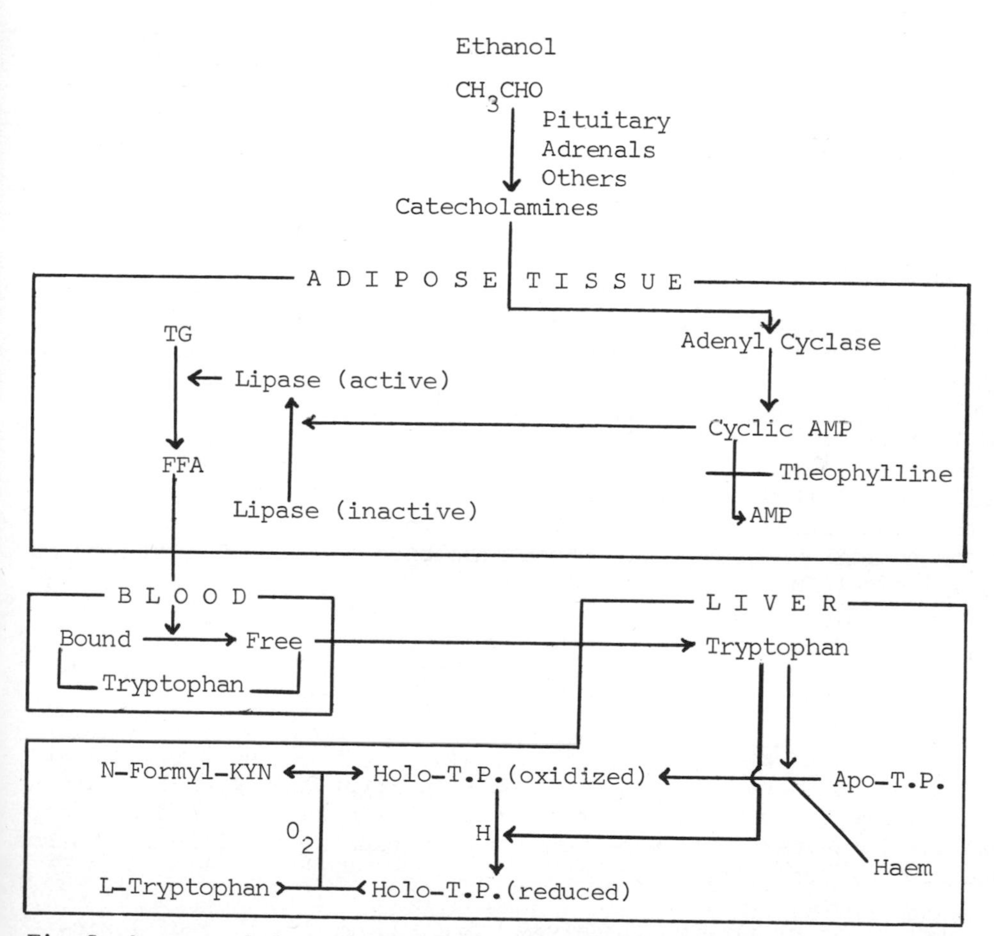

Fig.6. A summarizing diagram of the possible sequence of events leading to the activation of rat liver tryptophan pyrrolase by acute ethanol administration.

TG, Triglycerides; FFA, free fatty acids; T.P. tryptophan pyrrolase; KYN, kynurenine.

Phenobarbitone: Acute administration of phenobarbitone does not change the pyrrolase activity at 4h or 24h. This is in contrast with the finding (Seifert, 1973) of an increase at 24h after a single intraperitoneal injection. However, at 24h after administering the drug in drinking water (Seifert, 1973; Badawy & Evans, 1973 d,e) an increased haem saturation of the apoenzyme is observed (Fig.4). This effect is discussed in this section since the time-interval at which it occurs (24h) represents acute treatment. This increased haem saturation of the pyrrolase apoenzyme together with a similar effect on the liver microsomal cytochrome P-450 (De Matteis, 1972), may be due to the moderate stimulation of 5-aminolaevulinate synthetase activity, probably due to decreased degradation (Satyanarayana Rao et al., 1972).

Morphine: The moderate increase in the holoenzyme activity observed during acute morphine administration (Fig. 2) suggests that morphine may also increase haem utilization. The effect of morphine on other haem-requiring enzymes is not known. A simple direct stimulatory effect by morphine is unlikely since it does not alter the pyrrolase activities in vitro.

Comparison of the acute effects of drugs of dependence: While the activation of tryptophan pyrrolase is most marked after ethanol, a similar pattern of change is observed with the three drugs. This consists of increased haem saturation of the apoenzyme. With morphine and phenobarbitone, the increase in the holoenzyme activity is 100 and 75% respectively whereas with ethanol (at 3h) it is much larger, and is followed by an overall activation of both forms of the enzyme. It is of interest that the pronounced ethanol effect may increase the synthesis of NAD^+ from tryptophan thus compensating for the loss of this nucleotide during ethanol oxidation and correcting the ratio of concentrations of $NAD^+/NADH$ whose imbalance can affect a variety of physiological functions in the liver.

2. Chronic and Withdrawal Effects of Drugs on Tryptophan Pyrrolase Activity

The effects of chronic administration and subsequent withdrawal of ethanol, phenobarbitone and morphine on tryptophan pyrrolase are discussed together because of their general similarities and the fact that similar experiments were performed to examine possible mechanisms causing these effects.

Our previous finding (Badawy & Evans, 1972; 1973a,b) of inhibition of tryptophan pyrrolase activity in liver homogenates from rats chronically-treated with ethanol (Fig.3) has now been confirmed (Mørland, 1974) in the homogenate as well as in the perfusion system. The liver perfusion technique, however, if not carefully controlled, can introduce new problems, one of which is the lowering of pH which causes a decrease in the enzyme activity in the normal rat upon perfusion of the liver with ethanol (Mørland et al., 1972). Mørland (1974) found that the induction of tryptophan pyrrolase following perfusion with dexamethasone is inhibited in livers of rats chronically-treated with ethanol. The induction of tyrosine aminotransferase was also decreased although the basal level of this enzyme, unlike that of tryptophan pyrrolase, was not altered by chronic ethanol treatment. The author suggested that chronic ethanol treatment inhibits the pyrrolase by decreasing its synthesis. The author, however, did not report the effect of dexamethasone administration into intact chronically-alcoholized rats on the pyrrolase activity in liver homogenates. Our finding with cortisol in intact rats (Table 5) suggests that the synthesis of apotryptophan pyrrolase is not impaired by chronic ethanol treatment.

Chronic administration of phenobarbitone (apart from the 24h increase in the holoenzyme activity discussed earlier)(Fig.4), and of morphine (Fig.5) produce remarkably similar effects to those of ethanol (Fig.3) where the decrease in the total enzyme activity can be accounted for by inhibition of that of the apoenzyme by some 80-100%. Withdrawal of any of the three drugs causes a 2.5-fold increase in both holoenzyme and total pyrrolase activities although this increase is reached earliest during morphine, and latest during ethanol withdrawal. This withdrawal increase is not primarily cofactor-dependent, is associated with a holoenzyme/apoenzyme ratio of less than one (like the basal and cortisol-induced enzyme ratio) and may therefore be due to increased enzyme synthesis although such possibility cannot be confirmed without labelled amino acid incorporation experiments.

That the inhibition of tryptophan pyrrolase by chronic ethanol treatment is specific to the apoenzyme activity is also suggested by the finding (Table 4) of no inhibition by chronic ethanol administration into two other animal species that lack the apoenzyme. Moreover, the activation, in normal rats, of pre-existing apoenzyme by acute administration of tryptophan (Schimke, 1969), salicylate which acts by displacing protein-bound tryptophan (Badawy & Smith, 1971, 1972) and ethanol (Badawy & Evans, 1973a,b, and this paper), does not occur in rats chronically-treated with ethanol (Table 5), presumably because the apoenzyme there is already inactive (Fig.3). Inhibition of the apoenzyme activity in livers of rats treated with a single dose of the apoenzyme inhibitor allopurinol (Badawy & Evans, 1973c) also blocks the activation of the pyrrolase by tryptophan, salicylate or ethanol. The ethanol inhibition, unlike that by allopurinol which prevents the conjugation of the apoenzyme with its haem activator, is not reversed by excess haematin in vitro.

As is the case with ethanol (Table 5), the inhibition of apotryptophan pyrrolase activity by chronic phenobarbitone (Fig.4) and morphine (Fig.5) treatment blocks the activation by tryptophan, salicylate or ethanol, and the effectiveness of cortisol in inducing the enzyme (Table 7) suggests that apoenzyme synthesis is not affected by chronic drug treatment. Addition of excess haematin in vitro is also without effect in homogenates from rats chronically-treated with phenobarbitone or morphine.

The possibility that the inhibition of the pyrrolase activity by chronic ethanol administration may be due to increased liver tryptophol concentration was also excluded since although tryptophol causes inhibition (Madras & Sourkes, 1968) that is specific to the apoenzyme (see results section), the addition of excess haematin in vitro reverses it unlike the ethanol inhibition which is not reversed by the cofactor.

The exclusion by the above experiments of factors that might be responsible for the pyrrolase inhibition by chronic administration of ethanol, phenobarbitone or morphine leaves one with the question whether physiological inhibitors are involved. Among the physiologically-occurring inhibitors of tryptophan pyrrolase, NADH and NADPH, which act by an allosteric mechanism (Cho-Chung & Pitot, 1967) are known to be increased by chronic ethanol or phenobarbitone treatment. Ethanol increases both NADH and NADPH (Kalant et al., 1970) whereas phenobarbitone increases NADPH after oral administration in drinking water (Butcher, 1971) or in the rat's diet (McIntosh & Topham, 1972). The effects of chronic phenobarbitone administration on liver NADH, or of morphine on liver nicotinamide-adenine-dinucleotides are not reported as far as we could ascertain. Direct measurement of liver nucleotides was therefore made in rats chronically-treated with ethanol, phenobarbitone or morphine and was compared with values for normal rats of the same age, sex and body wt. (Table 6). Confirming previous findings (Kalant et al., 1970), we found the concentrations of liver NADH and NADPH to be increased by chronic ethanol treatment. Treatment with phenobarbitone increases NADPH only, and causes a decrease in both NAD^+ and NADH concentrations. The relatively larger decrease of NAD^+ compared with that of NADH levels causes phenobarbitone to decrease the ratio of concentrations of NAD^+/NADH. These decreases are interesting and require further investigation. Chronic morphine administration was found (Table 6) to increase the liver NADPH concentration only.

If these increased concentrations of NADH and/or NADPH are involved in the inhibition of tryptophan pyrrolase activity caused by chronic treatment with the above three drugs of dependence, then it should be possible to reverse such inhibition by modifying the reduced nucleotide concentrations. Both fructose and pyruvate are known to regenerate liver NAD^+, and this effect forms the basis of the use of the sugar in treating acute alcoholic intoxication (see Slater, 1972, for a review). An agent that rapidly reoxidizes both NADH and NADPH is phenazine methosulphate (Katz & Wals, 1970). The administration of fructose, pyruvate or phenazine methosulphate results in reversal of the inhibition of the total pyrrolase activity caused by chronic ethanol administration (Table 5). Also, the addition in vitro of NAD^+ or $NADP^+$ gives a similar effect, and these results together with those demonstrating increased reduced nucleotides (Table 6) suggest that the pyrrolase inhibition by ethanol may be due to both NADH and NADPH. In rats chronically treated with either phenobarbitone or morphine, the pyrrolase inhibition is not reversed by the administration of fructose nor pyruvate (Table 7) thus suggesting non-involvement of NADH in the inhibition; a conclusion also supported by the lack of increase in NADH concentration by chronic treatment with either drug (Table 6). The effectiveness of phenazine methosulphate in reversing the inhibition by these two drugs (Table 7) suggests that reversal may be due to the reoxidation of NADPH rather than NADH. These results with phenobarbitone and morphine therefore suggest that the pyrrolase inhibition is caused by NADPH whose concentration is also increased (Table 6). $NADP^+$ added in vitro to liver homogenates from rats chronically-treated with phenobarbitone or morphine effectively reverses the inhibition. But NAD^+ in vitro also reverses, although its regeneration in vivo by fructose or pyruvate does not. This discrepancy may be explained by NAD^+ being converted into NADH, concomitant with the reoxidation of NADPH present in excess, by various metabolic reactions involving endogenous substrates such as glutamate.

In conclusion, the similarities of effects of chronic administration and subsequent withdrawal of ethanol, phenobarbitone and morphine on rat liver tryptophan pyrrolase activity are remarkable. It is also interesting to observe the occurrence of acute withdrawal symptoms soon after administration of phenazine methosulphate to rats chronically-treated with all three drugs of dependence. These changes, their relevance to conditions of dependence and withdrawal and the influence they could have on the metabolism of tryptophan in the brain suggest that further work in this area may prove fruitful.

ACKNOWLEDGEMENTS

Permission by the Editorial Board of the Biochemical Journal to reproduce some previously published data is gratefully acknowledged. We thank the Medical Research Council (U.K.) and the Addiction Research Foundation for Wales for their support, Mr. C. J. Morgan for technical assistance, Mr. A. Dacey for animal maintenance and Mrs. Margaret Harrison for secretarial assistance. The following companies provided generous gifts of drugs: Burk Pharaceuticals Ltd. (disulfiram), Imperial Chemical Industries Ltd. (propranolol) and Smith Kline and French Laboratories Ltd. (dibenamine and dibenzyline).

REFERENCES

Badawy, A. A.-B. & Evans, M. (1972) Lancet, ii, 374-375
Badawy, A. A.-B. & Evans, M. (1973a) Biochem. Soc. Trans. 1, 193-195
Badawy, A. A.-B. & Evans, M. (1973b) Advan. Exp. Med. Biol. 35,105-123
Badawy, A. A.-B. & Evans, M. (1973c) Biochem. J. 133, 585-591
Badawy, A. A.-B. & Evans, M. (1973d) Biochem. J. 135, 555-557
Badawy, A. A.-B. & Evans, M. (1973e) Biochem. J. 136, 885-892
Badawy, A. A.-B. & Evans, M. (1974) Biochem. J. 138, 445-451
Badawy, A. A.-B. & Smith, M. J. H. (1971) Biochem. J. 123, 171-174
Badawy, A. A.-B. & Smith, M. J. H. (1972) Biochem.Pharmacol. 21, 97-101
Brodie, B. B., Butler, W. M. Jr., Horning, M. G., Maickel, R. P. & Maling, H. M. (1961) Amer. J. Clin. Nutr. 9, 432-435
Butcher, R. G. (1971) Biochem. J. 125, 22P-23P
Butcher, R. W., Robison, G. A., Hardman, J. G. & Sutherland, E. W. (1968) Advan. Enz. Regul. 6, 357-389
Cho-chung, Y. S. & Pitot, H. C. (1967) J. Biol. Chem. 242, 1192-1198
Curzon, G. (1969) Brit. J. Psychiat. 115, 1367-1374
Curzon, G., Friedel, J. & Knott, P. J. (1973) Nature New Biol. 242, 198-200
Curzon, G. & Knott, P. J. (1974) Brit. J. Pharmacol. 50, 197-204
De Matteis, F. (1972) Biochem. J. 130, 52P-53P
Denkla, W. D. & Dewey, H. K. (1967) J. Lab. Clin. Med. 69, 160-169
Druyan, R. & Kelly, A. (1972) Biochem. J. 129, 1095-1099
Ellis, S., Anderson, H. L., Jr. & Collins, M. (1953) Proc. Soc. Exp. Biol. Med. 84, 383-386
Feigelson, P. & Greengard, O. (1961) J. Biol. Chem. 236, 153-157
Greengard, O. & Feigelson, P. (1961) J. Biol. Chem. 236, 158-161
Kalant, H., Khanna, J. M. & Loth, J. (1970) Can. J. Physiol. Pharmacol. 48, 542-549
Katz, J. & Wals, P. A. (1970) J. Biol. Chem. 245, 2546-2548
Lapin, I. P. & Oxenkrug, G. F. (1969) Lancet, i, 132-136
Lester, D., Keokosky, W. Z. & Felzenberg, F. (1968) Q. J. Stud. Alochol, 29, 449-454
Madras, B. K. & Sourkes, T. L. (1968) Biochem. Pharmacol. 17, 1037-1047
McArthur, J. N. & Dawkins, P. D. (1969) J. Pharm. Pharmacol. 21, 744-750
McIntosh, D. A. D. & Topham, J. C. (1972) Biochem. Pharmacol. 21, 1025-1029
Mikac-Dević, D., Stanković, H. & Bošković, K. (1973) Clin. Chim. Acta, 45, 55-59
Mørland, J. (1974) Biochem. Pharmacol. 23, 21-35
Mørland, J., Christoffersen, T., Osnes, J. B., Seglen, P. O. & Jervell, K. F. (1972) Biochem. Pharmacol. 21, 1849-1859

Robison, G. A., Butcher, R. W. & Sutherland, E. W. (1967) Ann. N. Y. Acad. Sci. 139, 703–723
Sardesai, V. M. & Provido, H. S. (1972) Life Sci. 11, 1023–1028
Satyanarayana Rao, M. R., Malathi, K. & Padmanaban, G. (1972) Biochem. J. 127, 553–559
Schimke, R.T. (1969) Curr. Top. Cell. Regul. 1, 77–124
Seifert, J. (1973) Toxicology, 1, 179–186
Shanley, B. C., Zail, S. S. & Joubert, S. M. (1968) Lancet, i, 70–71
Slater, T. F. (1972) in Free Radical Mechanisms in Tissue Injury (Lagnado, J. R., ed.) pp. 171–197, Pion Ltd., London
Slater, T. F. & Sawyer, B. (1962) Nature (Lond.) 193, 454–456
Slater, T. F., Sawyer, B. & Straüli, U. (1964) Arch. Internat. Physiol. Biochim. 72, 427–447
Smith, H. G. & Lakatos, C. (1971) J. Pharm. Pharmacol. 23, 180–189
Tyrrell, D. L. J. & Marks, G. S. (1972) Biochem. Pharmacol. 21, 2077–2093

PITUITARY FUNCTION IN CHRONIC ALCOHOLISM

John Wright, Julius Merry, Denys Fry and Vincent Marks.
West Park Hospital, Epsom, Surrey, U.K. and University of Surrey, Guildford, Surrey, U.K.

Previous studies of pituitary function in alcoholics have in general concentrated on the hypothalamic-pituitary-adrenal axis and there have now been a considerable number of reports describing disturbances of this system. We have recently extended our investigations to include other aspects of pituitary function, in particular the hypothalamic-pituitary-gonadal axis. We have performed combined pituitary function tests in a group of chronic alcoholic patients admitted to West Park Hospital. All patients in the study had been drinking heavily right up to the time of admission to hospital and all tests were performed within the first 48 or 72 hours of admission. The tests performed included a short ACTH (Synacthen) stimulation test and a standard insulin stress test combined with stimulation tests using thyrotrophin releasing hormone (TRH) and luteinising hormone releasing hormone (LHRH).(1)

RESULTS

The Hypothalamic-pituitary-gonadal Axis (see accompanying table)

The mean basal LH value in a group of 13 male alcoholics was 9.2mU/ml compared with 5.4mU/ml in a group of normal control men ($p < 0.05$). Six of the alcoholics had basal levels of 12mU/ml or above, considerably higher than the basal levels for non-alcoholic subjects. Following intravenous administration of 100μg of LHRH (Roche), the mean maximal level of LH,

	Subject	Age (years)	17βOH Androgens (ng/ml)	LH (basal) (mU/ml)	LH (max) (mU/ml)
Alcoholics	1	33	8.6	12	>30
	2	53	10.0	5	12
	3	52	7.8	15	>30
	4	22	25.3	7	23
	5	51	7.8	12	>30
	6	35	13.2	6	18
	7	45	4.2	13	>30
	8	35	7.1	5	11
	9	51	13.3	9	17
	10	29	8.7	7	14
	11	43	8.3	4	11
	12	42	8.8	11	>30
	13	54	8.6	12	>30
Controls	1	25	5.2	4	25
	2	30	6.8	6	24
	3	46	5.5	6	24
	4	47	5.5	4	18
	5	37	6.1	7	13

Mean basal LH ($\pm$ S.D.) in alcoholics : 9.2 $\pm$ 3.6 mU/ml

Mean basal LH ($\pm$ S.D.) in controls : 5.4 $\pm$ 1.3 mU/ml

Results of hypothalamic - pituitary - gonadal function studies in alcoholics measured at either 20 or 60 minutes after the injection was higher in the alcoholics than in the control group and particularly high in the six alcoholics who had basal LH levels of 12mU/ml or more. This represents an exaggerated response to LHRH in six of the 13 alcoholics tested. There is no obvious explanation for the elevated LH levels and the exaggerated response to LHRH seen in these patients. There was no correlation between either the age of the patients or their clinical status and the LH values. There are a few recent reports of reduced plasma testosterone levels in alcoholics which might be consistent with elevated LH levels. However, the 17βOH androgens in this group of alcoholics were slightly higher than those found in normal non-alcoholic men. In addition, there was no correlation between LH and testosterone values for individual patients within the group of alcoholics.

Other pituitary function. Insulin stress tests were performed in 11 patients. Two patients showed an absent cortisol response to hypoglycaemia despite a normal growth hormone response and demonstration of adequate adrenal reserve. This confirms previous reports of "cortisol unresponsivness" in a proportion of chronic alcoholics (2). The majority of patients tested showed increased sensitivity to insulin with profound, prolonged hypoglycaemia. Growth hormone response to hypoglycaemia was normal in all patients. Two of 13 patients tested showed diminished TSH response to TRH despite normal conventional thyroid function tests (Thyroxine and T3 uptake).

Summary. Elevated basal LH levels and an exaggerated response to LHRH were found in a high proportion of male chronic alcoholics. This was not associated with reduced plasma testosterone values. In addition, previous reports of an absent cortisol response to hypoglycaemia in a proportion of chronic alcoholics are confirmed.

References

1. Harsoulis, P., Marshall, J.C., Kuku, S.F. et al Brit. Med.J. 4 326, 1973.

2. Merry, J. and Marks, V. Lancet 2 990, 1972.

BODY COMPOSITION IN CONTROL, ALCOHOLIC AND DEPRESSIVE INDIVIDUALS USING A MULTIPLE ISOTOPE TECHNIQUE AND WHOLE BODY COUNTING OF POTASSIUM

David MacSweeney

Academic Department of Psychiatry

Middlesex Hospital Medical School
London, W.1. U.K.

The most clinically useful classification of alcoholics currently available derives from the work of Winokur and his associates (Winokur, Reich, Rimmer and Pitts, 1970). They divided alcoholics into three main groups:-

(1) Primary alcoholics (those who had become addicted to alcohol without pre-existing depression, sociopathy, anxiety neurosis, hysteria, or schizophrenia).

(2) Depression-alcoholics who had clearcut episodes of depression without pre-existing alcoholism, sociopathy etc. and

(3) Sociopathy alcoholics, those who had histories of police problems "with early onset of excessive fighting, delinquency, job trouble, sexual promiscuity, wanderlust, or periods of being a 'runaway' ".

These workers have also demonstrated a genetic link between alcoholism and depression. In their careful study of 259 hospitalized alcoholic patients 507 of their first degree relatives were interviewed. Among the male primary alcoholics the morbid risk among first degree relatives for alcoholism was 22% and for depression 13%; 23% of the female relatives of this group suffered from depression. A total of 15% of the first degree relatives of male depression-alcoholics suffered from alcoholism and 7% from depression.

It is unnecessary to emphasise that when a genetic basis to an illness is postulated a significant biochemical component in the aetiology of that illness is implied. Thus genetic, constitutional or other similarities found between alcoholism and depression might be related to common aetiological mechanisms.

In addition, alcoholism might secondarily produce biochemical changes which are similar to those responsible for affective disorder or which are instrumental in lowering the individual's threshold to this group of illnesses; alternatively should these changes become established (over time) in certain predisposed individuals then the possibility of vicious circle loss-of-control drinking might occur.

BODY COMPOSITION IN DEPRESSION

Coppen & Shaw (1963) measured water and electrolyte distribution in severely depressed patients using an isotope dilution technique and 'whole body' counting of the naturally occurring radioactive potassium. Their most striking finding was that during depression residual sodium (that fraction of exchangeable sodium in cells and bone) was increased by 50%. It returned to normal after recovery. They also demonstrated a relative deficiency of total body potassium and the concentration of intracellular potassium was low when compared to values obtained on normal subjects.

BODY COMPOSITION IN ALCOHOLISM

Many studies suggest that alcoholism may give rise to abnormalities in various parameters and processes concerned with body composition; these include investigations of water and electrolytes (Kalant & Israel, 1967; Martin, McCuskey & Tupikova, 1959; Beard & Knott, 1968; Nielsen, 1963, observations on ATPases (Sun & Samorajski, 1970; Israel, Kalant, La Blanc, Bernstein & Salazar, 1970), and studies of protein synthesis (Kuriyama, Sze & Rauscher, 1971). Chronic alcoholism may also be a cause of myopathy (Ekbom, Hed, Kirstein & Astrom, 1964; Perkoff, Dioso, Bleisch & Klinkerfuss, 1967, and Faris & Reyes, 1971), and may be followed by loss of potassium from the brain (Shaw, Frizel, Camps & White, 1969; Shaw, Camps, Robinson, Short & White 1970).

PERSONALITY STRUCTURE OF PRIMARY ALCOHOLICS AND ENDOGENOUS DEPRESSIVES

MacSweeney (1973) has compared the personality structure of non-depressed primary alcoholics, endogenous depressives (ill and well) and normal subjects for Psychoticism, Extraversion, Neuroticism (Eysenck, 1972) and Validity or 'drive' (Coppen, 1966). He found both similarities and significant differences between the alcoholics and the recovered depressives. Significant differences between both patient groups and normal subjects were also found. These observations lend support to the genetic 'overlap' between alcoholism and depression as demonstrated by Winokur and his colleagues.

AIM OF PRESENT STUDY

This was to try and demonstrate in vivo whether or not abnormalities in body composition found in severe depression were also present in a homogeneous group of non-depressed alcoholics and whether or not both illnesses share similar abnormalities.

More specifically it attemps to find differences between the body composition of a group of primary alcoholics and a group of controls using a multiple isotope technique and whole body counting of potassium.

METHODS

POPULATIONS STUDIED

Alcoholic group: These were primary alcoholics who fulfilled two of three criteria which indicated physical dependence on alcohol; loss of control over drinking, frequent alcoholic amnesias and morning drinking. None were concurrently depressed at the time of their tests or had ever been depressed (Coppen & Shaw, 1963). They had been off alcohol for a minimum of 2 weeks before being tested and apart from the possible effects of alcoholism were in good physical health. Had these patients been studied in too close relation to their drinking there would have been the risk of merely showing temporary electrolyte changes. The purpose of the experiment was to try and demonstrate the more permanent and damaging effects of alcohol on body composition which might in the future be in some way instrumental in bringing about a resumption of uncontrolled drinking. They were therefore a carefully defined homogeneous group of alcoholics taken randomly in time. There were 21 males (mean age 41.4 years) and 15 females (mean age 42.3 years). (Had a population of Depression alcoholics been selected instead of Primary alcoholics then the experiment would have been loaded in favour of finding electrolyte abnormalities).

Control Group: These were volunteers from members of the laboratory, nursing and medical staff of West Park Hospital, Epsom, U.K. They had no personal or family history of depression, alcoholism or other mental disorder. Male controls who numbered 20 had a mean age of 41.7 years and female controls who numbered 21 a mean age of 40.9 years. The distributions of age were not significantly different in males and females and in controls and alcoholics (X^2).

BODY COMPOSITION AND RELATED MEASUREMENTS

The isotope-dilution principle is similar to that used in the measurement of plasma volume with dye. For estimated and derived variates see Table 3.

Total body potassium (K_T) was estimated by assay of the naturally occurring isotope of potassium (K_{40}) using a whole body scintillation counter at the M.R.C. Radiological Protection Service*. The Counter was calibrated by means of a polyethylene model of the human body (Bush, 1946) filled with a solution of potassium chloride of known concentration. The weight of the model was 75 kg. The heights and weights of the subjects were recorded at the time of measurement. The sensitivity of the counter was independent of height but dependent on weight. A factor was calculated using models simulating different weights for the same height to allow estimates of K_T to be corrected for variation in weight.

$^{24}NaCl$, $NH_4{}^{82}Br$ and $^{3}H_{2}0$ (Radiochemical Centre, Amersham) were diluted under sterile conditions and 50μCi, 15μCi and 100μCi respectively of each were transferred using Hamilton syringes (with 'Cheyne adaptors') to a single syringe. Simultaneously standards of appropriate dilutions were prepared from the dose solutions.

The mixture of radioactive isotopes was injected intravenously at 09.30 hrs. on the day following estimation of K_T. Thereafter the technique employed was as described previously (Shaw & Coppen, 1966) with the exception a) that the dilution of sample 'A' from passage through the column of resin was estimatef from the concentration of chloride in the eluate and plasma, and b) that assays of exchangeable sodium and total body water were from paired estimates from urine and plasma and urine and plasma or red cells respectively.

Abbreviations used

K_T	Total body potassium
Na_E	Exchangeable sodium
Cl_E	Exchangeable chloride
TBW	Total body water
ECW	Extracellular water
ICW	Intracellular water
K_{IN}	Mass of potassium in intracellular space

* Now National Radiological Protection Board, Belmont, Surrey.

Results

(1) Distribution of the data

All the variates examined - height, weight, TBW, K_T, Na_E, ECW, ICW and Cl_E - had skew distributions and therefore were transformed logarithmically before analysis to normalize their distributions.

(2) Validity of 'control group'

Because hospital personnel (the 'control group') may have been an unrepresentative sample of the population they were compared with a sample taken from the records of the RPS who were matched for height and weight.

There was no significant difference in the regressions of K_T on height or weight in the two groups, suggesting, therefore, that the hospital sample was as representative of the general population as were the more randomly selected group. Uncorrected values of K_T were used for this calculation only.

(3) Comparison of alcoholics and controls.

Table 1 gives the means of the main variates in controls, alcoholics and the two groups together. Alcoholics did not differ significantly from controls in any of the given parameters. A large number of relationships between parameters was calculated and again no significant differences were found between alcoholic and control subjects. In particular the regressions of K_T on weight (males), TBW and age and of K_{IN} on ICW did not differ significantly. The data on body composition for controls and alcoholics were considered as coming from one population and were analyzed together.

(4) Regression equations of control and alcoholic groups combined (Table 2)

Analysis of the data in control and alcoholic groups showed that Na_E, body weight or ECW did not regress on age, but there were significant regressions for K_T on age for both sexes. TBW and ICW did not regress significantly on age in females but such regressions were significant in the males.

In all regressions where equations have been given for both males and females there were significant differences between them in elevations but not in slopes. It is of interest that regressions of Na_E, TBW, ECW and ICW on weight were different in the sexes and this also was true for K_T on TBW and K_{IN} on ICW. The two regressions where there were no significant differences between males and females both involved Na_E (regressions on ECW and TBW).

Table 1. *Means of main estimated and determined variables*

	K_T (m.mol)			TBW (L)			ICW (L)		
Group	N	'mean'*	$R_1 - R_2$†	N	'mean'*	$R_1 - R_2$†	N	'mean'*	$R_1 - R_2$†
Male controls	20	3394	3243–3553	20	44.35	42.97–45.75	20	25.33	24.36–26.32
Male alcoholics	21	3247	3155–3342	21	42.86	41.75–44.00	21	24.22	23.34–25.11
All males	41	3318	3231–3406	41	43.58	42.69–44.49	41	24.75	24.11–25.42
Female controls	21	2210	2158–2262	21	32.97	32.21–33.75	21	18.57	17.90–19.25
Female alcoholics	15	2148	2068–2231	16	33.18	32.25–34.13	15	17.02	16.39–17.67
All females	36	2184	2140–2228	37	33.05	32.47–33.67	36	17.97	17.43–18.40

	ECW (L)			Na_E (m.mol)			Cl_E (m.mol)		
Group	N	'mean'*	$R_1 - R_2$†	N	'mean'*	$R_1 - R_2$†	N	'mean'*	$R_1 - R_2$†
Male controls	20	18.99	18.43–19.56	20	3099	3015–3186	20	2305	2238–2374
Male alcoholics	22	18.40	17.86–18.94	20	3015	2919–3113	20	2216	2147–2287
All males	42	18.67	18.28–19.08	40	3057	2993–3128	40	2260	2212–2310
Female controls	21	15.05	14.66–15.46	21	2425	2353–2500	21	1819	1774–1864
Female alcoholics	15	14.86	14.46–15.29	15	2487	2418–2558	15	1822	1766–1878
All females	36	14.97	14.69–15.27	36	2451	2404–2497	36	1820	1786–1855

*'mean' = antilog (log mean).
†R_1 = antilog (log mean − S.E.). R_2 = antilog (log mean + S.E.)
Analysis of variance showed no significant differences between control and alcoholic groups. In all variables mean values were significantly different in males and females ($p < 0.01$).

Table 2. *Regression Equations*

Group	N	Regression equations	Regression coefficient ± SE
M	39*	$\log(K_T) = 0.869 \log(\text{body weight}) + 1.906$	0.869 ± 0.120
M	40	$\log(K_T) = 1.078 \log(\text{TBW}) + 1.753$	1.078 ± 0.208
F	36	$\log(K_T) = 0.765 \log(\text{TBW}) + 2.181$	0.765 ± 0.203
M	40	$\log(K_T) = -0.003 (\text{age}) + 3.648$	−0.003 ± 0.0009
F	35	$\log(K_T) = -0.002 (\text{age}) + 3.424$	−0.002† ± 0.0008
M	40	$\log(Na_E) = 0.704 \log(\text{body weight}) + 2.176$	0.704 ± 0.074
F	35	$\log(Na_E) = 0.534 \log(\text{body weight}) + 2.436$	0.534 ± 0.106
M + F	75	$\log(Na_E) = 0.920 \log(\text{ECW}) + 2.311$	0.920 ± 0.039
M + F	76	$\log(Na_E) = 0.776 \log(\text{TBW}) + 2.213$	0.776 ± 0.092
M	40	$\log(\text{TBW}) = 0.650 \log(\text{body weight}) + 0.432$	0.650 ± 0.133
F	36	$\log(\text{TBW}) = 0.426 \log(\text{body weight}) + 0.754$	0.426 ± 0.113
M	40	$\log(\text{TBW}) = -0.002 (\text{age}) + 1.703$	0.002 ± 0.0007
M	40	$\log(\text{ECW}) = 0.550 \log(\text{body weight}) + 0.249$	0.550 ± 0.144
F	36	$\log(\text{ECW}) = 0.532 \log(\text{body weight}) + 0.227$	0.532 ± 0.128
M	40	$\log(\text{ICW}) = -0.003 (\text{age}) + 1.514$	−0.003 ± 0.0008
M	40	$\log(\text{ICW}) = 0.785 \log(\text{body weight}) - 0.065$	0.785 ± 0.168
F	36	$\log(\text{ICW}) = 0.363 \log(\text{body weight}) + 0.598$	0.363† ± 0.162
M	40	$\log(K_{IN}) = 0.916 \log(\text{ICW}) + 2.227$	0.916 ± 0.171
F	35	$\log(K_{IN}) = 0.551 \log(\text{ICW}) + 2.640$	0.551 ± 0.152

In none of the values above were there significant differences between alcoholics and controls.
Na_E body weight and ECW did not regress with age.
K_T did not regress on body weight in females.
M = male controls and alcoholics.
F = female controls and alcoholics.
*One marked aberrant result omitted.
†For these regression coefficients $p < 0.05$: for all others $p < 0.001$.

TABLE 3

ESTIMATED AND DERIVED VARIATES

Estimated variates	Abbreviation	Units	Method of calculation
Height	-	cm	-
Weight	-	kg	-
Total body potassium	K_T	mmol	-
24-hour exchangeable sodium	Na_E	mmol	$\frac{^{24}Na \text{ counts given} - ^{24}Na \text{ counts lost from body}}{\text{counts } ^{24}Na/\text{mmol Na in 'spot' urine (sample C) or diluted plasma (sample E)}}$
Distribution volume of bromide	DBr	1	$\frac{^{82}Br \text{ counts given} - ^{82}Br \text{ counts lost from body}}{\text{counts } ^{82}Br \text{ in sample A} \times \frac{[Cl] \text{ in plasma}}{[Cl] \text{ in sample A}}}$
Total body water	TBW	1	$\frac{^{3}H_2O \text{ counts given} - ^{3}H_2O \text{ counts lost from body}}{^{3}H_2O \text{ counts/ml in spot urine, plasma or red cells}}$
Concentrations of sodium, potassium, and chloride in plasma	[Na], [K], [Cl]	mmol/1	
Derived variates			
Extracellular water	ECW	1	DBR x 0.9
Intracellular water	ICW	1	TBW - ECW
Exchangeable chloride	Cl_E	mmol	DBr x [Cl]

Discussion

There are two parts to the discussion of the data presented in this paper - one concerned with the initial aim of comparing body composition in alcoholic and control individuals, and the other concerned with the absolute values of, and relationships between, the variates studied:-

(1) Comparison of alcoholic and control subjects.

The data showed that the alcoholics did not differ from the controls in any of the measures derived variates, nor in the various relationships between these variates. This was a surprising finding. As mentioned above, the acute changes in the distribution of water and electrolytes occurring in acute alcoholism and in delirium tremens have been well documented, but these are relatively transient phenomena. In the present study a peripheral dificiency of potassium to mirror the lack of this cation, which seems to be a feature of the brains of alcoholic subjects determined post mortem was being sought. A loss of this type might have been the accompaniment of 'silent' loss of muscle mass, a condition which has been described a number of times in chronic alcoholism. Alternatively, it might have occurred if there had been changes in the composition of the intracellular phase. There was no evidence of peripheral potassium loss, or, as stated above, of any changes in electrolytes or water compartments. It seemed therefore that either there were no changes or they were corrected during the short period of abstinence.

It seems relevant to mention here some recent work by Shaw and his team (Shaw 1974) on the kinetic behaviour of radioactive tryptophan (^{3}H-L-tryptophan) in depressives both while ill and after recovery in the context of MacSweeney's preliminary findings on tryptophan levels in a group of nondepressed primary alcoholics. In the former report depressives, whether ill or after recovery, showed a reduction in the tryptophan mass present in the body when compared with controls. In the second study the non-depressed (dry) alcoholic population showed significantly lower total tryptophan levels with low diffusible tryptophan when compared with a suitably matched control group.

Two double blind trials to test the possible prophylactic value of L-tryptophan both in preventing relapse in recurrent depressives and in preventing a return to alcohol abuse in dry primary alcoholics is currently being set up by the author. (The dose of L-tryptophan (Optimax) proposed is 6 tablets daily i.e. half the conventional dose).

(2) Relationships between the variates studied

The control data on body composition are not dissimilar to those collected by Moore, Olesen, McMurray, Parker, Ball & Boyden (1963). However, the values presented here were not normally distributed and the use of logarithmically transformed values made direct comparisons difficult.

The absence of regressions of Na_E and ECW on age probably indicated the relative lack of change in the extra-cellular compartment with ageing. The loss of potassium with age taken with the absence of a regression of weight on age may mean that there was cellular loss which was partly balanced by gains by other tissues, presumably in the form of fat. The significant decrease in ICW with age in males was noteworthy and suggested that loss of cellular mass with increasing age is prominent in men.

The significant differences between males and females in the elevations of the regressions of Na_E, TBW, ECW and ICW on weight probably reflected the different contributions of the fluid compartments to body weight in the two sexes, the contribution by these spaces to body mass being smaller in the females. Similarly, the differences in elevation of K_T on TBW was probably a function of the proportionately smaller cellular mass in women as a part of total body water.

SUMMARY

(1) Height, weight, total body potassium, exchangeable sodium, bromide space, total body water and concentrations of sodium, potassium and chloride in plasma were measured in control subjects and individuals suffering from alcoholism, with techniques which included body counting and a multiple isotope method using ^{24}Na, ^{82}Br and $^{3}H_{2}0$.

(2) No differences were found between control and alcoholic subjects so there was no evidence that chronic alcoholism altered body composition. In particular there was no evidence of cellular damage or loss which would have been reflected in changes in K_T or K_{IN}.

(3) The data were combined and were analysed to give information on the relationships of the variates.

(4) On-going work by the author on tryptophan metabolism in primary alcoholics is compared to Shaw's findings on the kinetic behaviour of tryptophan in affective disorder. The possible prophylactic value of L-tryptophan (Optimax) in preventing both recurrent depression and recurrent alcohol abuse is outlined.

(5) Data on body composition in normal subjects not hitherto available in the literature is provided.

ACKNOWLEDGEMENTS:

This work would not have been possible without the kind co-operation of Dr. Julius Merry and his patients. The author is also grateful to Dr. D.M. Shaw and Dr. Tony Johnson (M.R.C.) for much advice and to Mr. Norman Woolcock for technical assistance.

REFERENCES

Beard, J.D. & Knott, D.H. Fluid and electrolyte balance during acute withdrawal in chronic alcoholic patients. Journal of the Americal Medical Association, Vol. 204, 133-139, 1968.

Bush, F. Energy observation in radium therapy. British Journal of Radiology, Vol. 19, 14-21, 1946.

Coppen, A. The Marke-Nyman temperament Scale: an English translation, Brit. J of Med. Psychol. Vol 39, 55-59, 1966.

Coppen, A. & Shaw, D.M. Mineral metabolism in melancholia. British Medical Journal, Vol 2, 1439 - 1444, 1963.

Ekbom, K., Hed, R., Kirstein, L. & Astrom, K.E. Muscular affections in chronic alcoholism. Archives of Neurology, Vol. 10, 449 - 458, 1964.

Faris, A.A. & Reyes, M.G. Reappraisal of alcoholic myopathy. Journal of Neurology, Neurosurgery & Psychiatry, Vol. 34, 86 - 92, 1971.

Israel, Y., Kalant, H., Le Blanc, E., Bernstein, J.C. & Salazar, I. Changes in cation transport and (Na + K)-activated adenosene triphosphate produced by chronic administration of ethanol. Journal of Pharmacology, Vol. 174, 330 - 336, 1970.

Eysenck, S.13.G & Eysenck, H.J. The questionnaire measurement of psychoticism. Psychological Med. Vol. 2, 50 - 55, 1972.

Kalant, H. & Israel, Y. Effect of ethanol on active transport of cations. In: Biochemical Factors in Alcoholism (ed. R.P. Marckel), Pergamon Press, Oxford, 1967.

Kuriyama, K., Sze, P.Y. & Rauscher, G.E. Effect of acute and chronic ethanol administration on ribosomal protein synthesis in mouse brain and liver. Life Sciences, Vol. 10, 181 - 189, 1971.

MacSweeney, D.A. The personality structure of primary alcoholics and endogenous depressives. Proceedings of 7th Internat. Congress on Suicide prevention, Amsterdam, The Netherlands, 1973.

Martin, H.E., McCuskey, C. & Tupikova, N. Electrolyte disturbance in acute alcoholism with particular reference to magnesium. American Journal of Clinical Nutrition, Vol.7, 191 - 196, 1959.

Moore, F.D., Olesen, K.H., McMurray, J.D., Parker, H.U., Ball, H.R. & Boyden, C.M. The body cell mass and its supporting environment. W.B. Saunders Company, Philadelphia - London, 1963.

Nielsen, J. Magnesium metabolism in acute alcoholics. Danish Medical Bulletin, Vol. 10., 225 - 233, 1963.

Perkoff, G.T. Dioso, M.M., Bleisch, V. & Klinkerfuss, G. A spectrum of myopathy associated with alcoholism. Annals of International Medicine, Vol. 67, 481 - 492, 1967.

Shaw, D.M. The kinetic behaviour of ^{3}H-L-tryptophan in affective disorder Psychological Med. (in press).

Shaw, D.M., Camps, F.E., Robinson, A.E., Short, R. & White, S. Electrolyte content of the brain in alcoholism. British Journal of Psychiatry. Vol. 116, 185 - 193, 1970.

Shaw, D.M. & Coppen, A.J. Potassium and water distribution in depression. British Journal of Psychiatry, Vol. 112, 269 - 276, 1966.

Shaw, D.M. Frizel, D., Camps, F.E. & White, S. Brain electroytes in depressive and alcoholic suicides. British Journal of Psychiatry, Vol. 115, 69 - 79, 1969.

Sun, A.Y. & Samorajski, T. Effects of ethanol on the activity of adenosine triphosphate and acetylcholinesterase in synaptosomes isolated from guinea pig. Journal of Neurochemistry, Vol. 17, 1365 - 1372, 1970.

Winokur, G., Reich, T., Rimmer, J. & Pitts, Jnr., F.N. Alcoholisms III; Diagnosis and familial psychiatric illness in 259 alcoholic probands. Arch. Gen. Psychiat, Vol. 23, 104 - 111, 1970.

IMPAIRMENT OF MEMORY AFTER PROLONGED ALCOHOL CONSUMPTION IN MICE

G. Freund

Veterans Administration Hospital and Department of Medicine, University of Florida, Gainesville, Florida 32602, U. S. A. *

The association of chronic excessive ethanol consumption with anatomical and functional brain impairment has long been observed by clinicians. It is controversial whether this association in some or all patients is merely an unrelated coexistence of alcoholism with other degenerative brain disorders or whether an actual cause-effect relationship exists. If chronic alcohol consumption were truly related to mental impairment, the question arises if impaired brain function is a cause or a result of chronic excessive alcohol consumption (Goodwin and Hill, 1974). Assuming that chronic ethanol consumption is indeed the cause of CNS impairment, it can be postulated that malnutrition that may accompany chronic alcoholism, rather than alcohol per se, is the cause (Victor, Adams, and Collins, 1971). Alternatively, ethanol itself, alone or in conjunction with malnutrition, could cause CNS impairment (Freund, 1973).

POSSIBLE INTERPRETATIONS OF A CORRELATION OF "A" WITH "B"

[A] – Chronic alcohol consumption [B] – CNS impairment

1) Unrelated: "A" and "B" coexist

2) Related: "B" causes "A"

"A" causes "B":
(Malnutrition, Ethanol toxicity or both)

* Supported by the Veterans Administration (MRIS 2932) and PHS Grant ROI-AA00218

Only a clinical, controlled prospective study could definitely resolve this controversy. This would necessarily involve the consumption of large amounts of alcohol by the experimental subjects for many years and the induction of physical and perhaps psychological dependence. Such a study would obviously be unethical. The next best approach to the problem of potential CNS toxicity of ethanol is to approach alcohol like any other drug and to investigate it in animals with classical pharmacological techniques. Every drug is subjected to extensive chronic animal toxicity studies before it can be used in humans. In addition, investigations of chronic CNS toxicity of ethanol in animals may potentially yield leads applicable to humans regarding pathogenesis, prophylaxis, and therapy. The basic technology is available, and beyond the scope of neuropsychopharmacology an understanding of the biology of learning and memory owes much to animal studies (Barondes, 1970; Horn et al, 1973; McGaugh, 1973; Deutsch, 1973).

It has been demonstrated previously that chronic ethanol consumption impairs a variety of learning tasks in rodents (Freund, 1970; Walker and Freund, 1971, 1973), that the impairment is irreversible, dose-dependent, and cannot be attributed to malnutrition, stress, change in sensory perception (Freund and Walker, 1971), or sedation (Freund, in press). It is therefore of interest to determine whether chronic ethanol consumption affects memory as well as learning. The technique employed in this investigation is a one-trial, passive, avoidance-learning task that is widely used to study the effects of electroconvulsive shock, drugs, and other manipulations upon retention (Schneider et al, 1974).

METHODS

Female, 3-month-old C-57 BL 6/J mice were fed a liquid diet containing 35% of calories derived from ethanol, fortified with vitamins and minerals as described previously (Freund, 1970). After 4 weeks on this diet, the ethanol-derived calories were increased to 38%. Sucrose animals were pair-fed with the same basic diet, except that it contained isocaloric amounts of sucrose instead of ethanol. The second control group received laboratory chow (Ralston Purina Co., St. Louis, Missouri) and water ad libitum.

Mice were trained in the step-through apparatus developed by Jarvik and Kopp (1967), which consists of a small, clear plexiglass chamber attached to a large, dark compartment. The two chambers are separated by a partition with a 2.5 x 2.5 cm hole. Two lights (Tensor lamps set at low intensity) on both sides of the clear chamber illuminate the small compartment.

Mice are placed into the lighted, small compartment and will explore the dark, large compartment through the hole. Inside the dark chamber a 0.5 mA foot-shock was administered through the metal floor as soon as the mouse activated a touch sensitive electrical switch 8 cm from the entry into the dark compartment. Latencies were recorded automatically by the activation of touch relays connected to electrical timers. The time interval between placement of a mouse into the small, lighted compartment and the entry into the large, dark compartment was measured for each animal (response latency at acquisition trials). After various time intervals the trials were repeated under identical conditions (retention trials). Retention trials were terminated if a mouse remained in the clear chamber for 480 seconds. The acquisition latencies were subtracted from the retention trial latencies (latency differences). Mann-Whitney U tests were used to analyze differences in the latency difference scores (retention minus acquisition latencies) between the various groups. An increase in step-through latency (the time a mouse stays in the small compartment) at retention testing was assumed to represent increased retention ("memory") for the foot-shock punishment when entering the large, dark compartment.

Two basic experimental designs were employed: a) prolonged ethanol consumption preceded both acquisition and retention trials (table 1); b) alcohol was fed between acquisition and retention trials (table 2). The experimental details are listed in the tables.

RESULTS

The step-through retention latencies in mice fed the stock diet decreased with the passage of time (Figure 1) with almost no observable retention left at 24 weeks. The number of mice in each experiment is indicated below the abscissa. Based on these data, it was considered possible to demonstrate impaired retention 8 to 10 weeks after acquisition when control mice still had appreciable retention.

Results of studies using design a): Ethanol and sucrose-containing diets were fed for 4 months, then laboratory chow for 2 weeks, followed by acquisition trials and 1 or 8 weeks later by retention trials. It is apparent from Table 1 and Figure 2 (left side) that the latency times of the alcohol-treated mice were shorter than those of control mice.

Figure 1

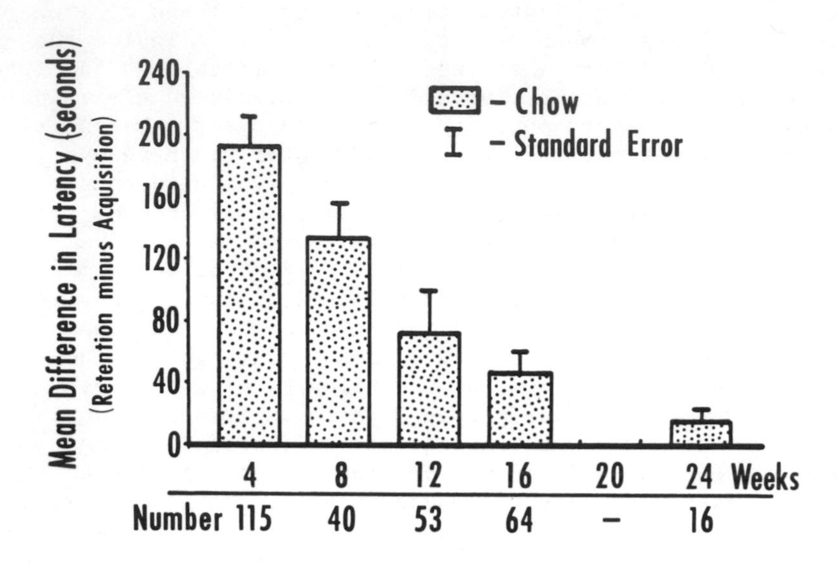

Figure 2

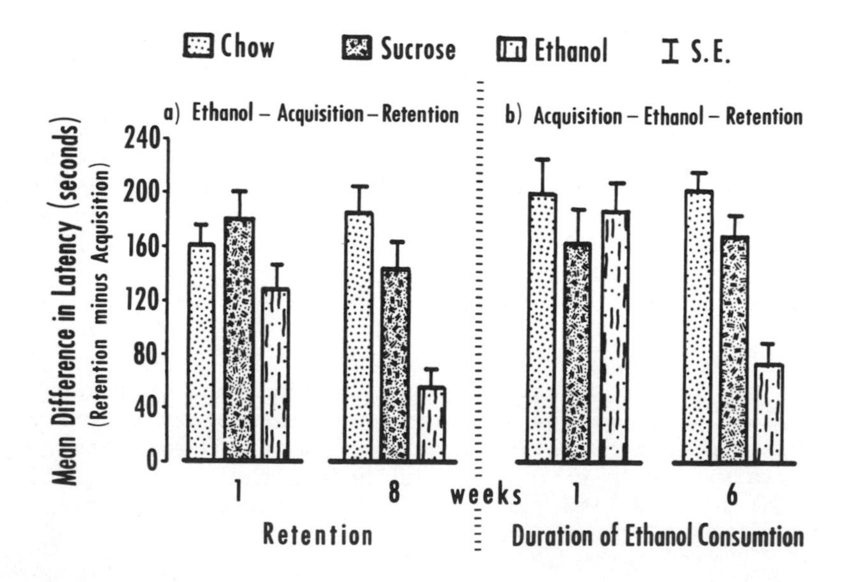

Table 1

Order of Procedures					$\overline{X}$ Latencies: Retention-Acquisition					
N*	Ethanol	Chow	Acquisition Trials	Retention Trials **	Eth - Sucr		Eth - Chow		Sucr - Chow	
					Sec	p	Sec	p	Sec	p
250	4 months	14 days	✓	1 week	-50	0.026	-32	0.019	18	N.S.
312	4 months	14 days	✓	8 weeks	-87	0.0002	-127	0.0001	-24	N.S.

Table 2

Order of Procedures						$\overline{X}$ Latencies: Retention-Acquisition					
N*	Acquisition Trials	Chow	Ethanol	Chow	Retension Trials **	Eth - Sucr		Eth - Chow		Sucr- Chow	
						Sec	p	Sec	p	Sec	p
221	✓	1 week	6 weeks	1 week	8 weeks	-92	0.002	-133	<0.001	-41	N.S.
168	✓	1 week	1 week	6 weeks	8 weeks	24	N.S.	- 12	N.S.	-36	N.S.
85	✓	1 day	1 week	7 weeks	8 weeks	50	N.S.	13	N.S.	-37	N.S.

*N: Total number of mice/experiment
**Time interval between acquisition and retention trials.

Results of studies using design b): Ethanol consumption interposed between acquisition and retention trials (8 weeks after acquisition trials) are shown in Table 2 and Figure 2 (right side). The latencies of ethanol-treated mice were shorter than control mice if they consumed alcohol diets for 6 weeks, but not significantly different if ethanol consumption lasted only 1 week. The acquisition latencies of all control and ethanol-treated groups were statistically indistinguishable.

DISCUSSION

Mice that have consumed alcohol over prolonged periods of time, rapidly step from a small, lighted compartment into a large, dark compartment where they have previously received a foot-shock. Control mice seem to have retained the foot-shock experience because they do not enter the large chamber at all or only after long hesitation. Such behavior has generally been interpreted as impaired retention of information, and impaired retrieval is less likely (Schneider et al, 1974). Similar results have been reported when electroconvulsive seizures were administered, usually within seconds after the learning trials or by inhibitors of protein synthesis at the time of learning or within minutes thereafter (v.s.). In addition, certain prerequisites must be fulfilled for the demonstration of retention deficits, such as appropriate arousal and a minimum degree of functional and anatomical integrity of the CNS. Many processes, at all levels of biological organization, can therefore potentially interfere with the chain of events beginning with sensory registration of external input leading to the storage and retrieval of information as is evident from the animal experimental literature (Deutsch, 1973). In analogy to experiments using convulsive or protein synthesis inhibition treatments, the impairment described here is probably a result of decreased retention per se, although decreased retrieval of information has not been excluded. The possibility that impaired retention performance in alcohol-treated animals is a result of changes in sensitivity of visual perception or a decreased emotional reaction to shock is very remote because the extensive clinical literature has never documented such effects. The large majority of clinical disorders of memory cannot be explained at present, at least not on the basis of localized brain damage (Whitty and Lisham, 1966).

In contrast to the findings reported here, Walker and Hunter (1974) found no impairment of retention of shuttle-box, avoidance learning in rats when ethanol was fed for 4 1/2 months between acquisition and retention trials. They attributed the failure to demonstrate a retention deficit to the possibilities of overlearning (daily sessions for 30 days) or the long acquisition-retention interval of 7 1/2 months, which may represent remote, rather than more short-term memory. The conditions for the present experiment were selected so that control animals had retention latencies intermediate between perfect retention and complete loss of retention.

CONCLUSION

In conclusion, this study indicates that under the conditions of this experiment, prolonged but not short-term alcohol ingestion results in a deficit of retention performance. This deficit occurs under nutritionally controlled conditions in mature, adult animals. It occurs when alcohol administration precedes both learning and retention trials. Retention deficits are also demonstrable when ethanol is administered between learning and retention trials but is begun after long-term memory consolidation has generally thought to have taken place, that is 24 hours after learning trails (Barondes, 1970). Possibly the mechanisms whereby alcohol consumption interferes with retention performance is different under the two sets of experimental conditions. The results of the present investigation do not define the mechanism whereby chronic ethanol consumption impairs retention performance. But a better understanding of this mechanism could lay the foundation for rational prophylaxis or therapy of the chronic brain syndrome in alcoholic patients.

REFERENCES

Barondes, S. H. Multiple steps in the biology of memory. In I. O. Schmitt (Ed.), The Neurosciences. Second Study Program, pp. 272-278. New York, Rockefeller University Press, 1970.

Deutsch, J. A. (Ed.). The Physiological Basis of Memory, New York, Academic Press, 1973.

Freund, G. Impairment of shock avoidance learning after long-term alcohol ingestion in mice. Science, 168:1599-1601, 1970.

Freund, G., and Walker, D. W. Impairment of avoidance learning by prolonged ethanol consumption in mice. J. Pharmacol. Exp. Ther., 179:284-292, 1971.

Freund, G. Chronic central nervous system toxicity of alcohol. Ann. Rev. Pharmacol., 13:217-227, 1973.

Freund, G. Normal shuttle box avoidance learning after chronic phenobarbital intoxication in mice. (in press)

Goodwin, D. W., and Hill, S. Y. Chronic effects of alcohol and other psychoactive drugs on intellect, learning and memory. In J. Rankin (Ed.), Effects of Chronic Use of Alcohol and Other Psychoactive Drugs on Cerebral Function, Addiction Research Foundation, Toronto, 1974. (in press)

Horn, G., Rose, S. P. R., and Bateson, P. P. G. Experience and plasticity in the central nervous system. Science, 181:506-514, 1973.

Jarvik, M. E., and Kopp, R. An improved one-trial passive avoidance learning situation. Psycholog. Rep., 21:221-224, 1967.

McGaugh, J. L. Drug facilitation of learning and memory. Ann. Rev. Pharmacol., 13:229-241, 1973.

Schneider, A. M., Tyler, J., and Jinich, D. Recovery from retrograde amnesia: a learning process. Science, 184:87-88, 1974.

Victor, M., Adams, R. D., Collins, G. H. In F. Plum and F. H. McDowell (Eds.) The Wernicke-Korsakoff Syndrome, Philadelphia, F. A. Davis Company, 1971.

Walker, D. W., and Freund, G. Impairment of shuttle box avoidance learning following prolonged alcohol consumption in rats. Physiol. Behav., 7:773-778, 1971.

Walker, D. W., and Freund, G. Impairment of timing behavior after prolonged alcohol consumption in rats. Science, 182:597-599, 1973.

Walker, D. W., and Hunter, B. E. Prolonged consumption in the rat: absence of retrograde amnesia for an avoidance response. Pharmacol. Biochem. Behav., 2:63-66, 1974.

Whitty, C. W. M., and Lisham, W. A. Amnesia in cerebral disease. In C. W. M. Whitty and O. L. Zangwill (Eds.) Amnesia, pp. 36-76. New York, Appleton-Century-Crofts, 1966.

MAXIMIZATION OF ETHANOL INTAKE IN THE RAT

Roy A. Wise, Center for Research on Drug Dependence

Department of Psychology, Sir George Williams University

Montreal, Quebec, Canada*

Over the last few decades there has been a great deal of effort invested in attempts to produce lower animal analogues for the human alcoholic. The literature on this problem is substantial and has been extensively reviewed; the recurrent conclusion of successive reviews has been that no viable lower animal model has been found (Lester, 1966; Mello, 1973; Myers and Veale, 1972).

The pursuit of a lower animal model of alcoholism has been guided by several criteria set forth by various authors (see e.g., Lester and Freed, 1973). Of these criteria the minimal generally accepted requirements are two critical characteristics: the animal must be physically dependent on ethanol, and must maintain this physical dependence by its own voluntary ethanol intake. Both criteria have provided difficulties when rodents were used as subjects; it took many years to produce physical dependence, and it has not yet been possible to produce voluntary intake sufficient to either maintain or establish physical dependence. Most investigators have focussed their attention to the problem of producing dependence, and recently some have been successful in producing it. The thrust of this work will be summarized below, and the critical difference between successful and unsuccessful attempts will be discussed. A few other investigators have attempted to produce high levels of voluntary intake, ignoring, for the most part, the problem of physical dependence. These latter studies are the main focus of the present paper.

---

* This work was supported by grants from the Non-Medical Use of Drugs to Z. Amit and from the Medical Research Council and United States Public Health Service (AA00203) to Roy A. Wise.

PHYSICAL DEPENDENCE

The basic problem is that lower animals generally do not voluntarily drink enough ethanol to intoxicate themselves; they usually do not drink enough to even approach intoxication. While some strains of rat and mouse have unusual willingness to drink ethanol solutions (Brewster, 1969; Eriksson, 1968; Mardones, 1960; Rodgers, 1966), most rodents prefer water to ethanol solution when ethanol concentrations higher than 8 or 10% are offered. When rodents drink more highly concentrated solutions than this they usually drink small amounts.

Thus only man seems to become alcoholic spontaneously, when ethanol is simply made readily available. Attempts to produce an animal analogue have consequently attempted to experimentally augment the low initial ethanol intake of lower animals. The rationale for continuing to seek an animal model for alcoholism has not been made explicit, but what seems to have been a common general assumption was that while animals did not provide a good model for the _development_ of alcoholism, they still might provide a good model for _later stages_ of alcoholism. Thus many investigators continued a line of study which began to include manipulations designed to augment the initially low ethanol intake in lower animals.

The immediate goal of much of this research became the production of ethanol physical dependence. Again it might be suggested that there was a widely held implicit assumption guiding many studies: here the assumption seems to have been that while naive animals would not drink significant amounts of ethanol solution, physically dependent animals might.

The problem here was seen as one of getting a lot of ethanol into uncooperative animals. This has been accomplished in a number of ways. The taste of ethanol has been masked by other tastes to encourage drinking (Lester and Greenberg, 1952). Ethanol intake has been rewarded by food in hungry animals (Keehn, 1969; Mello and Mendelson, 1965; Senter, Smith and Lewin, 1967), and animals have been maintained on ethanol-diluted liquid diets as their only source of nutrients (Branchley, Rauscher and Kissin, 1971; Freund, 1969; Ogata, Ogata, Mendelson and Mello, 1972). Ethanol solutions have also been offered as an only source of fluids (Carey, 1972; Gibbons, Kalant, Le Blanc and Clark, 1971; Ratcliffe, 1972). Animals have been shocked if they failed to drink ethanol at a determined rate (Cicero, Myers and Black, 1968; Mello and Mendelson, 1971). Animals have also been non-contingently shocked on the assumption that this would produce anxiety, which in turn would produce dependence (Casey, 1960). The schedule-induced polydipsia technique has been used to induce ethanol consumption (Falk, Sampson and Winger, 1972; Lester,

1961; Ogata et al., 1972). Ethanol vapor has been put in animals' cages so that they must take ethanol whenever they breathe (French and Morris, 1972; Goldstein, 1972; Goldstein and Pal, 1971). Electrical stimulation of the hypothalamic drinking circuitry has been used to make animals drink ethanol solutions (Amit, Stern and Wise, 1970; Wayner, Greenberg, Carey and Nolley, 1971). All of these methods have successfully caused animals to take in ethanol, but the intake can scarcely be termed "voluntary." Moreover, with only recent exceptions (Branchley et al., 1971; Falk et al., 1972; French and Morris, 1972; Freund, 1969; Goldstein, 1972; Goldstein and Pal, 1972; Ogata et al., 1972; Ratcliffe, 1972), the animals have not become physically dependent, and with even fewer exceptions (Amit et al., 1970), the animals have not sustained an elevated voluntary intake of ethanol when the experimental restrictions were lifted and animals were given a free choice between ethanol solutions and water.

Stronger measures have also been taken. Since the animals will not take in enough ethanol on their own, experimenters have injected it for them. It has been injected into brain (Myers and Veale, 1969), vein (Deneau, Yanagita and Seevers, 1969; Woods, Ikomi and Winger, 1971), and stomach (Ellis and Pick, 1970; Essig and Lam, 1968), again with mostly discouraging results. There have been signs of dependence in recent studies, but no signs of subsequent sustained voluntary intake.

Four of the paradigms which have been successful in producing dependence have involved active ingestion of ethanol solution by the animals: these are ethanol solution as an only source of fluid (Ratcliffe, 1972), ethanol liquid diets as an only source of calories (Branchley et al., 1971; Freund, 1969; Ogata et al., 1972), ethanol vapor as an only source of oxygen (French and Morris, 1972; Goldstein, 1972; Goldstein and Pal, 1971), and schedule-induced polydipsia (Falk et al., 1972). The schedule-induced polydipsia technique seems to best illustrate what appears to be the essential difference between the large number of failures to produce dependence, and the few recently successful studies.

Schedule-induced polydipsia has been used with ethanol by a number of investigators (Everett and King, 1970; Falk et al., 1972; Freed, 1972; Lester, 1961; Meisch and Thompson, 1972; Ogata et al., 1972; Senter and Sinclair, 1967). The phenomenon was first explored by Falk (1961), who discovered that animals on well-spaced food-reward schedules tend to do a good deal of post-pellet drinking of water. Animals given a small food pellet once every minute or once every two minutes will drink up to half their body-weight in water during a single 3 hour session. This technique will cause excessive drinking of ethanol solution as well, if low-concentration ethanol solutions are used. However, when animals are intoxicated by this method on a daily basis, they do not usually develop dependence. What Falk et al. (1972) have done to successfully produce dependence is to give his animals several ethanol polydipsia sessions per day, one every four hours. In this way the blood ethanol levels are maintained at high concentration on a continuous basis throughout the day. Continuously high blood ethanol also seems to typify the other studies which have produced dependence; in all cases the animals take in ethanol every few hours. In the inhalation method (Goldstein and Pal, 1971) the animals take in ethanol every few seconds. Thus it appears that if ethanol is maintained in the system at high levels on a continuous basis, ethanol dependence can be produced in rodents. The continuous maintenance of high systemic ethanol levels seems to be a necessary condition, since in studies where ethanol has been given less often - usually once per day - there has been a uniform failure to produce dependence.

MAXIMIZATION OF VOLUNTARY INTAKE

In none of the studies reporting ethanol dependence is there any sign that the animals have an increased motivation to drink ethanol voluntarily: when tested during subsequent ethanol withdrawal the animals show their normal aversion to concentrated alcohol solutions (see, e.g., Begleiter, 1974).

Several years ago, Drs. Amit, Stern and I (1970) discovered a method which did produce a permanent, voluntary preference for concentrated ethanol solutions--solutions that were initially aversive to our rats. At first we thought that we saw signs of physical dependence in our animals, but it is now clear that we did not. For the last few years we and our students have been exploring various aspects of our finding, with an aim to maximizing voluntary intake and preference, and ignoring, for the time, the question of physical dependence.

Our initial study was really in the tradition of studies based on a dependence model, however, in that we assumed that if we got enough alcohol into the animals for a long enough period, the animals might begin to drink on their own. We used hypothalamic electrical stimulation to make the animals drink 5-7 ml of aversive (about 20%) ethanol solutions once per day for 30 days, while giving the animals free access to the solutions in their home cages on every other day. At the end of each half hour session, especially at the beginning of the experiment, the animals were clearly intoxicated; they had a good deal of trouble getting to the spout to continue drinking when stimulation was on.

At the end of 30 days, the animals had begun to drink in their home cages, and even though the brain stimulation treatment was now terminated, the animals continued to increase their intake until they were drinking the ethanol exclusively. This preference continued for the rest of the animal's lives--over another year in several cases.

In this study there were a number of factors which had been designed to maximize our animals' voluntary intake, and we have been exploring the importance of each of them. Amit and Stern (1971) went on to look at the brain stimulation itself; was it the stimulation *per se* which caused increasing drinking? Was it the association of the (pleasant, rewarding) brain stimulation with drinking or with ethanol? They found that it was apparently the stimulation itself which was important; stimulation of animals in a barren box was as effective as the same stimulation used to induce ethanol consumption. Our control animals also drank unexpected (but lesser) amounts; because of this finding we have also explored several factors of our paradigm which were independent of stimulation.

Genetic Factors

It is well known that genetic differences contribute to the variability of voluntary ethanol intake as measured in a variety of paradigms (Brewster, 1969; Eriksson, 1968; Mardones, 1960; Rodgers, 1966). We have found that our paradigm is no exception. In both stimulated and unstimulated control animals given ethanol on the every-other-day, free-choice schedule used in most of our studies, males drink more than females, Wistars drink more than Sprague-Dawleys or hoodeds, and Tryon maze brights drink more than Tryon maze dulls (Z. Amit and M.H. Stern, personal communication).

Moreover, there can be significant differences in ethanol intake between animals of the same strain purchased from different breeders. For example, we (Wise, 1974) compared the intake of unstimulated animals of the Sprague-Dawley and Wistar strains purchased from two different breeders. The Wistars drank more than the Sprague-Dawleys, but the difference was statistically significant only for the animals from one of the breeding farms. The intake of the Sprague-Dawleys from one farm was almost double that of the Sprague -Dawleys from the other farm. The strains can be distinguished by slight differences in head size and shape, and there was no question about having received the wrong strain of animals. To ensure that we had not simply seen differences due to sampling variation we replicated this entire study, and found almost identical intake levels from each replication group. The initial study was run in the winter, and the replication run in the spring, so the intake levels seem not sensitive to difference in seasonal stresses, and our hypothesis is that genetic drift between the breeding stocks of different suppliers can be sufficient to produce these dramatic differences in voluntary ethanol ingestion.

Schedule of Presentation

The schedule on which ethanol is offered to rats proves to be an extremely important variable in studies of voluntary intake. It has been known for some time that the type of ethanol exposure influences voluntary intake, and our original paradigm took this into consideration. First, we always offered our animals water when we offered them ethanol solutions. Periods of forced ethanol (ethanol solution available as an only source of fluid) tend to produce reluctance to drink ethanol under subsequent free choice conditions (Carey, 1972; Richter, 1956; Veale and Myers, 1969). Thus we always offered ethanol in free choice with water. Second, we first offered our animals weak ethanol solutions, and only gradually raised the concentration to normally aversive levels. This is termed acclimation (Veale and Myers, 1969), and is known to lead to the animals' willingness to drink more concentrated solutions than would be accepted otherwise. Third, we gave our animals ethanol solutions only on alternate days. This was because of the finding that periods of ethanol withdrawal led to elevated subsequent voluntary intake (Sinclair and Senter, 1967). We consequently withdrew ethanol on an every-other-day basis.

We have subsequently assessed the importance of these factors in our paradigm. We find that acclimation gives animals an initially higher intake level, but that this effect produces only temporary differences between animals over the course of our 72-day paradigm (Wise, 1973). Animals given 20% solutions on the first day of exposure come to drink as much as animals previously acclimated to 20% solutions within the first three weeks of exposure. Thus acclimation gives animals an initial head-start, but the advantage is lost in a few weeks.

Free choice conditions do not seem to be particularly important in our paradigm (Wise, 1973). The same subseqeunt voluntary intake levels are seen after 30 days where 20% ethanol solution is an only source of fluid and after 30 days where 20% ethanol solutions are always offered in free choice with water. This is true regardless of whether ethanol solutions are offered every day or every other day.

Intermittent exposure turns out to have a very dramatic influence on subsequent intake (Wise, 1973). We find that animals given 20% ethanol solutions on an every-other-day basis develop a mild preference for these solutions, where animals given the same solutions on an every day basis do not. Thirty days of intermittent exposure produces voluntary intake levels on the order of 10g/kg/day, where the same period of continuous exposure produced less than 2g/kg/day. This finding holds regardless of whether the animals are given their ethanol in free choice with water or whether they are given it as an only source of fluid. That is, if animals are given only water to drink on even-numbered days, and only 20% ethanol solutions on odd-numbered days, they will subsequently show high intake levels under free-choice conditions. This effect is seen both in hypothalamically stimulated animals and also in unstimulated animals (Engel, 1972; Wayner and Greenberg, 1972; Wayner, Greenberg, Tartaglione, Nolley, Fraley and Cott, 1972; Wise, 1973).

Hypothalamic Stimulation Treatment

In our hands, lateral hypothalamic stimulation treatment has consistently produced elevated ethanol intake in the home cage. This has not been true for other investigators, and we have had occasions when our stimulated groups did not drink more than control animals, although we have never had a case where control animals drank more that stimulated animals. A good deal of our attention has been focussed on the possible procedural differences which could contribute to the variability of findings between investigators and between replications. Thus far, two groups of investigators have published reports concluding that hypothalamic stimulation has no effect on intake. The first was Wayner's group. They originally reported confirmation of our results (Wayner *et al.*, 1971a); they then reported negative results (Wayner *et al.*, 1971b); and most recently they have again concluded that lateral hypothalamic stimulation does elevate ethanol intake (Wayner and Greenberg, 1972), but to a lesser degree in their hands than in ours. The other group is Myers' group; they (Martin and Myers, 1972) concluded that there was no effect of hypothalamic stimulation in a self-stimulation paradigm.

Neither the Wayner nor the Myers studies were true replications of our work, and there are several of the procedural differences between our studies and theirs which might be expected to be important. In the case of the Wayner studies, an explanation of the negative conclusion reported in their second study (Wayner *et al.*, 1971b) is clear from their own work (Wayner and Greenberg, 1972) as well as from our work (Engel, 1972; Wise, 1973) discussed under schedule of presentation (above). In their study (Wayner *et al.*, 1971b) only four stimulated animals were tested, and no control animals were provided for comparison. The animals were given ethanol exposure on an every day basis, and their negative conclusion regarding the effect of stimulation was based on a comparison of the drinking levels after stimulation treatment to the levels before stimulation. As confirmed by subsequent studies in their lab (Wayner *et al.*, 1972; Wayner and Greenberg, 1972), the low levels of drinking that they saw were primarily due to the fact that they offered ethanol every day: when they subsequently compared the intake of animals given ethanol on an intermittent basis the baselines were higher, and differences between the stimulation and control conditions were seen (Wayner and Greenberg, 1972). This is not to suggest, however, that hypothalamic stimulation only facilitates intake when ethanol is intermittently available. It is just that the baseline is so low with animals exposed every day that differences between stimulated and unstimulated animals are small and not easily detected. Stern, however, has data indicating that stimulation is effective in animals given ethanol every day as well as every other day (M.H. Stern, personal communication).

In the case of the Myers finding (Martin and Myers, 1972) the paradigm was quite different from ours. They used a self-stimulation paradigm where animals received very brief pulses of square-wave stimulation, while we (Amit, et al., 1970) used a stimulation-induced drinking paradigm where animals received long trains of sine-wave stimulation. They used a low drinking strain of animals and only three or four animals in each group. Thus there are a number of possible explanations of their failure to see elevated intake in their stimulated animals. One aspect of their procedure warrants particular attention, however, since it bears on a problem which we have studied in connection with variability in our own studies, and in connection with differences between the levels of intake of Wayner's (Wayner and Greenberg, 1972) animals and our own. This has to do with electrode placement: the Myers electrode coordinates were intentionally different from ours.

We have come to suspect that precise electrode placement may be very important, despite the fact that intake levels are not obviously correlated with individual electrode placements. As mentioned above, we have had some groups of stimulated animals that did not differ from controls. Histological examination of these animals revealed placements more dorsal than those in our earlier studies, and the placements of Wayner and Greenberg (who reported a weaker effect of stimulation than that which we have seen) were also dorsal to ours (Compare Amit et al., 1970 to Wayner and Greenberg, 1972). Amit and his colleagues (1974) have lesion data suggesting that it is the ventral and not the dorsal aspects of the lateral hypothalamus which are important for ethanol intake. Thus we have recently compared animals with electrode placements differing in the dorsal-ventral plane. Since the difference in placement between our best and poorest groups and between our best groups and the animals of Wayners group is on the order of a few tenths of a millimeter, the target coordinates for the dorsal and ventral placements under discussion here was 0.5 mm; the intention was to have one group of placements just slightly dorsal to the level of the fornix, and one group of placements just at or just ventral to the level of the fornix. The actual placements obtained are represented in Figure 1; they were quite nicely clustered as intended. Figure 1 also shows the intake levels of the two groups of animals and a group of unstimulated control animals. Both stimulated groups drank more than control animals, but the ventral group drank significantly more than the dorsal group. Thus our

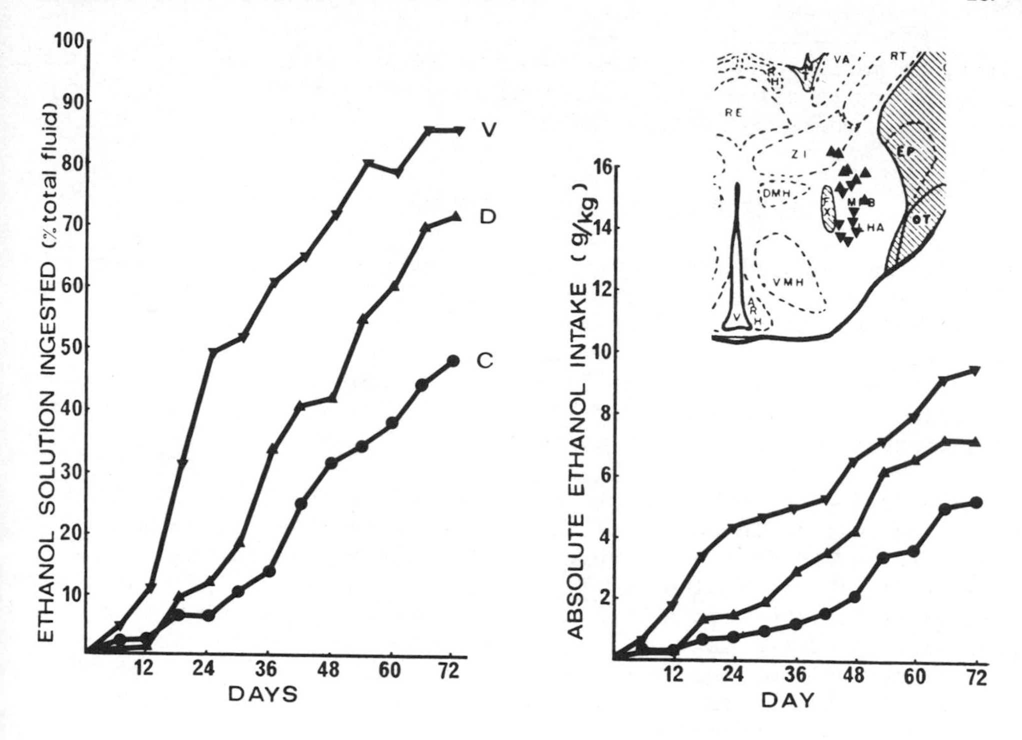

Figure 1. Ethanol intake and electrode placements for animals given dorsal (D) or ventral (V) lateral hypothalamic brain stimulation treatment and for unstimulated control animals (C).

recent work supports the view that small differences in electrode placement may be important, and that the portion of the lateral hypothalamus just ventro-lateral to the fornix may be the focus of the neural circuitry which must be reached to produce elevated intake by hypothalamic stimulation.

A cautionary note should be added for those attempting to maximize ethanol intake using the hypothalamic stimulation paradigm. While the stimulation itself increases intake, the surgical procedure involved in electrode implantation has an inhibitory influence. We have found (Wise and James, 1974) that intracranial surgery inhibits subsequent intake for up to several weeks, and thus two things should be kept in mind with the stimulation paradigm. First, the comparison baseline must be a group of animals which received surgery along with the experimental animals. We initially thought that this was not important, but it now appears that it seemed not to make a difference earlier only because there was a long recovery period following surgery in our early work. This is the second point; to maximize the absolute level of intake with stimulation it is necessary to allow a long surgical recovery period. A period of four to six weeks would seem safest.

DEPENDENCE IN ETHANOL-PREFERING RATS

Having produced high levels of voluntary intake in our animals we have looked closely for signs of physical dependence. We have tested animals to see if they would have audiogenic seizures in response to key jangling; while this is a poorly controlled stimulus, it reliably produces seizures in ethanol dependent rats during periods of ethanol withdrawal (Falk *et al.*, 1972). We have never seen a case of audiogenic seizure in our animals.

We have also looked to see if there is a quantitative tendency toward dependence in our animals. Dependent rats have a heightened jump response to foot shock which can be reliably quantified (French and Morris, 1972; Gibbins *et al.*, 1971). In our studies the animals were given footshock in a light cage supported by a strain gauge. The gauge indicated the maximum force exerted by the jumping rat after each of several footshocks at each of a variety of shock intensities. It has been shown with a similar paradigm that dependent rats have a heightened jump response which reaches its maximum intensity about four days after ethanol withdrawal (French and Morris, 1972). Data from ten animals that had been drinking at rates of more than 10g/kg/day in our paradigm are compared to data from non-drinking animals in Figure 2. There is clearly no difference between the drinkers and the non-drinkers. Thus we see no signs of either an early (peak sensitivity of key jangling occurs in the first day of withdrawal) or a late (four day) withdrawal syndrome in our animals.

ETHANOL WITHDRAWAL AND ETHANOL MOTIVATION

It is not surprising that we do not see ethanol dependence in our animals; there are two good reasons why we should not. First, while our animals did show the strongest voluntary ethanol intake that we have seen reported, they did not drink enough to be intoxicated for any significant period. The 10g/kg/day of intake that our animals reach after several weeks of exposure does not exceed their ability to metabolize ethanol, and while there are significant ethanol concentrations in the blood of our animals after the main drinking bouts (the most dramatic drinking in our animals comes just after the ethanol tubes are put in the cages), significant blood levels are not sustained, and our animals do not show measureable signs of intoxication. Rather the animals seem to drink just to the level of intoxication, but never higher (except when we are forcing them to drink with hypothalamic stimulation). Thus even on ethanol exposure days, our animals are probably not drinking at the

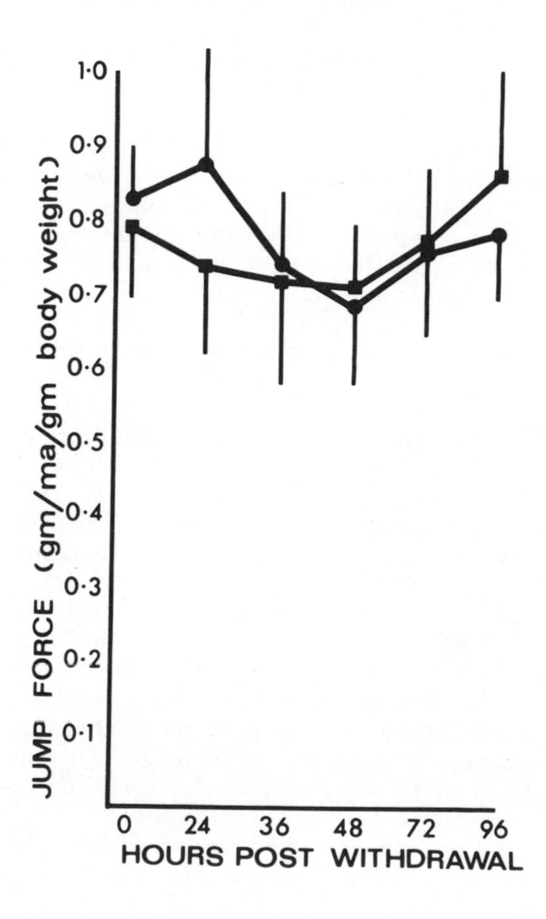

Figure 2. Strength of jump response to footshock in high drinking (squares) and low drinking (circles) animals as a function of ethanol withdrawal.

rate necessary to produce dependence.

More importantly, our animals are drinking intermittently where it is apparent that continuous exposure to ethanol is needed to produce dependence, as mentioned earlier. Involuntary intake exceeding that which we produce has not caused ethanol dependence in animals that have only intermittent ethanol intake. Conversely, dependence has been produced only when ethanol intake has been frequent enough to produce and maintain continuously high blood levels of ethanol or its metabolites.

Thus our data suggest an incompatibility between the methods for maximizing ethanol dependence (continuous or frequent ethanol intake) and the methods for maximizing voluntary ethanol intake levels (intermittent ethanol exposure). In order to increase dependence signs in our animals we would have to give ethanol continuously; this would reduce the intake in our animals. In order to produce higher intake in their animals, other investigators might be required to disrupt ethanol exposure; this would eliminate dependence in their animals.

Our studies do indicate that ethanol motivation varies with ethanol withdrawal, but it does not seem to vary with the stress which accompanies withdrawal in dependent animals (Begleiter, 1974). This is consistent with reports that humans (Mello and Mendelson, 1972) and rhesus monkeys (Woods _et al_., 1971) frequently fail to voluntarily initiate ethanol drinking during periods of maximal withdrawal stress, and also that they frequently do initiate drinking during periods of minimal withdrawal stress. Thus our data are not the first to suggest a lack of relationship between ethanol motivation and ethanol withdrawal. However our data suggest the further possibility that in the laboratory rat maximal motivation to consume ethanol may not accompany any aspect of ethanol dependence. If this suggestion is valid, then it may be more fruitful to use the rat to study components of the human condition rather than to continue to attempt to satisfy all of the proposed requirements (Lester and Freed, 1973) for a lower animal analogue for alcoholism.

REFERENCES

Amit, Z., Meade, R.G. and Singer, J. The effects of lesions, experience and ventral-dorsal variation on ethanol preference induced by hypothalamic stimulation. Paper presented at the meeting of the Canadian Psychological Association, Windsor, 1974.

Amit, Z., Stern, M.H. and Wise, R.A. Alcohol preference in the laboratory rat induced by hypothalamic stimulation. _Psychopharmacologia (Berl.)_, Vol. 17, 367-377, 1970.

Amit, Z. and Stern, M.H. A further investigation of alcohol preference in the laboratory rat induced by hypothalamic stimulation. Psychopharmacologia (Berl.), Vol. 21, 317-327, 1971.

Begleiter, H. Ethanol consumption subsequent to physical dependence. (This volume.)

Branchley, M., Rauscher, G. and Kissin, B. Modifications in the response to alcohol following establishment of physical dependence. Psychopharmacologia (Berl.), Vol. 22, 314-322, 1971.

Brewster, D.J. Ethanol preference in strains of rats selectively bred for behavioral characteristics. J. genet. Psychol., Vol. 11, 217-227, 1969.

Carey, R.J. A decrease in ethanol preference in rats resulting from forced ethanol drinking under fluid deprivation. Physiol. Behav., Vol. 8, 373-375, 1972.

Casey, A. The effect of stress on the consumption of alcohol and reserpine. Quart. J. Stud. Alcohol, Vol. 21, 208-216, 1960.

Cicero, T.J., Myers, R.D. and Black, W.C. Increase in volitional ethanol consumption following interference with a learned avoidance response. Physiol. Behav., Vol. 3, 657-660, 1968.

Deneau, G., Yanagita, T. and Seevers, M.H. Self administration of psychoactive substances by the monkey. Psychopharmacologia (Berl.), Vol. 16, 30-48, 1969.

Ellis, F.W. and Pick, J.R. Evidence of ethanol dependence in dogs. Fed. Proc., Vol. 29, 649, 1970.

Engel, L. Role of schedule of alcohol presentation in development of alcohol preference by brain stimulation. Paper presented at the meeting of the Canadian Psychological Association, June, 1972.

Eriksson, K. Genetic selection for voluntary alcohol consumption in the albino rat. Science, Vol. 159, 739-741, 1968.

Essig, E.F. and Lam, R.C. Convulsions and hallucinatory behavior following alcohol withdrawal in the dog. Archs. Neurol., Vol. 18, 626-632, 1968.

Everett, P.B. and King, R.A. Schedule-induced alcohol ingestion. Psychonom. Sci., Vol. 18, 278-279, 1970.

Falk, J.L. Production of polydipsia in normal rats by an intermittent food schedule. Science, Vol. 133, 195-196, 1961.

Falk, J., Sampson, H.H. and Winger, G. Behavioral maintenance of high concentrations of blood ethanol and physical dependence in the rat. Science, Vol. 177, 811-813, 1972.

Freed, E.X. Alcohol polydipsia in the rat as a function of caloric need. Quart. J. Stud. Alcohol, Vol. 33, 504-507, 1972.

French, S.W. and Morris, J.R. Ethanol dependence in the rat induced by non-intoxicating levels of ethanol. Res. Comm. Chem. Pathol. Pharmacol., Vol. 4, 221-233, 1972.

Freund, G. Alcohol withdrawal syndrome in mice. Archs. Neurol., Vol. 21, 315-320, 1969.

Gibbins, R.J., Kalant, H., LeBlanc, A.E. and Clark, W. The effects of chronic administration of ethanol on startle threshold in rats. Psychopharmacologia (Berl.), Vol. 19, 95-104, 1971.

Goldstein, D.B. An animal model for testing effects of drugs on alcohol withdrawal reactions. J. Pharmac. exp. Ther., Vol. 183, 14-22, 1972.

Goldstein, D.B. and Pal, N. Alcohol dependence produced in mice by inhalation of ethanol: Grading the withdrawal reaction. Science, Vol. 172, 288-290, 1970.

Keehn, J.D. "Voluntary" consumption of alcohol in rats. Quart. J. Stud. Alcohol, Vol. 30, 320-329, 1969.

Lester, D. Self-maintenance of intoxication in the rat. Quart. J. Stud. Alcohol, Vol. 22, 223-231, 1961.

Lester, D. Self-selection of alcohol by animals, human variation and the etiology of alcoholism: A critical review. Quart. J. Stud. Alcohol, Vol. 27, 395-438, 1966.

Lester, D. and Freed, E.X. Criteria for an animal model of alcoholism Pharmac. Biochem. Behav., Vol. 1, 103-107, 1973.

Lester, D. and Greenberg, L.A. Nutrition and the etiology of alcoholism: The effect of sucrose, fat and saccharine on the self-selection of alcohol by rats. Quart. J. Stud. Alcohol, Vol. 13, 553-560, 1952.

Mardones, J. Experimentally induced changes in the free selection of ethanol. Int. Rev. Neurobiol., Vol. 2, 41-76, 1960.

Martin, G.E. and Myers, R.D. Ethanol ingestion in the rat induced by rewarding brain stimulation. Physiol. Behav., Vol. 8, 1151-1160, 1972.

Meisch, R.A. and Thompson, T. Ethanol intake during schedule-induced polydipsia. Physiol. Behav., Vol. 8, 471-475, 1972.

Mello, N.K. A review of methods to induce alcohol addiction in animals. Pharmac. Biochem. Behav., Vol. 1, 89-101, 1973.

Mello, N.K. and Mendelson, J. Operant drinking of alcohol on a rate-contingent ratio schedule of reinforcement. J. Psychiat. Res., Vol. 3, 145-152, 1965.

Mello, N.K. and Mendelson, J. The effects of drinking to avoid shock on alcohol intake in primates. In M.K. Roach, W.M. McIsaac and P.J. Creaven (Eds.) Biological Aspects of Alcohol. Austin: University of Texas Press, 1971, p. 317-340.

Mello, N.K. and Mendelson, J.H. Drinking patterns during work-contingent and non-contingent alcohol acquisition. Psychosom. Med., Vol. 34, 139-164, 1972.

Myers, R.D. and Veale, W.L. Alterations in volitional alcohol intake produced in rats by chronic intraventricular infusions of acetaldehyde, paraldehyde or methanol. Arch. Int. Pharmacodyn., Vol. 180, 100-113, 1969.

Myers, R.D. and Veale, W.L. The determinants of alcohol preference in animals. In B. Kissin and H. Begleiter (Eds.) The Biology of Alcoholism, Vol. 11. New York: Plenum, 1972.

Ogata, H., Ogata, R., Mendelson, J.H. and Mello, N.K. A comparison of techniques to induce alcohol dependence in mouse. J. Pharmac. exp. Ther., Vol. 180, 216-230, 1972.

Ratcliffe, F. Ethanol dependence in the rat: Its production and characteristics. Arch. Int. Pharmacodyn., Vol. 196, 146-156, 1972.

Richter, C.P. Production and control of alcoholic cravings in rats. In Abramson (Ed.) Neuropharmacology. New York: Josiah Macy Foundation, 1956.

Rodgers, D.A. Factors underlying differences in alcohol preference among inbred strains of mice. Psychosom. Med., Vol. 28, 408-513, 1966.

Senter, R.J. and Sinclair, J.D. Self-maintenance of intoxication in the rat: A modified replication. Psychonom. Sci., Vol. 9 291-292, 1967.

Senter, R.J., Smith, F.W. and Lewin, S. Ethanol ingestion as an operant response. Psychonom. Sci., Vol. 8, 291-292, 1967.

Sinclair, J.D. and Senter, R.J. Increased preference for ethanol in rats following alcohol deprivation. Psychonom. Sci., Vol. 8, 11-12, 1967.

Wayner, M.J. and Greenberg, I. Effects of hypothalamic stimulation, acclimation and periodic withdrawal on ethanol consumption. Physiol. Behav., Vol. 9, 737-740, 1972.

Wayner, M.J., Gawronski, D., Roubie, C. and Greenberg, I. In N. Mello and J. Mendelson (Eds.) Recent Advances in Studies of Alcoholism, Washington, D.C. U.S. Government Printing Office, 1971, p. 219-273. (a)

Wayner, M.J., Greenberg, I., Carey, R.J. and Nolley, D. Ethanol drinking elicited during electrical stimulation of the lateral hypothalamus. Physiol. Behav., Vol. 7, 793-795, 1971. (b)

Wayner, M.J., Greenberg, I., Tartaglione, R., Nolley, D., Fraley, S., and Cott, A. A new factor affecting the consumption of ethyl alcohol and other sapid fluids. Physiol. Behav., Vol. 8, 345-362, 1972.

Wise, R.A. Voluntary ethanol intake in rats following exposure to ethanol on various schedules. Psychopharmacologia (Berl.), Vol. 29, 203-210, 1973.

Wise, R.A. Strain- and supplier-related differences in rat ethanol intake. Quart. J. Stud. Alcohol, 1974 (in press).

Woods, J.H., Ikomi, F. and Winger, G. The reinforcing properties of ethanol. In M.K. Roach, W.M. McIsaac and P.J. Creaven (Eds.) Biological Aspects of Alcoholism. Austin: University of Texas Press, 1971, p. 371-388.

ALTERATION OF ETHANOL PREFERENCE IN RATS; EFFECTS OF β-CARBOLINES

Irving Geller and Robert Purdy

Southwest Foundation for Research and Education

San Antonio, Texas 78228

Our interest in the β-carbolines as influencing factors in alcohol preference stems from the early work of McIsaac (1961) who reported the *in vitro* formation of 6-methoxytetrahydroharman as a condensation product of 5-methoxytryptamine and acetaldehyde. When rats were administered these agents in conjunction with their respective metabolic inhibitors, iproniazed and disulfiram, small amounts of 6-methoxytetrahydroharman were found in the urine. These findings were recently confirmed by Dajani and Sehab (1973). They reported the isolation of labelled β-carbolines in 24 hour urine samples of rats administered labelled ethanol and 5-hydroxytryptophan or 5-methoxytryptophan.

Similar reports of the formation of biologically active alkaloids through condensation reactions (Cohen and Collins, 1970; Davis and Walsh, 1970) and the suggestion of a possible relationship of physical dependence and addiction in alcoholism to the *in vivo* formation of these agents, prompted us to investigate the effects of 6-methoxytetrahydroharman on ethanol preference in the rat. We found that chronic administration of this drug reduced ethanol intake in rats and produced a significant increase in whole brain serotonin (Geller et al., 1973).

METHODS

The present investigation represents an extension of these

* This research, conducted in part at Texas Tech University School of Medicine in the Department of Psychiatry, was supported by P.H.S. grants AA-01245 and DA-00747.

efforts. The structurally related β-carbolines shown in Figure 1 were compared for their effects on ethanol drinking in the rat. These compounds were prepared by condensation in aqueous solution at physiological pH and temperature and the products purified and crystallized as the water soluble hydrochloride salts. Since 6-hydroxytetrahydroharman proved to be inactive in reducing alcohol preference in doses up to 150 mg/kg, it is not described further in this report.

Male Sprague-Dawley rats were housed individually in cages 9 X 15 X 18 inches (Wahman L C-28). They were kept in a laboratory with ambient temperatures of 21° - 24° C and were maintained on an unrestricted diet of Wayne Lab Blox. Water and an ethanol solution were available at all times in 100 ml, calibrated drinking tubes mounted on the back or on either side of the cages so that the drinking spouts protruded into the cages approximately 1 1/4 inches above the cage floor level. The two fluid, three-bottle choice method was used to prevent rats from selecting a fluid based on position preference (Myers and Holman, 1966). The cages contained a tube of water, a tube of ethanol an an empty tube. Each day at 10 A.M., the amounts of fluids consumed during the preceding 24 hour period were recorded. The drinking tubes were washed, refilled and put back on the cages and their positions were rotated randomly from day to day. Starting with a 2% solution, ethanol concentrations were increased every other day until a level of rejection was

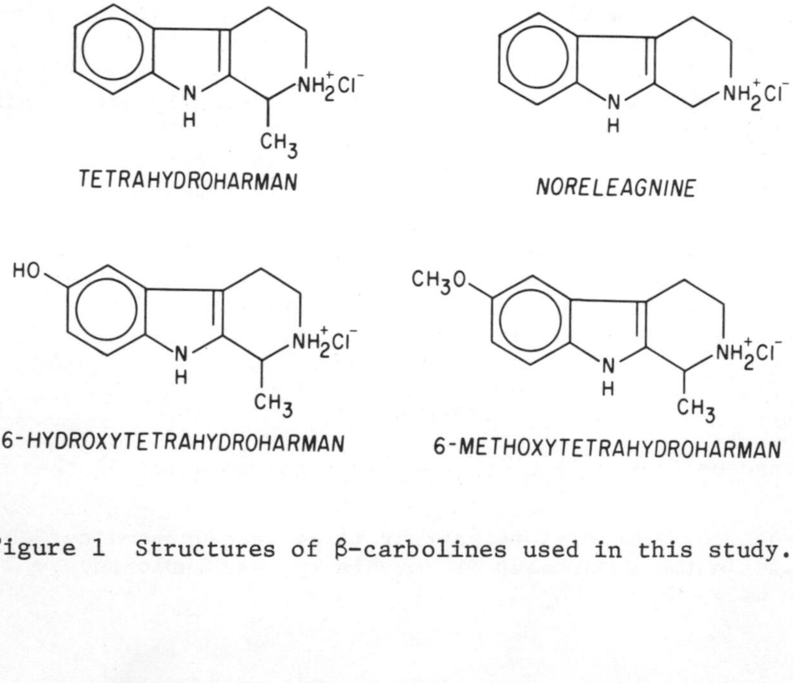

TETRAHYDROHARMAN

NORELEAGNINE

6-HYDROXYTETRAHYDROHARMAN

6-METHOXYTETRAHYDROHARMAN

Figure 1 Structures of β-carbolines used in this study.

attained. Rats were then maintained throughout the experiment at a concentration at least 3 percentage points below the rejection threshold. Those animals that rejected even the lowest ethanol concentration were placed in total darkness in order to induce drinking (Geller, 1971). Tetrahydroharman, 6-methoxytetrahydroharman and noreleagnine were administered in a mixed order in doses of 10 to 50 mg/kg. Drugs were given intraperitoneally at intervals at least a week apart or in many cases much longer when baseline drinking values did not return to control levels soon after drug administration.

RESULTS

The effects of acute administration of each compound at 25, 40 and 50 mg/kg are shown in Figure 2. The solid lines represent alcohol and the broken lines represent water intake. The arrows indicate readings obtained 18 hours after injection of the drug. The 25 and 40 mg/kg doses of 6-methoxytetrahydroharman produced no observable effect on alcohol intake while a rather extensive increase of water intake occurred three days post-drug. The 50 mg/kg dose produced a slight reduction of alcohol drinking with a concomitant increased water intake.

All doses of tetrahydroharman produced a reduction of alcohol intake accompanied by an increased water intake. The greatest reduction of alcohol intake below control levels was produced by noreleagnine at all doses tested. Increased water intake occurred only at the 50 mg/kg dose where the reduced alcohol intake persisted over a 48 hour period.

In Figure 3 are shown data for an individual rat (2 L-D) as well as data averaged four four rats when the drugs were administered at 40 mg/kg. For the individual rat only the 40 mg/kg dose of noreleagnine reduced ethanol intake. The data averaged for four animals under each drug condition demonstrate increasing potency in the order of 6-methoxytetrahydroharman, tetrahydroharman, and noreleagnine.

DISCUSSION

Like 5-hydroxytryptophan, the serotonin precursor, and unlike parachlorophenylalanine, the tryptophan hydroxylase inhibitor (Geller, 1973) the β-carbolines reduced alcohol intake by the rat. Consideration of these findings coupled with those previously reported, strongly suggest a serotonergic involvement in the action of these β-carbolines on alcohol preference in the rat. Other investigators have reported a preferential increase in brain serotonin over norepinephrine in mice administered the 6-methoxy analogue of noreleagnine, 6-methoxy-1,2,3,4-tetrahydro-β-carboline.

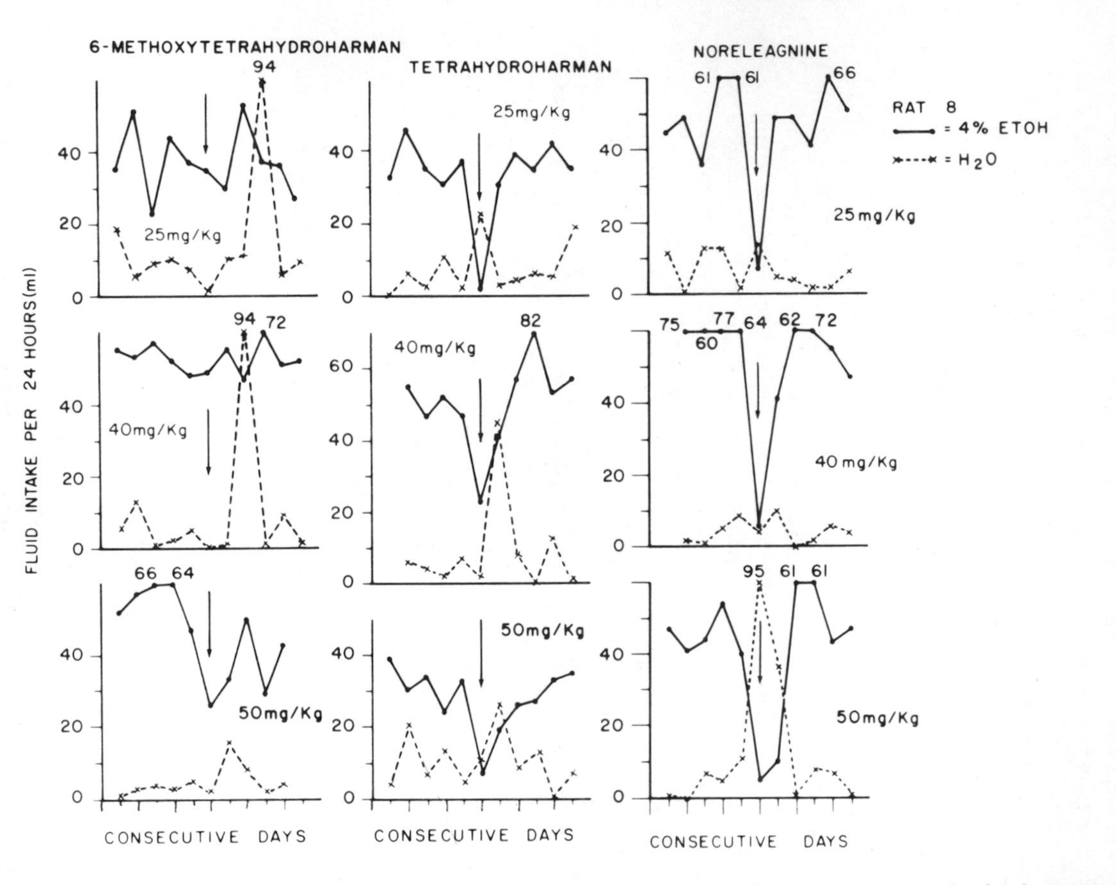

Figure 2 Effects of acute administration of tetrahydroharman, 6-methoxytetrahydroharman and noreleagnine on ethanol intake in the rat. Solid lines show ethanol intake; broken lines water intake. Drugs were injected intraperitoneally at 4 P.M. at 25, 40 or 50 mg/kg. Arrows represent readings taken 18 hours after injection.

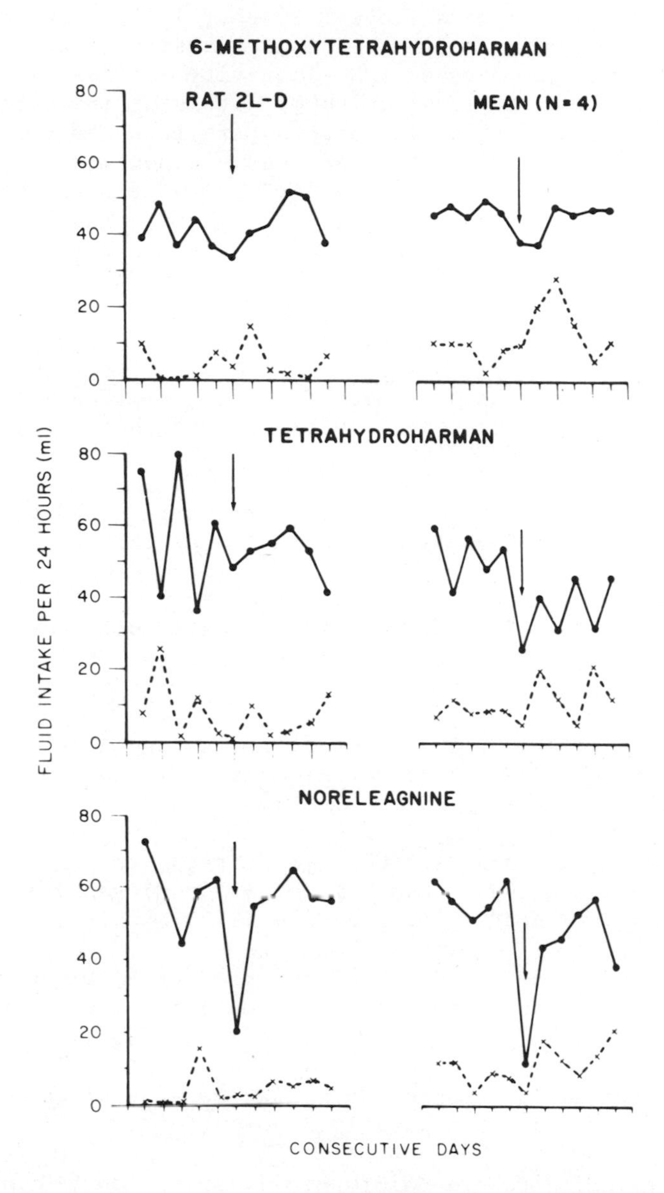

Figure 3 Data for an individual rat and data averaged for four rats showing the effects of 40 mg/kg of tetrahydroharman, 6-methoxy-tetrahydroharman and noreleagnine on ethanol intake. Solid lines show ethanol intake; broken lines water intake. Drugs were injected intraperitoneally at 4 P.M. Arrows indicate readings taken 18 hours after injection.

Similarly, administration of 6-methoxytetrahydroharman increased whole brain serotonin of rats (Geller et al., 1973). The β-carbolines employed in this study have been demonstrated to cause an inhibition of monoamine oxidase (MAO) activity (McIsaac and Estevez, 1966; Ho et al., 1968; Ho et al., 1969). It is of interest that their order of potency as MAO inhibitors parallels their order of potency in reducing alcohol intake of rats. Reduction of alcohol drinking in rats by the β-carbolines may involve the serotonergic system insofar as MAO inhibition would necessarily result in a higher level of brain serotonin, a condition which appears to be related to reduced alcohol intake in the rat (Geller, 1973).

REFERENCES

McIsaac, W. M. "Formation of 1-methyl-6-methoxy-1,2,3,4-tetrahydro-2-carboline Under Physiological Conditions." Biochim. Biophys. Acta 52: 607, 1961.

Dajani, R. M. and Saheb, S. E. "A Further Insight into the Metabolism of Certain β-Carbolines." Annals New York Acad. Sci. 215: 120, 1973.

Cohen, G. and Collins, M. "Alkaloids from Catecholamines in Adrenal Tissue: Possible Role in Alcoholism." Sci. 167: 1749, 1970.

Davis, V. E. and Walsh, M. J. "Alcohol Amines and Alkaloids. A Possible Biochemical Basis for Alcohol Addiction." Sci. 167: 1005, 1970.

Geller, I., Purdy, R., and Merritt, J. "Alterations in Ethanol Preference in the Rat: The Role of Brain Biogenic Amines." Annals New York Acad. Sci. 215: 54, 1973.

Myers, R. D. and Holman, R. B. "A Procedure for Eliminating Position Habits in Preference-Aversion Tests for Ethanol and Other Fluids." Psychon. Sci. 6: 235, 1966.

Geller, I. "Ethanol Preference in the Rat as a Function of Photoperiod." Sci. 173: 456, 1971.

Geller, I. "Effects of Para-Chlorophenylalanine and 5-Hydroxytryptophan on Alcohol Intake in the Rat." Pharmacol. Biochem. and Behavior 1: 361-365, 1973.

Ho, B. T., Taylor, D. and McIsaac, W. M. "Studies on the Mechanism of action of 6-methoxytetrahydro-β-Carboline in Elevating Brain Serotonin." In Brain Chemistry and Mental Disease. Beng T. Ho and William M. McIsaac, Editors. 1: 97-112, 1971.

McIsaac, W. M. and Estevez, V. "Structure-action Relationships of β-carbolines as Monoamine Oxidase Inhibitors." Biochem. Pharmacol. 15: 1625-1627, 1966.

Ho. B. T., McIsaac, W. M. and Tansey, L. W. "Inhibitors of Monoamine Oxidase IV: 6 (or 8) Substituted Tetrahydro-β-carbolines and Their 9-Methyl Analogues." J. Pharmac. Sci. 58: 998-1001, 1969.

Ho, B. T., McIsaac, W. M., Walker, K. R. and Estevez, V. "Inhibitors of Monoamine Oxidase: Influence of Methyl Substitution on the Inhibitory Activity of β-carbolines." J. Pharmac. Sci. 57: 269-273, 1968.

EVIDENCE OF A CENTRAL CHOLINERGIC ROLE IN ALCOHOL PREFERENCE

Andrew K. S. Ho and Benjamin Kissin

College of Pharmacy, Wayne State University, Detroit, Mich. and Dept. of Psychiatry, State University of New York, Downstate Medical Center, Brooklyn, New York, U.S.A. *

Specific genetic strains of mice which are alcohol preferrers and others which are alcohol non-preferrers have been differentiated. For example, the C_{57}Bl strains have been shown to be alcohol preferrers and the DBA strains have been found to reject alcohol (Rodgers and McClearn, 1962). Thus far, the major biochemical differentiating factor between preferrer and non-preferrer strains has been in the rate of acetaldehyde metabolism where the DBA strains have been shown to metabolize acetaldehyde less rapidly and thus to show greater accumulation of acetaldehyde after ethanol ingestion with consequent greater associated signs of toxicity (Schlesinger et al., 1966). Other neurochemical parameters to differentiate the alcohol preferrers have been focused on the role of serotonin (5-HT). Alcohol selecting strain of rats have been shown to have a higher content of brain serotonin; chronic alcohol consumption further increases the brain content of serotonin in the alcohol preferrers but not in the non-preferrers (Ahtee and Eriksson, 1972). Para-chloro-phenylalanine (pCPA), a selective inhibitor of tryptophan hydroxylase, has been reported to reduce alcohol preference by some investigators (Myers and Veale, 1968) but not by others (Nachman et al., 1970).

In this report, we present some preliminary data in support of a possible direct cholinergic involvement in the differential selection of alcohol by different strains of mice. Our findings showed that the brain content of acetylcholine and the uptake of ^{14}C-choline by the whole brain homogenate were significantly higher in the C_{57}Bl/6J strain, whereas the brain acetylcholinesterase

* Supported by Grant MH-16477

activity was greater in the DBA strain. Furthermore, we found that the alcohol preference of the C_{57}Bl/6J strain could be significantly reduced by 4-(1-Naphthylvinyl) pyridine (NVP), a putative inhibitor of choline transferase (ChAc) as well as by lithium, a putative inhibitor of choline uptake.

METHODS

Adult male C_{57}Bl/6J and DBA/2J mice weighing about 20 to 25 g were used in this study. The animals were kept individually or in groups of three in a constant temperature room. To measure alcohol consumption in a free choice situation, two small drinking tubes were fitted onto each of the cages, one filled with water and the other with a 5% solution of ethanol. Food, water and alcohol solution were available ad libitum and the daily consumptions were recorded at 10 a.m. The position of the tubes were changed and different tubes were used each day. The baseline levels of water and alcohol consumptions were established for at least 4 days; alcohol and water consumption levels were established for each of the animals. NVP was given intraperitoneally every 12 hours for two days; the dose used was 2 mg/kg. Lithium was administered i.p. at the dose of 0.5 mEq/kg twice daily for a period of 5 days. Volitional consumption of alcohol was recorded daily before, during, and for four days following cessation of medication.

To determine whether there was a significant difference in the cholinergic system between the alcohol-selecting and the water-selecting strains of mice, we studied the levels of brain acetylcholine, acetylcholinesterase, choline acetylase, and the uptake of ^{14}C-choline in both the C_{57}Bl/6J and the DBA/2J mice. Acetylcholine in the brain was estimated by a bioassay method using the isolated guinea pig ileum preparation (Bentley and Shaw, 1952). The enzyme activities of both cholinesterase and choline acetylase were assayed by methods described previously by Ho et al. (1965). Uptake of ^{14}C-choline using the S_1 fraction was carried out according to a modified method of Coyle and Snyder (1969).

RESULTS

Results obtained showed that the level of brain acetylcholine was significantly higher in the C_{57}Bl/6J than in the DBA/2J mice. Mean values of 3.30 ± 0.18 μg/g (n=6) and 2.20 ± 0.12 μg/g (n=6) were obtained for the C_{57}Bl/6J and the DBA/2J mice respectively ($p < .01$). These values represent a difference of approximately 45%. The uptake of ^{14}C-choline by brain homogenates for the C_{57}Bl/6J mice was significantly higher ($p < .05$) than that for the DBA/2J mice by about 20%. On the other hand, the data obtained for cholinesterase activity

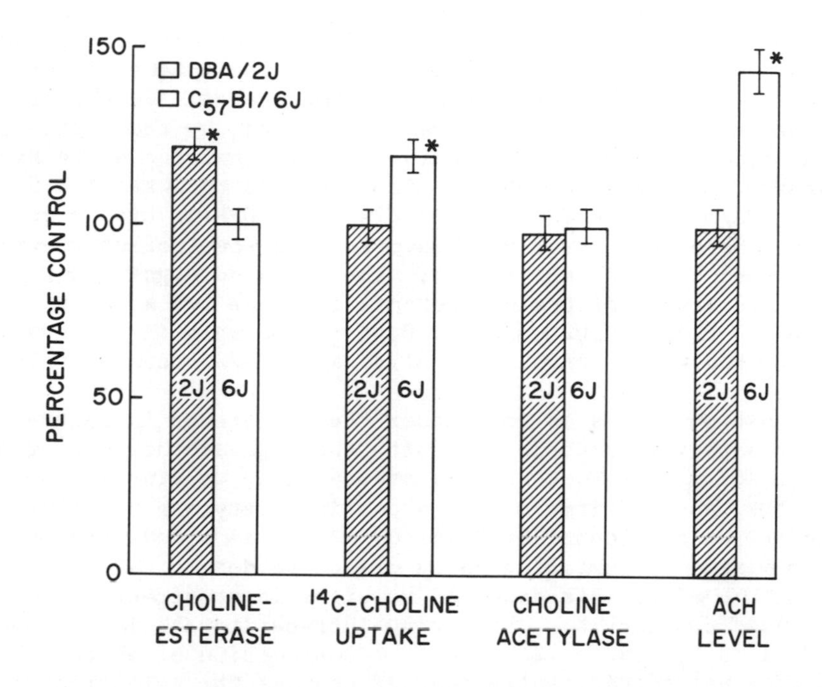

Figure 1. Differences in various aspects of brain acetylcholine metabolism between DBA/2J and C_{57}Bl/6J mice strains.

* p ∠.05

showed that the DBA/2J mice had a significantly higher level than did the C_{57}Bl/6J mice ($p < .05$). There were no significant differences between the two strains of animal in their brain choline acetylase activities (Figure 1). These results suggest that the synthesis rate and storage of acetylcholine is greater in the C_{57}Bl/6J mice whereas the rate of metabolism of acetylcholine is more rapid in the DBA/2J mice. These findings may be interpreted to indicate that brain cholinergic activity in the C_{57}Bl/6J mice is higher.

On the other hand, preliminary studies in our laboratory showed that there were no significant differences in the uptake of ^{3}H-norepinephrine and ^{3}H-dopamine by brain homogenates. Uptake of ^{3}H-norepinephrine and ^{3}H-dopamine was carried out by the method of Coyle and Snyder (1969). Data obtained showed that the whole brain homogenates of C_{57}Bl/6J and DBA/2J mice had mean uptakes of $4.59 \pm 0.23 \times 10^3$ CPM/mg tissue and $4.54 \pm 0.24 \times 10^3$ CPM/mg tissue respectively. Serotonin level was measured by a spectrofluorometric assay described by Ho et al. (1971). There was no significant difference in the levels of brain serotonin in these two strains. Mean values of 0.44 ± 0.02 µg/g and 0.46 ± 0.02 µg/g (n=6) were obtained for the whole brain of C_{57}Bl/6J and DBA/2J respectively.

The question arises as to whether the difference in central cholinergic activity is correlated with the high alcohol preference of the C_{57}Bl/6J mice. To test this proposition, we studied the effect of NVP, an inhibitor of brain choline acetylase activity, on the selection of alcohol by C_{57}Bl/6J mice. In a dose-response study, the minimum pharmacologically effective dose of NVP was found to be 10 mg/kg. This dose had no significant effect on water or total fluid consumption. Following intraperitoneal injections of this dose of NVP every 12 hours, the consumption of alcohol by C_{57}Bl/6J mice was significantly reduced whereas the selection for water was significantly increased (Figure 2). No gross behavioral changes, such as sedation or increased psychomotor activities, appeared. Determination of brain choline acetylase activity, however, showed only a small reduction in enzyme activity of about 10%. The possibility that a more significant regional change occurred in areas such as the lateral hypothalamus could not be excluded.

A second presumed method of reducing central cholinergic activity involved the use of i.p. lithium in doses of 0.5 mEq/kg given twice daily. Here, again, as was the case with NVP, there was a sharp decrease in the voluntary consumption of alcohol and an increase in the consumption of water (Figure 3). The decrease in alcohol consumption persisted as long as the lithium continued to be administered. When it was stopped, there was a rapid return to the original pattern of alcohol preference.

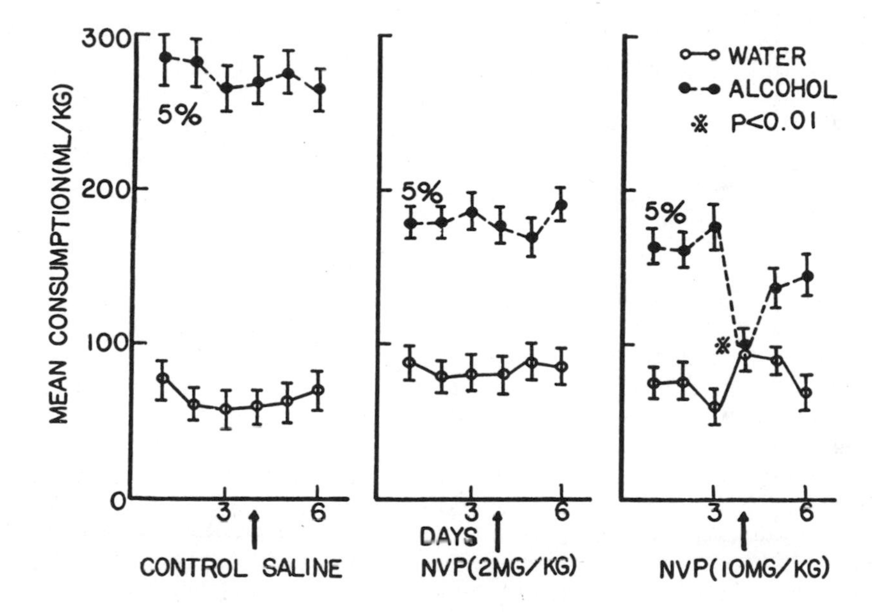

Figure 2. Changes in alcohol and water consumption in C_{57}Bl/6J mice after NVP injection. Values were expressed as ml/kg in mean ± S.E.M. (n=6). Drug was given on the days indicated by (↑).

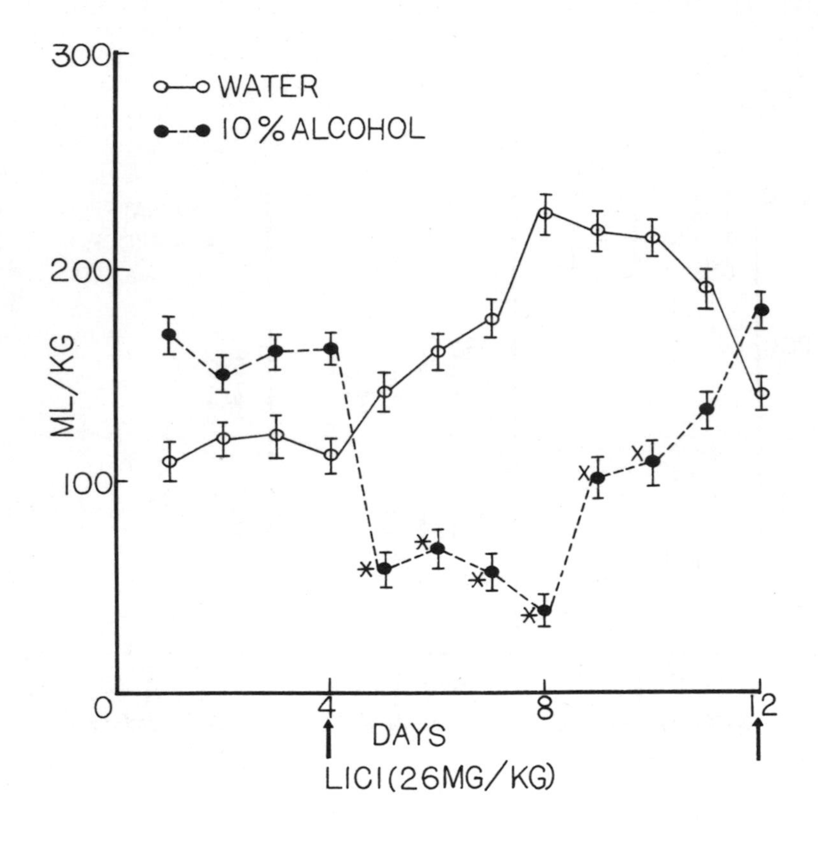

Figure 3. Changes in alcohol and water consumption in $C_{57}Bl/6J$ mice after lithium injection.

DISCUSSION

The findings indicate that there is a correlation between the level of cholinergic activity in the brain and preference of alcohol in C_{57}Bl/6J mice as compared to the DBA/2J strain. This is evident in the fact that C_{57}Bl/6J mice have a greater central cholinergic activity and a higher preference for alcohol than do animals of the DBA/2J strain. When presumed central anticholinergic substances were administered to C_{57}Bl/6J mice - either NVP or lithium - there was a sharp drop in the consumption of alcohol and a corresponding rise in the consumption of water. In neither instance did the animals appear toxic. The fact that these two substances reduce central cholinergic activity through two entirely different mechanisms strongly suggests that central cholinergic activity is the significant variable. NVP has been shown to be an active inhibitor of choline acetylase in the brain (Goldberg et al., 1971), although in our present study that enzyme activity was reduced only by some 10%. Lithium has been reported by Snyder et al. (1973) to inhibit the uptake of choline by neuronal tissues and thus to interfere with acetylcholine formation. The widely different modes of action of these two drugs which appear to have in common mainly the reduction of central cholinergic activity, tends to reduce the probability of other operative effects such as a disulfiram-like effect.

However this hypothesis is somewhat marred by the nonspecificity of the two putative anticholinergic substances used, i.e., NVP and lithium. NVP is said to be a specific choline transferase inhibitor (Goldberg et al., 1971), yet on actual determination, we found the choline transferase level to have decreased only 10%. Since choline transferase is in any event not the rate-limiting enzyme in acetylcholine formation, it is questionable whether the intended effect of the drug was the real one. Similarly, lithium is a known inhibitor of choline uptake but is also inhibitory of the uptake of precursors of other neurotransmitters (Snyder et al., 1973). Consequently, the reduction of alcohol preference associated with the use of these two substances can only be hypothesized as being related to their anticholinergic effect; by no means can this hypothesis be considered as proven. On the other hand, the demonstrated increased central cholinergia in the C_{57}Bl/6J strain as opposed to that in the DBA/2J strain tends to support this hypothesis.

REFERENCES

Ahtee, L. and Eriksson, K. 5-Hydroxytryptamine and 5-hydroxyindoleacetic acid content in the brain of rat strains selected for preferring alcohol to water. Physiol. Behav., 1972, 8, 123-126.

Bentley, G. A. and Shaw, F. H. The separation and assay of acetylcholine in tissue extracts. J. Pharmac. exp. Ther., 1952, 106, 193-199.

Coyle, J. T. and Snyder, S. H. Catecholamine uptake by synaptosomes in homogenates of rat brain: Stereospecificity in different areas. J. Pharmac. exp. Ther., 1969, 170, 221-231.

Goldberg, M. E., Salama, A. I., and Blum, S. W. Inhibition of choline acetyltransferase and hexobarbitone-metabolizing enzymes by naphthylvinyl pyridine analogues. J. Pharm. Pharmac., 1971, 23, 384-385.

Ho, A. K. S., Freeman, S. E., Freeman, W. P., and Lloyd, H. J. Action of tricyclic anti-depressant drugs on central processes involving acetylcholine. Biochem. Pharmac., 1966, 15, 817-824.

Ho, A. K. S., Singer, G., and Gershon, S. Biochemical evidence of adrenergic interaction with cholinergic function in the central nervous system of the rat. Psychopharmacologia, 1971, 21, 238-246.

Nachman, M., Lester, D., and LeMagren, J. Alcohol aversion in the rat: Behavioral assessment of noxious drug effects. Science, 1970, 168, 1244-1246.

Rodgers, D. A. and McClearn, G. E. Mouse strain differences in preference for various concentrations of alcohol. Quart. J. Stud. Alc., 1962, 23, 26.

Schlesinger, K., Kakehana, R., and Bennett, E. L. Effects of tetraethylthiuramidisulfide (Antabuse) on metabolism and consumption of ethanol in mice. Psychosom. Med., 1966, 28, 514-520.

Snyder, S. H., Yamamura, H. I., Pert, C. B., Logan, W. J., and Bennett, J. B. Neuronal uptake of neurotransmitters and their precursors. Studies with "transmitter" amino acids and choline, in New Concepts in Neurotransmitter Regulation (A. J. Mandell, ed.), Plenum Press, New York, 1973, pp. 195-222.

THE LATERAL HYPOTHALAMUS, CATECHOLAMINES AND ETHANOL SELF-ADMINISTRATION IN RATS*

Zalman Amit, Robert G. Meade and Michael E. Corcoran**

Center for Research on Drug Dependence, Department of Psychology, Sir George Williams University, Montreal

The possible involvement of the medial forebrain bundle and the lateral hypothalamus in the maintenance, regulation and control of ethanol and morphine intake in animals received a fair amount of attention over the past five years. In what we believe to be the first study of this kind, Segal, Nerobkova and Rybalkina (1969) reported that electrical stimulation of the ventro-medial hypothalamic nucleus of rats resulted in a temporary increase in ethanol consumption. Stimulation parameters were similar to those commonly used in self-stimulation experiments. This elevated intake of ethanol returned to pre-stimulation baseline after three days of stimulation sessions. In contrast with this study, Marfaing-Jallat, Larue and LeMagnen (1969) reported that lesions of the ventro-medial nucleus also resulted in an increase in the consumption of 8% ethanol solution. Martin and Myers (1972) found that rats that were reinforced for drinking ethanol with lateral hypothalamic stimulation, ingested voluminous amounts of ethanol to the point of intoxication. However, Martin and Myers reported that this electrically reinforced ethanol drinking did not have any effect on home-cage preference levels for ethanol. Thus in spite of the large amounts of ethanol consumed by these rats during the electrical stimulation sessions, they continued to prefer water over ethanol in the home cage. My

*Supported in part by a grant to Dr. Z. Amit from the Non-Medical use of Drugs Directorate, Department of National Health and Welfare, Canada

**Kinsmen Laboratories of Neurological Research, Faculty of Medicine, University of British Columbia, Vancouver, Canada.

colleagues and I have reported (Amit, Stern and Wise, 1970; Amit and Stern, 1971; Stern and Amit, 1972; Amit and Cohen, 1974; Corcoran and Amit, 1974) that repeated electrical stimulation of the lateral hypothalamus (LH) resulted in the development of a permanent home-cage preference for ethanol over water in rats that previously re-jected these ethanol solutions when presented to them. The rats continued to prefer ethanol for as long as they were allowed to live and even when the ethanol was adulterated with high doses of quinine (0.05%). Wayner and Greenberg, (1972) confirmed these findings although they argued that the elevated ethanol preference observed in these studies was due more to the schedule of ethanol presentation and less to hypothalamic stimulation. We also reported that lesions produced in ethanol prefering rats through the electrode previously used to stimulate the LH, caused a partial reduction in ethanol intake while lesions produced contralateral to the side previously stimulated did not have any effect on ethanol drinking (Amit and Stern, 1972; Levitan, 1971). However, in subsequent tests we found it difficult to replicate this finding. In this context, it is interesting to note that Wise and James (1974) recently reported that surgical insult to the LH (such as that produced by implanting a chronic hypothalamic electrode) reduced ethanol intake in rats placed on an alternate day method of ethanol presentation. They further reported that this reduction in ethanol intake was elim-inated by electrical stimulation of the LH through the implanted electrode.

More recently several investigators focussed their attention on the involvement of the hypothalamus in morphine intake and dependence. Kerr and Pozuello (1971) found that lesions of the ventro-medial hypothalamic nucleus (VMH) attenuated the withdrawal symptoms which typically occur after terminating morphine admin-istration to dependent rats. They also reported that lesions of the LH had no effect on morphine withdrawal symptoms. Kerr and Pozuello's findings are somewhat difficult to interpret since their VMH lesions clearly encroached on the more medial aspects of the LH.

Glick and Charap (1973) reported that bilateral lesions of the posterior lateral hypothalamus (apparently induced in one-stage), resulted in an increase in oral consumption of morphine in rats. Amit and Corcoran (1972) demonstrated that two-stage bilateral lesions of the more anterior aspects of the lateral hypothalamus blocked completely oral consumption of a 0.05% morphine hydrochloride solution in rats offered morphine as the only source of fluids. These rats refused to drink morphine to the point of death. The authors induced the lesions in two stages with 30 days separating the two stages, thus avoiding the onset of the aphagic-adipsic syndrome which typi-cally occurs in rats when bilateral hypothalamic lesions are induced in one stage (Fass, Jordan, Rubman, Seibel and Stein, 1974; Amit, Meade, Corcoran and Teichman, 1974). In a more recent report, Amit, Corcoran, Amir and Urca (1973) found that only lesions of the ventral portion of the lateral hypothalamus were effective in blocking morphine intake in rats. This finding suggested that the involvement of the lateral hypothalamus in morphine intake may be differentiated along a dorsal-ventral dimension of this limbic structure. This finding also led us to re-examine the histological data of several of the experiments in which we first succeeded and later failed to induce changes in ethanol consumption by hypothalamic lesions. This examination revealed that most of the lesions which were effective in reducing ethanol consumption were located at or below the level of the fornix while the ineffective lesions were located above the level of the fornix. We thus decided to investigate this question more systematically. We (Meade, Amit and Singer, 1974), implanted electrodes in two groups of male Wistar rats. In one group the electrodes were aimed at the dorsal aspects of the LH (7.5 mm below the superior surface of the skull) while in the second group the electrodes were aimed at the ventral LH (8.2 mm below the superior surface of the skull). Both groups were then lesioned through the implanted electrode. After a seven day recovery period, we began stimulating the area adjacent to the lesion through the chronically implanted lesioning electrode. Stimulation sessions were maintained for thirty days. During the 30 days of stimulation and for 40 days following the stimulation, the animals were offered a free choice between ethanol and water in their home cage. We found that while the dorsally lesioned animals developed ethanol preference similar to that previously observed in several LH stimulation experiments (e.g., Amit and Stern, 1971; Amit and Cohen, 1974; Corcoran and Amit, 1974), the ventrally lesioned group did not display any increase in ethanol intake and preference. There was little difference in ethanol intake and the preference pattern between the dorsally lesioned group and a non-lesioned stimulated group. On the other hand, there was a significant difference between the intake of the non-lesioned group and the ventrally lesioned group which displayed an intake pattern similar to that of non-stimulated control animals. We interpreted these results as indicating that the integrity of the ventral aspects of the lateral hypothalamus must be protected in order for LH stimulation to be effective in inducing ethanol preference. The development of ethanol preference in the dorsal group was ascribed to the spread of stimulation current from the dorsal location of the electrode into the adjacent intact ventral tissue. It is of interest that the same ventral lateral hypothalamic lesions induced after the animals acquired ethanol preference produced only a small decrement in the established preference. This and the previous findings seem to be in agreement with studies on morphine intake which show that ventral LH lesions can block consumption of morphine in naive rats (Amit *et al.*, 1973). Rats that were given experience in drinking morphine did not stop drinking morphine after ventral LH lesions (Amir, 1974).

In an attempt to explain the involvement of the ventral LH in ethanol and morphine intake we considered the possibility that brain catecholamines are involved in the regulation of ethanol and morphine

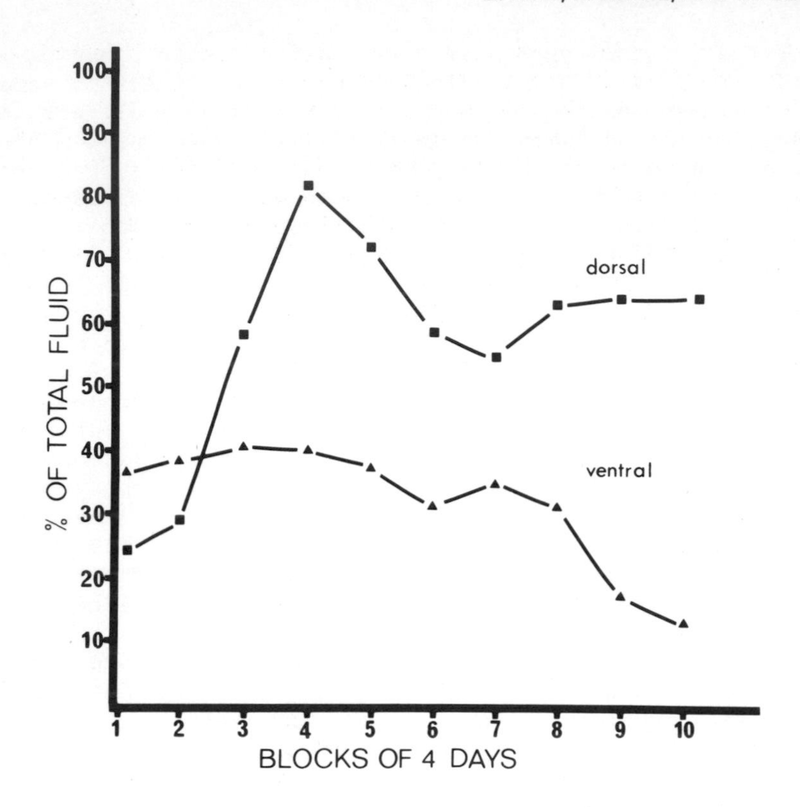

Figure 1. Ethanol intake patterns in animals lesioned in the ventral and dorsal aspects of the lateral hypothalamus.

self-administration. Several investigators (e.g., Olson and Fuxe, 1972; Ungerstedt, 1971) demonstrated that the ventral noradrenergic bundle and to some extent also the mesolimbic dopamine system course through the ventral aspects of the LH. Thus it is possible that the lesions described above (Amit et al., 1973; Amit, Meade and Singer, 1974; Amir, 1974) damaged one or even two of the catecholaminergic systems and in this way interfered with some of the mechanisms underlying ethanol and morphine self-administration.

The possibility that catecholamines are involved in ethanol and morphine intake was discussed by several investigators (e.g., Cohen and Collins, 1970; Davis and Walsh, 1970; Amit and Stern, 1972; Cohen, 1973; Pozuello and Kerr, 1972; Glick, Zimmerberg and Charap, 1973;

Davis and Smith, 1973). Adopting this possibility as a working hypothesis, one would predict that experimental manipulation of catecholamines should produce a significant change in self-administration of ethanol or morphine or both. We recently completed two studies which were designed to examine this question. In the first study (Meade and Amit, 1974) we examined the effects on oral morphine consumption of intraventricular infusions of 6-hydroxydopamine (6-OHDA). Two infusions of 6-OHDA, which is known to selectively deplete brain catecholamines (Uretsky and Iverson, 1970) by destroying catecholamine-containing neurons resulted in a significant blockade of morphine intake in rats. As can be seen from Figure 2, this blockade was dose-dependent. In order to control for the possibility that the reduction in morphine intake resulted from hypersensitivity to the bitter taste,

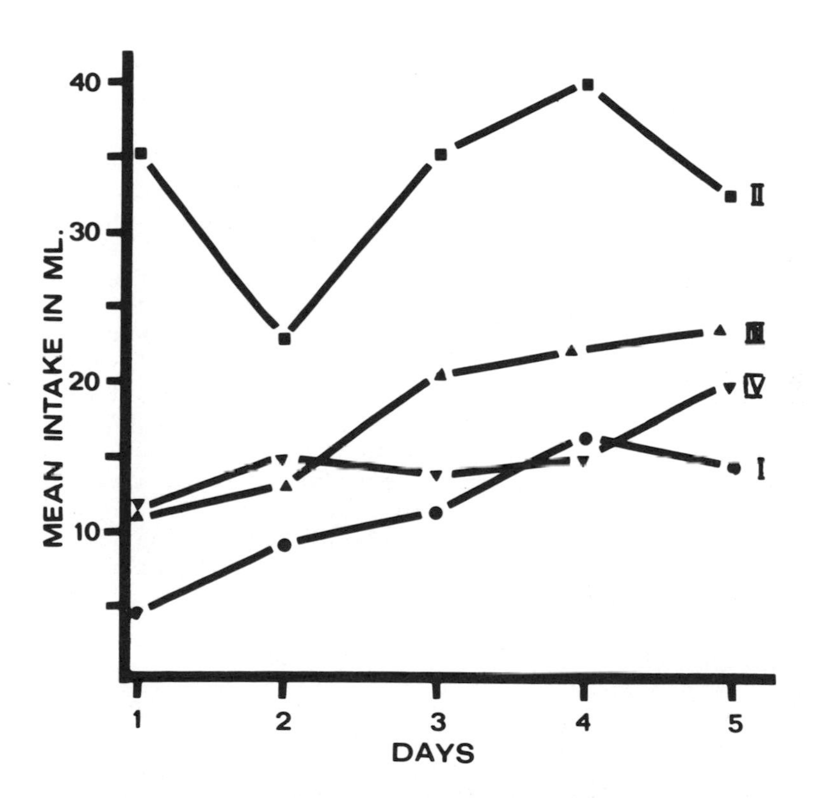

Figure 2. Intake patterns of several fluids after the following treatments. (I) 2 x 350 μg 6-OHDA (drinking morphine); (II) 2 x 350 μg 6-OHDA (drinking water); (III) 2 x 350 μg 6-OHDA; (IV) 2 x 250 μg 6-OHDA (drinking morphine).

or because of adipsia induced by the infusions, two control groups were run. One of them had available a .025% quinine hydrochloride solution as their only source of fluid while the other group had water. Both groups drank readily both water and quinine solution. Mean intake for both fluids after the first twenty-four hours of presentation exceeded 13 mls. In the second study (Meade, Amit and Singer, 1974) we have examined the effects of disulfiram injections on ethanol intake in rats made to prefer ethanol by LH stimulation. Disulfiram, which is frequently used for preventing alcohol intake in alcoholics, partially inhibits aldehyde dehydrogenase thus preventing the metabolism of acetaldehyde. It also inhibits dopamine beta hydroxylase (DBH) the enzyme which catalyses the conversion of dopamine to noradrenaline. We found that injections of 25 mg/kg of disulfiram reduced intake of ethanol in ethanol prefering rats from a mean of 82% of total fluid intake to a mean of 21% for a period of 3 days, after which, ethanol intake gradually returned to baseline. A second group of ethanol prefering rats was subjected to injections of FLA-63 (17 mg/kg) which also inhibits DBH but has no known effects on aldehyde dehydrogenase. To our surprise, FLA-63 effected the same reduction in ethanol intake. Rats, which previously ingested over 80% of their total fluid intake in ethanol, when injected with FLA-63 (17 mg/kg) reduced their intake to around 20% of total fluid intake. A third group of ethanol drinking rats received injections of calcium cyanamide citrated (temposil) which inhibits aldehyde dehydrogenase but has no effect on DBH. These animals showed only a small and insignificant reduction in ethanol intake. A fourth group of animals which was given injections of Haloperidol, a dopamine receptor blocker, displayed a pattern of decreased ethanol intake. However, the ethanol intake of these animals was significantly higher than in the animals exposed to disulfiram and FLA-63. These findings support Collier's (1972) suggestion that the action of disulfiram in preventing ethanol intake in humans may occur via this drug's action on DBH. It also strengthens the argument that brain norepinephrine is involved in the control of ethanol intake. It is important to point out in this context that both disulfiram and FLA-63 while depleting norepinephrine have no effect on dopamine. Figure 3 summarizes the data obtained in this experiment.

The picture that seems to emerge from these studies suggests that catecholamines are involved in the control of ethanol and morphine self-administration. However, it seems that while dopamine is primarily (but not perhaps exclusively) involved in morphine intake, norepinephrine is the major catecholamine involved in ethanol intake. Our data concerning the involvement of brain catecholamines in self-administration of morphine and ethanol seems to be in agreement with several reports in the literature. Myers and Veale (1968) reported that alpha-methyl-para-tyrosine (aMpT) which blocks the synthesis of catecholamines decreased intake of ethanol when injected to rats.

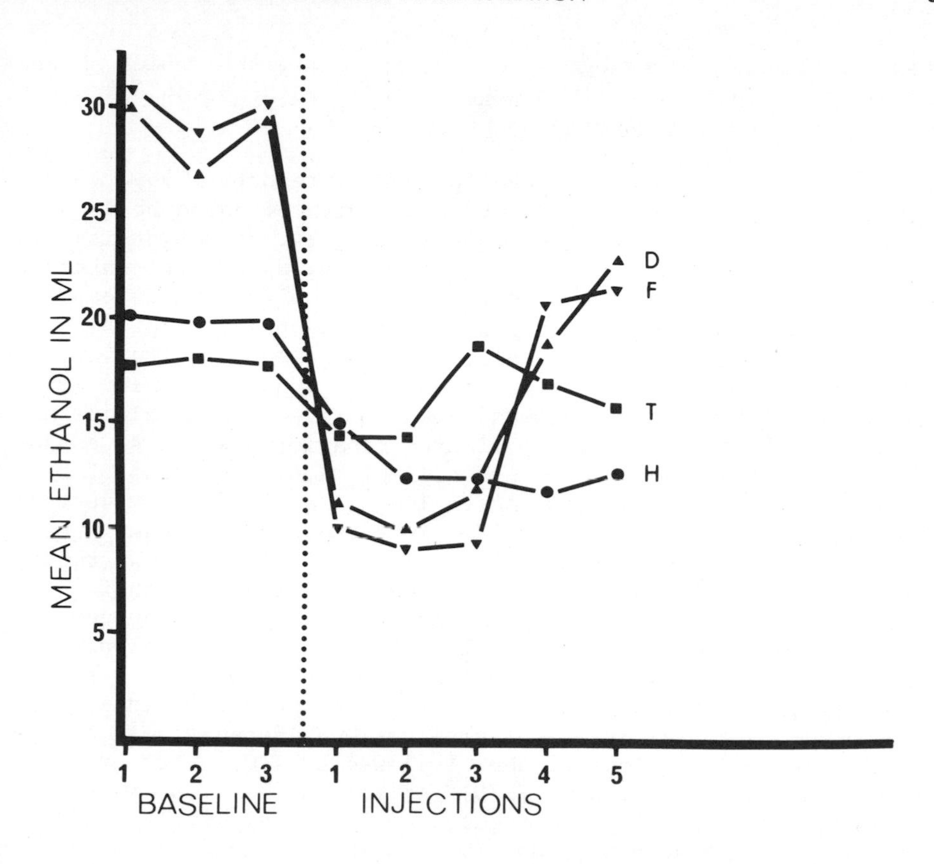

Figure 3. Ethanol intake patterns after inhibition of dopamine beta hydroxylase or aldehyde dehydrogenase or both. D: Disulfiram; F: FLA-63; T: Temposil; H: Haloperidol.

Glick et al., (1973) and Davis and Smith (1973) reported that aMpT also decreased morphine self-administration in rats. Similarly, Pozuello and Kerr (1972) reported that this compound decreased morphine self-administration in monkeys. We feel that these data support a modified version of the notion initially put forth by Davis and Walsh (1970). In its modified form, it suggests that at the level of self-administration (which seems to be at least in part unrelated to physical dependence) there may be some overlap in the catecholaminergic mechanisms which may underlie intake of morphine and ethanol. This overlap may be reflected in the fact that while both norepinephrine and dopamine may effect self-administration of both ethanol and morphine, the involvement of dopamine may be primarily related to intake of morphine and the involvement of norepinephrine is primarily related to intake of ethanol. A recent study by Sinclair (1974) which showed that intake of ethanol in rats is greatly modified by morphine injections seems to lend further support to this notion.

Cohen and his associates argued in several reports (1969, 1970, 1971, 1973) that acetaldehyde biosynthesizes with norepinephrine and that the product of this biosynthesis, 1,2,3,4, tetrahydroisoquinoline alkaloids (TIQ), may play a role in regulating ethanol intake. They reported among other things (1973) that TIQ alkaloids were taken up by rat iris tissue after depletion of norepinephrine from this tissue by aMpT. In an attempt to elucidate the nature of the involvement of catecholamines in intake of ethanol, we (Duby and Amit, 1974) examined the question whether catecholamine derived TIQ alkaloids may play a functional role in ethanol oriented behavior. We first established an intake baseline for 10% ethanol solution in a group of rats. After intake had stabilized we injected 1 μg of aMpT into the ventral LH. Four hours after injecting aMpT we injected, via the same cannula, 5 μg of 1,2,3,4, tetrahydroisoquinoline (TIQ) into the LH. We repeated these injections for 10 days. We found that injections of aMpT followed by TIQ reduced intake of ethanol in those rats (mean per cent change from baseline of 27%). Injections of vehicle alone had no effect on ethanol intake and injections of aMpT alone had a small and insignificant effect on ethanol intake. The reduction in ethanol intake observed after injections of aMpT was considerably smaller than that observed after infusions of aMpT followed by TIQ. These data both support Cohen and Collins, (1970) contention that TIQ may be playing an important role in the regulation of ethanol intake, and are also in agreement with Collins' (1974) findings that TIQ alkaloids (e.g., Salsolinol) were identified in rats after prolonged intubation with alcohol. These data are also strengthened by Sandler's (1973) finding that TIQ alkaloids were isolated in Parkinson patients treated with L-dopa who drank alcohol.

In summary, several recent experiments carried out in our laboratory suggest that the ventral LH is involved in the self-administration of ethanol and morphine. This involvement seems to be related to the fact that two of the main catecholamine pathways (norepinephrine and dopamine) in the brain pass through this limbic structure. While both catecholamines seem to be involved in self-administration of the two drugs, dopamine seems to be primarily involved in morphine self-administration and norepinephrine in ethanol self-administration. These data support the notion that there may be some overlap in the mechanisms underlying the self-administration of both drugs. Finally, TIQ alkaloids may be involved in ethanol intake as a function of a possible biosynthesis of brain norepinephrine (and to a lesser extent also dopamine) and acetaldehyde.

REFERENCES

Amir, S. The effect of experience on blockade of drinking of morphine by hypothalamic lesions. Paper presented at the meeting of the Eastern Psychological Association, Philadelphia, 1974.

Amit, Z. and Cohen, J. The effect of hypothalamic stimulation on oral ingestion of diazepam in rats. Behavioral Biology 10, 223-229, 1974.

Amit, Z. and Corcoran, M.E. Blockade of drinking of a morphine solution by hypothalamic lesions and 6-hyrdoxydopamine infusions in rats. Paper presented at the meeting of the Neuroscience Society, San Diego, 1973.

Amit, Z., Corcoran, M.E., Amir, S. and Urca, G. Ventral hypothalamic lesions block the consumption of morphine in rats. Life Sciences 13, 805-816, 1973.

Amit, Z., Meade, R.G., Corcoran, M.E. and Teichman, M. Elimination of aphagia and adipsia by temporal separation of surgical and neurochemical manipulations in rats. Paper presented at the meeting of the 5th International Conference on Physiology of Food and Fluid Intake, Jerusalem, 1974.

Amit, Z. and Stern, M.H. A further investigation of alcohol preference in the laboratory rat induced by hypothalamic stimulation. Psychopharmacologia 21, 317-327, 1971.

Amit, Z. and Stern, M.H. Electrochemical interaction in the medial forebrain bundle and ethanol preference in rats. In O. Forsander and E.K. Eriksson (Eds.) Biological Aspects of Alcohol Consumption, Finish Foundation, 225-232, 1972.

Amit, Z., Stern, M.H. and Wise, R.A. Alcohol preference in the laboratory rat induced by hypothalamic stimulation. Psychopharmacologia 17, 367-377, 1970.

Brezenoff, H.E. and Cohen, G. Hypothermia following intraventricular injection of a dopamine-derived tetrahydroisoquinoline alkaloid. Neuropharmacology 12, 1033-1038, 1973.

Cohen, G. The role of tetrahydroisoquinolines as false adrenergic neurotransmitters in alcoholism. In M.M. Gross (Ed.), Alcohol Intoxication and Withdrawal: Experimental Studies I, pp. 33-44. Plenum, New York, 1973.

Cohen, G. and Barrett, R. Fluorescence microscopy of catecholamine-derived tetrahydroisoquinoline alkaloids formed during methanol intoxication. Federal Proceedings 28, 288, 1969.

Cohen, G. and Collins, M. Alkaloids from catecholamines in adrenal tissue: Possible role in alcoholism. Science 167, 1749-1752, 1970.

Collins, M. Biosynthesis of tetrahydroisoquinoline alkaloids in brain and other tissues of ethanol intoxicated rats. (This volume.)

Corcoran, M.E. and Amit, Z. Reluctance of rats to drink hashish suspensions: Free-choice and forced consumption, and the effects of hypothalamic stimulation. Psychopharmacologia 35, 129-147, 1974.

Davis, W.M. and Smith, S.G. Blocking of morphine based reinforcement by alphamethyltyrosine. Life Sciences 12, 185-191, 1973.

Davis, V.E. and Walsh, M.J. Alcohol, amines and alkaloids: A possible biochemical basis for alcohol addiction. Science 167, 1005-1006, 1970.

Duby, S. and Amit, Z. Tetrahydroisoquinolines and ethanol intake in rats. Manuscript in preparation, 1974.

Fass, B., Jordan, H., Rubman, A., Seibel, S. and Stein, D. Recovery of function after serial or one-stage lesions of the lateral hypothalamus in rats. Paper presented at the meeting of the Eastern Psychological Association, Philadelphia, 1974.

Glick, S.D. and Charap, A.D. Morphine dependence in rats with medial forebrain bundle lesions. Psychopharmacologia 30, 343-348, 1973.

Glick, S.D., Zimmerberg, B. and Charap, A.D. Effects of Methyl-p-Tyrosine on morphine dependence. Psychopharmacologia 32, 365-371, 1973.

Heikkila, R., Cohen, G. and Dembiec, D. Tetrahydroisoquinoline alkaloids uptake by rat brain homogenates and inhibition of catecholamine uptake. Journal of Pharmacology and Experimental Therapeutics 179, 250-258, 1971.

Kerr, F.W.L. and Pozuello, J. Suppression or reduction of morphine dependence in rats by discrete stereotaxic lesions in the hypothalamus. Journal of Mayo Clinic Proceedings 46, 653-665, 1971.

Levitan, D. Blockade of ethanol intake by hypothalamic lesions in rats. Unpublished Honours Thesis, McGill University, 1971.

Marfaing-Jallat, J., Larue, C. and LeMagnen, J. Alcohol intake in hypothalamic hyperphagic rats. Physiology and Behavior 8, 1151-1160, 1972.

Martin, G.E. and Myers, R.D. Ethanol ingestion in the rat induced by rewarding brain stimulation. Physiology and Behavior 8, 1151-1160, 1972.

Meade, R.G. and Amit, Z. Blockade of morphine intake by catecholamine depletion in rats. Paper presented at the meeting of the American Psychological Association, New Orleans, 1974.

Meade, R.G., Amit, Z. and Singer, J. Reduction in ethanol intake induced by dopamine beta hydroxelase in ethanol prefering rats. Manuscript in preparation, 1974.

Meade, R.G., Amit, Z. and Singer, J. The effects of lesions, experience and ventral-dorsal variation on ethanol preference induced by hypothalamic stimulation. Paper presented at the meeting of the Canadian Psychological Association, Windsor, 1974.

Myers, R.D. and Veale, W.L. Alteration in volitional alcohol intake produced in rats by chronic intraventricular infusions of acetaldehyde, paraldehyde or methanol. Archives of International Pharmacodynamics and Therapeutics 180, 100-113, 1969.

Olson, L. and Fuxe, K. Further mapping out of central noradrenaline neuron systems: Projections of the "subcoeruleus" area. Brain Research 43, 289-295, 1972.

Pozuello, J. and Kerr, F.W. Suppression of craving and other signs of dependence in morphine addicted monkeys by administration of alpha-methyl-para-tyrosine. Journal of Mayo Clinic Proceedings 47, 621-628, 1972.

Sandler, M., Carter, S.B., Hunter, K.R. and Stern, G.M. Tetrahydroisoquinoline alkaloids: In vivo metabolites of L-dopa in man. Nature 241, 439, 1973.

Segal, B.M., Nerobkova, L.N. and Rybalkina, S.V. "Drive" for alcohol-stimulation of hypothalamic nuclei in rats. Journal of Higher Nervous Activity - I.P. Pavlov, 688-691, 1969.

Sinclair, J.D. and Senter, R.J. Development of an alcohol deprivation effect in rats. Quarterly Journal of Studies on Alcohol 29, 863-867, 1968.

Wayner, M.J. and Greenberg, I. Effects of hypothalamic stimulation acclimation and periodic withdrawal on ethanol consumption. Physiology and Behavior 9, 737-740, 1972.

Wise, R.A. and James, L. Rat ethanol intake: Suppression by intracranial surgery and facilitation by intracranial stimulation. Paper presented at the meeting of the Eastern Psychological Association, Philadelphia, 1974.

ESTABLISHMENT OF ETHANOL AS A REINFORCER FOR RHESUS MONKEYS VIA THE ORAL ROUTE: INITIAL RESULTS

Richard A. Meisch, Jack E. Henningfield and Travis Thompson
Psychiatry Research Unit
University of Minnesota
Minneapolis, Minnesota 55455

Ethanol functions as a reinforcer for rhesus monkeys when it is self-administered intravenously (Deneau, Yanagita and Seevers, 1969; Winger and Woods, 1973; Woods, Ikomi and Winger, 1971) or intragastricly (Yanagita and Takahashi, 1973). However, it has been reported that ethanol is not drunk by rhesus monkeys in either large quantities or at high concentrations except under special circumstances. For example, if rhesus monkeys are totally restricted to a liquid diet containing ethanol, physiological dependence develops (Pieper and Skeen, 1972). In three of four rhesus monkeys the chronic intraventricular infusion of ethanol resulted in abrupt and intermittent increases in ethanol intake (Myers, Veale and Yaksh, 1972), but such increases were not found in another study (Koz and Mendelson, 1967). Following a six-day interruption in the monkeys' continuous access to ethanol, reintroduction of access to ethanol resulted in a pronounced but transient increase in intake (Sinclair, 1971).

In a number of studies it has been found that rhesus monkeys drink very little ethanol, and these findings have been attributed to ethanol's aversive taste properties. When offered a choice between water and different ethanol concentrations from 3 to 30% (W/V), rhesus monkeys consistently preferred water (Myers, Stoltman and Martin, 1972). A requirement to drink ethanol in order to avoid an electric shock did not eventuate in the drinking of large quantities of ethanol, and the results were interpreted as indicating that rhesus monkeys have a strong aversion to ethanol (Mello and Mendelson, 1971a). In another study, during schedule-induced polydipsia rhesus monkeys ingested volumes of 4% (W/V) ethanol at levels equal to or exceeding water control values, but sustained intake of higher concentrations was not observed, nor was there an increase in ethanol intake within the polydipsia situation (Mello and Mendelson, 1971b).

The present report described results obtained with two rhesus monkeys that indicate it is possible to establish ethanol as a reinforcer using the oral route. The experimental strategies employed are similar to those used successfully in this laboratory with rats (Meisch and Thompson, 1971, 1972a, 1974a), and the results obtained using the monkeys confirm and extend previous findings made using rats (Meisch and Thompson, 1971, 1972a, 1972b, 1974a).

METHODS

Subjects: Two adult male rhesus monkeys (Macaca mulatta) served as subjects. At 80% of free-feeding weight, monkey M-P weighed 5.4 kg, and monkey M-S weighed 6.2 kg. The monkeys were individually housed in experimental chambers which were constantly illuminated and maintained at a temperature of 20.5°C. Both monkeys were experimentally naive at the beginning of Experiment 1.

Apparatus: Housing - Monkey M-P was housed in a Plexiglas restraining chair (Foringer #1206M2) which restricted body movement by means of a loosely-fitting neck yoke. The chair was mounted in a sound-attenuated isolation booth which was constantly ventilated by an electric fan. Monkey M-S was housed in a cubical stainless steel primate cage (Labco #ME1305) having three solid walls and one barred wall (inside dimensions 66.6 cm x 66.6 cm x 66.6 cm). For both subjects, operanda and stimulus lights were mounted on one wall as illustrated in Figure 1.

Liquid operanda: A prerequisite to any oral liquid self--administration experiment is a liquid-delivery device that allows both accurate measurement of liquid actually consumed and a means of recording discrete responses. The standard drinkometer spout is deficient in that leakage frequently occurs (particularly as the concentration of the ethanol solution is increased). With the lip-lever-operated spout, liquid can be spilled. The spout designed for the present study required at least one centimeter of the spout to be inside a monkey's mouth to complete a drinkometer circuit. The mouth-contact response opened a solenoid valve for the duration of lip contact. Visual inspection revealed virtually no liquid spillage.

Food operanda: The restraining chair was equipped with a stainless steel mouth operated feeder tube (Thompson, Schuster. Dockens and Lee, 1964) which dispensed 1-g Noyes banana pellets via an automatic pellet dispenser (Foringer #1282). The open cage

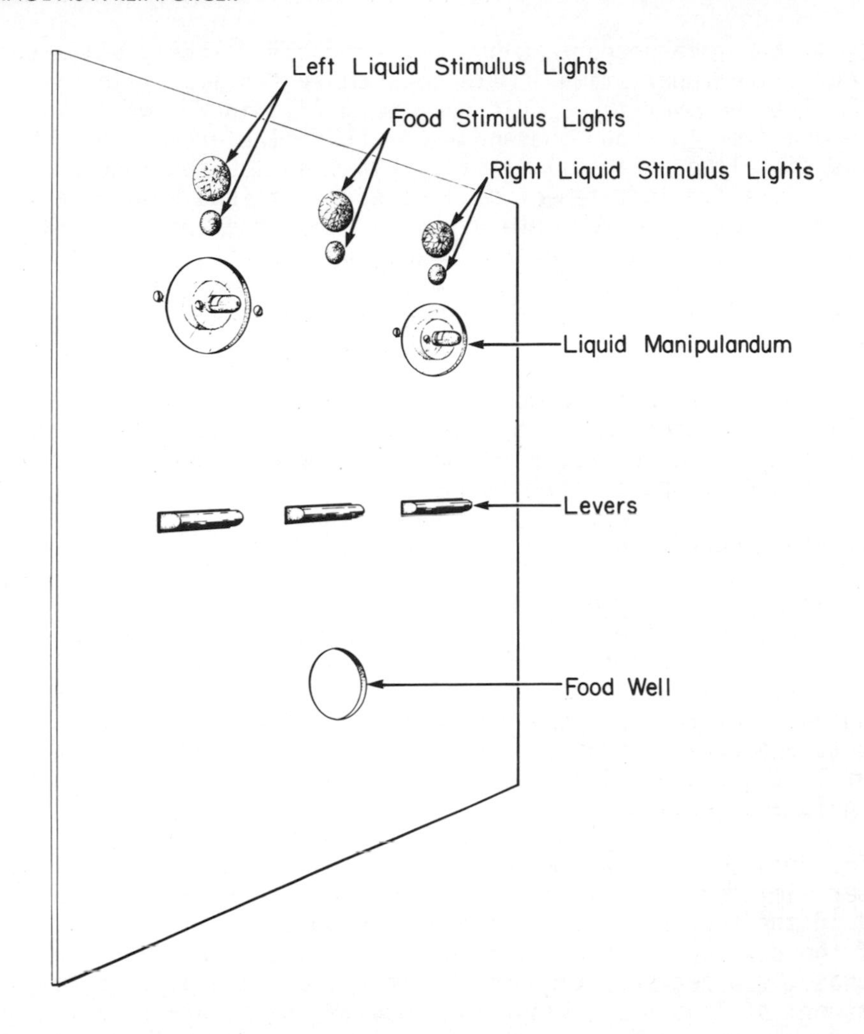

Figure 1. Layout of the monkey work panel. The following dimensions specify distances between centers of components. The levers are on a line 12.0 cm above the food well (5.0 cm diameter) and are separated by 15.5 cm. Each liquid manipulandum is 15.5 cm above one of the outer levers. The smaller stimulus lights are 5.5 cm above the liquid manipulanda, and the larger stimulus lights are 3.75 cm above the smaller lights.

was equipped with an automatic pellet dispenser (Foringer #1282) that delivered pellets to a small tray recessed in the cage wall. The feeder operandum was a commercial primate lever (BRS/LVE #PRL-001/121-07).

Stimuli: In both chambers, two stimulus fixtures, illuminated by 4.76-W incandescent bulbs, were mounted directly above each operandum. The upper stimuli were jeweled lenses 1.3 cm in diameter. The one over the liquid operandum (liquid S^D) was green and the one over the food operandum was red. The green liquid stimulus signalled availability of liquid, and the red food stimulus signalled instatement of a schedule of food reinforcement. The lower stimuli, mounted more directly above the operanda, had clear lenses 1.4 cm in diameter. The clear lensed liquid stimulus was paired with delivery of liquid and the clear lensed food stimulus (food S^D) signalled the immediate availability of a food pellet.

Control equipment: Commercial electromechanical control equipment was located in a room adjacent to the experimental rooms. Electromechanical counters, printout counters and cumulative recorders recorded experimental events.

Experimental procedures:

Experiment 1: Ethanol intake during schedule-induced polydipsia. Daily experimental sessions were 3 hours in duration. A 1-hour stimulus blackout preceded each session and a 3-hour stimulus blackout followed each session. Water was available on a continuous reinforcement schedule (CRF) except during stimulus blackouts and during ethanol sessions. A session consisted of 1 hour of liquid availability followed by 2 hours of concurrent food and liquid availability.

The food schedule was a mult (EXT 120 sec)(FR 1):food pellet. An upper limit of 45 pellets for M-P and 50 pellets for M-S resulted in the food schedule's terminating up to 30 min before the end of the session, thus completing the session in a CRF, liquid-only phase. To decrease the probability of accidental food reinforcement of liquid drinking, the food S^D was never presented within 8 seconds following a liquid response. The maximum number of food pellets available was chosen for each subject so as to provide a daily quantity of food that would maintain the monkey at 80% of its free-feeding weight when supplemented by a fruit and multiple-vitamin pill.

Ethanol sessions were signalled by blinking the green liquid S^D at 2 flashes per second. Solutions of 95% ethanol and tap water were mixed at least 20 hours prior to ethanol sessions. For example, the 8% solution was prepared by adding 53.0 ml of ethanol to a volumetric flask with sufficient tap water to make a total volume of 500 ml. Water was also stored for at least 20 hours prior to control sessions. Four concentrations of ethanol solutions were used in Experiment 1. Expressed in grams percent, they were instated in the following sequence: 1%, 2%, 4% and 8%.

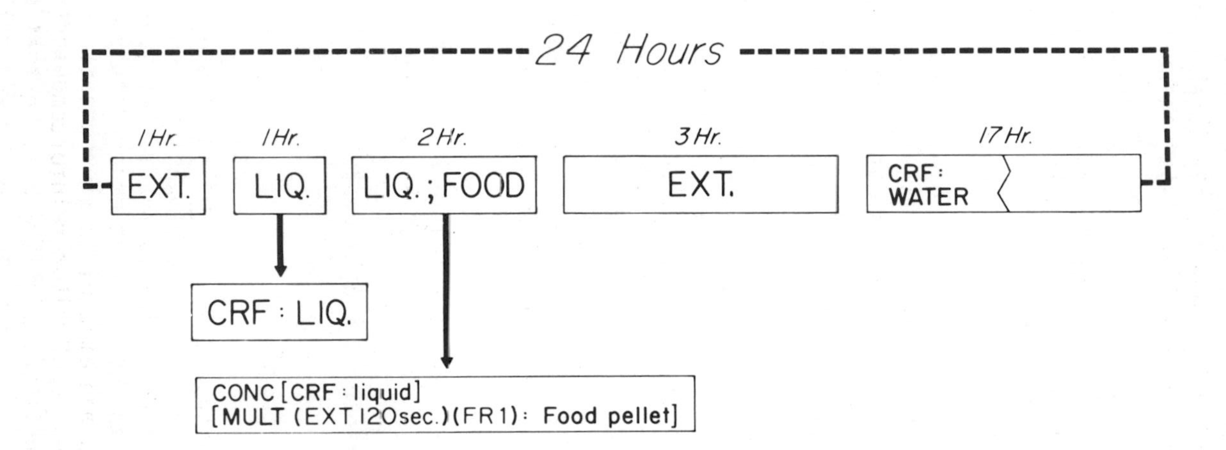

Figure 2. Scheduled activities in daily 24-hour sequence. The insert labelled CRF:LIQ. indicates that each liquid response was reinforced, and the insert labelled CONC [CRF: liquid] [MULT (EXT 120 sec)(FR 1):food pellet] indicates that each liquid response was reinforced, and that each food response was reinforced when it occurred in the presence of a discriminative stimulus. This stimulus had its onset 120 sec following the last food--reinforced response, and its offset occurred following the next reinforced response.

Individual ethanol sessions were separated by at least 2 water control sessions; water control sessions continued until water drinking returned to baseline values. Concentrations of ethanol were not manipulated until a criterion of 5 consecutive stable ethanol sessions had been attained, as indicated by visual inspection of the data.

The dependent variables of experimental interest during Experiment 1 were total number of liquid reinforcements per 3-hour session, volume consumed per 3-hour session, and the number of liquid reinforcements that occurred during the first hour of the session.

Experiment 2: Ethanol intake following termination of schedule-induced polydipsia. Concurrent availability of food during the second and third hours of the session was discontinued, and liquid only was available on a CRF schedule. The daily allotment of food pellets was available on a CRF schedule immediately following the post-session stimulus blackout.

For monkey M-S, 4% (W/V) ethanol was present during every third session while water was available during the intervening sesions. For M-P, three concentrations were studied in the sequence of 8, 16 and 32% (W/V). Expressed in terms of proof, these ethanol concentrations were 20, 40 and 80 proof, respectively. Again, ethanol sessions were separated by at least 2 water control sessions, and as before, a criterion of 5 consecutive stable ethanol sessions was attained before the ethanol concentration was changed.

The dependent variables of experimental interest were total number of liquid reinforcements per 3-hour session, total volume consumed, and the time course of liquid reinforcements, indicated by a printout counter that printed out data every 10 minutes of the session.

RESULTS

Experiment 1. Ethanol intake during schedule-induced polydipsia. The relative intakes of water and ethanol varied depending on whether or not food was concurrently available. During the first hour of each session when food was not concurrently available, little water was drunk, and ethanol reinforcements at all concentrations consistently exceeded water reinforcements (Figures 3 and 4). These data are consistent with the premise that ethanol was functioning as a reinforcer. During the second and third hours of each 3-hour session when food was concurrently available on a mult (EXT 120 sec)(FR 1), the two monkeys exhibited schedule-induced polydipsia, with monkeys M-P and M-S drinking 1140 ml and 1192 ml of water, respectively, over a 2-hour period. In

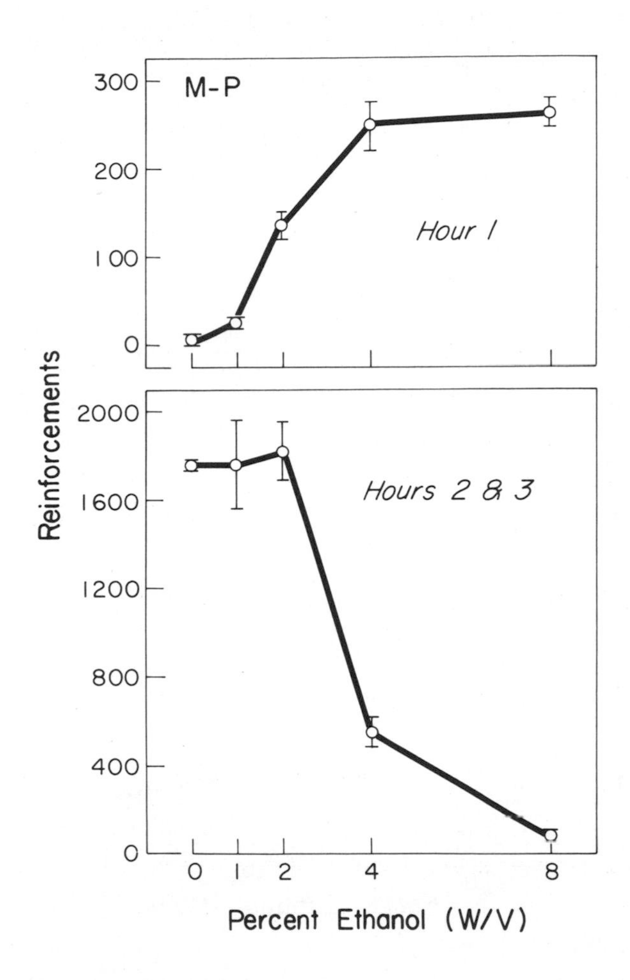

Figure 3. Number of liquid reinforcements as a function of ethanol concentration for monkey M-P. Upper frame: Number of reinforcements during Hour 1 when food was not available. Lower frame: Number of reinforcements during Hours 2 and 3 when 1-g food pellets were concurrently available on a mult (EXT 120 sec)(FR 1) schedule, and schedule-induced polydipsia occurred. Each point at 0% is the mean of 40 observations. Each point at the ethanol concentrations is the mean of 5 observations. Brackets indicate the standard error of the mean.

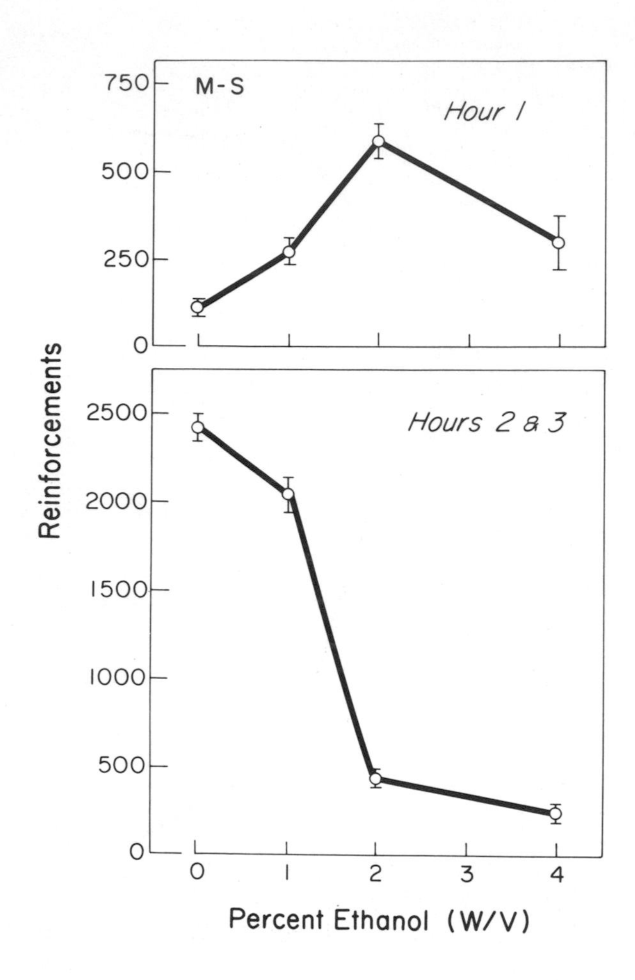

Figure 4. Number of liquid reinforcements as a function of ethanol concentration for monkey M-S. Upper frame: Number of reinforcements during Hour 1 when food was not available. Lower frame: Number of reinforcements during Hours 2 and 3 when 1-g food pellets were concurrently available on a mult (EXT 120 sec)(FR 1) schedule, and schedule-induced polydipsia occurred. Each point at 0% is the mean of 30 observations. Each point at the ethanol concentrations is the mean of 5 observations. Brackets indicate the standard error of the mean.

Table 1

Quantity of Ethanol Consumed in g/kg/hr

	Monkey	Percent Ethanol (W/V)			
		1	2	4	8
Hour 1	M-P	0.03	0.30	0.87	2.66
	M-S	0.28	1.21	1.08	----
Hours 2 & 3	M-P	0.98	1.78	0.96	0.42
	M-S	1.01	0.41	0.46	----

contrast to the results obtained during the first hour, ethanol reinforcements were equal to or less than water reinforcements (Figures 3 and 4). The decrease in ethanol reinforcements below water levels was especially marked at the highest concentrations used with this procedure. Figure 5 shows representative cumulative records from monkey M-S which illustrate the main findings in that during the first hour water-reinforced responding was low and less than for ethanol, whereas during the second and third hours, when food was concurrently available, water-reinforced responding was high and above that for ethanol. These records also show the initial burst of ethanol-reinforced responding at the beginning of the session. Table 1 presents the quantity of ethanol consumed (g/kg/hr) for each concentration.

<u>Experiment 2.</u> <u>Ethanol intake following termination of schedule-induced polydipsia.</u> Ethanol, but not water intake persisted following the permanent discontinuation of concurrent food availability. Figure 6 shows that monkey M-S's ethanol reinforcements consistently and substantially exceeded water values when 4% (W/V) ethanol was available every third day for 3 hours and water was available on intervening days. At this concentration, monkey M-S obtained 91 percent of its reinforcements in the first 30 minutes, and over the 3-hour session it consumed a mean quantity of 0.56 g/kg/hr.

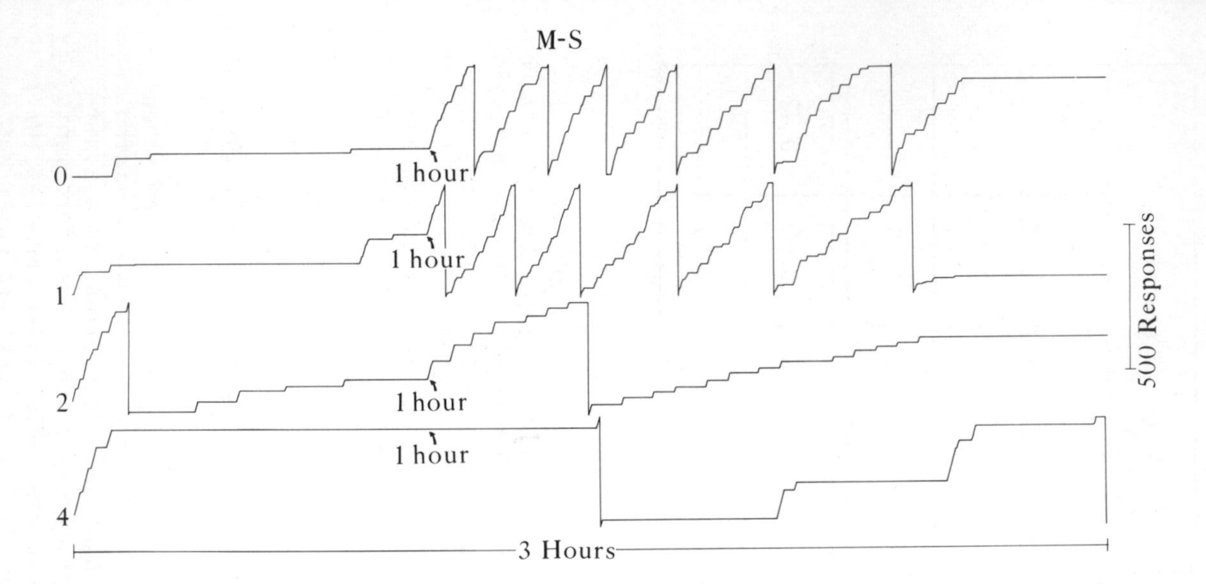

Figure 5. Representative cumulative records of monkey M-S's liquid responding during the 3-hour sessions of Experiment 1. Numbers along the left side specify the ethanol concentration. Responses or reinforcements (at FR 1 responses equal reinforcements) are cumulated along the ordinate, and time is indicated along the abscissa. The rate of intake is indicated by the slope of the record. Records selected are closest to representing mean performance at a given concentration. Note that with zero percent (water only) little responding occurred during Hour 1 and most responding occurred during Hours 2 and 3; whereas with ethanol present, responding exceeded the water values during Hour 1 but was less than the water values during Hours 2 and 3.

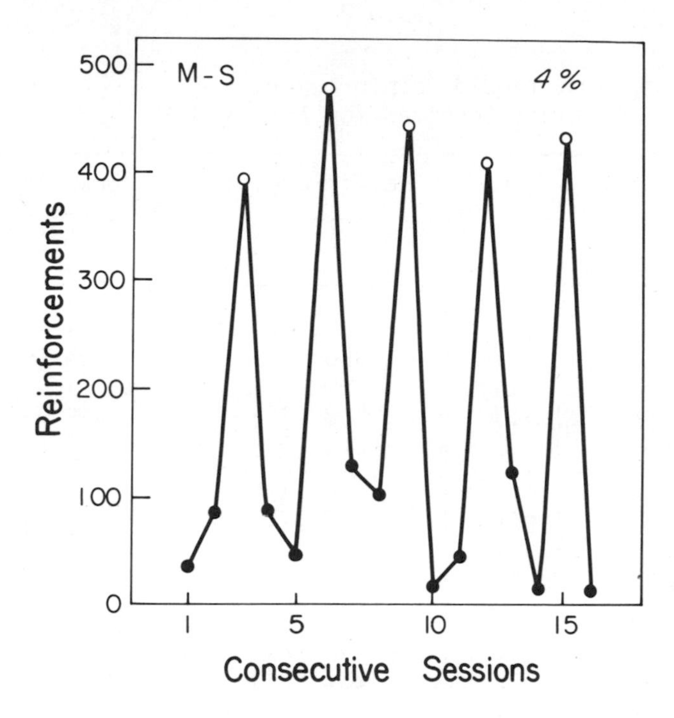

Figure 6. Liquid reinforcements per 3-hour session for monkey M-S when food was no longer concurrently available. Ordinate: Reinforcements (or responses) per 3-hour session. Abscissa: Consecutive sessions. Open circles: 4% (W/V) ethanol sessions. Filled circles: Water control sessions.

With monkey M-P drinking was studied at concentrations of 8, 16 and 32% (W/V). At these concentrations reinforcements substantially exceeded water control values (Figure 7 and Table 2). As the concentration was increased, the number of reinforcements decreased, but the quantity of ethanol consumed remained relatively constant (Table 2). When the ethanol solution was returned to 8% (W/V) after a series of sessions at 32% (W/V), it was possible to replicate the initial 8% (W/V) values (Figure 7 and Table 2). Figure 8 shows that most of the ethanol drinking occurred at the beginning of the session. These findings are also illustrated by cumulative records of the monkey's lip responding (Figure 9). At concentrations of 8% (W/V) and above, the monkey was markedly intoxicated. By the end of an experimental session, the monkey was

Table 2

Mean Liquid Reinforcements and
Quantity Consumed for Monkey M-P

Concentration	n	Reinforcements $\bar{x}$ (S.E.)	Quantity* $\bar{x}$ (S.E.)
0%	10	37 (10)	--- ---
8%	5	350 (23)	2.94 (0.17)
0%	10	58 (22)	--- ---
16%	5	211 (5)	3.38 (0.24)
0%	10	8 (3)	--- ---
32%	5	108 (11)	3.69 (1.34)
(retest)			
0%	10	12 (2)	--- ---
8%	5	327 (24)	3.78 (0.38)

* g/kg of body weight/hr

often slumped down in its restraining chair. Its movements were diminished and passive in nature. On one occasion the animal became only slightly responsive to stimuli and had to be removed from the apparatus and placed on its side. Intoxication of this magnitude was not seen in previous studies with rats.

Liquid intake in the presence and absence of concurrent food reinforcement. The effects of concurrent availability of food pellets on water and ethanol intake may be evaluated by comparing some results of Experiments 1 and 2. Specifically, water and ethanol intake during concurrent food reinforcement (i.e., during schedule-induced polydipsia) may be compared with water and ethanol intake in the absence of concurrent food reinforcement (i.e., following termination of schedule-induced polydipsia). Figure 10 shows that for monkey M-P ethanol intake did not differ in the presence and absence of concurrent food reinforcement, whereas water intake was far greater when food was available. Figure 11 shows similar results for monkey M-S. It should be noted that these figures compare results obtained under conditions

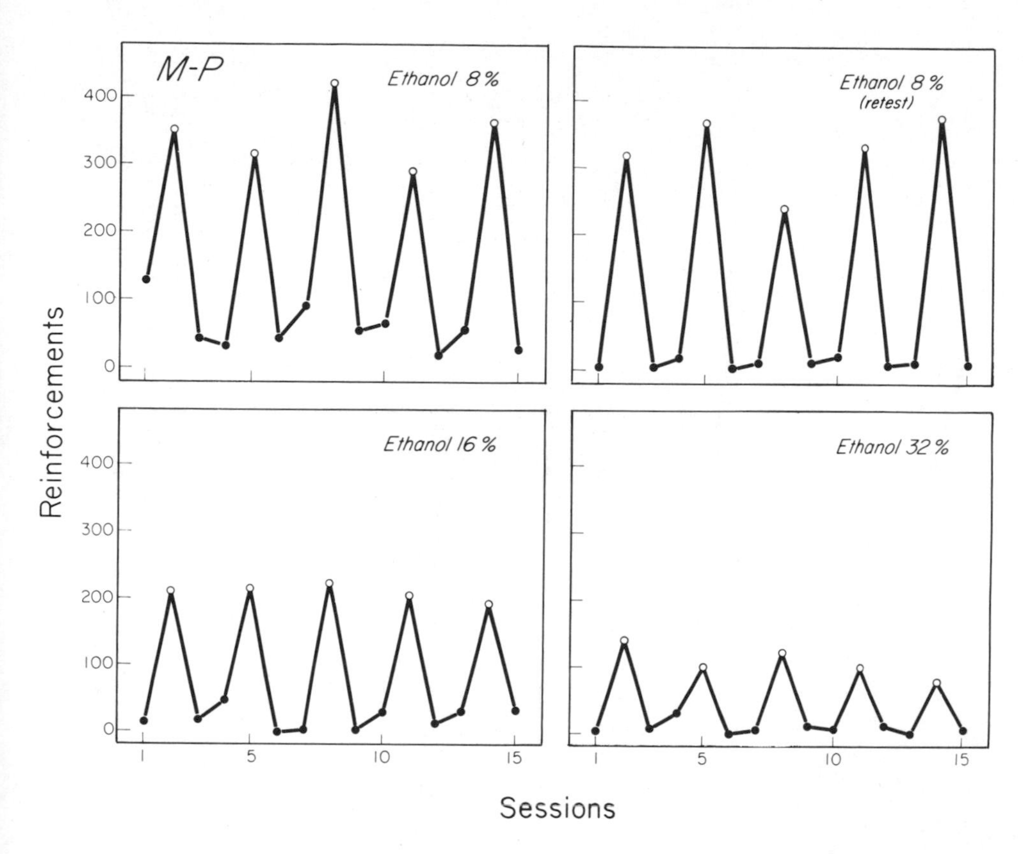

Figure 7. Reinforcements per 3-hour session for monkey M-P when food was no longer concurrently available. Ordinate: Reinforcements (or responses) per 3-hour session. Abscissa: Sessions. Open circles: Ethanol sessions. Filled circles: Water control sessions.

of stabilized responding. The high water intake when food was available indicates the occurrence of schedule-induced drinking (Schuster and Woods, 1966). The lack of effect of food availability on ethanol intake was surprising, indicating that once ethanol-reinforced responding stabilized it probably occurred independently of concurrent access to food.

DISCUSSION

That the number of ethanol reinforcements and the volume consumed consistently exceeded water control values indicates that ethanol was functioning as an oral reinforcer for the rhesus monkey. These data supplement previous findings that ethanol can serve as

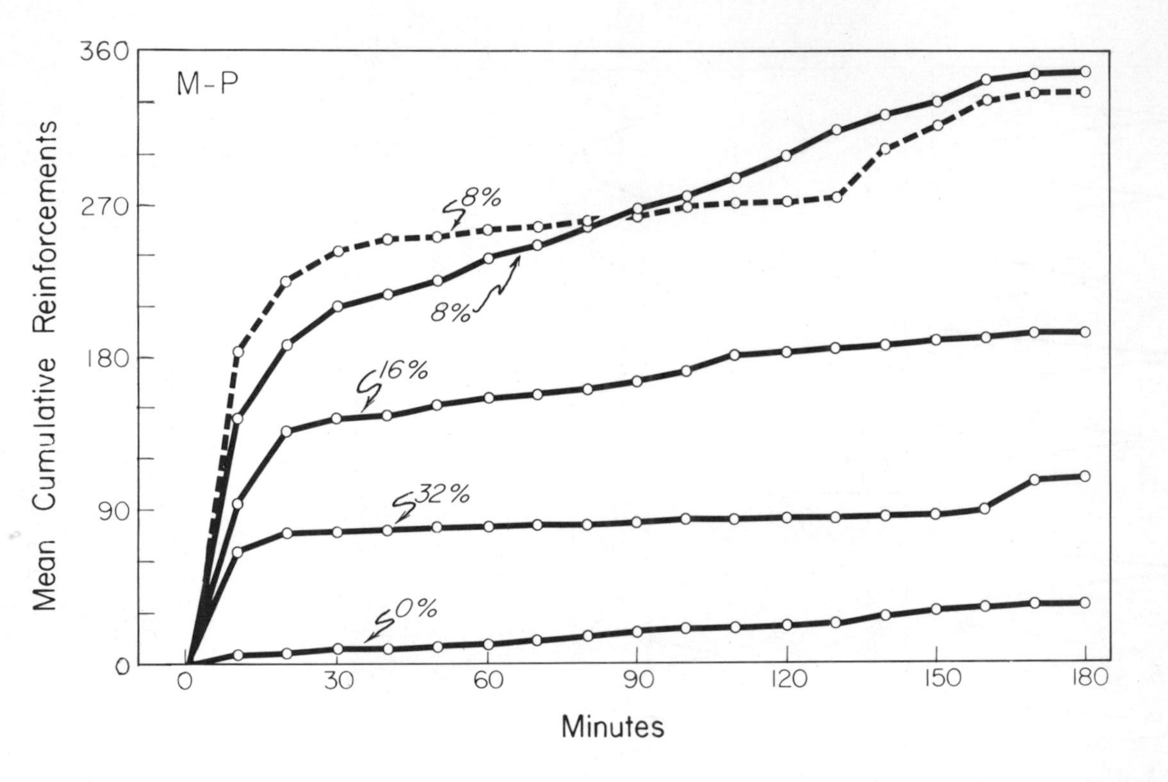

Figure 8. Mean cumulative reinforcements over 3-hour sessions when food was not present as a function of concentration for monkey M-P. Ordinate: Mean cumulative reinforcements. Abscissa: Time within sessions. Each point on the 0% curve is the mean of 40 observations. Each point on the ethanol curves is the mean of 5 observations. The dotted line connects the values for 8% (W/V) that were obtained when the monkey was returned to this concentration following the series of sessions at 32% (W/V). Note that the highest rate of intake occurred at the beginning of the session, and that ethanol values exceeded water control values.

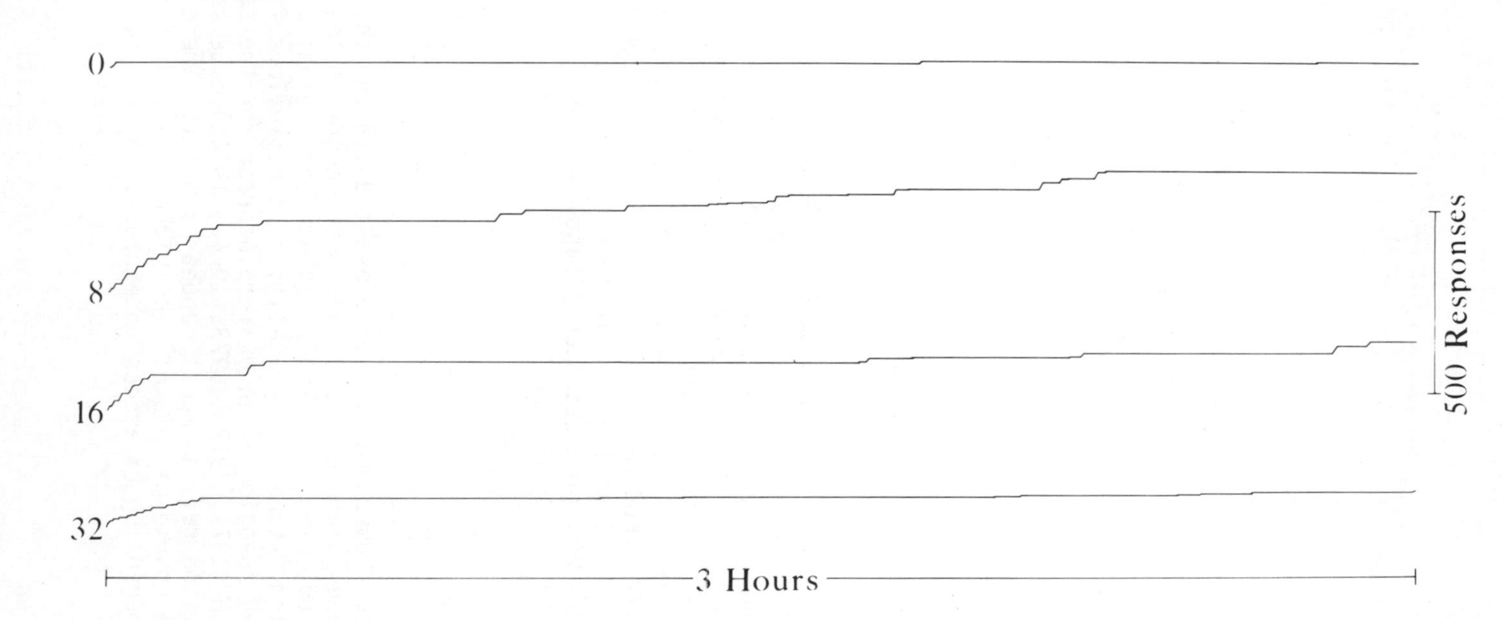

Figure 9. Representative cumulative records for monkey M-P's responding showing the pattern of intake at each concentration. Responses (or reinforcements) are cumulated along the ordinate, and time is indicated along the abscissa. Numbers along the left side specify the ethanol concentration. Each record was selected on the basis of being the closest to the mean performance at a concentration. Note that responding usually occurred in sustained bursts and at a high rate at the beginning of a session.

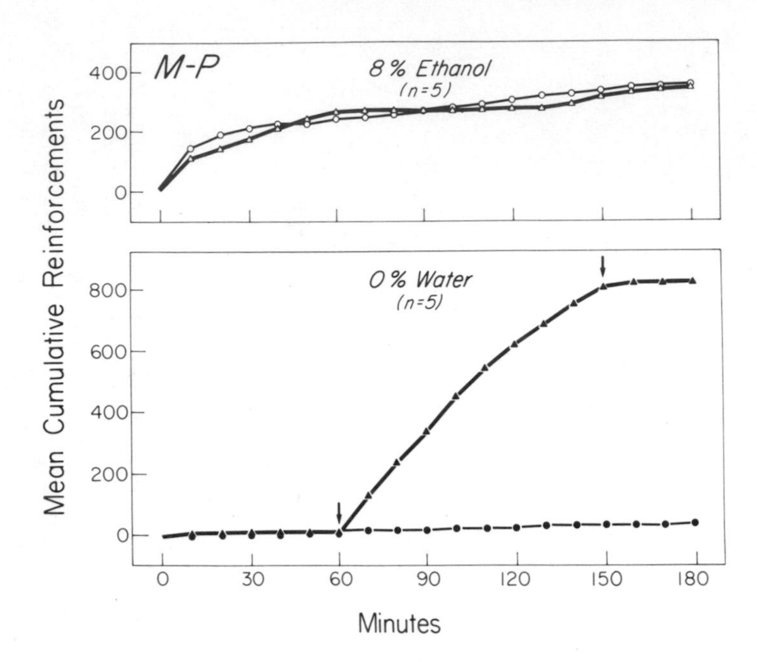

Figure 10. Mean cumulative water and 8% (W/V) ethanol reinforcements for monkey M-P from 3-hour sessions with and without concurrent food. Ordinate: Mean cumulative reinforcements. Abscissa: Time within sessions. Each point is a mean of observations from 5 sessions. Triangles represent data points from sessions when concurrent food was present. The arrow at 60 minutes indicates the onset of concurrent food reinforcement, while the subsequent arrow indicates the offset of food reinforcement. Water sessions are those that immediately preceded each ethanol session.

a reinforcer for the rhesus monkey when presented via the intragastric (Yanagita and Takahashi, 1973) and intravenous routes (Deneau, et al., 1969; Winger and Woods, 1973; Woods, et al., 1971). The data also extend the generality of the previous finding that ethanol can function as an oral reinforcer for the rat (Meisch and Thompson, 1971, 1972a, 1973, 1974a, 1974b). Certain of the findings for the rat and monkey appear similar. There is a characteristic time course of intake: The highest rate of responding occurs at the beginning of access to ethanol. Also, the

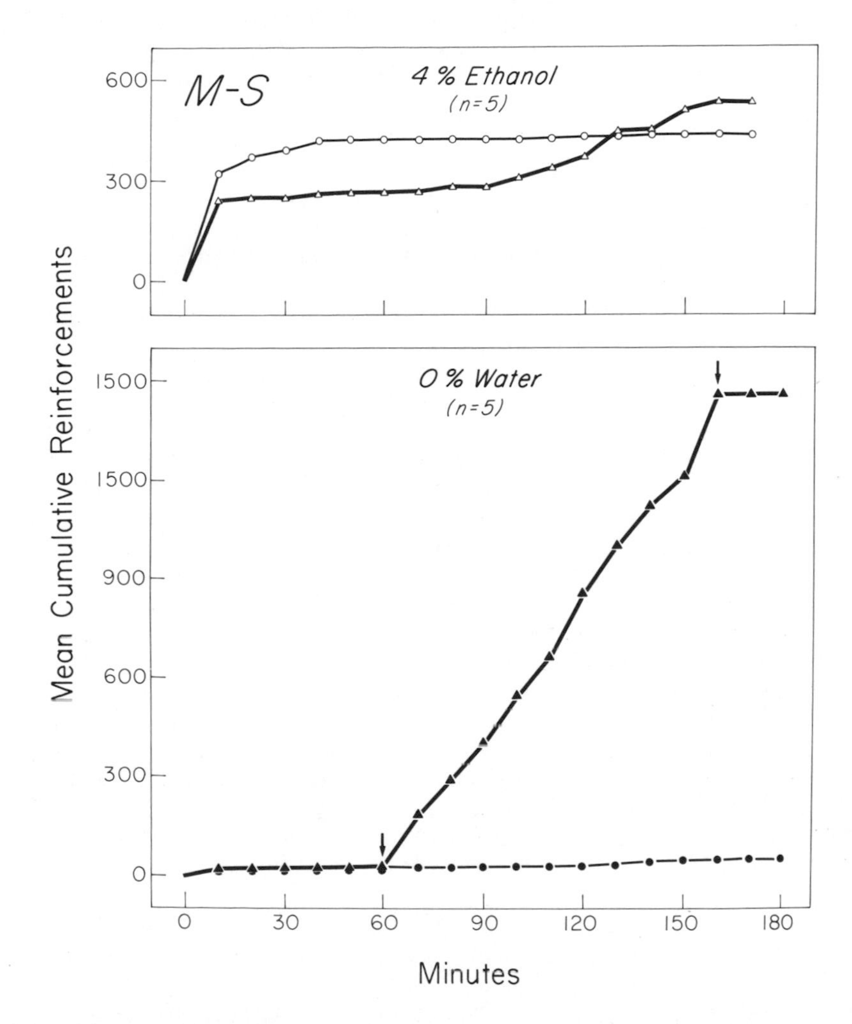

Figure 11. Mean cumulative water and 4% (W/V) ethanol reinforcements for monkey M-S from 3-hour sessions with and without concurrent food. See Figure 10 for further details.

quantity consumed increases with the concentration. Both rats and monkeys consume more ethanol than water when food is not present, and both do just the opposite when food is concurrently available on an intermittent schedule.

In order to establish ethanol as a reinforcer via the oral route two difficulties must be circumvented: its aversive taste, and the delay in reinforcement between drinking and experiencing its effects following absorption (Mello and Mendelson, 1971a, 1971b). The aversiveness of ethanol's taste properties is indicated by the fact that the monkey does not prefer even low concentrations to water (Myers, Stoltman and Martin, 1972). Thus, the ethanol that _is_ consumed is often insufficient to produce a discriminable effect; even if large quantities were to be consumed, the delay in reinforcement caused by the time required for absorption would hinder association of drinking behavior with subsequent ethanol effects. In the present study these problems were overcome by starting with a low concentration, viz. 1% (W/V), and then slowly increasing the concentration as the monkeys became adapted to ethanol's taste and to drinking large volumes. The use of food-deprived monkeys helped to minimize the delay in absorption, and the use of schedule-induced polydipsia permitted the repeated and reliable drinking of large volumes. Thus, the taste of the ethanol solutions was repeatedly paired with the effects that follow ingestion of substantial volumes.

The rhesus monkey may have several advantages over rats in the study of ethanol dependence. The monkey has a longer life span and a more complex behavioral repertoire. Additionally, the monkey's behavior may be less affected by the delay of reinforcement due to absorption and the monkey is phylogenetically closer to man than is the rat. Thus, the use of the procedure described in the present paper will hopefully result in an animal preparation that permits an investigation of factors controlling human ethanol dependence.

Acknowledgements: This research was supported by a grant from the Medical Foundation of the Minneapolis Clinic of Psychiatry and Neurology and in part by USPHS Grant MH 20919. Portions of these data were reported to the Committee on Problems of Drug Dependence, 1974. We thank Dr. Carol Iglauer for her helpful comments concerning the manuscript, and we also thank Pat Beardsley, Linda Stark and Dale Kliner for assistance in conducting the experiments.

REFERENCES

Deneau, G., Yanagita, T. and Seevers, M.H. Self-administration of psychoactive substances by the monkey. A measure of psychological dependence. Psychopharmacologia, 16:30-48, 1969.

Koz, G. and Mendelson, J.H. Effects of intraventricular ethanol infusion on free choice alcohol consumption by monkeys. In (Ed.) R.P. Maickel, Biochemical Factors in Alcoholism, New York: Pergamon, pp. 17-24, 1967.

Meisch, R.A. and Thompson, T. Ethanol intake in the absence of concurrent food reinforcement. Psychopharmacologia, 22:72-79, 1971.

Meisch, R.A. and Thompson, T. Ethanol reinforcement: Effects of concentration during food deprivation. International Symposium Biological Aspects of Alcohol Consumption, 27-29 September 1971, Helsinki. The Finnish Foundation for Alcohol Studies, 20:71-75, 1972a.

Meisch, R.A. and Thompson, T. Ethanol intake during schedule--induced polydipsia. Physiol. Behav., 8:471-475, 1972b.

Meisch, R.A. and Thompson, T. Ethanol as a reinforcer: Effects of fixed-ratio size and food deprivation. Psychopharmacologia, 28:171-183, 1973.

Meisch, R.A. and Thompson, T. Rapid establishment of ethanol as a reinforcer for rats. Psychopharmacologia, 1974a, in press.

Meisch, R.A. and Thompson, T. Ethanol intake as a function of concentration during food deprivation and satiation. Pharmac. Biochem. Behav., 1974b, in press.

Mello, N.K. and Mendelson, J.H. The effects of drinking to avoid shock on alcohol intake in primates. In (Eds.) M.K. Roach, W.M. McIsaac and P.J. Creaven, Biological Aspects of Alcohol, Austin: University of Texas Press, pp. 313-332, 1971a.

Mello, N.K. and Mendelson, J.H. Evaluation of a polydipsia technique to induce alcohol consumption in monkeys. Physiol. Behav., 7:827-836, 1971b.

Myers, R.D., Stoltman, W.P. and Martin, G.E. Effects of ethanol dependence induced artifically in the rhesus monkey on the subsequent preference for ethyl alcohol. Physiol. Behav., 9:43-48, 1972.

Myers, R.D., Veale, W.L. and Yaksh, T.L. Preference for ethanol in the rhesus monkey following chronic infusion of ethanol into the cerebral ventricles. Physiol. Behav., 8:431-435, 1972.

Pieper, W.A. and Skeen, M.J. Induction of physical dependence on ethanol in rhesus monkeys using an oral acceptance technique. Life Sci., 11:989-997, 1972.

Schuster, C.R. and Woods, J.H. Schedule-induced polydipsia in the rhesus monkey. Psychol. Rep., 19:823-828, 1966.

Sinclair, J.D. The alcohol-deprivation effect in monkeys. Psychon. Sci., 25:21-22, 1971.

Thompson, T., Schuster, C.R., Dockens, W. and Lee, R. Mouth--operated food and water manipulanda for use with monkeys. J. exp. Anal. Behav., 7:171-172, 1964.

Winger, G.D. and Woods, J.H. The reinforcing property of ethanol in the rhesus monkey: I. Initiation, maintenance and termination of intravenous ethanol-reinforced responding. Ann. N. Y. Acad. Sci., 215:162-175, 1973.

Woods, J.H., Ikomi, F. and Winger, G.D. The reinforcing property of ethanol. In (Eds.) M.K. Roach, W.M. McIsaac and P.J. Creaven, Biological Aspects of Alcohol, Austin: University of Texas Press, pp. 371-388, 1971.

Yanagita, T. and Takahashi, S. Dependence liability of several sedative-hypnotic agents evaluated in monkeys. J. Pharmac. exp. Ther., 185:307-316, 1973.

ALCOHOL WITHDRAWAL CONVULSIONS IN GENETICALLY DIFFERENT POPULATIONS OF MICE*

Dora B. Goldstein and Ryoko Kakihana

Departments of Pharmacology and Psychiatry

Stanford University School of Medicine, Stanford, CA, USA

The question whether alcoholism is inherited is a complex one. Some human populations have much higher rates of alcoholism than others. Genetic differences between populations might explain their alcoholism rates but it is equally likely that cultural habits determine the extent of drinking in a given population and thus secondarily account for the incidence of alcoholism. When the heritability of alcoholism has been examined by methods that rule out cultural differences, the results suggest that there is indeed a genetic component. Studies of twins (Kaij, 1960; Partanen et al, 1966) and of adopted sons of alcoholic biological parents (Schuckit, 1972; Goodwin et al, 1973) show a high incidence of alcoholism in relatives of alcoholics.

Studies in humans leave unanswered the question of what is being inherited. Is it the propensity to drink or is it a likelihood of becoming physically dependent at a given level of alcohol intake? Animal models can deal with these two questions separately. The first question has already been examined; genetic factors are known to affect voluntary alcohol intake in rats and mice (Eriksson, 1968; Mardones, 1960; McClearn and Rodgers, 1959). This paper addresses the second question, whether there is a genetic variability in the response to a given amount of chronic alcohol intake.

The response we would like to measure is physical dependence, which we assume to be a biochemical alteration of the brain. Since

* Supported by United States Brewers Association, Inc. and by USPHS, grant no. AA0498.

we do not know its nature, however, we have no way at all to estimate its magnitude. What we can do is put the question in terms of the withdrawal reaction, which is the overt peripheral expression of the underlying physical dependence. This paper shows that genetic factors contribute to the intensity of withdrawal reactions. That is as far as we can go until we know what to look for biochemically.

METHODS

Mice were made physically dependent on alcohol by a standard "cycle" of alcohol administration, a 3-day period of continuous intoxication with blood alcohol levels of about 2 mg/ml. The previously described inhalation technique was used (Goldstein, 1972). Mice were given a priming dose of ethanol i.p. and then housed in a vapor chamber containing ethanol vapor, 10 to 12 mg/L. Pyrazole, an inhibitor of alcohol dehydrogenase, was administered daily at a dose of 1 mmole/kg, intraperitoneally, in order to stabilize the blood alcohol concentration. No pyrazole was given on the day of withdrawal.

To measure the intensity of withdrawal reactions, we used only one sign, which we call "convulsions on handling". We have previously shown that this sign is a good measure of the overall withdrawal reaction in our model (Goldstein, 1972). Seizures were elicited by picking the mouse up by the tail during the withdrawal period. The seizures were scored repeatedly (hourly for the first 14 hr) on the basis of their severity, using a rating system with a range of 0 to 4 points. Seizure scores were plotted against time after removing the mice from the vapor chamber. The area under such curves, up to 30 hr after withdrawal, was computed for individual animals and designated as the "withdrawal score", an estimate of the intensity of the withdrawal reaction.

SELECTIVE BREEDING

In a standard 3-day cycle of alcohol intoxication, the withdrawal scores of individual mice vary widely. Selective breeding experiments (Goldstein, 1973c), with high-scoring and low-scoring mice, are summarized here.

The first point to establish was whether individual mice behaved consistently in two cycles of intoxication and withdrawal. Male and female Swiss-Webster mice were subjected to two cycles separated by an interval of 6 weeks. This was more than enough time to insure that the physical dependence developed during the first cycle would not carry over into the second (Goldstein, 1974) Analysis of variance showed that individual mice differed

Table 1: High-Scoring and Low-Scoring Pairs Selected for Breeding

Group	Sex	Withdrawal Score		
		Cycle 1	Cycle 2	Group Mean
High	Male	62	76	
	Female	50	64	
				55 ± 5.3
High	Male	45	49	
	Female	46	47	
Low	Male	30	14	
	Female	16	9	
				15 ± 4.9
Low	Male	23	23	
	Female	2	2	

One score per mouse (mean of the two cycles) was used to compute the group mean and its standard error. The means of the two groups differed significantly. $P < 0.01$. Student's t-test with 6 degrees of freedom.

---

significantly in withdrawal scores. Some scored high in both cycles, others low.

From the original group of 37 mice, four breeding pairs were selected - two pairs of high-scoring mice and two low-scoring. Their withdrawal scores were consistent in the two cycles, as shown in Table 1. The means of the high and low groups differed significantly, $P < 0.01$.

All the mice of the four litters were tested in the vapor chamber and scored for convulsions on handling after withdrawal. The results are shown in Table 2. The offspring clearly behaved like their parents with respect to withdrawal scores. Analysis of the data was slightly complicated by the fact that female mice developed significantly lower blood alcohol levels than males, at the same alcohol vapor concentration. Withdrawal reactions vary with blood alcohol levels maintained during the intoxication phase (Goldstein, 1972) so it was necessary to correct for this discrepancy. Multiple regression analysis was used to deal with each independent variable separately, and the results showed that withdrawal reactions in the F1 generation varied with the score-group

Table 2: F1 Generation

	Parental Group	Mean ± S.E.M. (N)	
		Males	Females
Blood Alcohol (mg/ml)	High	2.6 ± .12 (5)	1.7 ± .10 (6)
	Low	2.4 ± .20 (3)	1.6 ± .06 (6)
Withdrawal Score	High	41.0 ± 6.6	23.0 ± 4.1
	Low	21.0 ± 8.3	8.0 ± 2.2

Multiple regression analysis of withdrawal scores

By groups	P < .01
By blood alcohol	P < .01
By sex	N. S.

of the parents (P < 0.01) and (as expected) with the blood alcohol levels of individual mice (P < 0.01) but not by sex. That is, there was no sex difference in intensity of withdrawal reactions when the blood alcohol level was taken into account.

The selective breeding was continued for one more generation, which produced a slightly greater separation of the lines. The experiment confirmed the lack of sex difference in withdrawal scores when corrected for blood alcohol levels.

This analysis (including the statistical treatment by individual mice rather than by litters) does not take into account possible maternal effects. Studies with a larger number of mating pairs, including cross-fostering experiments, would be necessary to establish a genetic effect conclusively.

INBRED STRAINS OF MICE

Among inbred strains of mice, the C57BL strain is of particular interest in alcohol research because mice of this strain will voluntarily consume most of their fluid in the form of ethanol, even when water is freely available. By contrast, strains such as DBA or BALB almost completely refuse to drink alcohol in a similar two-bottle choice situation. Furthermore, Kakihana et al (1966) have shown that C57BL mice have a lower brain sensitivity to ethanol than BALB mice, a nondrinker strain.

We compared the withdrawal reactions of DBA/2J, BALB/cJ and C57BL/6J male mice after a standard cycle of alcohol exposure (Goldstein and Kakihana, in press). Figure 1 shows a plot of withdrawal scores for C57BL, DBA and the previously tested Swiss-Webster mice. Mean blood alcohol levels during the inhalation period were 2.4, 1.8 and 2.0 mg/ml, respectively, for these three strains. The withdrawal reactions of DBA mice (and BALB mice, not shown) resembled that of Swiss-Websters. But the C57BL mice showed almost no convulsions on handling.

How to interpret this failure of C57BL mice to show withdrawal seizures? It does not necessarily mean that mice of this strain do not become physically dependent on ethanol. Indeed, Freund and Walker (1971) have reported that female C57BL mice undergo convulsions, both spontaneous and audiogenic, after alcohol withdrawal*. We did see some mild signs of alcohol dependence in this strain, including a slight loss of body weight and a tendency for hyperactive startle reactions. C57BL mice are known to be relatively resistant to audiogenic and electroshock convulsions under some conditions (Schlesinger and Griek, 1970). Perhaps they are also resistant to withdrawal convulsions. As explained above, we have no way of knowing to what extent they were physically dependent in a biochemical sense.

The response of the inbred strains to reserpine was examined, because previous work had suggested that alcohol withdrawal reactions and certain reserpine effects had something in common. Swiss-Webster mice given single injections of reserpine had been shown to display the typical convulsions on handling (Goldstein, 1973b). The seizures were evoked by doses of 1 to 5 mg/kg, they could be elicited over a period of many hours and they appeared to be identical in form to the convulsions elicited by handling during alcohol withdrawal reactions. It was therefore of interest to see

---

* Studies in progress in one of our laboratories (R. K.) show that DBA/2J and BALB/cJ male mice had convulsions on handling but C57BL/6J mice did not, after alcohol administration in the Freund liquid diet.

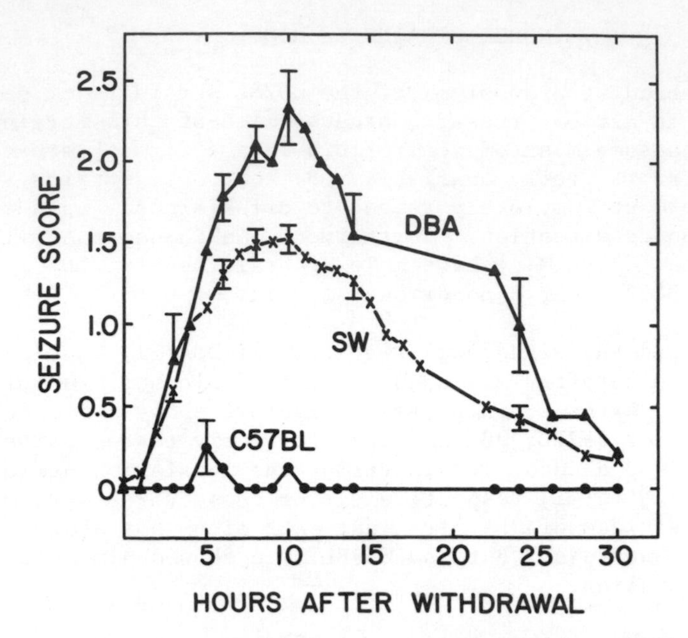

Figure 1. Scores for convulsions on handling after withdrawal from alcohol. ▲, DBA mice, N = 9. X, Swiss-Webster mice, N = 95, from previous experiments. ●, C57BL mice, N = 8. Points are means; vertical bars indicate S. E. M.

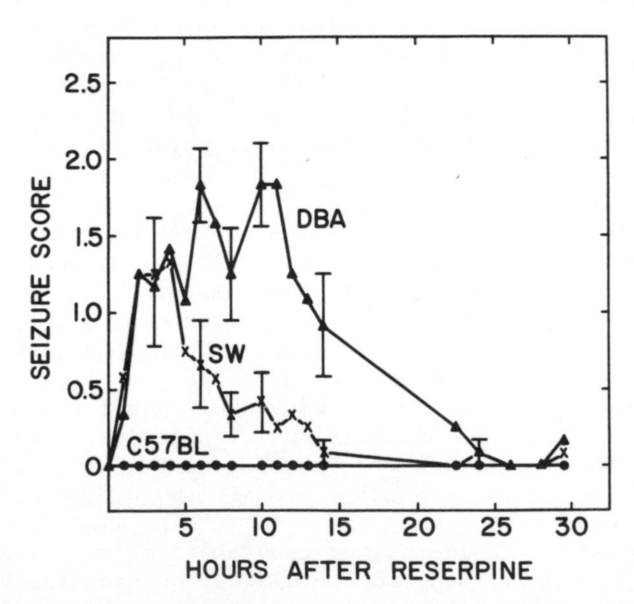

Figure 2. Scores for convulsions on handling after reserpine, 5 mg/kg. Symbols as in Figure 1. N = 12 for each strain.

whether C57BL mice would be resistant to reserpine as well as to alcohol withdrawal convulsions. Mice of the C57BL/6J and DBA/2J strains were treated with reserpine and then scored for convulsions on handling. As shown in Figure 2, the strain differences seen during alcohol withdrawal were exactly paralleled in the response to reserpine.

We had other evidence suggesting the involvement of catecholamines (but not serotonin) in alcohol withdrawal reactions (Goldstein, 1973a) so this strain correlation between sensitivity to reserpine and to alcohol withdrawal seizures seemed meaningful. However, it turned out to be fortuitous, as the next set of experiments showed.

MICE OF A HETEROGENEOUS STOCK

A correlation of two traits in a few inbred strains looks more important than it actually is. For this type of experiment, one strain represents one individual, because the mice within an inbred strain are homozygous for practically all loci. They can be considered as genetically identical. What is necessary in order to see whether the two traits correlate is to test the association in a large number of animals with genetic variability. In a genetically heterogeneous population, the same individual mice that have high alcohol withdrawal scores should have high seizure scores after reserpine, if the same biochemical mechanism determines both responses. We tested this hypothesis in HS mice, a reproducibly heterogeneous stock of mice derived from a balanced blend of 8 inbred strains (McClearn et al, 1970).

Thirty-two male HS mice were obtained from the Institute for Behavioral Genetics, University of Colorado, Boulder. Their alcohol sleep times and elimination rates were measured. Six weeks later, they were given a standard 3-day cycle of alcohol exposure in the vapor chamber, and withdrawal scores were recorded for each mouse. Finally, 3 to 4 weeks after the inhalation treatment, the same mice were observed for convulsions on handling following an injection of reserpine. Correlation coefficients between the variables were calculated and are shown in Table 3. There was only one significant correlation, an expected inverse correlation between alcohol elimination rates and sleep times. Sleep times did not correlate with withdrawal scores, suggesting that susceptibility to acute and chronic alcohol effects do not have the same mechanism. Withdrawal reactions did not correlate significantly with blood alcohol levels in the immediately preceding intoxication phase. (Previous experiments, mentioned above, where there was more variability in blood alcohol levels did show such a correlation). Finally, there was no significant correlation between alcohol withdrawal scores and seizures evoked by reserpine. This result makes it unlikely that the same biochemical characteristic determines

Table 3: Correlation of Alcohol and Reserpine Effects

Variable	Mean ± s.d.	Correlation Coefficient, r 1	2	3	4	5
1. Sleep Time (minutes)	50 ± 31	1.0	-0.52*	-0.10	-0.06	-0.18
2. Elimination Rate (mg/ml per hr)	0.62 ± 0.08		1.0	0.04	0.20	0.09
3. Mean Blood Alc. (mg/ml)	2.17 ± 0.22			1.0	0.10	0.06
4. Withdrawal Score (area)	34.6 ± 18.4				1.0	0.20
5. Reserpine Score (area)	16.0 ± 15.3					1.0

The data are from 3 experiments with the same 32 HS male mice. In Exp. I, sleep times and alcohol elimination rates were measured after single injection of ethanol, 3.5 g/kg, i.p. Exp. II was an inhalation experiment; blood alcohol levels were assayed daily and the withdrawal aeizures were scored. In Exp. III, convulsions on handling were scored after a single injection of reserpine (Serpasil, Ciba), 5 mg/kg, i.p. $^{*}P < 0.05$

sensitivity to alcohol withdrawal reactions and to reserpine convulsions.

SUMMARY

Individual Swiss-Webster mice differed in the intensity of their withdrawal reactions after a standard regimen of alcohol administration, and offspring resembled their parents in this respect. Inbred strains of mice also differed in the severity of their alcohol withdrawal reactions. C57BL mice showed almost no convulsions on handling. These results indicate that a genetic factor participates in determining the severity of alcohol withdrawal seizures in mice.

The sensitivity to excitant actions of reserpine paralleled the sensitivity to alcohol withdrawal reactions in three strains of mice but not among individual mice of a genetically heterogeneous stock. The strain correlation between these traits was therefore fortuitous and does not demonstrate a common biochemical lesion in alcohol-withdrawn and reserpine-treated mice.

REFERENCES

Eriksson, K. Genetic selection for voluntary alcohol consumption in the albino rat. Science 159: 739-741, 1968.

Freund, G. and Walker, D. W. Sound-induced seizures during ethanol withdrawal in mice. Psychopharmacologia 22: 45-49, 1971.

Goldstein, D. B. Relationship of alcohol dose to intensity of withdrawal signs in mice. J. Pharmacol. Exp. Ther. 180: 203-215, 1972.

Goldstein, D. B. Alcohol withdrawal reactions in mice: Effects of drugs that modify neurotransmission. J. Pharmacol. Exp. Ther. 186: 1-9, 1973a.

Goldstein, D. B. Convulsions elicited by handling: A sensitive method of measuring CNS excitation in mice treated with reserpine or convulsant drugs. Psychopharmacologia 32: 27-32, 1973b.

Goldstein, D. B. Inherited differences in intensity of alcohol withdrawal reactions in mice. Nature 245: 154-156, 1973c.

Goldstein, D. B. Rates of onset and decay of alcohol physical dependence in mice. J. Pharmacol. Exp. Ther., in press.

Goldstein, D. B. and Kakihana, R. Alcohol withdrawal reactions and reserpine effects in inbred strains of mice. Life Sci., in press.

Goodwin, D. W., Schulsinger, F., Hermansen, L., Guze, S. B. and Winokur, G. Alcohol problems in adoptees raised apart from alcoholic biological parents. Arch. Gen. Psychiat. 28: 238-243, 1973.

Kaij, L. Alcoholism in Twins. Almqvist and Wiksell, Stockholm, 1960.

Kakihana, R., Brown, D. R., McClearn, G. E. and Tabershaw, I. R. Brain sensitivity to alcohol in inbred mouse strains. Science 154: 1574-1575, 1966.

Mardones, J. Experimentally induced changes in the free selection of ethanol. Int. Rev. Neurobiol. 2: 41-76, 1960.

McClearn, G. E. and Rodgers, D. A. Differences in alcohol preference among inbred strains of mice. Quart. J. Stud. Alc. 20: 691-695, 1959.

McClearn, G. E., Wilson, J. R. and Meredith, J. E. The use of isogenic and heterogenic mouse stock in behavioral research. Contributions to Behavior-Genetic Analysis. G. Lindzey and D. D. Thiessen, Eds. Appleton-Century-Crofts, New York, 1-22, 1970.

Partanen, J., Bruun, K. and Markkanen, T. Inheritance of Drinking Behavior. Finnish Foundation for Alcohol Studies, Helsinki, 1966.

Schlesinger, K. and Griek, B. J. The genetics and biochemistry of audiogenic seizures. In Contributions to Behavior-Genetic Analysis. G. Lindzey and D. D. Thiessen, Eds. Appleton-Century-Crofts, New York, 219-257, 1970.

Schuckit, M. A. Family history and half-sibling research in alcoholism. Ann. N. Y. Acad. Sci. 197: 121-125, 1972.

Figures reprinted, with permission, from Goldstein and Kakihana, 1974, Pergamon Press.

A BEHAVIORAL AND ELECTROPHYSIOLOGICAL ANALYSIS OF ETHANOL DEPENDENCE IN THE RAT

Don W. Walker, Bruce E. Hunter and Joseph Riley

Veterans Administration Hospital and Department of Neuroscience, University of Florida, Gainesville, Florida, U. S. A.*

The identification of the neurobiological concomitants of ethanol dependence and withdrawal would not only provide a better understanding of the basic nature of such phenomena, but would also facilitate the development of techniques of detection, prevention, and treatment of alcoholic disease and its psychobiological consequences.

The development and application of animal models is necessary in order to provide information concerning the neurophysiological, neuropharmacological, and neurochemical concomitants of ethanol dependence. During the past five years significant progress has been made in the development of animal models of ethanol dependence, as evidenced by reports of behavioral signs of ethanol withdrawal in a variety of species (C.F. Mello, 1973).

The rat would appear to be a particularly suitable species for use in the delineation of the neurobiological concomitants of ethanol dependence. Experiments concerning neurophysiological (Hunter et al., 1973) and neurochemical (Hunt and Majchrowicz, 1974; Littleton, Griffiths, and Ortiz, 1974; Pohorecky, 1974; Pohorecky, Jaffe, and Berkeley, 1974) alterations during ethanol

---

* Supported by the Veterans Administration (MRIS 9183) and PHS Grants AA00200 and MH10320

The authors thank Pat Burnett, Larry Ezell, Dot Robinson, and Joe Welcome for technical assistance.

withdrawal require large groups of subjects, practical only with small laboratory animals. The rat has several advantages over the mouse, particularly in studies of regional neurochemical or neurophysiological alterations (or both) in the brain. These advantages include larger brain size, availability of comprehensive and detailed stereotaxic brain atlases, and a relatively greater background literature of behavioral, neuroanatomical, neurophysiological, and neurochemical information.

Although a rat model of physiological dependence was desirable, only recently have several laboratories reported the development of physiological dependence in the rat, as reflected by overt behavioral signs of ethanol withdrawal. These reports of ethanol withdrawal reactions in rats have involved several techniques of ethanol exposure, including oral administration via liquid diets (Branchey, Rauscher, and Kissin, 1971; Hunter et al., 1973, 1974; Lieber and DeCarli, 1973; Pohorecky, 1974) or schedule-induced polydipsia (Falk, Samson, and Winger, 1972), gastric intubation (Hunt and Majchrowicz, 1973, 1974; Wallgren, 1973), and inhalation (Roach et al., 1973). Although the above experiments have provided valuable information, they have been characterized either by small numbers of animals, incomplete descriptions of the nature and time course of the behavioral indices of withdrawal, or marked differences in the dose and duration of exposure to ethanol. A behavioral analysis of the rat withdrawal syndrome, including a rating scale of severity, would greatly facilitate the study of the neurobiological correlates of ethanol dependence. Such a rating scale would allow correlation of the severity of the withdrawal reaction with the extent of alteration in the measured physiological variable of interest for individual animals, thus providing a more powerful data analysis.

Recently we reported the successful adaptation to the rat (Hunter et al., 1973, 1974) of the liquid-diet technique previously used for mice (Freund, 1969, 1973; Freund and Walker, 1971; Walker and Zornetzer, 1974). Rats were maintained on nutritionally fortified liquid diets containing 35 to 40% of total calories from ethanol. Removal of ethanol resulted in the time-dependent appearance of a variety of withdrawal signs, including tail signs, tremors, and convulsions. The present experiment was undertaken to examine carefully some parameters and characteristics of the ethanol withdrawal syndrome in rats by using the liquid diet technique. Specifically, we sought to 1) examine the relationship between the duration of ethanol exposure and the intensity of the ensuing withdrawal signs, 2) provide a more detailed description of the time course and characteristic nature of the alcohol withdrawal syndrome than has been previously available, and 3) develop a reliable behavioral rating scale of withdrawal severity. Also included in the present paper will be a presentation of preliminary data from a series of experiments in which we are using the liquid diet technique with rats to investigate the neurophysiological concomitants of ethanol dependence and withdrawal. In the neurophysiological series of experiments, we are monitoring electroencephalographic (EEG) activity from chronically implanted depth electrodes in a variety of brain structures during ethanol withdrawal.

METHODS

The animals in these experiments were 60-day-old, male Long Evans hooded rats purchased from Charles River. The rats were individually housed in stainless steel cages located in a colony room having an automatic 7:00 a.m. to 7:00 p.m. light cycle.

Ethanol was administered via a liquid diet in which ethanol was incorporated as 35 to 40% of total calories (Walker and Freund, 1971). The diets were prepared from a 63.3% (v/v) stock ethanol solution (prepared from 95% ethanol and distilled water) mixed with Metrecal Shape (Mead Johnson Company), so that the final ethanol concentration in the diet was 8.1 to 9.7% (v/v). Control diets were prepared in an identical fashion, and isocaloric sucrose solution was substituted for ethanol. The diets were additionally fortified with Vitamin Diet Fortification Mixture, 3.0 g/liter of diet, and salt mixture XIV, 5.0 g/liter of diet (Nutritional Biochemicals Company). The diets were prepared fresh daily and administered in calibrated bottles.

Rats were reduced to 75 to 80% of their free-feeding weight by restricting food consumption to 5 g of pelleted laboratory food per day for 7 to 8 days and then started on the liquid diets. The percentage of ethanol calories was gradually increased from 35 to 40% (8.1 to 9.7% v/v ethanol) by increases of one percentage point at intervals of five days. Blood samples (30 to 50 μl) from the tail were collected in heparinized capillary tubes for subsequent determination of ethanol concentration by a gas chromatography procedure described previously (Freund, 1967).

On the day of withdrawal the ethanol diets were removed at 8:00 a.m. Only one or two rats were withdrawn on any one day in order to allow continuous observation. Each rat was placed in an observation chamber (12" x 18" x 12") that had a Plexiglass^R front. Both tap water and a liquid diet, isocalorically substituted for ethanol, were available at all times during withdrawal. Blood samples were collected at hourly intervals for at least 6 hours post-withdrawal (PW). Each rat was continuously observed for behavioral signs of alcohol withdrawal for 8 to 10 hours PW.

Eight hours PW the susceptibility to sound-elicited convulsions was assessed by a jangling of keys near the top of the observation chamber for a maximum of 15 seconds.

Four weight-matched groups (10 rats/group) were maintained on the ethanol-containing diets for 10, 15, 20, or 30 days. Blood samples were taken after 1, 5, 10, 15, 20, 25, and 30 days of ethanol consumption at 8:00 a.m.

A separate group of eight rats was used in the determination of diurnal variations in blood ethanol concentration (BEC). The rats were treated as described above, and blood samples were collected at 2:00 a.m., 8:00 a.m., 2:00 p.m., and 8:00 p.m. on days 14 and 16 of ethanol exposure.

Twenty-seven rats were used in an additional experiment designed to investigate the EEG alterations recorded from forebrain structures during ethanol withdrawal. Stereotaxic surgery was performed under Nembutal anesthesia (50 mg/kg). Three monopolar depth electrodes, one bipolar electrode, and two monopolar cortical screw electrodes were implanted in each rat. The monopolar depth electrodes (00 stainless steel insect pins insulated with epoxylite to within 0.5 mm of the tip) were implanted in the right amygdala and bilaterally in the ventral hippocampus. A twisted, bipolar, platinum-iridium electrode (125 μ) was implanted in the left amygdala. Two stainless steel screws (1/8 in x 080) were placed in the skull overlying anterior cortex. A similar screw placed in the frontal air sinus served as ground. Electrodes were connected to a nine pin ITT Cannon connector, and the entire assembly was fixed to the skull with dental cement.

Following recovery from surgery (9 to 16 days), the rats were divided into three weight-matched groups of nine rats each. Two of the groups--alcohol withdrawal (AW) and alcohol control (AC)--received the ethanol liquid diet, and the remaining control group (SC) was pair-fed the sucrose diet.

After 15 days of diet consumption, groups AW and SC were withdrawn. Group AC continued to have access to the ethanol diet. Each AW rat was connected to a flexible, low-noise, shielded cable (Microdot, Inc.) and placed in the observation chamber located in a shielded room. EEG activity from all electrode placements was continuously monitored on a Grass model 7 polygraph. Behavior was simultaneously observed. AC and SC rats were also observed, and sample periods of EEG activity were recorded.

RESULTS

Parametric Behavioral Results

Prior to weight reduction, the mean free-feeding weights of the 10-day (10D), 15-day (15D), 20-day (20D), and 30-day (30D) ethanol exposure groups ranged from 254 to 264 g. All rats gained weight during the alcoholization period and had reattained predeprivation weight after 15 to 20 days of liquid diet consumption. Mean daily ethanol consumption was 15.3 g/kg/rat. Daily observation revealed signs of gross intoxication, including docility, ataxia, and loss of motor coordination.

Generally the level of ethanol consumption was stable; however, spontaneous abstinence periods were observed, characterized by a sudden decrease in ethanol consumption, together with the appearance of signs of withdrawal. Representative examples of some individual ethanol consumption patterns of rats from each group are shown in Figure 1. When ethanol consumption fell below 8 g/kg (indicated by dottled line), signs of ethanol dependence, including tail stiffening, tremors, ataxia, and hyperreactivity were invariably observed. The duration of the spontaneous abstinence periods never extended beyond 24 hours, as shown in Fig. 1. This phenomenon, while

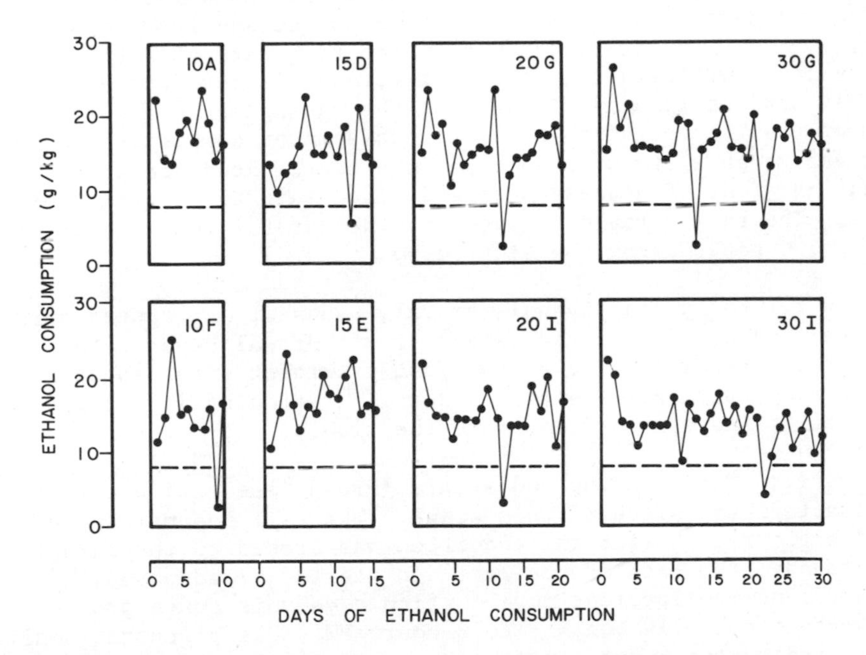

Figure 1. Individual patterns of ethanol consumption. Letters represent individual animal designations within each group. The dotted line represents an ethanol consumption of 8.0 g/kg/day.

pervasive, did not occur in all rats. Both rat 10A and 15E (see Figure 1), for example, exhibited stable patterns of consumption throughout the duration of ethanol exposure. Note that the occurrence of spontaneous abstinence periods during ethanol exposure did not reduce the intensity of the subsequent withdrawal reaction, as long as there were 48 hours or more between the last spontaneous abstinence period and the scheduled withdrawal period.

The rats maintained consistently elevated blood ethanol concentrations (BEC) during the alcoholization period. The 8:00 a.m. BEC ranged between approximately 150 and 250 mg/100 ml. Numerous reports have provided evidence indicating that the nervous system must be exposed continuously over a 24-hour period before physiological dependence can develop (Freund, 1970; Ogata et al., 1972; Seevers and Deneau, 1963). In the present experiment, mean 24-hour BEC, determined in a separate group of rats was: at 8:00 a.m., 264 mg % (range 235-281); at 2:00 p.m., 192.2 mg % (range 78-308); at 8:00 p.m., 240 mg % (range 84-298); and at 2:00 a.m., 297 mg % (range 202-379). These results are consistent with results reported for mice when the liquid diet technique was used (Freund, 1970), since BEC was elevated throughout the day and night with a peak during the dark phase of the light cycle.

The removal of ethanol resulted in the ordered appearance of a variety of withdrawal symptoms. The time course and types of symptoms observed will be discussed within the framework of a behavioral rating scale of withdrawal intensity developed in our laboratory. While only semiquantitative in nature, the rating scale arranges a variety of easily recognizable signs into symptom complexes (stages) of progressive severity that develop in a coordinated and time-dependent fashion. Convulsions were divided into three stages of severity. The rating scale is based on the careful observation of more than 100 rats during the past year.

Stage I. Following ethanol removal, behavior may appear normal for several hours. The initial signs of withdrawal began 2 to 5 hours postwithdrawal (PW). The two most prominent signs associated with this stage were piloerection and tail stiffening (the tail appeared stiff and was carried above the floor).

Stage II. Tail arching and ataxia (broad-based gait) were the most characteristic signs of this stage. The tail became extended at a 45° angle from horizontal and often was arched to the extent that it touched the back of the rat. An ataxic, broad-based, and rigid gait rendered movement more difficult as this stage progressed. These symptoms normally began 3 to 6 hours PW. Sleep occurred only in short, fragmented bursts interspersed with periods of activity. We have verified this sleep fragmentation electrographically as well as behaviorally (unpublished observations).

Stage III. Signs of withdrawal severity increased markedly during this stage, beginning 5 to 8 hours PW. This stage was characterized by the additional appearance of hypoactivity, extensor rigidity, and fasciculations of axial musculature. Extensor rigidity often resulted in the complete extension of the limbs with the abdomen flush against the floor of the observation chamber. Muscular fasciculations often resembled tremors but actually consisted of bursts of rhythmic jerking of the back muscles. The rats also began to exhibit increases in hyperreactivity to various visual and auditory stimuli during this stage. A gentle tap on the side of the observation chamber was often sufficient to elicit an exaggerated startle response and vocalization.

Stage IV. The primary new symptom that developed during this stage was sudden sprawling movements. A sprawling movement began with a series of rapid steps followed by complete tonic forelimb and hindlimb extension, opisthotonus, vocalizations, and severe tremors. These sprawling movements usually occurred spontaneously but could be elicited by normal laboratory noises. Motor activity consisted almost entirely of these sprawling episodes. Stage IV signs were not observed in all rats. In those rats that did exhibit stage IV signs, the time course of withdrawal symptoms was compressed with stages I to III collapsed in time, so that stage IV began as early as 4 to 6 hours PW. Stage IV was additionally characterized by a rapid decline in excitability as measured by susceptibility to sound-induced convulsions. Spontaneous convulsions, when observed, usually occurred subsequent to a sprawling episode, and sound-induced convulsions were often elicited during the early portions of stage IV. Thus the early portions of stage IV were usually characterized by an increased level of hyperexcitability. However, when stage IV symptoms continued for several hours, hyperexcitability declined rapidly. This dissociation between behavioral symptoms and level of hyperexcitability was often particularly dramatic, since sprawling movements and severe tremors were evident, yet auditory stimuli failed to elicit a convulsion.

Three convulsive stages were differentiated: stage V, audiogenic convulsion; stage VI, spontaneous convulsion; and stage VII, convulsion ending in death.

Audiogenic convulsions were characteristically induced during stage III to IV preconvulsive withdrawal severity. They were invariably preceded by running episodes ranging from 5 to 15 seconds in duration. The behavioral components of audiogenic convulsions were of four types: 1) whole-body tonus, 2) whole-body clonus, 3) forelimb tonus and hindlimb clonus, and 4) forelimb clonus and hindlimb tonus.

Table 1

CRITERIA FOR BEHAVIORAL STAGE OF WITHDRAWAL

Pre-Convulsive Stages	Symptoms	Persisting Symptoms	Onset (Hours PW)
I	Tail stiffening Piloerection	-	2-5 Hr.
II	Tail arching Broad-based Gait	Piloerection	3-6 Hr.
III	Hypoactivity Extensor Rigidity Muscular fasciculations Hyperreactivity Susceptibility to Audiogenic Convulsion	Piloerection Tail arching Broad-based Gait	5-8 Hr.
IV	Sprawling episodes Spontaneous vocalizations Whole-body Rigidity Severe tremors	Piloerection Tail arching Broad-based gait Extensor rigidity Hyperreactivity Susceptibility to Audiogenic Convulsions	4-8 Hr.
Convulsive Stages			
V	Audiogenic convulsion	-	-
VI	Spontaneous convulsion	-	-
VII	Convulsion ending in death	-	-

Spontaneous convulsions were relatively rare and characteristically milder than audiogenic convulsions. Spontaneous convulsions usually developed after a sprawling episode and began with automatisms of the head and vibrissae, culminating in loss of righting and whole-body, tonic-clonic limb movements (10 to 25 sec. duration). The stages of withdrawal severity and associated symptoms are summarized in Table 1.

Table 2 shows the maximum stages of withdrawal intensity displayed by rats in the 10D, 15D, 20D, and 30D groups. As little as 10 days of ethanol exposure was sufficient for the development of severe withdrawal signs. As shown in Table 2, 50% of the rats from the 10D group were judged to have stage III to IV withdrawal signs. Further increases in ethanol exposure did not markedly alter the intensity of preconvulsive withdrawal symptoms (stage I to IV), but instead tended to increase the percentage of rats exhibiting the most severe abstinence signs (stages III-IV). The major feature that distinguished rats receiving varying durations of ethanol exposure was observed in the convulsive stages. An examination of Table 2 reveals an increased incidence of convulsions as a function of duration of ethanol exposure, reaching asymptote at approximately 20 days. Based on these results, it appears that 10 days of ethanol exposure is sufficient for the development of ethanol dependence, including behavioral evidence indicative of aberrations in autonomic and somatic nervous system function. Further increases in ethanol exposure did not appear to result in concomitant increments in the intensity of these behavioral signs, but instead resulted in a marked growth in hyperexcitability as indicated by susceptibility to audiogenic convulsions.

Table 2

NUMBER OF RATS IN EACH GROUP

DISPLAYING EACH BEHAVIORAL STAGE OF WITHDRAWAL

	MAXIMUM STAGE OF WITHDRAWAL								
GROUP	0	I	II	III	IV	V	VI	VII	NUMBER OF CONVULSIONS
10 DAY	0	1	3	3	2	1	0	0	1
15 DAY	0	1	2	4	0	3	0	0	3
20 DAY	0	0	1	1	1	3	2	2	7
30 DAY	0	1	3	1	0	4	0	1	5

Withdrawal severity among the four groups was further examined in relation to several other parameters assessed during the present experiment. Table 3 compares values for each group obtained for withdrawal intensity, ethanol consumption, BEC at time of withdrawal and the rate of ethanol elimination during the PW period.

The mean BEC for each group at the beginning of withdrawal ranged from 133 to 224 mg % as shown in Table 3. A Spearman rank correlation calculated using all rats revealed a significant ($r = .38$, $Z = 2.27$, $p < .012$) relationship between BEC at the beginning of this correlation between BEC and withdrawal severity is obscured by the fact that BEC also increased as a function of duration of ethanol exposure. However, if it is assumed that the 8:00 a.m. BEC provides an indication of the level of blood ethanol maintained over a 24 hour-period, then it might be expected that those rats

TABLE 3

AVERAGE MAXIMUM WITHDRAWAL SEVERITY, ETHANOL CONSUMPTION,

BLOOD ETHANOL CONCENTRATION AND ETHANOL ELIMINATION

FOR EACH GROUP

GROUP	MEDIAN WITHDRAWAL STAGE	MEAN* ETHANOL CONSUMPTION	MEAN** BLOOD ETHANOL CONCENTRATION AT WITHDRAWAL	MEAN RATE*** OF ETHANOL ELIMINATION
10 DAY	III	15.0 (0.8)	133.0 (37.0)	61.5 (5.8)+
15 DAY	III	15.0 (0.6)	156.0 (26.8)	58.7 (5.4)
20 DAY	V	15.6 (0.5)	224.0 (23.6)	67.0 (3.9)
30 DAY	IV	13.6 (0.5)	203.0 (28.8)	61.0 (2.8)

* g/kg/rat/day during the five days prior to withdrawal
** mg/100 ml
*** mg/100 ml/hour
+ Numbers in parenthesis are standard errors of the mean

maintaining higher 8:00 a.m. BEC would develop the most severe dependence.

Although the induction of metabolic tolerance has been assumed, at best, to account partially for behavioral tolerance after prolonged ethanol exposure (Kalant, LeBlanc, and Gibbins, 1971), no role has been specified with respect to physiological dependence. Since the appearance of withdrawal symptoms was contingent upon a decline in BEC, we reasoned that the nervous system might be sensitive to the rate of decline of blood ethanol. The rate of ethanol elimination was determined for each rat from the descending limb of the ethanol disappearance curve. The results of this analysis are shown in Table 3. The hypothesis generated above would predict a direct relationship between ethanol elimination and withdrawal intensity. However, a Spearman rank correlation failed to show a statistically significant relationship. The values obtained for ethanol elimination in the present experiment were substantially higher than those previously reported in the rat (Wallgren and Barry, 1970). We have subsequently replicated and extended these findings (unpublished observations). Preliminary results have indicated that the values for ethanol elimination obtained in the present experiment represent an increase of approximately 50 to 100% over control levels. Metabolic tolerance appears to develop gradually and is nearly complete after 10 days of ethanol exposure. Lieber and DeCarli (1973) have recently reported a 40% increase in the rate of ethanol elimination rate by using a similar liquid diet procedure.

Electrophysiological Correlates

As previously mentioned, we are investigating the EEG concomitants of the ethanol withdrawal reaction in the rat in a separate series of experiments. The results of one of these experiments (Hunter et al., 1973) in which EEG was recorded from several forebrain structures will be summarized here. Preliminary results from ongoing experiments in which EEG is being recorded from diencephalic, mesencephalic, and forebrain limbic structures will be discussed briefly.

The ethanol consumption, weight gain during ethanol exposure, and the time course and nature of the behavioral manifestations of withdrawal were virtually identical to those described previously for the parametric behavioral experiment. Sound-induced, tonic-clonic convulsions were elicited in all of the AW group rats. No behavioral signs of withdrawal were observed in either the AC or SC control groups. Furthermore, auditory stimulation in AC and SC rats (shortly following elicitation of convulsions in AW rats) produced no convulsions.

EEG recording from AW rats during the withdrawal period indicated the widespread development of abnormal forebrain epileptiform activity. The abnormalities usually began with the development of synchronous activity (1-5 HZ), together with transient spikes resembling interictal epileptiform events. The appearance of spiking roughly coincided with the initial appearance of behavioral symptoms. The epileptiform spike events increased in an amplitude and frequency, becoming more organized during the latter stages of withdrawal. This increased organization consisted of brief bursts of spike activity or sustained epileptiform episodes.

Cortical EEG activity during withdrawal was continuously evaluated up to the time that sound-induced convulsions were elicited; then it was classified into one of the following four stages: Stage 1--synchronized high amplitude EEG, together with transient spike events (a peak-to-peak amplitude of at least twice background activity was judged as a spike event) occurring with a frequency of less than 1/min; stage 2--increased occurrence and amplitude of spike events with a frequency of 1 to 10/min; stage 3--organized bursts of spike activity consisting of an envelope of 3 to 10 spikes within a 3 to 5 sec epoch; or stage 4--sustained seizure-like activity.

The results of this cortical EEG analysis for five AW rats are shown in Table 4. Also shown in Table 4 is the individual ethanol consumption for the last five days of ethanol exposure. Although the final stage of severity differed among AW rats, organized cortical epileptiform activity developed progressively during the withdrawal period. The cortical EEG abnormalities appeared to progress in a correlated fashion with the behavioral symptoms. However, cortical epileptiform activity did not appear to be directly related to specific behavioral signs of withdrawal. For example, a specific tremor or behavioral automatism did not necessarily coincide with a spike event or other EEG anomaly.

A similar temporal development of epileptiform activity to that described above was also observed in recordings from amygdala and hippocampus. Note, however, that although cortex, amygdala, and hippocampus all exhibited abnormal EEG activity, there was substantial independence among these structures with respect to the occurrence of specific spike events. These considerations suggest that widespread areas of forebrain develop similar patterns of paroxysmal activity during ethanol withdrawal.

Table 4

ANALYSIS OF CORTICAL EEG ACTIVITY DURING ETHANOL WITHDRAWAL* AND INDIVIDUAL ETHANOL CONSUMPTION

Hours Postwithdrawal	AW 25	AW 21	AW 15	AW 8	AW 20
1	-	-	-	-	-
2	-	1	-	-	1
3	1	2	1	-	1
4	2	3	1	-	1
5	3	4	1	-	1
6	4	5	3	1	1
7			3	1	
8				3	
9				3	
10				3	
Alcohol Consumption†	14.7	13.1	15.8	17.0	13.0

*See text for details of Stage 1, 2, 3, and 4
†g/kg/day during the last five days of the ethanol treatment period

Examples of the abnormal EEG activity associated with the ethanol withdrawal syndrome are shown in Figure 2. Figure 2A and 2B are examples of transient spiking and high-voltage slow waves that were typically observed during the early stages of ethanol withdrawal. Clearcut bursts of paroxysmal activity were typically observed during the later stages of withdrawal as shown in Figure 2C. Simultaneous recordings from a control rat having continued access to ethanol are shown in Figure 2D. Figure 2E illustrates a forebrain seizure with a latency of 30 seconds following the onset of a sound-induced, whole-body, tonic-clonic convulsion. Brain seizure activity during this behavioral convulsion never appeared in the cortex. Control recordings during auditory stimulation in both AC and SC groups resulted in neither convulsions nor evidence of forebrain paroxysmal activity.

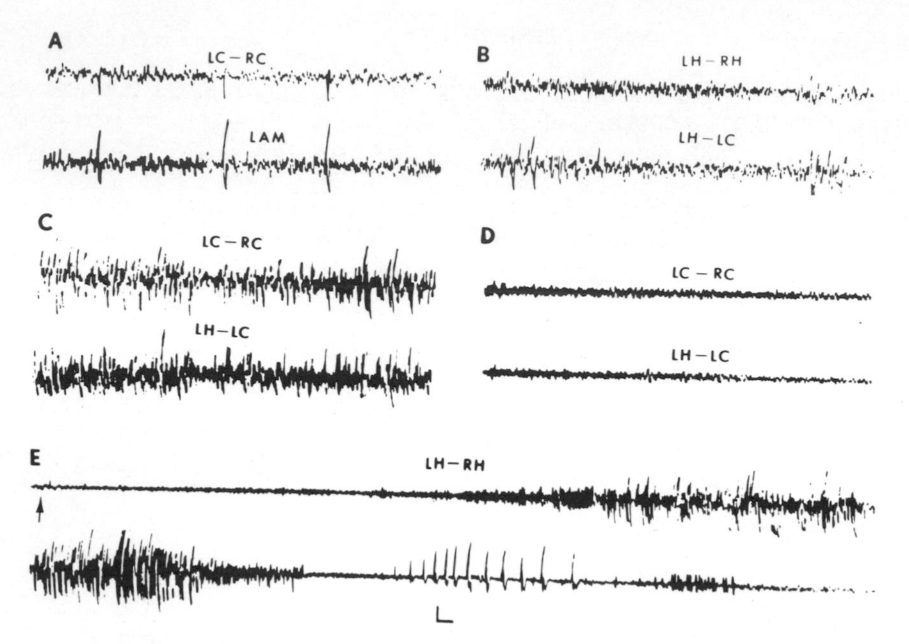

Figure 2. Examples of abnormal EEG activity associated with the ethanol withdrawal syndrome: (A) Transient spiking seen 3 hours after withdrawal; calibration 1 sec, LC-RC 100 µV, LAM 200 µV. (B) High amplitude slow waves seen 3 hours after withdrawal; calibration 1 sec, 150 µV. (C) Organized paroxysmal activity seen 5 hours after withdrawal and prior to auditory stimulation; calibration 2 sec, LC-RC 200 µV, LH-LC 250 µV. (D) Control record from AC rat taken at same time as that in (C); calibration 2 sec, LC-RC 100 µV, LH-LC 250 µV. (E) Seizure activity associated with sound-induced convulsion. Note latency from the onset of convulsion (indicated by arrow) to the beginning of brain seizure; calibration 1 sec, 300 µV (bottom trace is a continuation of top trace). Abbreviations: LC-Left Cortex, RC-Right Cortex, LAM-Left Amygdala, LH-Left Hippocampus, RH-Right Hippocampus.

During ethanol withdrawal in the rat, a dissociation between forebrain seizures and severe behavioral convulsions often occurs; this fact suggests that seizure activity need not spread over the entire brain for a convulsion to occur and, more importantly, that seizures observed in forebrain areas may result from propagated seizure activity originating in more caudal brain regions. Note also that on a number of occasions we have observed brain seizures in the absence of behaviorally observable convulsions during ethanol withdrawal. These brain seizures usually occur in limbic structures during behavioral arrest. It would be impossible, based on only behavioral observation, to surmise that the rat was experiencing a brain seizure.

DISCUSSION

Results of the present parametric study of behavioral characteristics of ethanol withdrawal confirm and extend previous reports of induction of ethanol dependence in the rat using a liquid diet technique (Branchey et al., 1971; Hunter et al., 1973; Pohorecky, 1974). In the present experiment, rats consuming ethanol-containing liquid diets for 10 to 30 days maintained substantial ethanol intake leading to behavioral intoxication concomitant with substantial and relatively stable blood ethanol concentrations. The replacement of ethanol by sucrose in the liquid diets resulted in evidence of physiological dependence indicated by the occurrence of a number of behavioral abnormalities and spontaneous and sound-elicited convulsions. The withdrawal symptoms developed in a characteristic order that was related to the decline in blood ethanol concentration.

A direct relationship was found between the duration of ethanol exposure and the severity of the ethanol withdrawal reaction, supporting similar observations made in other species (Ellis and Pick, 1971; Goldstein, 1972a). Ten days of ethanol exposure was sufficient for the development of signs of ethanol dependence reflecting autonomic and somatic nervous system dysfunction. Further increases in the duration of ethanol exposure generally resulted in increments in the proportion of rats exhibiting severe preconvulsive withdrawal symptoms and spontaneous or sound-elicited convulsions. The development of physiological dependence appeared to have reached an asymptotic level at 20 days, since further increases in the severity of the withdrawal reaction, as measured here, were not apparent after 30 days of ethanol exposure.

The behavioral rating scale for ethanol withdrawal intensity described within the present report is the first to be developed for the rat, although similar rating scales have been developed for mice (Freund, 1969; Goldstein, 1972b). While only semi-quantitative, this rating system has proved to be very reliable in our laboratory. The scale has been useful in the evaluation of maximum withdrawal intensity and in the examination of temporal characteristics of withdrawal symptoms. Withdrawal signs appeared in complexes of overt symptoms that form the basis for a portion of the rating scale. These symptom complexes developed in a consistent, time-dependent order following ethanol removal. Some of the symptoms described in the present report become obvious only during periods of movement. Careful and frequent observations are thus required for the reliable evaluation of ethanol withdrawal intensity.

The potential significance of the development of a rat model of ethanol dependence has led to considerable experimental effort.

Recent reports of the successful induction of ethanol dependence in rats have involved various techniques of ethanol administration, including inhalation (Roach et al., 1973), intubation (Hunt and Majchrowicz, 1974; Wallgren, 1973), ethanol-containing liquid diets (Branchey et al., 1971; Hunter et al., 1973, 1974; Lieber and DeCarli, 1973; Pohorecky, 1974), or use of a modified schedule-induced polydipsia technique (Falk et al., 1972). The use of varied techniques of ethanol administration is desirable, in that it provides an experimental control for any possible confounding variables inherent in any single model. In this regard it is important to note that many of the withdrawal symptoms described in the present report appear to be identical to those reported by other investigators using intubation (Wallgren, 1973), the polydipsia technique (Falk et al., 1973) or the liquid diet technique (Branchey et al., 1971; Pohorecky, 1974). Each of the above models appears to have its advantages and disadvantages, with the final choice of a model contingent upon the experimental question to be investigated. Use of the liquid diet technique provides for the rapid induction of dependence in large groups of animals, a noncomplicated route of ethanol administration, and convenient pair-feeding to control for the empty calories provided by ethanol. Thus, the liquid diet technique appears to be particularly suitable for the investigation of the neurophysiological and neurochemical correlates of ethanol dependence and withdrawal and of prolonged ethanol exposure.

In the second part of the present report, we presented data concerning the forebrain EEG correlates of ethanol withdrawal in the rat. These results indicated a progressive development of forebrain epileptiform activity during the first few hours following ethanol withdrawal. EEG abnormalities characteristically began with slowing and increased amplitude coupled with the appearance of transient spiking. The spiking usually progressed in severity and became organized into bursts of spikes or sustained seizure activity. A similar progression of forebrain epileptiform activity during ethanol withdrawal has recently been reported in mice (Walker and Zornetzer, 1974).

The occurrence and characteristics of this forebrain epileptiform activity during ethanol withdrawal suggests two interpretations: 1) independent loci of neural hyperexcitability develop throughout forebrain, or 2) a structure, or set of structures, serves to organize and propagate epileptiform activity to forebrain areas. The present results can be interpreted to support both of these possibilities. Epileptiform activity appeared to develop independently in anterior cortex, ventral hippocampus, and amygdala during the early postwithdrawal hours. However, forebrain seizure activity during sound-induced convulsions appeared to be propagated from more caudal brain loci, since seizure activity occurred in the forebrain with a significant delay after the onset of the tonic-clonic convulsion. Thus, as ethanol concentration dissipates throughout the brain following withdrawal, epileptiform activity appears to develop in an independent fashion in many brain regions. The most severe behavioral symptoms, especially convulsions, may be contingent upon the influence of a structure, or set of structures, that organizes the epileptiform events into sustained patterns of seizure activity.

There are several candidates for such organizing regions. The limbic system, particularly the hippocampus and amygdala, is known to have a low seizure threshold. Thus, brain seizure activity during ethanol withdrawal might be initiated in the hippocampus or the amygdala because of the low threshold for seizure in these areas. On the other hand, the recruitment phenomena seen during spike bursts in both the mouse (Walker and Zornetzer, 1974) and the rat (this study; unpublished observations) suggest that the midline thalamus or the brain stem reticular formation may be responsible for the organizing and propagation of brain seizure activity during ethanol withdrawal. Interestingly, Pohorecky (1974) has recently reported a time-dependent acceleration of norepinephrine turn-over in the brainstem, relative to other brain regions, during ethanol withdrawal in the rat. The peak effect occurred 8 hours postwithdrawal, a time coincident with the peak level of behavioral and neural excitability reported in the present paper. We have generally favored the brain stem reticular formation or midline thalamus as the most likely candidate for such an organizing region, rather than the hippocampus or amygdala (Hunter et al., 1973; Walker and Zornetzer, 1974). The direct involvement of forebrain limbic structures, such as the hippocampus and amygdala, seemed unlikely, since we observed seizure activity in those structures only after a significant delay following the onset of the behavioral convulsion.

Preliminary results from ongoing experiments in our laboratory, however, indicate that both mesencephalic reticular formation (MRF) and limbic structures may be involved in the initiation and propagation of seizure activity during alcohol withdrawal convulsions under different, but yet unspecified, conditions. We have regularly observed brain seizure activity during alcohol withdrawal to begin in MRF and subsequently propagate to diencephalic and limbic areas. That is not to say that the MRF is the locus of origin of the brain seizure activity associated with convulsion, since we have yet to investigate many brain regions. We have occasionally observed brain seizure activity to begin in the limbic system (amygdala and anterior limbic field) and subsequently to propagate to other brain regions. It remains to be determined under what conditions these various patterns of brain seizures occur. It may be that a specific type of convulsion is associated with a specific pattern of brain seizure propagation. These questions are presently under investigation.

REFERENCES

Branchey, M., Rauscher, G., and Kissin, B. Modifications in the response to alcohol following the establishment of physical dependence. Psychopharmacologia (Berl.), 22:314-322, 1971.

Ellis, F. W., and Pick, J. R. Dose- and time-dependent relationships in ethanol-induced withdrawal reactions. Fed. Proc., 30:568, 1971.

Falk, J. L., Samson, H. H., and Winger, G. Behavioral maintenance of high concentrations of blood ethanol and physical dependence in the rat. Science, 177:811-813, 1972.

Falk, J. L., Samson, H. H., and Tang, M. Chronic ingestion techniques for the production of physical dependence of ethanol. In M. M. Gross (Ed.) Alcohol Intoxication and Withdrawal: Experimental Studies I, pp. 197-212. New York, Plenum Press, 1973.

Freund, G. Exchangeable injection port cartridge for gas chromatographic determination of volatile substances in aqueous fluids. Anal. Chem., 39:545-546, 1967.

Freund, G. Alcohol withdrawal syndrome in mice. Arch. Neurol., 21:315-320, 1969.

Freund, G. Alcohol consumption and its circadian distribution in mice. J. Nutr., 100:30-36, 1970.

Freund, G. Alcohol, barbiturate, and bromide withdrawal syndrome in mice. Ann. N. Y. Acad. Sci., 215:224-234, 1973.

Freund, G., and Walker, D.W. Impairment of avoidance learning by prolonged ethanol consumption in mice. J. Pharmacol. Exp. Ther., 179:284-292, 1971.

Goldstein, D. B. Relationship of alcohol dose to intensity of withdrawal signs in mice. J. Pharmacol. Exp. Ther., 180:203-215, 1972a.

Goldstein, D. B. An animal model for testing effects of drugs on alcohol withdrawal reactions. J. Pharmacol. Exp. Ther., 183:14-22, 1972b.

Hunt, W. A., and Majchrowicz, E. Turnover rates and steady-state levels of brain serotonin in alcohol-dependent rats. Brain Research, 72:181-184, 1974.

Hunter, B. E., Boast, C. A., Walker, D. W., and Zornetzer, S. F. Alcohol withdrawal syndrome in rats: Neural and behavioral correlates. Pharmac. Biochem. Behav., 1:719-725, 1973.

Hunter, B. E., Walker, D. W., and Riley, J. N. Dissociation between physical dependence and volitional ethanol consumption: Role of multiple withdrawal episodes. Pharmac. Biochem. Behav., 1974. (in press)

Kalant, H., LeBlanc, A. E., and Gibbins, R. J. Tolerance to and dependence on, ethanol. In Y. Israel and J. Mardones (Eds.) Biological Basis of Alcoholism, pp. 235-269. New York, Wiley-Interscience, 1971.

Lieber, C. S., and DeCarli, L. M. Ethanol dependence and tolerance: A nutritionally controlled experiment model in the rat. Res. Commun. Chem. Pathol. Pharmacol., 6:983-991, 1973.

Littleton, J. M., Griffiths, P. J., and Ortiz, A. The induction of ethanol dependence and the ethanol withdrawal syndrome. J. Pharm. Pharmacol. 26:81-91, 1974.

Mello, N. K. A review of methods to induce alcohol addiction in animals. Pharmac. Biochem. Behav., 1:89-101, 1973.

Ogata, H., Ogata, F., Mendelson, J. H., and Mello, N. K. A comparison of techniques to induce alcohol dependence and tolerance in the mouse. J. Pharmac. Exp. Ther., 180:216-230, 1972.

Pohorecky, L. A. Effects of ethanol on central and peripheral noradrenergic neurons. J. Pharmacol. Exp. Ther., 189:380-391, 1974.

Pohorecky, L. A., Jaffe, L. S., and Berkeley, H. A. Effects of ethanol on serotonergic neurons in the rat brain. Res. Commun. Chem. Pathol. Pharmacol., 8:1-11, 1974.

Roach, M. K., Khan, M. M., Coffmann, R., Pennington, W., and Davis, D. L. Brain (NA^+ + K^+)-activity and neurotransmitter uptake in alcohol-dependent rats. Brain Research, 63:323-329, 1973.

Seevers, J. H., and Deneau, G. A. Physiological aspects of tolerance and physical dependence. In W. S. Root and F. G. Hoffmann (Eds.) Physiological Pharmacology, pp. 565-640. New York, Academic Press, 1963.

Walker, D. W., and Freund, G. Impairment of shuttlebox avoidance learning following prolonged alcohol consumption in rats. Physiol. Behav., 7:773-778, 1971.

Walker, D. W., and Zornetzer, S. F. Alcohol withdrawal in mice: Electroencephalographic and behavioral correlates. Electroencephalogr. Clin. Neurophysiol., 36:233-243, 1974.

Wallgren, H. Neurochemical aspects of tolerance to and dependence on ethanol. In M. M. Gross (Ed.) Alcohol Intoxication and Withdrawal: Experimental Studies, pp. 15-32. New York, Plenum Press, 1973.

Wallgren, H., and Barry, H., III. Actions of alcohol. New York: Elsevier, 1970.

ETHANOL CONSUMPTION SUBSEQUENT TO PHYSICAL DEPENDENCE

Henri Begleiter, Department of Psychiatry

Downstate Medical Center, State University of New York

Brooklyn, New York, U.S.A. *

A number of elegant techniques have been developed to induce pharmacological dependence on alcohol in animals. However, our understanding of the behavioral aspects of alcohol ingestion or self-administration is still quite limited. It is only quite recently that the relationship between physical dependence and subsequent alcohol intake has been investigated. Unfortunately, the few findings obtained by different investigators appear to be quite contradictory. An early study by Freund (1969) concluded that mice do not change their preference for ethanol subsequent to physical dependence. A study by Myers, Stoltman and Martin (1972) dealt with the effects of ethanol dependence induced artificially in the Rhesus monkey on the subsequent preference for alcohol. Their findings show that each monkey rejected the ethanol solution offered even at low concentrations, in spite of the fact that symptoms of physical dependence on alcohol were quite manifest. In a more recent experiment, Deutsch and Koopmans (1973) demonstrated a large and lasting enhancement of alcohol consumption over control levels after direct infusion of 10% alcohol into the stomach of rats for six days.

Because of the inconsistency in past findings, we undertook to study ethanol consumption subsequent to physical dependence on ethanol.

EXPERIMENT 1

The subjects were 40 male Long Evans hooded rats, 10 weeks old and weighing 250-300 gm. at the start of the experiment. The

* Supported by NIAAA Grant AA01231

animals were housed individually in rat cages. The room temperature was kept at 76° F. and a 12-hour day and night cycle was maintained by an automatic timer. The animals were always given food and water ad lib.

Alcohol preference was measured with the use of a two-bottle method, which is a modification of a previously described procedure (Myers and Veale, 1972). The animals were simultaneously offered a choice between plain water and a solution of 95% ethanol prepared volumetrically with tap water. The fluids were contained in two 100 ml. inverted graduated bottles which were fitted with steel spouts that protruded into the cage. The bottles were placed randomly and their positions were interchanged daily to eliminate error due to position or bottle preference. Fresh solutions of alcohol and water were placed in the bottles daily.

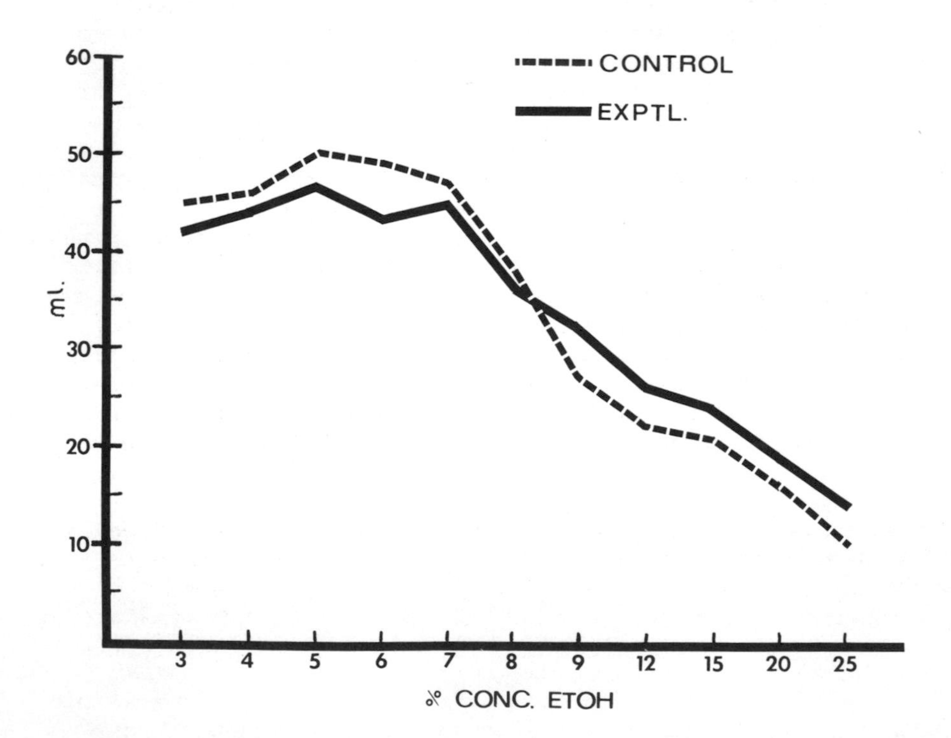

Figure 1

For the baseline period, measurement of alcohol preference and consumption was established by increasing the concentration of alcohol solution each day in a stepwise fashion as follows: 3, 4, 5, 6, 7, 9, 12, 15, 20 and 25%. Only one concentration of this ascending series was offered per day for the 10 days of baseline. The daily fluid consumption was recorded at 9 a.m., 11 a.m. and 2 p.m. Subsequent to the baseline period, half the animals (N = 20) were randomly assigned to the experimental group, the other half to the control group. In the experimental group, animals were intubated daily with 4 g/kg of alcohol for the first 10 days, 5 g/kg for the second 10 days, and with 6 g/kg for the final 10 days. The control animals were intubated with an equivalent amount of water.

After 30 days of intubation, withdrawal symptoms were precipitated by total removal of alcohol. We observed withdrawal signs, e.g., hyperactivity, body tremors, spasticity, hyperreflexia, piloerection, episodes of generalized convulsions with prominent clonic components in 87% of the experimental animals. After the last intubation alcohol preference was again established with the method described before, that is by increasing the concentration of alcohol solution each day in a stepwise fashion.

A statistical comparison of alcohol preference and consumption between the experimental and control animals did not yield a significant difference. (See Figure 1)

EXPERIMENT 2

It is possible that the negative results obtained in our previous experiment might be due to the fact that a long-term pattern of physical dependence had not been established. Consequently, we carried out an experiment in which 15 experimental animals were intubated with ethanol as described before and 15 control animals were intubated with water. All animals were subjected to three different periods of intubation. Each 20-day period of intubation was followed by withdrawal symptoms described before.

Subsequent to these three separate epochs of withdrawal symptoms, all animals were again tested for alcohol preference with the method described before.

A statistical comparison of alcohol preference and consumption between the experimental and control animals did not yield a significant difference. (See Figure 2)

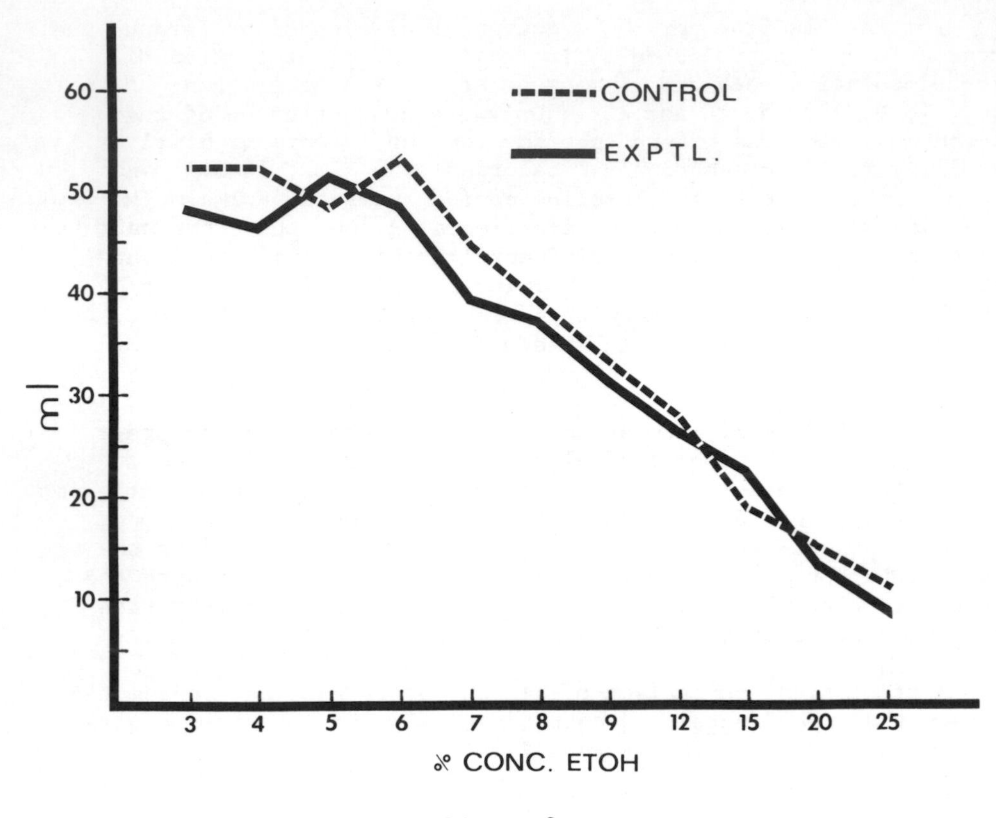

Figure 2

EXPERIMENT 3

It is quite conceivable that our failure to obtain greater alcohol preference subsequent to withdrawal was caused by the inability of the animals to associate the intake of ethanol with the attenuation of the stress of withdrawal. Consequently, we carried out an experiment in which the experimental animals were administered alcohol to relieve signs of withdrawal.

In this experiment, alcohol preference was again established for 12 experimental and 12 control animals. The experimental animals were intubated with 6 g/kg of ethanol for a period of 20 days, and withdrawal symptoms were induced by the removal of alcohol.

While the experimental animals were manifesting signs of withdrawal, they were again intubated with 1 g/kg of alcohol. The intake of alcohol was quite efficacious in relieving all observable signs of withdrawal in all experimental animals. The animals were allowed a period of 5 days subsequent to withdrawal, during which time they were given just food and water ad libitum. Then, once again, the process to induce physical dependence was reinitiated. The same animals were intubated for a period of 20 days followed by total withdrawal from alcohol. During withdrawal all animals were again intubated with 1 g/kg of alcohol which brought about substantial relief from the withdrawal syndrome. Finally, after another 5-day period of rest, the same process was reinstated for a period of 20 days.

Subsequent to removal from alcohol, alcohol preference was established by the previously described method.

A statistical comparison of alcohol preference and consumption between the experimental and control animals did not yield a significant difference. (See Figure 3)

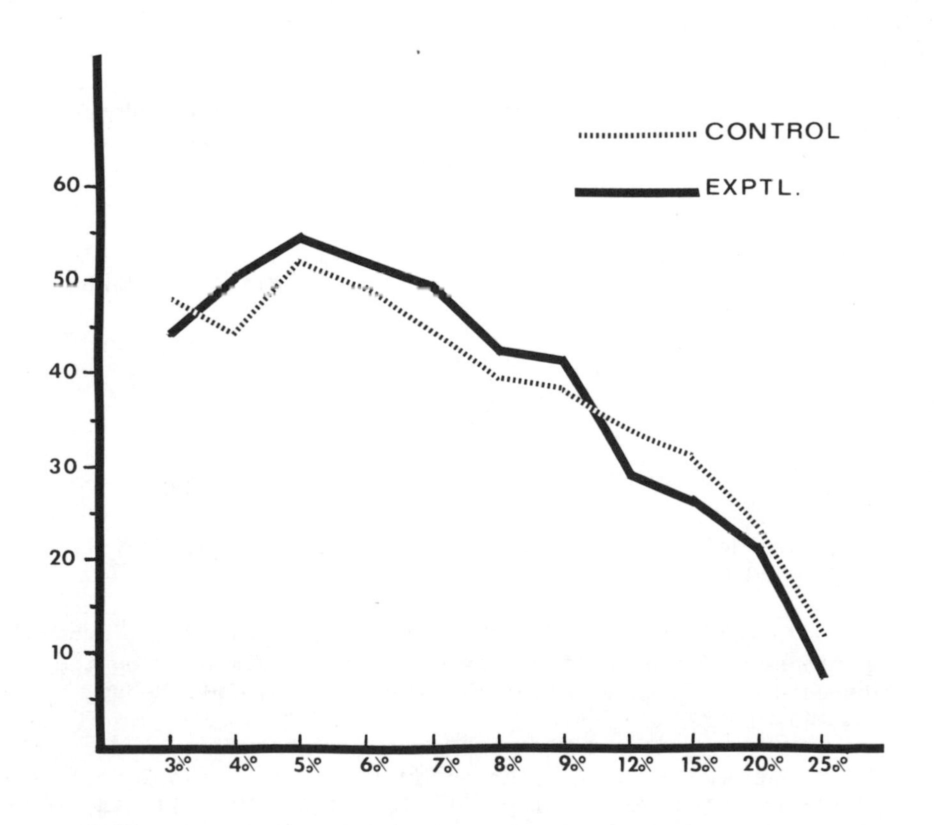

Figure 3

DISCUSSION

The most significant and consistent finding of our experiments is that rats did not show an increase in alcohol preference, in spite of the fact that some symptoms of physical dependence on this drug were strikingly obvious. In recent years a number of elegant methods have been developed to artificially induce physical dependence on ethanol. It has been shown that intravenous self-administration, intragastric intubation, forced feeding of a liquid diet and inhalation of ethanol vapor can be used rather successfully to induce symptoms of withdrawal. However, it is quite possible that while these methods appear quite efficacious in eliciting withdrawal symptoms, they are somewhat ineffective in producing a complete state of physical dependence. It is possible that withdrawal symptoms represent only a partial aspect of the physical dependence phenomena. While the experimental methods may be useful for investigating the biological symptoms of withdrawal, the results of our experiments imply that a theoretical basis for examining the development of an abnormal consumption of ethanol requires a more inclusive experimental procedure than the systematic administration of alcohol, though this appears to be quite sufficient in the case of morphine intake in animals.

It is probable that the development of physical dependence on ethanol can only take place with the use of a procedure which involves voluntary ingestion of a palatable solution.

While our results indicate that in rats ethanol consumption does not increase subsequent to signs of withdrawal, it is quite probable that the etiology of a long term high intake of alcohol is specific to man.

REFERENCES

Deutsch, J. A. and Koopmans, H. S. Preference enhancement for alcohol by passive exposure. Science 179: 1242-1243, 1973.

Freund, G. Alcohol withdrawal syndrome in mice. Arch. Neurol. 21: 315-320, 1969.

Myers, R. D., Stoltman, W. P. and Martin, G. E. Effects of ethanol dependence induced artificially in the Rhesus monkey on the subsequent preference for ethyl alcohol. Physiol. Behav. 9: 43-48, 1972.

Myers, R. D. and Veale, W. L. The determinants of alcohol preference in animals. In: The Biology of Alcoholism, Vol. II, edited by B. Kissin and H. Begleiter, New York: Plenum Press, 1972, pp. 131-168.

ALCOHOLIC LIVER INJURY: EXPERIMENTAL MODELS IN RATS AND BABOONS*

Charles S. Lieber and Leonore M. DeCarli

Laboratory of the Section of Liver Disease and Nutrition,

Veterans Administration Hospital, Bronx, New York and

Department of Medicine, Mount Sinai School of Medicine

of the City University of New York, New York, N.Y.

ABSTRACT

A model has been developed for the administration to rats and baboons of ethanol as part of a nutritionally adequate liquid diet. With this regimen, ethanol intake was much higher than with conventional procedures. All animals gained or maintained their body weight, and liver morphology was normal in the controls. Isocaloric substitution of carbohydrate by ethanol (36% of total calories in rats and 50% in baboons) resulted in the production of fatty liver in all animals, while the baboons also developed alcoholic hepatitis and cirrhosis with increased activities of serum glutamic oxaloacetic transaminase. Inebriation and manifestation of dependence upon withdrawal of the diet were observed in baboons and quantitated in the rat. Chemical alterations produced by ethanol at the fatty liver stage were characterized by hyperlipemia, striking triglyceride accumulation in the liver and enhanced activities of microsomal drug metabolizing enzymes, including the microsomal ethanol oxidizing system (MEOS). Ultrastructural changes of the mitochondria and the endoplasmic reticulum were already present at the fatty liver stage and persisted throughout the hepatitis and cirrhosis. The lesions were similar to those observed in alcoholics (including the inflam-

*Most of the original studies were supported, in part, by grants from the National Institute of Alcohol Abuse and Alcoholism and the National Institute of Arthritis, Metabolism and Digestive Diseases and projects from the Veterans Administration.

mation and the central sclerosis), and differed strikingly from the alterations produced by other models of liver injury. In showing that all aspects of liver injury observed in alcoholics can be reproduced in animals by the feeding of pure ethanol with an adequate diet, this study incriminates ethanol itself as a cause for the hepatic complications. This new experimental model is proposed as a tool for the study of the pathogenesis and treatment of alcoholic liver injury and dependence.

I. INTRODUCTION

With increasing alcohol consumption, the incidence of related complications has been rising steadily, particularly that of associated liver disease, to the extent that at the present time, cirrhosis of the liver, the most severe hepatic complication of alcoholism, is the third cause of all deaths between the ages of 25 and 65 in the city of New York. In addition to cirrhosis of the liver characterized by diffuse hepatic scarring, alcohol abuse is also associated with hepatic inflammation and necrosis (alcoholic hepatitis) and excess fat accumulation (alcoholic fatty liver). The relationship of these various liver injuries to each other however has been questioned. Furthermore, since not all alcoholics develop liver injury, there has been considerable debate concerning the question whether alcohol itself or some associated factor such as dietary deficiency is the main cause for the liver disease. The question has both theoretical and practical implications. The classic belief that liver injury could be prevented in the alcoholic by merely controlling the diet was challenged by evidence that alcohol might exert direct toxic effects upon the liver; it was indeed shown that the fatty liver, the most benign stage of the disease, could be produced in volunteers given alcohol in association with adequate or enriched diets (Lieber et al., 1963, 1965; Lane and Lieber, 1966, Lieber and Rubin, 1968; Rubin and Lieber, 1968a). Fatty liver however is still a fully reversible lesion and the question remained whether alcoholic hepatitis, associated with a high morbidity, and irreversible cirrhosis could also be linked directly to alcohol ingestion itself rather than to a deficient diet. This problem could not be studied in volunteers, in view of the severity of the lesions involved. Previous attempts to produce these lesions in animals failed because of the reluctance of all species used to consume enough alcohol. Indeed, when ethanol is given as part of the drinking water, rats usually refuse to take a sufficient amount of alcohol to develop liver injury, if the diet is adequate (Best et al., 1949). This aversion of rats to ethanol was surmounted by the introduction of the new technique of feeding of ethanol as part of a nutritionally adequate totally liquid diet (Lieber et al., 1963,

Key Words: Alcohol, Ethanol, Fatty Liver, Cirrhosis, Liquid Diet, Baboon.

1965; DeCarli and Lieber, 1967). With this procedure, ethanol consumption was sufficient to produce a fatty liver despite adequate dietary intake. This technique is now widely adopted for the study of the pathogenesis of the fatty liver in the rat. In addition to the fatty liver, ethanol dependence develops in these rats, as witnessed by typical withdrawal seizures after cessation of alcohol intake (Lieber and DeCarli, 1973). Rats fed alcohol in liquid diets develop a fatty liver but not the more severe forms of liver injury seen in alcoholics, namely hepatitis and cirrhosis. We wondered whether this failure might be due to the fact that even when alcohol is given as part of a liquid diet, the rat will not consume more than 36% of total calories as ethanol and maintain an adequate diet. Of potential importance was also the fact that, whereas in man, development of cirrhosis requires five to twenty years of steady drinking, the rat only lives about two years. To overcome this difficulty, we turned to the baboon, a species which is long-lived and phylogenetically close to man. We adapted for the baboons (Lieber and DeCarli, 1974) the liquid diet which we had developed for the rat. With this technique, the alcohol intake was increased to 50% of total calories, and the entire spectrum of liver lesions observed in the alcoholic was produced, namely not only the fatty liver but also the hepatitis and the cirrhosis.

II. RAT MODEL OF ETHANOL DEPENDENCE, FATTY LIVER AND ASSOCIATED ADAPTIVE CHANGES IN ETHANOL, DRUG AND LIPOPROTEIN METABOLISM

Our original diet (Lieber et al., 1963) comprised amino acids and sucrose, whereas our new formula consists of casein (supplemented with methionine and cystine) and dextrin-maltose (DeCarli and Lieber, 1967), with 18% of total calories as protein, 35% as fat and 47% as carbohydrates in the control diet, and 11% carbohydrate in the diet containing 36% of total calories as ethanol (table 1). Rat littermates weighing 120-150 grams (Sprague-Dawley strain purchased from Charles River Breeding Laboratories, Inc., Wilmington, Mass.) are pair-fed the isocaloric diets once or twice daily in Richter graduated drinking tubes. Ethanol is introduced in the diet gradually to reach the final concentration on the fifth day. The ethanol fed animal is usually rate limiting and determines the amount of liquid diet given to the control. Actually, when the liquid diet is fed ad libitum to the control animals, the food intake is excessive and obesity develops. Conceivably, this could serve as a model for the study of obesity in the rat. In any event, consumption of the liquid diet is sufficient to assure continued growth in all animals and normal livers in the controls, whereas in the rats fed ethanol, fatty liver develops, which is evident both morphologically and on chemical analysis. Hepatic triglycerides increase progressively with an average 6-fold rise after one month, associated with a 5-fold increase of hepatic cholesterol ester and a small but significant rise in phospholipids.

TABLE 1

COMPOSITION OF CONTROL RAT LIQUID DIET (Per Liter)

Micropulverized Casein	41.4 g	Ethyl Linoleate	2.7 g
L-Cystine	0.5 g	Vitamin A Acetate	2.0 mg
DL-Methionine	0.3 g	Calciferol	0.01 mg
Vitamin Mixture*	5.0 g	DL-α-tocopherol Acetate	30.0 mg
Hegsted Salt Mixture	10.0 g	Dextri-Maltose	150.9 g**
Corn Oil	8.5 g	Sodium Carrageenate	2.5 gm
Olive Oil	28.5 g	Distilled Water	to 1 L.

*Contains (in mg) Thiamine 0.725, Riboflavin 1.25, Pyridoxine HCl 0.725, Ca Panthothenate 5.0, Nicotinamide 3.75, Choline Chloride 250.0, Biotin 0.025, Folic Acid 0.25, Inositol 25.0, 2 Methyl-1,4 Naphthoquinone 0.25, Vitamin B_{12} 0.025, PABA 12.5 and Glucose 4700.50.

**Replaced by 49.7 g dextri-maltose and 50.0 g ethanol in the ethanol formula.

After four to five weeks, the degree of steatosis reaches a plateau and persists for at least 22 weeks (Lieber and DeCarli, 1970a). The major source for the fatty acids accumulating in this experimental model is dietary fat (Lieber et al., 1966), most likely because ethanol blocks lipid oxidation in the liver (Lieber et al., 1967). A reduction in the fat content of the liquid formula to 25% of total calories does indeed reduce the degree of steatosis (Lieber and DeCarli, 1970a). Below that level, no further significant gains are achieved, however, and even with a diet practically devoid of fat, ethanol produces a significant increase in hepatic triglycerides which comprise endogenously synthesized fatty acids, probably made in the liver itself (Lieber et al., 1966). A reduction in the protein and lipotrope content of the diet increases the fat accumulation (Lieber et al., 1969). In addition to the fatty liver, ethanol dependence developed in these rats, as witnessed by typical withdrawal seizures after cessation of alcohol intake (Lieber and DeCarli, 1973).

The degree of steatosis observed in the rat is comparable to that seen in volunteers after a similar moderate ethanol intake (Lane and Lieber, 1966; Lieber and Rubin, 1968). Associated ultrastructural changes (giant distorted mitochondria, proliferation of the smooth endoplasmic reticulum) are similar in both species (Iseri et al., 1966; Lane and Lieber, 1966; Lieber and Rubin, 1968; Rubin and Lieber, 1968a). Both in man and in rats, the proliferation of the smooth endoplasmic reticulum is accompanied by increased activities of microsomal ethanol and drug detoxifying enzymes and enhanced rates of ethanol and drug metabolism (Rubin et al., 1968; Misra et al., 1971; Joly et al., 1973). The human fatty liver is often associated with mild hyperlipemia (Lieber et al., 1963; Losowski et al., 1963). In our rats with an alcoholic fatty liver, administration of ethanol with a fat containing diet also results in hyperlipemia due, at least in part, to enhanced lipoprotein production (Baraona and Lieber, 1970; Baraona et al., 1973).

III. BABOON MODEL OF HIGH ETHANOL INTAKE, FATTY LIVER, ALCOHOLIC HEPATITIS AND CIRRHOSIS

A. Experimental Procedure

In order to avoid complications due to hepatocystic parasitic disease, the adolescent or young adult animals used in this study were almost exclusively hamadryas, or olive and yellow baboons, the latter mostly born and raised in the USA. The animals imported from Africa were studied after prolonged quarantine. They were housed in individual cages at the Laboratory for Experimental Medicine and Surgery in Primates (LEMSIP), Tuxedo, N.Y. Until the actual study period, they were given a routine regimen of Purina Monkey Chow (Ralston Purina, St. Louis, Mo.) ad libitum, supplemented with a daily vitamin preparation. The animals entered the study after both prolonged observation and repeated hematological and stool examinations had indicated absence of disease.

Surgical biopsies of the liver were performed at regular intervals. Samples were taken for analysis of total lipids and triglycerides, light and electron microscopy and alcohol dehydrogenase activity in the cytosol, as described before (Lieber et al., 1972). The activity of the microsomal ethanol oxidizing system (MEOS) was determined (Lieber and DeCarli, 1970b) as well as that of aniline hydroxylase (Imai et al., 1966) in liver microsomes. Blood samples were taken for the measurement of ethanol (Bonnichsen, 1963) and of cholesterol, albumin, bilirubin, alkaline phosphatase, glutamic oxaloacetic transaminases (SGOT), creatinine, urea and glucose in an Autoanalyzer (Technicon Instrument Co., Tarrytown, N.Y.). Absence of hepatitis-associated antigen in the blood was verified by radioimmunoassay.

The overall composition of the control and and ethanol-containing diets is shown in table 2. The protein content of 18% of total calories corresponds to that of commonly used commercial diets considered to be satisfactory for the baboon. It is almost twice the amount recommended for man (National Research Council, 1974). The carbohydrate was 61% of total calories in the control diet. Even when carbohydrate was replaced by ethanol to the extent of 50% of total calories, the remaining carbohydrate was still 11% of total calories. The mineral composition of the liquid diet is given in table 3, and the vitamin content in table 4; both exceeded the requirement for the monkey as formulated by Foy et al. (1964), Portman (1970) and Hummer (1970) The present formulation (table 4) includes some of the vitamins given originally as a separate tablet.

The diets were prepared at least once or twice a week (by Bio Serv, Inc., Frenchtown, N.J.) stabilized with a seaweed extract similar to the stabilizer used in our rat liquid diet (DeCarli and Lieber, 1967), and were kept at +4°C until their use. They were given to the baboons twice a day in standard drinking bottles equipped with a valve-containing drinking tube. Except for a daily small carrot, the animals received no food but the liquid diet. The diet consumed was measured daily. Each of the alcohol-fed animals was matched with a control according to species, sex and weight. The dietary intake of the control was limited daily to that of its mate. This technique of daily pair feeding was adopted to assure a strictly equal caloric intake in both ethanol-treated animals and in their individual pair-fed controls. The diet formula presented here was arrived at after a number of trial experiments had solved problems observed with other formulas. The difficulties encountered included lack of appetite and therefore inadequate food intake to maintain body weight, diarrhea, food spoilage, insufficient alcohol consumption, and instability of the diet. These problems were successfully overcome in the present model.

TABLE 2
COMPOSITION OF BABOON LIQUID DIETS (g/L)

	ALCOHOL DIET	CONTROL DIET
Micropulverized casein	41.4	41.4
L-cystine	0.5	0.5
DL-Methionine	0.3	0.3
Corn oil	5.1	5.1
Olive oil	17.1	17.1
Ethyl linoleate	1.6	1.6
Dextrin-maltose	28.0	156.4
Ethanol	71.0	0

TABLE 3

MINERAL CONTENT OF BOTH CONTROL AND ETHANOL-CONTAINING BABOON LIQUID DIETS (per liter)

Calcium	3.0 g	Iron	80.0 mg
Phosphorus	1.4 g	Sulfur	50.0 mg
Potassium	2.8 g	Zinc	3.0 mg
Magnesium	0.4 g	Manganese	1.4 mg
Sodium	1.0 g	Copper	1.5 mg
Chlorine	1.3 g	Iodine	8.4 mg

B. Results

Nine pairs of animals were pair-fed alcohol containing or the control liquid diet for 8 to 22 months. The average duration of the treatment was 15 months and the mean intake was 80.0 ± 2.24 ml per kg per day. The alcohol fed baboons and their controls had average initial weights of 10.6 ± 0.35 and 10.6 ± 0.33 kg respectively.

TABLE 4

VITAMIN CONTENT OF CONTROL AND ETHANOL-CONTAINING BABOON LIQUID DIETS (per liter)

Thiamine hydrochloride	3.0 mg	P-aminobenzoic acid	200 mg
Pyridoxine hydrochloride	3.0 mg	Riboflavin	5.0 mg
Folic acid	1.5 mg	Cyanocobalamin	10.0 ug
Calcium pantothenate	10.0 mg	Vitamin K	2 ug
Nicotinic acid	30.0 mg	Biotin	100 ug
Ascorbic acid	200 mg	Vitamin A	8000 iu
Choline chloride	100 mg	DL-α-tocopherol acetate	30 mg
Inositol	200 mg	Calciferol	1000 iu

The baboons fed alcohol maintained their weight (10.2 ± 0.46 kg) throughout the study, whereas the controls increased their weight to an average of 11.8 ± 0.34 kg ($P<0.01$). No abnormalities developed in the controls. Inebriation was commonly observed in the animals drinking the alcohol containing diet. Blood ethanol concentration in inebriated animals were 262 and 258 mg/100 ml on two occasions in one baboon and 358 and 376 on one occasion in two other animals. Alcohol consumption resulted in the development of a fatty liver, with an average triglyceride content of 144.2 ± 36.2 compared to 10.4 ± 1.7 mg/g in the control ($P<0.01$). Four pairs of animals were biopsied sequentially after 8.5 and 21 months; whereas the triglyceride was only 62.4 ± 19.5 mg/g after 8.5 months of alcohol treatment, this value increased to a mean of 165.3 ± 39.8 mg/g of liver at the end of 21 months. A case of obvious steatosis is shown in Fig. 1. In addition to the fat, mild inflammation, cellular degeneration and some fibrosis were noted. In three animals fed ethanol for 9 months and one animal fed ethanol for 12 months, alcoholic hepatitis developed, as defined by cell degeneration, inflammation and central sclerosis.

In four animals which developed hepatitis, triglyceride acumulation was more pronounced (180.4 ± 55.77 mg/g) than in the five animals which had only a fatty liver (33.0 ± 5.7 mg/g). Two of the four animals which had developed hepatitis after 9 months were again subjected to biopsy after 20 months. Each showed the development

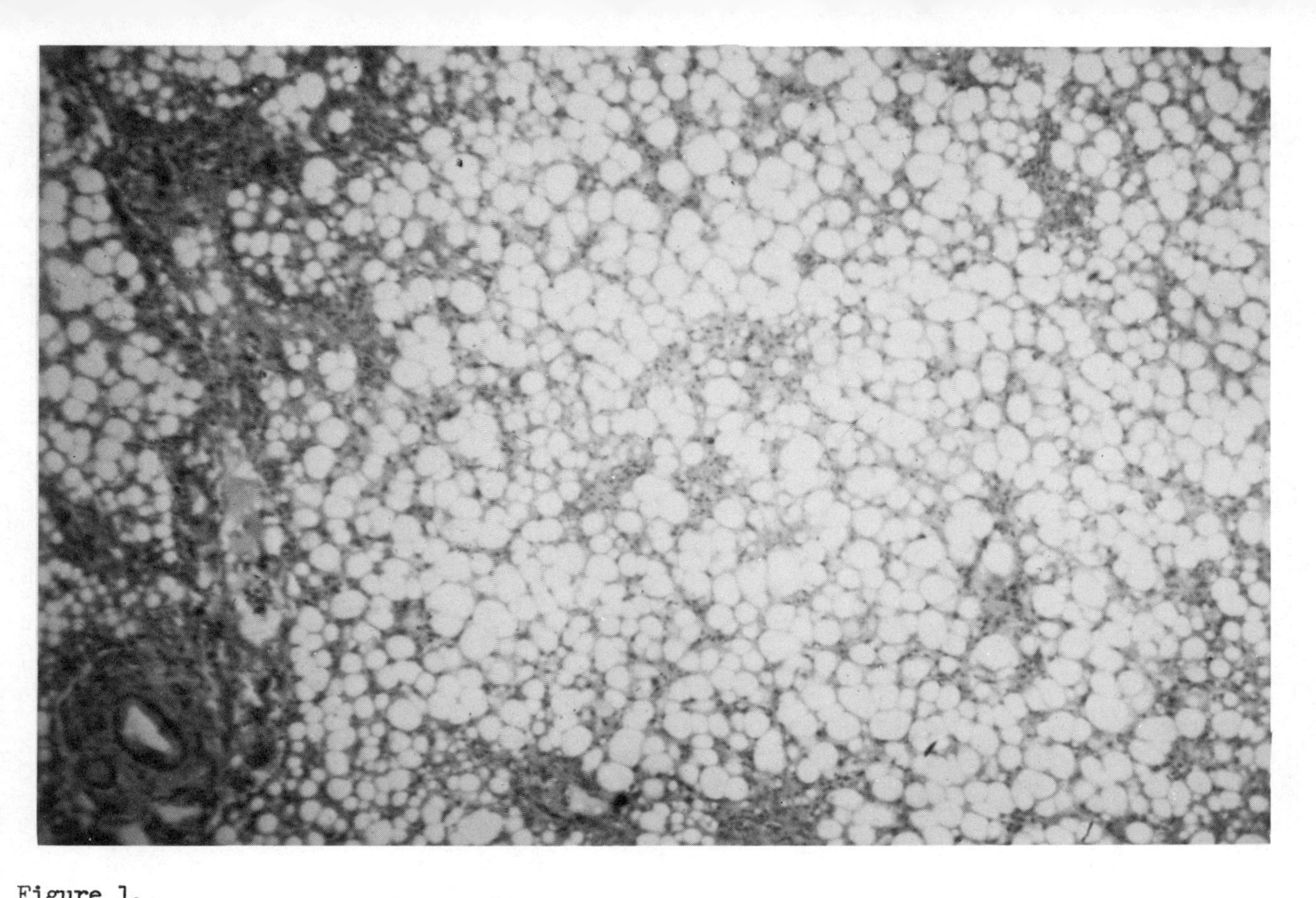

Figure 1.
Fatty liver in a baboon fed the ethanol-containing liquid diet. Hematoxylin eosin stain (X 150). From Lieber and DeCarli (1974).

of extensive fibrosis corresponding to a diagnosis of incomplete cirrhosis. One of the animals died after 2 years of treatment because of withdrawal symptoms (convulsions) which developed when the intake of ethanol decreased because of an intercurrent infection. The autopsy revealed typical alcoholic (Laennec's) cirrhosis of the liver (Fig. 2). Another animal which showed alcoholic hepatitis also died after 18 months of treatment because of a similar complication. The autopsy revealed no pathological changes other than those of the liver. When alcohol intake was decreased for reasons of intercurrent upper respiratory infection, withdrawal symptoms (such as seizures) were observed in at least four animals. Ultrastructual changes were pronounced already in the fatty livers; they remained present throughout the stages of hepatitis and cirrhosis. The mitochondrial lesions were characterized by enlargement, irregular forms and disoriented cristae. The rough endoplasmic reticulum was decreased, and the smooth endoplasmic reticulum was vesicular and proliferated.

There was an increase in the activity of the microsomal ethanol oxidizing system in the animals fed ethanol. MEOS activity after alcohol feeding was 23.0 $\pm$ 2.5 nmoles/min/mg protein vs. 13.7 $\pm$ 1.5 in the controls ($P<0.01$). It is noteworthy however that of the four animals which had the hepatitis, only two had a significant increase in MEOS activity, whereas the two others had values comparable to that of the corresponding controls. We have reported previously that other microsomal enzymes also increase in activity after ethanol feeding, both in rats (Rubin and Lieber, 1968b; Joly et al., 1973; Ishii et al., 1973) and in baboons (Lieber et al., 1972). This was confirmed for microsomal aniline hydroxylase activity measured in 4 pairs of baboons: whereas the mean value in the animals fed ethanol was 0.874 $\pm$ 0.42 nmoles/min/mg microsomal protein the corresponding controls had activities of 0.299 $\pm$ 0.12.

Serum cholesterol was moderately increased in the alcohol treated baboons (214.1 $\pm$ 22.2 vs 153.5 $\pm$ 11.0 mg per 100 ml, $P<0.05$). In all pairs of animals, the values of SGOT were higher in the ethanol fed animals than in the corresponding controls. In the five animals with a simple fatty liver, however, the increase was small. By contrast, in the four animals which had hepatitis, there were striking elevations of 2650, 227, 123 and 75 mU/ml as compared to values of 40, 37, 35 and 38 in the corresponding controls. No significant abnormalities were noted in albumin, bilirubin, alkaline phosphatase, creatinine, urea and glucose. Hemoglobin and hematocrit values had a tendency to be lower in the alcohol fed animals but no meaningful decrease has been noted thus far.

An additional group of 12 animals which had been given a solid diet with either alcohol or carbohydrates in the drinking water as described before (Lieber et al., 1972) for a period varying from 17 to 34 months were then changed to the liquid diet for an average

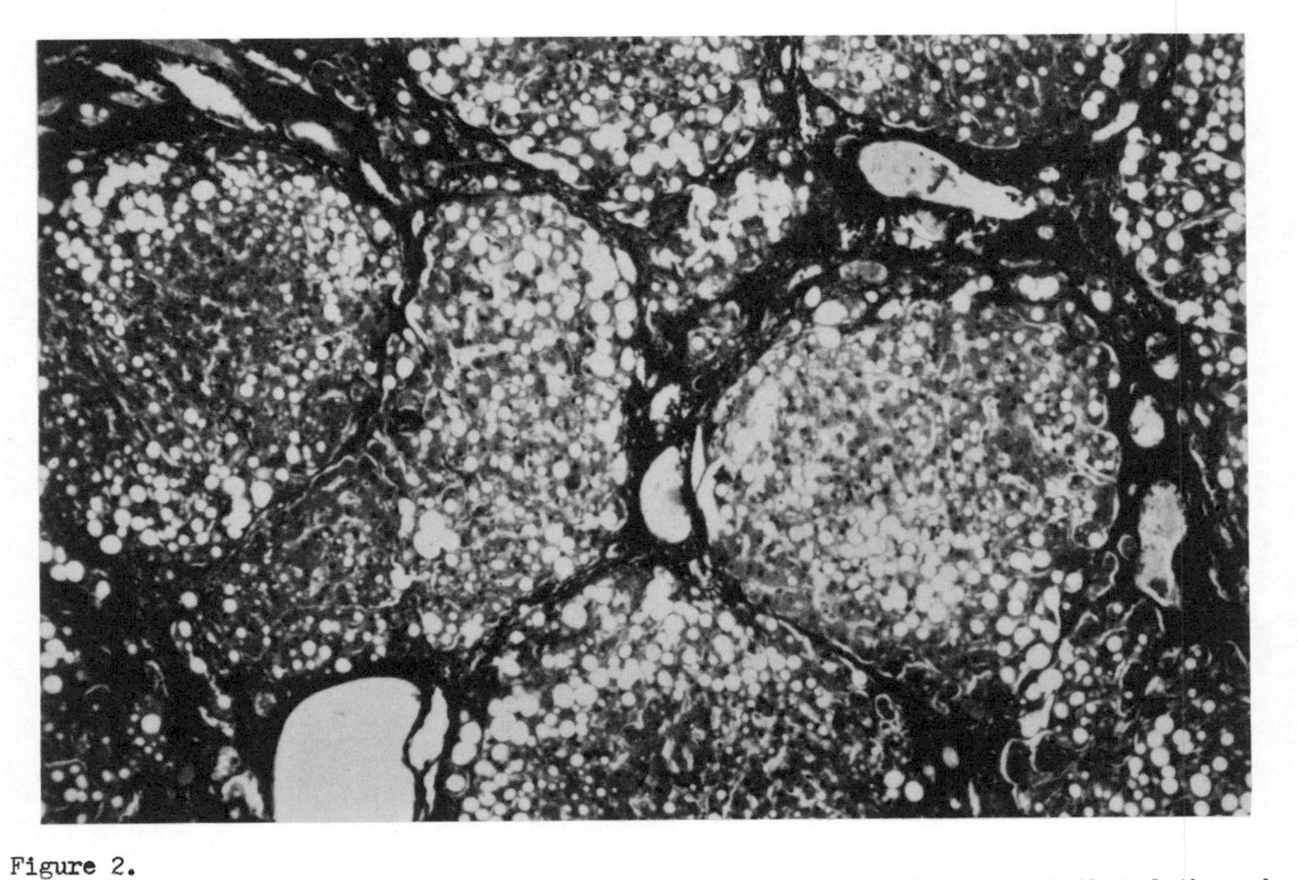

Figure 2.
Cirrhosis in a baboon fed alcohol. Fat is regularly distributed through nodules surrounded by connective tissue septa. Chromotrope-aniline blue. (X 60). From Lieber and DeCarli (1974).

of 17 months. Whereas when alcohol was given with the solid diet no lesions more severe than fatty liver had developed (Lieber et al., 1972), with this new regimen, four of the six animals fed alcohol progressed to a more severe stage: one to alcoholic hepatitis (after 29 months of the solid diet and 19 months on the liquid diet), two to incomplete cirrhosis (after 30 months on the solid diet and 15 months on the liquid diet) and one to complete cirrhosis (after 34 months on the solid diet and 19 months on the liquid diet). These preliminary findings were previously reported (Rubin and Lieber, 1974).

C. Discussion

Our experimental model in the baboon reproduces all the liver lesions observed in the alcoholic, namely the fatty liver, the hepatitis and cirrhosis. Fatty liver had been described before after short-term ethanol administration in man (Lieber et al., 1963, 1965; Lane and Lieber, 1966; Lieber and Rubin, 1968) and in rats (Lieber et al., 1963, 1965; DeCarli and Lieber, 1967). However, no experimental production of hepatitis has been reported before: fibrosis or cirrhosis only has been produced in primate animals after the feeding of severely deficient diets, lacking protein and/or choline (Wilgram, 1959; Hoffbauer and Zaki, 1965; Ruebner et al., 1969). The adequacy of our diet was shown by the fact that the liver morphology remained normal in the controls. The animals fed ethanol maintained their weight, whereas the controls had a slight weight gain. This finding is consistent with our previous observations that both in rats (Lieber et al., 1965; Saville and Lieber, 1965) and in man (Pirola and Lieber, 1972) ethanol promotes less weight gain than isocaloric carbohydrates.

The preliminary data gathered thus far with this new model of alcoholic liver injury bring already an answer to some questions which have been the subject of long-standing debate. The most important one is the demonstration that hepatitis and cirrhosis can be produced by prolonged alcohol ingestion, even in the absence of a deficient diet. This finding has immediate practical applications since it disproves the widespread belief that adequate diets prevent the development of cirrhosis in alcoholics. Another debate centered on the question of the relationship of the fatty liver stage to that of hepatitis and cirrhosis. Some believe that the fatty liver, though benign and reversible, is nevertheless a precursor form of the hepatitis and cirrhosis, whereas others hold the view that fatty liver has no relationship to the more severe forms of liver disease. It is relevant to this question that the animals which had developed the more severe lesions (hepatitis or cirrhosis) had significantly higher liver triglycerides. Moreover, it was found that already at the fatty liver stage, collagen, the characteristic protein of the fibrous bands, is increased in amount and that the activity of the enzyme peptidylproline hydroxylase involved in the production of collagen is increased after alcohol feeding prior to the development of hepatitis and cirrhosis (Feinman and Lieber, 1972). Therefore, already at the fatty liver stage, changes occur which can be considered as precursors for the more severe forms of the disease. This view is supported by the observation that the ultrastructural changes seen in the alcoholic fatty liver (Iseri et al., 1966, Lane and Lieber, 1966; Lieber and Rubin, 1968; Lieber et al., 1972) are like those of the alcoholic hepatitis (Svoboda and Manning, 1964; Rubin and Lieber, 1974).

In rats and man, ethanol administration results in increased activity of microsomal enzymes (Rubin and Lieber, 1968b; Ishii et al., 1973; Joly et al., 1973) including that of the MEOS (Lieber and DeCarli, 1968, 1970b; Mezey and Tobon, 1971). The present model reproduces these 'inductive' effects of ethanol on microsomal enzymes. The moderate hypercholesterolemia observed in the baboons fed ethanol is also comparable to a similar effect previously described in man (Lieber et al., 1963) and in the rat (Lefevre et al., 1972).

We hope that the present model will enable us to define the transition between the adaptive response of the liver due to the metabolic overload produced by alcohol (Lieber, 1973) and the development of injury, as well as to recognize a possible precursor lesion of the cirrhosis. The fact that not all the baboons fed ethanol developed either hepatitis or cirrhosis raises the challenging question of the cause for the enhanced susceptibility to alcohol in some of them. It would be of obvious clinical importance to learn to recognize this enhanced susceptibility at an early stage of the disease. We are also seeking to determine whether the baboons will eventually develop not only the morphologic lesions of hepatitis and cirrhosis, but also the secondary complications asssociated clinically with the full-blown disease including hypoalbuminemia, portal hypertension, ascites and collateral circulation. Our model might also contribute to elucidation of the mechanism whereby alcohol produces these liver lesions, thereby allowing for development of a rational form of prophylaxis and therapy. Furthermore, since our animals fed ethanol develop signs of physical dependence upon withdrawal of the drug, this model could be used for the verification of various theories proposed for the explanation of dependency, as well as for the assessment of the efficacy of new treatment modalities to alleviate the complications associated with ethanol withdrawal.

REFERENCES

Baraona, E. and Lieber, C.S. 1970, J. Clin. Invest. 49:769.

Baraona, E., Pirola, R.C. and Lieber, C.S. 1973, J. Clin. Invest. 52:296.

Best, C.H., Hartroft, W.S., Lucas, C.C. and Ridout, J.H. 1949, Brit. Med. J. ii:1001.

Bonnichsen, R. Methods of Enzymatic Analysis. edited by H.U. Bergmeyer, New York, Academic Press, 1963, p. 285.

DeCarli, L.M. and Lieber, C.S. 1967, J. Nutr. 91:331.

Feinman, L. and Lieber, C.S. 1972, Science 176:795.

Foy, H., Kondi, A. and Mbaya, V. 1964, Brit. J. Nutr. 18:307.

Hoffbauer, F.W. and Zaki, F.G. 1965, Arch. Path. 79:364.

Hummer, R.L. Feeding and Nutrition of Nonhuman Primates. edited by R.S. Harris, New York, Academic Press, 1970, p. 183.

Imai, Y., Ito, A. and Sato, R. 1966, J. Biochem. 60:417.

Iseri, O.A., Lieber, C.S. and Gottlieb, L.S. 1966, Amer. J. Path. 48:535.

Ishii, H., Joly, J.-G. and Lieber, C.S. 1973, Biochim. Biophys. Acta 291:411.

Joly, J.-G., Ishii, H., Teschke, R., Hasumura, Y. and Lieber, C.S. 1973, Biochem. Pharmacol. 22:1532.

Lane, B.P. and Lieber, C.S. 1966, Amer. J. Path. 49:593.

Lefevre, A.F., DeCarli, L.M. and Lieber, C.S. 1972, J. Lipid Res. 13:48.

Lieber, C.S. 1973, New Eng. J. Med. 288:356.

Lieber, C.S. and DeCarli, L.M. 1968, Science 162:917.

Lieber, C.S. and DeCarli, L.M. 1970a, Amer. J. Clin. Nutr. 23:474.

Lieber, C.S. and DeCarli, L.M. 1970b, J. Biol. Chem. 245:2505.

Lieber, C.S. and DeCarli, L.M. 1973, Res. Commun. Chem. Path. Pharmacol. 6:983.

Lieber, C.S. and DeCarli, L.M. 1974, J. Med. Primatology 3:153.

Lieber, C.S. and Rubin, E. 1968, Amer. J. Med. 44:200.

Lieber, C.S., Jones, D.P., Mendelson, J. and DeCarli, L.M. 1963, Trans. Ass. Amer. Physicians 76:289.

Lieber, C.S., Jones, D.P. and DeCarli, L.M. 1965, J. Clin. Invest. 44:1009.

Lieber, C.S., Spritz, N. and DeCarli, L.M. 1966, J. Clin. Invest. 45:51.

Lieber, C.S., Lefevre, A., Spritz, N., Feinman, L. and DeCarli, L.M. 1967, J. Clin. Invest. 46:1451.

Lieber, C.S., Spritz, N. and DeCarli, L.M. 1969, J. Lipid Res. 10:283.

Lieber, C.S., DeCarli, L.M., Gang, H., Walker, G. and Rubin, E. Medical Primatology-1972. edited by E.I. Goldsmith and J. Moor-Jankowski, Basel, S. Karger, Part 3, 1972, p. 270.

Losowsky, M.S., Jones, D.P., Davidson, C.S. and Lieber, C.S. 1963, New Eng. J. Med. 268:651.

Mezey, E. and Tobon, F. 1971, Gastroenterology 61:707.

Misra, P.S., Lefevre, A., Ishii, H., Rubin, E. and Lieber, C.S. 1971, Amer. J. Med. 51:346.

National Research Council. Food and Nutrition Board. Recommended Dietary Allowances. 8th Edition. Washington, D.C. Natl. Acad. Sci. 1974.

Pirola, R.C. and Lieber, C.S. 1972, Pharmacology 7:185.

Portman, O.W. Feeding and Nutrition of Nonhuman Primates. edited by R.S. Harris, New York, Academic Press, 1970, p. 87.

Rubin, E. and Lieber, C.S. 1968a, New Eng. J. Med. 278:869.

Rubin, E. and Lieber, C.S. 1968b, Science 162:690.

Rubin, E. and Lieber, C.S. 1974, New Eng. J. Med. 290:128.

Rubin, E., Hutterer, F. and Lieber, C.S. 1968, Science 159:1469.

Ruebner, B.H., Moore, J., Rutherford, R.B., Seligman, A.M. and Zuidema, G.D. 1969, Exp. Molec. Path. 11:53.

Saville, P.D. and Lieber, C.S. 1965, J. Nutr. 87:477.

Svoboda, D.J. and Manning, R.T. 1964, Amer. J. Path. 44:645.

Wilgram, G.F. 1959, Ann. Intern. Med. 51:1134.

HEPATIC AND METABOLIC EFFECTS OF ETHANOL ON RHESUS MONKEYS

Boris H. Ruebner, Robert I. Krieger, Jeffrey L. Miller,

Makepeace Tsao and Marie Rorvik

Departments of Pathology and Surgery, School of Medicine and Department of Environmental Toxicology, College of Agriculture, University of California, Davis, California 95616*

I. INTRODUCTION

Human alcoholics may suffer from a great variety of hepatic morphologic and biochemical abnormalities such as fatty liver, alcoholic hepatitis and cirrhosis. In addition, the metabolism of alcohol and of drugs is accelerated in human alcoholics (Shah, et al, 1972). The mechanisms of these changes are still disputed. Hepatic fatty change has been attributed to nutritional imbalance (Porta et al, 1970) and to a direct toxic effect of ethanol (Lieber et al, 1971). The mechanism of the accelerated metabolism of alcohol also remains uncertain.

We have studied these problems in non-human primates because they resemble man more closely in some aspects of metabolism than rats. Our model consists of Rhesus monkeys force-fed a liquid diet containing 41 percent of calories in the form of alcohol (Ruebner et al, 1972). In the diet of our control animals glucose was substituted isocalorically for alcohol. This model also has the advantage over most previous studies that the dietary and alcoholic intakes of our animals are known more accurately than in experiments

* Supported by Grants RR06138, RR0163, and ES00054 from the National Institutes of Health, Bethesda, Maryland. Mr. Miller is a Shell Fellow in the Department of Environmental Toxicology.

utilizing spontaneous alcohol intake. In a previous study our alcohol animals, but not our controls, developed a fatty liver after 10 days on a diet based on Metrecal (Ruebner et al, 1972). Since Metrecal may not be an adequate basal diet, we have substituted for it a more accurately defined diet based on casein (De Carli and Lieber, 1967). In this investigation the duration of the experiment has also been extended to one month. The effects of this diet on the morphology and fat content of the liver, some biochemical parameters and on body weights have been studied. In addition, we have investigated the effect of alcohol on its own metabolism and on that of antipyrene, a substance metabolized by the microsomal drug metabolizing enzymes. Conversely we have studied the effects of phenobarbital, a well-known inducer of drug metabolizing enzymes, on the metabolism of alcohol.

II. MATERIALS AND METHODS

Animals

Groups of Rhesus monkeys of approximately 4-5 kg. body weight were used. Each experiment had a group of three animals of the same sex (males have been used predominantly) as controls. The duration of the experiments was 28 days. After a rest period of one month the animals reverted to normal liver histology and serum biochemistry and were used again.

Diets

The liquid diet was based on that of DeCarli and Lieber (1967). Ethyl linoleate was added because primates require more linoleate than rats (Portman, 1970). The diet was given by nasogastric tube (8 French, 42 inch) in three meals between 9 a.m. and 4 p.m. to monkeys in specially designed restraining chairs which allow maximum freedom without interfering with the experiments. After addition of grain alcohol the composition of the diet was

Protein........................15%
Carbohydrate...................23%
Fat............................21%
Ethanol........................41%

This diet contained one calorie per ml. and 7.32% ethanol vol/vol.

Detailed Composition of the Alcohol Diet

Casein	3.67 g
Methionine	0.03 g
Cystine	0.05 g

Glucose	3.52 g
Alcohol	7.32 ml
Olive Oil	1.6 g
Corn Oil	0.67 g
Ethyl Linoleate	0.35 g
Vitamin Mixture	4 g
Vitamin A	3600 I.U.
Vitamin D	400 I.U.
Vitamin E	0.02 g
Vitamin B_1	0.004 g
Vitamin B_6	0.004 g
Niacin	0.018 g
Panthothenic Acid	0.012 g
Ascorbic Acid	0.18 g
Choline Chloride	0.3 g
Folic Acid	0.3 g
Vitamin B_{12}	0.0054 mg
Biotin	0.08 mg
Salt Mixture	1 g (HEGSTED)
Sodium Carageenate	0.2 g
Make up to 100 ml with water	

The animals were given 100 calories per kg per day which contained 5.78 g (7.32 ml) of alcohol per kg per day. In the control diet glucose was substituted isocalorically for the ethanol.

In the phenobarbital experiments the animals were given the standard primate center diet (Purina Monkey Chow) containing 15 percent protein and 5.2 percent fat.

Clinical and Liver Biopsy Studies

The animals were weighed at the beginning and end of the experiments. Biochemical and morphologic studies were done before and at the conclusion of the experiments.

Liver biopsies were done 24 hours before the start of each experiment and 72 hours after its conclusion under light Surital anesthesia with an adult-size Menghini needle which provided 10-20 mg of tissue.

For light microscopy liver biopsies were fixed in buffered formalin, paraffin enbedded and sectioned. Sections were stained with hematoxylin and eosin, the periodic acid Schiff reaction with diastase digestion for glycogen by Mallory's trichrome stain for connective tissue and by the Prussian blue reaction for hemosiderin. Pieces 1 mm in diameter were processed for electron microscopy after fixation in 4% phosphate buffered glutaraldehyde followed by Millonig's phosphate buffered osmium (1962), and embedded in an araldite epon mixture. One micron thick sections were stained by Toluidine blue for orientation and identification of lipid after cutting by a Porter Blum II microtome and stained by uranyl acetate and lead citrate using the method of Venables and Coggershall (1965). These sections were studied with AEI electron microscopes.

Total proteins (Gornall 1949), and triglycerides (Eggstein 1966; Irsigler and Hrabal 1968), were measured on the tissues obtained at liver biopsy.

Serum Biochemistry

The following were measured by SMA 12 autoanalyzer: LDH, SGOT, Alk phosphatase, K, Na, BUN, glucose, uric acid, total protein, calcium and phosphate and cholesterol.

Blood alcohols were determined on heparinized plasma by gas chromatography. Hourly measurements were taken during the day and next morning after an overnight fast.

Alcohol Tolerance Tests. Forty-eight hours before the start of each experiment and 24 hours after its conclusion the animals were given one gram per kg of ethanol intraperitoneally (Makar and Mannering, 1970). Alcohol levels were measured half hourly from one to three hours after administration. The results were expressed as mg of ethanol metabolized per 100 ml of blood per hour and differences in metabolism between the alcohol animals and controls before and after the experiments were evaluated by regression analysis.

Antipyrene Metabolism. Seventy-two hours before the start of each alcohol experiment and 48 hours after its completion ^{14}C-antipyrene (15mg/kg) was administered intravenously. Plasma antipyrene levels were monitored after 1/2, 1, 1 1/2, and 2 hours. Unmetabolized antipyrene was extracted using 12% *iso*-amyl alcohol from plasma made alkaline with NaOH. Portions of the extract were analyzed for ^{14}C-antipyrene using a scintillation counter. The antipyrene half-life was determined using regression analysis.

Phenobarbitone Experiments. Sodium phenobarbitone was given at 30 mg per kg per day (i.m.) for four days to animals on the standard primate center diet.

III. RESULTS

Ethanol Experiments

Both the alcohol animals and the controls tolerated this regime well.

The mean body weight of nine alcohol animals decreased from 5.5 kg to 5.3 kg. This was a statistically significant weight loss ($t= 3.2$; $p<.02$).

The mean body weight of nine carbohydrate animals increased from 5.0 to 5.25 kg. This was a statistically significant weight gain ($t= 4.8$; $p<.01$).

Hepatic Changes

Light Microscopy. Control biopsies prior to the intubations showed virtually no fat vacuoles in the hepatocytes. After administration of the alcohol diet the biopsies consistently showed moderate fatty change, more marked in centrilobular areas. The animals on the carbohydrate diet did not develop any fatty change.

Electron Microscopy. Compared with controls the most striking changes in the alcohol animals were fatty metamorphosis and mitochondrial swelling with disarray of the cristae and focal cytoplasmic degradation. The carbohydrate animals showed no significant alterations.

Liver Biochemistry. The liver triglycerides of four of the animals on the alcohol diet increased significantly from a mean of 66.5 mg percent of dry weight to 121 mg percent ($t= 8.5$; $p<0.01$). The carbohydrate control animals showed no change (from 78.9 to 84.1 mg percent).

Blood Changes

Blood alcohol levels during the infusions rose from zero at 8 a.m. to approximately 200 mg per 100 ml at 6 p.m.

Serum, S.G.O.T., alkaline phosphatase, sodium, potassium, cholesterol, calcium, phosphate, glucose, BUN, uric acid, and total proteins showed no change in the experimental or control groups.

The animals on the alcohol diets, however, did develop a significant increase in serum lactic dehydrogenase (LDH). In four animals there was a rise from 190 to 400 units ($t= 38$; $p<.01$). The glucose diet had no significant effect on the LDH. In four animals the mean LDH fell from 200 to 170 units.

Alcohol Tolerance Tests. Ethanol metabolism in three alcohol animals increased from a mean of 17.4 mg per 100 ml per hour to 26.6 mg. The data were analyzed by regression analysis and the difference between controls and treated animals was statistically significant ($t = 5.29$; $p < 0.05$). Two carbohydrate animals showed no significant change. Their alcohol metabolism decreased from 20.0 per 100 mg per hour to 17.0 mg.

Antipyrene Metabolism. The half life of antipyrene in eight ethanol animals decreased significantly from a mean of 61.0 minutes to 49.9 minutes ($t = 3.4$; $p =< .025$). Six control animals had a mean half life of 60.6 minutes which was unchanged after 28 days on the carbohydrate diet.

Phenobarbital Experiments

Hepatic Changes

Light Microscopy. Both before and after the phenobarbital administration hepatocellular morphology was normal.

Electron Microscopy. After phenobarbital there was a striking increase in the smooth endoplasmic reticulum which affected virtually all hepatocytes. No other changes were noted.

Blood Changes

Alcohol Tolerance Tests. Ethanol prior to phenobarbital administration was metabolized at a mean rate of 16 mg per 100 ml per hour in twelve animals. After phenobarbital administration the rate was 22.5 mg per hour. This difference was statistically significant ($t = 3.25$; $p < .005$).

Antipyrene Tolerance Tests. The half life of antipyrene in six phenobarbitone animals decreased significantly from a mean of 76.5 minutes to 33.6 minutes ($t = 7.27$; $p < 0.0005$).

IV. DISCUSSION

Our observation that the body weight of the alcohol animals decreased while that of the carbohydrate animals increased was unexpected. Dehydration or diarrhea did not appear to be responsible for the loss of weight of the alcohol animals. Whether the loss of weight of the alcohol animals was the result of malabsorption (Lindenbaum et al. 1972) or of altered protein metabolism (Pirola and Lieber 1973; Rodrigo et al 1971) requires further investigation.

There is general agreement that alcohol causes hepatic fatty change and mitochondrial alterations. This was confirmed in the present experiment. However, even after one month fatty change and mitochondrial swelling remained relatively mild and similar in severity to the changes observed previously in our monkeys after 10 days of an alcohol diet based on Metrecal (Ruebner, 1972). The diet employed here, based on casein (De Carli and Lieber 1967), thus appeared less effective in producing fatty change than the Metrecal diet employed previously. Since both diets were similar in protein, fat and carbohydrate contents this may well indicate that the casein diet employed here is of better quality nutritionally than our previous Metrecal diet. It might also be argued that the hepatic fatty change in our animals is transient and would disappear if the duration of our experiment is prolonged. This can only be settled by further experiments.

If it is accepted that our diet is nutritionally adequate and that it is adequately absorbed, then the effects on the liver which we observed are a direct effect of alcohol rather than the result of nutritional imbalance. Our observations do not conflict with those of Rubin and Lieber in baboons consuming spontaneously amounts of alcohol similar to those which we force-fed our Rhesus monkeys. During the first month of their experiment their results were similar to ours. Alcoholic hepatitis was first seen by them after nine months and cirrhosis after four years.

There is almost universal agreement that ethanol metabolism is accelerated in human alcoholics and in experimental animals after ethanol administration (Shal et al., 1972; Pieper et al., 1973; Misra et al., 1971). One exceptional result (Belfrage et al) can probably be explained by a relatively low alcohol intake, 16 percent of calories, compared with approximately 40 percent in most other studies. However, the mechanism of the acceleration of ethanol metabolism remains disputed. Rubin, Lieber and their collaborators (Misra et al., 1971) have suggested that a microsomal ethanol oxidizing system (Meos) can be induced by ethanol and is responsible for increased ethanol metabolism while others (Carter and Isselbacher, 1972; Thurman, 1973; Vatsis and Schulman, 1973; Oshino et al., 1973) believe that catalase is involved rather than an enzyme related to the microsomal drug metabolizing enzymes.

Our observations support those who believe that ethanol can be metabolized via a microsomal system related to the drug metabolizing enzymes. In agreement with the observations of Misra et al., (1971) in the rat and in man we found acceleration of drug metabolism (antipyrene) in Rhesus monkeys after chronic ethanol administration.

In preliminary assays using biopsy material we have found hepatic p-chloro-n-methyl aniline demethylase, a drug metabolizing enzyme, to be elevated but not aldrin epoxidase. Several previous studies have shown increases in various hepatic drug metabolizing enzymes after ethanol administration (Ideo, et al., 1971; Singlevich and Barboriak, 1971; Joly et al., 1973).

The effects of known inducers of drug metabolizing enzymes on alcohol metabolism have been more controversial (Lieber, Ethanol and the Liver in Bourne; Watkins). Like Lieber and De Carli, (1972) in rats and Mezey and Robles (1974) in man we found an acceleration of ethanol metabolism after phenobarbital administration.

There was, thus, induction by both alcohol and phenobarbital of alcohol metabolism and of drug metabolism. However, alcohol was a more powerful inducer of its own metabolism than phenobarbital. Phenobarbital, on the other hand was a better inducer of antipyrene metabolism than alcohol. Thus, while our observations do not prove that alcohol is metabolized by a pathway related to the microsomal drug metabolizing enzymes, they strongly suggest it.

V. SUMMARY

Rhesus monkeys were tube-fed 100 calories per kg of a liquid diet based on casein in which 41% of the calories were derived from grain alcohol. The alcohol intake was 5.8 g per kg per day. Control diets contained isocaloric amounts of glucose. The protein content of the diet was 15% and fat supplied 21% of the calories. After 28 days the animals which had been fed ethanol developed hepatic fatty change and serum L.D.H. levels were elevated. The most striking electron microscopic changes in the alcohol animals were mitochondrial swelling, focal cytoplasmic degradation, and dilatation of the rough endoplasmic reticulum. In the monkeys which had received ethanol the metabolism of alcohol increased from 17.4 mg per 100 ml per hour to 26.6 mg and antipyrene half-life decreased from 61.0 minutes to 49.9 minutes. The carbohydrate animals showed no significant change in alcohol metabolism or antipyrene half life. The ethanol animals lost weight significantly while the carbohydrate animals gained significantly. The metabolic effects of alcohol thus were not reproduced by glucose. Administration of phenobarbital at 30 mg per kg for 5 days increased alcohol metabolism from 16.5 mg per hour to 22.5 mg per hour and shortened antipyrene half life from 76.5 minutes to 33.6 minutes. Alcohol and phenobarbital both induced enhanced drug metabolism but alcohol was a more powerful inducer of its own metabolism than phenobarbital. Phenobarbital on the other hand was a better inducer of antipyrene metabolism than alcohol.

REFERENCES

1. Belfrage, P., Berg, B., Chronholm, T., Elmquist, D., Hagerstrand, I., Johansson, B., Nilsson-Ehle, P., Norden, G., Sjovall, J., Wiebe, T.: Prolonged Administration of Ethanol to Healthy Volunteers: Effects on Biochemical, Morphological and Neurophysiological Parameters. Acta. Med. Scand. Suppl 552.

2. Carter, E.A., Isselbacher, K.J.: Hepatic Microsomal Ethanol Oxidation, Mechanism and Physiologic Significance. Laboratory Investigation. 27:283-286, 1972.

3. De Carli, L.M., Lieber, C.S.: Fatty Liver in the Rat After Prolonged Intake of Ethanol with a Nutritionally Adequate New Liquid Diet. J. Nutr., 91:331-336, 1967.

4. Eggstein, M.: Eine neue Bestimmung der Neutralfette im Blutserum und Gewebe II. Klin. Wschr , 44:267, 1966.

5. Gornall, A.G., Bardawill, C.J., David, M.M.: Determination of Serum Proteins by Means of the Biuret Reaction. J. Biol. Chem., 177:751, 1949.

6. Ideo, G., de Franchis, R., Del Ninno, E., Corucci, C., Dioguardi, N.: Increase of Some Rat Liver Microsomal Enzymes as a Consequence of Prolonged Alcohol Intake: Comparison with the Effect of Phenobarbitone. Enzyme 12(4): 473-480, 1971.

7. Irsigler, K. and Hrabal, I.: Zur Neutralfettbestimmung im Biopsie-material der Menschlichen Leber. Klin. Wschr , 46:432, 1968.

8. Joly,J-G, Ishii, H.,Teschke, R., Hasumura, Y., Lieber, C.S.: Effect of Chronic Ethanol Feeding on the Activities and Submicrosomal Distribution of Reduced Nicotinamide Adenine Dinucleotide Phosphate-Cytochrome P-450 Reductase and the Demethylases for Aminopyrine and Ethylmorphine. Biochemical Pharmacology, Vol 2: 1532-1535, 1973.

9. Lieber, C.S., De Carli, L.M.: The Role of the Hepatic Microsomal Ethanol Oxidizing System (Meos) for Ethanol Metabolism in Vivo. J. Pharmacol. and Exptl. Therap. 181:279-287, 1972.

10. Lieber, C.S., "Ethanol and the Liver" in Alcoholism, Progress in Research and Treatment. Edited by Bourne, P.G. and Fox, R. Academic Press Inc., New York and London, 1973

11. Lieber, C.S., Rubin, E., De Carli, L.M., Gang, H., Walker, G.: Hepatic Effects of Long Term Ethanol Consumption with High or Low Protein Diets in Primates. Gastroenterology, 60:, 1971.

12. Lindenbaum, J., Shea, N., Saha, J.R., Lieber, C.S.: Alcohol Induced Impairment of Carbohydrate Absorption. Clin. Res., 20:459, 1972.

13. Makar, A.B., Mannering, G.J.: Kinetics of Ethanol Metabolism in the Intact Rat and Monkey. Biochemical Pharmacology. Vol. 19, 2017-2022, 1970.

14. Mezey, E., Robles, E.A.: Effects of Phenobarbital Administration on Rats of Ethanol Clearance and on Ethanol-Oxidizing Enzymes in Man. Gastroenterology. Vol. 66, 248-253, 1974.

15. Millonig, G.: Further Observations on a Phosphate Buffer for Osmium Solutions in Fixation in Electron Microscopy. Fifth International Congress for Electron Microscopy: ed. Breese, S.S.: Academic Press, New York 2:, 1962.

16. Misra, P.S., Lefevre, A., Ishii, H., Rubin, E., Lieber, C.S.: Increase of Ethanol, Meprobamate and Pentobarbital Metabolism After Chronic Ethanol Administration in Man and in Rats. J. Med. 51:346-351, 1971.

17. Oshino, N., Oshino, R., Chance, B.: The Characteristics of the 'Peroxidatic' Reaction of Catalase in Ethanol Oxidation. Biochemical Journal. Vol 131:555-567, 1973.

18. Pieper, W.A., Skeen, M.J.: Changes in Rate of Ethanol Elimination associated with Chronic Administration of Ethanol to Chimpanzees and Rhesus Monkeys. Drug Metab. Disposition 1(4):634-641, 1973.

19. Pirola, R.C., Lieber, C.S.: Energy Cost of Ethanol Metabolism. Clin. Res., 21:719, 1973.

20. Porta, E.A., Koch, O.R., Hartroft, W.S.: Recent Advances in Molecular Pathology: A Review of the Effects of Alcohol on the Liver. Exper. and Mole. Path., 12:104-132, 1970.

21. Portman, O.W.: Nutritional Requirements of Non-Human Primates in "Feeding and Nutrition of Non-Human Primates". Edited by Harris, R.S.; Academic Press, 1970.

22. Rodrigo, C., Antezana, C., Baraona, E.: Fat and Nitrogen Balances in Rats with Alcohol-Induced Fatty Liver. (E.) J. Nutr., 101: 1307-1310, 1971.

23. Rubin, E., Lieber, C.S.: Fatty Liver, Alcoholic Hepatitis, and Cirrhosis Produced by Alcohol in Primates. New Eng. J. Med., 290:128-135, 1974.

24. Ruebner, B.H., Brayton, M.A., Freedland, R.A., Kanayama, R., Tsao, M.: Production of a Fatty Liver by Ethanol in Rhesus Monkeys. Lab. Invest., 27:71-75, 1972.

25. Shah, M.N., Clancy, B.A., Iber, F.L.: Comparison of Blood Clearance of Ethanol and Tolbutamide and the Activity of Hepatic Ethanol-Oxidizing and Drug-Metabolizing Enzymes in Chronic Alcoholic Subjects. Amer. J. of Clinical Nutrition. Vol. 25 135-139, 1972.

26. Singlevich, T.E., Barboriak, J.J.: Ethanol and Induction of Microsomal Drug-Metabolizing Enzymes in the Rat. Toxicol. Appl. Pharmacol. 20(3):284-290, 1971.

27. Thurman, R.G., Induction of Hepatic Microsomal Reduced Nicotinamide Adenine Dinucleotide Phosphate-Dependent Production of Hydrogen Peroxide by Chronic Prior Treatment with Ethanol. Mol. Pharmacol. 9(5):670-675, 1973.

28. Vatsis, R.P.,Schulman, M.P.: Absence of Ethanol Metabolism in Acatalatic Hepatic Microsomes That Oxidize Drugs. Biochemical Biophysical Research Communications, 52:588-594, 1973.

29. Venables, J.H., Coggeshall, R.: A simplified Lead Citrate Stain for Use in Electron Microscopy. J. Cell Biol., 25:407, 1965.

30. Watkins, W.D.: Studies on the Toxicologic Role of Mammalian Alcohol Dedydrogenase. Dissertation Abstracts International. 32:1744-B, 1971.

RESPIRATORY DEPRESSANT EFFECTS OF ETHANOL: MEDIATION BY SEROTONIN

Alfred A. Smith, Charles Engelsher and Marsha Crofford

Departments of Psychiatry and Pharmacology

New York Medical College, N. Y. U.S.A. *

The present study demonstrates that ethanol and related alcohols possess a unique pharmacological property which distinguishes these compounds from other narcotics such as the barbiturates and opioids. Respiratory depression was used as the major pharmacological parameter since the changes induced by opioids, alcohols or barbiturates on pH and pCO_2 of mouse blood can be readily compared. Our results indicate that while adrenergic agonists can diminish respiratory depression induced by ethanol, a serotonergic mechanism appears to be the major mediator for this depressant effect. Serotonergic mediation of respiratory depression seems restricted to a series of homologous alcohols. Opioids and barbiturates do not display this requirement.

METHODS

Male or female Swiss-Webster mice were used in all studies. Ethanol (95%) was diluted to a 20% concentration (W/V) and injected intraperitoneally (i.p.) 30 min before obtaining a blood sample. To avoid clotting of the blood, heparin (500 u) was injected subcutaneously about 5 min before obtaining a blood sample. A superficial incision was made across the lower third of the tail. Free-flowing capillary blood was then drawn (with little or no exposure to air) into a 100 mm heparinized tube. The blood sample was then transferred to the blood-gas analyzer (BMS-3, Radiometer) and the pH and pCO_2 were determined. Diacetylmorphine, dl-methadone hydrochloride, and the barbiturates, sodium pentobarbital and phenobarbital, were dissolved in distilled

*Supported by Grants AA-00257 and DA-00020

water and also administered i.p. The pO_2 values ranged between 75 mm and 100 mm Hg and bore no relation to the moderate dosages of the administered narcotic. The high pO_2 levels suggest that mixed venous and arterial bloods were obtained. The data are presented as means ± S.E. Differences between means were calculated by Student's "t" test. At least 10 mice were tested for each treatment. Controls were done for each treatment series.

Effects of catecholamine depletors on responses to ethanol, methadone or pentobarbital.

The effects of ethanol and other narcotics on pH and pCO_2 of capillary bloods of mice are shown in Table 1. Dosages of 2-5 g/kg produce log-dose related increments in pCO_2 with concomitant decreases in pH. The injections of dl-methadone in dosages of 5 mg/kg, 10 mg/kg and 15 mg/kg, also produce similar log-dose related changes in pCO_2 and pH. Sodium pentobarbital injection elicits a very sharp linear dose response curve within the narrow limits of 30 mg/kg and 40 mg/kg.

To determine the effect of catecholamine depletion, a depletor of noradrenergic neurons, 6-hydroxydopamine (6-OHDA) was injected intracerebrally 24 hours before administration of either ethanol, methadone or pentobarbital. No significant change in respiratory parameters was noted for ethanol or for pentobarbital whereas a significant increase in pCO_2 occurred in mice tested with dl-methadone. This effect by 6-OHDA on response to dl-methadone is surprising. Schneider (1956) showed that reserpine treatment blocked the analgesic response to morphine and the respiratory activity of morphine might likewise be inhibited. As shown in Table 1, the opposite effect was obtained.

Alphamethyltyrosine methyl ester given in divided dosage 4 h and 2 h before the narcotic drugs was subsequently tested. The drug failed to alter the increase in pCO_2 or the decrease in pH produced by any of the drugs. Evidently catecholamine activity plays no important role in modulating respiratory depression induced by ethanol, methadone or pentobarbital.

Effects of Serotonergic Depletors. Reserpine, 2 mg/kg which depletes catecholamines as well as serotonin, was injected i.p. 4 h before administration of alcohol. The short time interval was selected to avoid later electrolyte disturbances which produce abnormal pH and pCO_2 values. The interval is however long enough to allow serotonin and catecholamine release without repletion of stores. Mice treated in this manner failed to show a rise in pCO_2 when injected with a large dosage (3 g/kg) of ethanol (Table 2). The difference in response compared with control is highly significant. The fall in blood pH was not blocked and is probably secondary to the profound metabolic effect of ethanol. Reserpine

Table 1

Drug	Dosage	Treatment	pH	pCO_2
-	-	-	7.400 ± .005	25.9 ± .49 (65)
Ethanol	2 g/kg	-	7.346 ± .006	28.6 ± .97 (10)
Ethanol	3 g/kg	-	7.337 ± .012	32.7 ± .60 (10)*
Ethanol	4 g/kg	-	7.209 ± .008	36.0 ± 1.2 (17)*
Ethanol	5 g/kg	-	7.233 ± .020	37.6 ± 1.9 (9)*
Ethanol	3 g/kg	6-OHDA,30 µg i.c. [a]	7.303 ± .008	34.0 ± .92 (11)
Methadone	10 mg/kg	6-OHDA,30 µg i.c. [a]	7.310 ± .005	35.0 ± .53 (10)
Pento-barbital	35 mg/kg	-	7.398 ± .027	34.0 ± 1.5 (9)
Pento-barbital	35 mg/kg	6-OHDA, 30 µg i.c.	7.324 ± .024	34.4 ± 1.5 (6)
Ethanol	3 g/kg	α-m-t. 100 mg/kg [b]	7.232 ± .020	34.5 ± .95 (10)
Methadone	10 mg/kg	α-m-t. 100 mg/kg [b]	7.338 ± .014	33.9 ± .76 (10)
Pento-barbital	35 mg/kg	α-m-t. 100 mg/kg [b]		

* $p < .01$

treatment did not prevent the respiratory depression produced by dl-methadone or pentobarbital.

In order to implicate serotonin more specifically, the inhibitor of tryptophan hydroxylase, p-chlorphenylalanine methyl ester (PCPA) was used. In the experiments listed below, the drug was injected i.p. 4 h and 24 h before testing blood pH and pCO_2. It can be seen (Table 2) that PCPA treatment was as effective as reserpine in blocking the rise in pCO_2 normally induced by ethanol but like reserpine failed to correct the metabolic acidosis. The drug was also ineffective in reversing the respiratory depression induced by either dl-methadone or pentobarbital.

Table 2. Effect of reserpine, 2 mg/kg or p-chlorphenylalanine, 300 mg/kg on the respiratory depression induced by ethanol 3 g/kg, methadone 10 mg/kg, or sodium pentobarbital 40 mg/kg. Figure in parenthesis indicates number of animals. Values given as mean ± S.E.M.

Drug	Reserpine	p-CPA	pH	pCO_2 (mm Hg)
-	-	-	7.400 ± .005	25.9 ± .34 (67)
Ethanol 3 g/kg	-	-	7.277 ± .004	33.6 ± .39 (70)
Ethanol 3 g/kg	2 mg/kg	-	7.255 ± .008	25.3 ± ,68 (27) *
Ethanol 3 g/kg	-	300 mg/kg	7.299 ± .017	25.0 ± .84 (10) *
-	-	300 mg/kg	7.396 ± .007	25.6 ± .99 (10)
Methadone 10 mg/kg	-	-	7.286 ± .010	38.1 ± 1.05 (16)
Methadone 10 mg/kg	2 mg/kg	-	7.249 ± .016	39.1 ± 1.60 (20)
Methadone 10 mg/kg	-	300 mg/kg	7.304 ± .016	37.1 ± 1.05 (10)
Pentobarbital 40 mg/kg	-	-	7.283 ± .014	33.6 ± 2.43 (9)
Pentobarbital 40 mg/kg	2 mg/kg	-	7.212 ± .020	37.1 ± 1.77 (10)
Pentobarbital 40 mg/kg	-	300 mg/kg	7.242 ± .017	36.0 ± 2.31 (10)

* $p < .01$ as compared to untreated experimental group.

PCPA blocked ethanol-induced increases in pCO_2 (Fig. 1) throughout the dose ranges of 2 g/kg to 4 g/kg. The apparent ability of PCPA to almost entirely block respiratory depression even at an anesthetic dosage (4 g/kg) of ethanol suggests that serotonin actually mediates this respiratory response. Sleeping time, was however, found unaffected by prior treatment with PCPA. Control sleeping time of mice given ethanol in a dosage of 4.5 g/kg i.p. was 29.6 ± 3.9 (29) whereas mean sleeping time for PCPA-treated mice was 24.0 ± 2.6 (36). The differences are insignificant. Evidently respiratory depression and narcosis are dissociable phenomena.

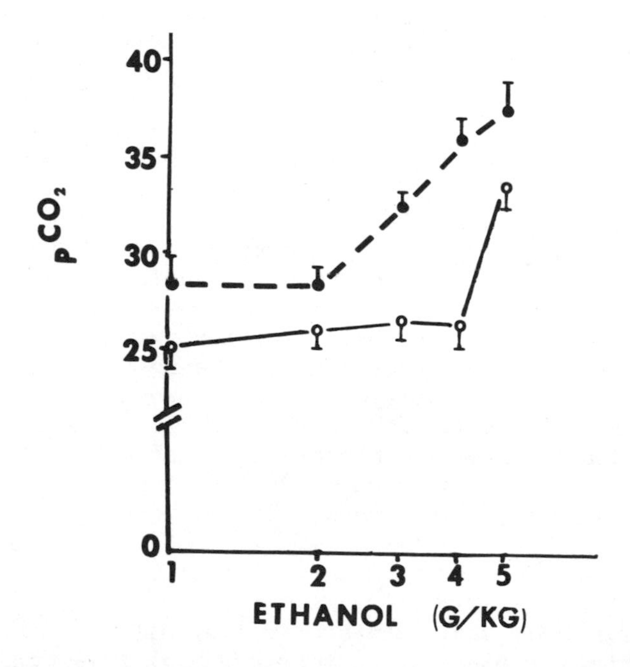

Figure 1. Effect of PCPA treatment on the respiratory depressive response to graded dosages of ethanol. Dotted line represents control.

Effect of PCPA on the Respiratory Depression Induced by an Homologous Series of Alcohols.

Do other alcohols or related compounds also require 5-HT for their respiratory depressant action? A study was undertaken using methanol, ethanol, isopropranol, n-propanol and chloral hydrate. The last is rapidly reduced *in vivo* to trichlorethanol. Dosages of the above compounds were selected to produce respiratory depression comparable to ethanol at a dosage of 3 g/kg. As shown in Table 3, the dosage required for this effect varied inversely with the chain length of the alcohol with the exception of chloral hydrate. After treatment with PCPA, the respiratory depressant effect of each of the alcohols with the exception of chloral hydrate was significantly attenuated. The reason for the exceptional behavior of chloral hydrate is not clear.

Two findings seem remarkable: First, isopropranol is not metabolized initially to an aldehyde but to a ketone suggesting that aldehydes are not involved in serotonin-mediated respiratory depression, secondly, the response to chloral hydrate on aldehyde quickly reduced to an alcohol is unaffected by PCPA treatment. Furthermore, the ability of PCPA to attenuate the alcohol effect seems to lessen with increasing chain length. Perhaps the 2-carbon length fits some special structural requirement which activates serotonergic systems.

Reversal by serotonin or 5-HTP: Reserpine or the inhibitor PCPA blocks the rise in pCO_2 induced by ethanol. The administration of serotonin or its precursor 5-hydroxytryptophan to animals treated with a serotonin depletor thus may restore the ability of ethanol to increase pCO_2. The results of these experiments are listed in Table 4. It can be seen that the intracerebral administration of 5-HT to mice 15 min. after ethanol injection reversed the blockade. The effect was enhanced with increased amounts of serotonin. By itself serotonin had no effect on respiratory parameters, with or without reserpine.

The precursor 5-HTP was administered i.p. one hour before ethanol injection, in place of an intracerebral injection of serotonin. This compound was also effective in overcoming the block in serotonin synthesis caused by PCPA. For reasons that are presently unclear 5-HTP produced a significant respiratory acidosis which makes interpretation of these data somewhat uncertain. Interestingly, 10 μg of serotonin injected i.c. did not enhance the respiratory depression elicited by ethanol.

Table 3. Effect of PCPA treatment on respiratory responses to a homologous series of alcohols and to chloral hydrate. Methanol was injected i.p. 15 min before obtaining capillary blood. All other drugs were administered 30 min prior to testing. The results are listed as means ± S.E. The figure in parentheses indicates number of mice used in each test.

Drug	Dosage	pCPA (300 mg/kg)	pH	pCO_2
Methanol	7 g/kg	-	7.304 ± .008	34.1 ± .56(28)
Methanol	7 g/kg	+	7.341 ± .008**	28.5 ± .69(30)**
Ethanol	3 g/kg	-	7.337 ± .01	32.7 ± .60(10)
Ethanol	3 g/kg	+	7.305 ± .02	26.7 ± .78(10)**
Isopropanol	2 g/kg	-	7.272 ± .015	36.5 ± 1.5(10)
Isopropanol	2 g/kg	+	7.289 ± .018	30.1 ± .90(9)**
n-Propanol	1 g/kg	-	7.278 ± .01	32.9 ± 1.0(10)
n-Propanol	1 g/kg	+	7.290 ± .01	30.0 ± .02(10)*
Chloral hydrate	450 mg/kg	-	7.280 ± .009	34.8 ± .53(22)
Chloral hydrate	450 mg/kg	+	7.248 ± .009	34.9 ± .92(20)

* p .05 as compared to group not treated with PCPA

** p .01

Table 4. Restoration of ethanol-induced respiratory depression by treatment with serotonin (5-HT) or dl-5-hydroxytryptophan methyl ester (5-HTP). Mice were pretreated with reserpine, 2 mg/kg 4 h before ethanol, or PCPA, 2 h before ethanol. Serotonin creatinine sulfate was injected intracerebrally 15 min after injection of ethanol. 5-HTP, 100 mg/kg was injected intraperitoneally 1 h before ethanol injection.

3 g/kg Ethanol	Reserpine	pCPA	5-HTP	5-HT	pH	pCO_2 (mm Hg)
-	-	-	-	-	7.400 ± .005	25.9 ± .34 (67)
+	-	-	-	-	7.196 ± .014	33.2 ± .48 (10)
+	+	-	-	-	7.255 ± .008	25.3 ± .68 (27)
+	+	-	-	10 μg	7.265 ± .016	33.3 ± .62 (10)*
+	-	+	-	-	7.299 ± .017	25.0 ± .84 (10)
+	-	+	+	-	7.191 ± .015	32.6 ± .67 (10)*
-	-	-	+	-	7.311 ± .010*	28.3 ± .75 (10)*
-	-	+	-	-	7.373 ± .010	27.2 ± .51 (11)

*$p < .01$ as compared with reserpine or PCPA treated groups.

Responses to Intracerebral Injections of Norepinephrine.

Fifteen minutes after injection of ethanol or other narcotic drugs, 250 ng of l-norepinephrine (N.E.) was administered intracerebrally. By itself, l-norepinephrine did not cause any change in blood gas parameters. However, the response to ethanol was attenuated significantly whereas the converse was true for methadone. No changes were observed in pH or pCO_2 when N.E. was injected after administration of sodium pentobarbital. The ability of N.E. to partially reverse the respiratory depression produced by ethanol suggests that this response might be mediated by an α-receptor. A β-receptor agonist, such as dl-isoproterenol (ISO) might be expected to exaggerate the depressant effect of ethanol. Intracerebral injection of ISO in a dose of 2 μg significantly decreased pCO_2 to 29.6 ± .94 from the control value of 34.0 ± .92.

This unexpected response was clarified by a subsequent experiment. Mice were treated with 6-hydroxytyramine in a dose of 30 µg. The injection was made intracerebrally. When ethanol was injected 24 h later and then the mice given ISO in the same dose as before dopamine, the average value of pCO_2 was found to be 32.9 ± .80, not significantly different. The ISO effect occurs only in mice with intact catecholamines stores and suggests that ISO causes catecholamine release. Some years ago we found that ISO potentiated the ability of opioids to produced transient cataracts in mouse lenses. The effect was virtually abolished in mice pretreated by injection with reserpine, evidence again for an indirect, tyramine-like effect for ISO (Smith, 1970).

Interestingly, the β-adrenergic antagonists such as sotalol or propranolol possess the isopropylethanolamine side chain of ISO. The reported (Hayashida and Smith, 1970) beneficial effect by sotalol on respiratory depression induced by ethanol may have been produced by some tyramine-like action of these drugs.

Table 5. Effect of catecholamines injected intracerebrally(i.c.) on the responses to ethanol methadone or pentobarbital. The i.c. injections were made 15. min before sampling blood. The 250 ng dose of l-norepinephrine was injected into the right hemisphere in a volume of 10 µl. Values given as means ± S.E.M.

Drug	pH	pCO_2
-	7.400 ± .005	25.9 ± .43 (65)
Ethanol 3 g/kg	7.277 ± .010	34.6 ± .88 (12)
Noreprinephrine (N.E.)	7.369 ± .010	26.1 ± 1.4 (10)
Ethanol 3 g/kg + N.E.	7.321 ± .009	29.5 ± .74 (10)*
Methadone 10 mg/kg	7.302 ± .015	36.6 ± 1.7 (10)
Methadone 10 mg/kg + N.E.	7.307 ± .020	41.0 ± 1.5 (10)*
Pentobarbital 40 mg/kg	7.285 ± .016	38.9 ± 2.4 (10)
Pentobarbital + N.E.	7.271 ± .023	38.4 ± 2.9 (10)

* p .05

DISCUSSION

Experiments with the catecholamine depletors, α-methyltyrosine and 6-hydroxydopamine gave no evidence that respiratory depressant effects of 3 representative narcotic drugs were dependent on the activity of catecholamine systems. Studies with disulfiram (unpublished) were attempted in order to evaluate the role of dopamine in modulating respiratory depressant activity of ethanol. Disulfiram treatment unfortunately caused considerable toxicity making any interpretation of the results impossible. However, dopamine, injected intracerebrally even in a dose of 50 µg failed to restore the respiratory depressant effects in mice treated with reserpine.

It would appear that catecholamines do not significantly modulate ethanol's respiratory effect except when given in pharmacological doses. On the other hand, treatment with drugs that either prevent storage of serotonin or block hydroxylation of tryptophan selectively prevent the respiratory depressant response to ethanol and other alcohols. Loss of serotonergic activity did not affect pentobarbital or methadone. Respiratory depression followed administration of pharmacologically adequate dosages of these drugs despite treatment of mice with blockers of serotonin. Additional studies were done with diacetylmorphine and with phenobarbital. Both compounds were given in dosages which produced respiratory depression. In neither instance did pretreatment with PCPA influence the rise in pCO_2 caused by these narcotics.

Thus serotonergic activity in as yet an unknown site in the brain appears to modulate alcohol's respiratory depressant effect. That this effect may be more than modulation is suggested by the results illustrated in Figure 1. Almost total blockade by PCPA of the effect of ethanol was seen at dosages from 2 g/kg to 4 g/kg. The last dosage produces anesthesia which was not blocked, indicating dissociation between anesthetic and respiratory depressant actions of ethanol.

The unusual serotonin requirement that ethanol demands for the mediation of respiratory depression parallels the finding (Myers and Veale, 1968) that PCPA treatment blocks the preference for drinking ethanol solutions by animals. More recent data (Myers and Martin, 1973; Schecter, 1973), using behavioral testing in rats support a significant role for serotonin in adaptation to the drinking of ethanol solutions or the loss of preference for such solutions. Our data now indicates a pharmacological role for serotonin in mediation of the respiratory depressant effect of ethanol.

References

Hayashida, K. and A. Smith, Reversal by sotalol of the respiratory depression induced in mice by ethanol. J. Pharm. Pharmacol Vol 23; 718-719, 1971.

Myers, R.D. and G.E. Martin, The role of serotonin in the ethanol preference of animals. Ann. N.Y. Acad. Sci. Vol. 215; 135-144, 1973.

Myers, R.D. and W.L. Veale, Alcohol preference in the rat: Reduction following depletion of brain serotonin. Science, Vol. 160; 1469-1471, 1968.

Schecter, M.D. Ethanol as a discriminative cue: Reduction following depletion of brain serotonin. Europ. J. Pharmacol. Vol. 24: 278-281, 1973.

Schneider, J.A. Reserpine antagonism of morphine analgesia in mice. Vol 87; 614-615, 1954.

Smith, A., Hayashida, K. and Y. Kim. Inhibition by propranolol of ethanol-induced narcosis. J. Pharm. Pharmacol. Vol. 22; 644-645, 1970.

CORRELATION OF BRAIN AMINE CHANGES WITH ETHANOL-INDUCED SLEEP-TIME IN MICE

C.K. Erickson* and J.A. Matchett

Department of Pharmacology and Toxicology
School of Pharmacy
University of Kansas
Lawrence, Kansas 66045

Earlier studies in our laboratories (Erickson and Graham, 1973; Graham and Erickson, 1974) have pointed out our interest in correlating central amine changes with certain aspects of ethanol intoxication. In the earlier studies we saw changes in acetylcholine (ACh) release from the cerebral cortex and reticular formation after various intravenous doses of ethanol in rabbits. Available literature concerning the effects of ethanol on catechol- and indolealkylamines is voluminous, but incomplete and sometimes conflicting. For example, ethanol has been reported to deplete serotonin (5HT) and norepinephrine (NE) in the brain stem of rabbits (Gursey and Olson, 1960), and to raise brain 5HT of rats (Bonnycastle *et al.*, 1962). Many workers, such as Häggendal and Lindqvist (1961), however, have seen no effect of ethanol on cerebral NE, DA or 5HT in rats or rabbits. Wallgren and Barry (1970) have concluded that, in general, ethanol appears to change brain levels of catecholamines and 5HT little, if at all. With regard to the effects of ethanol on the central amine ACh, Kalant *et al.* (1967) have seen decreased release of ACh in rat cortical slices *in vitro*, while we have seen a short-lived increase in total brain ACh after ethanol in rats (Erickson and Graham, 1973).

It thus appears that the reports of ethanol's actions on major central neurotransmitters to date are a blend of differing experimental designs, and the studies include differences in species used

---

*The acetylcholine and choline analyses reported in this paper were carried out by C.K. Erickson while he was a visiting researcher in the Toxicology Department at Karolinska Institutet, Stockholm, Sweden, under the direction of Professor Bo Holmstedt.

and differences in parts of the brain analyzed. In addition, the studies vary according to whether they measure free, bound, or total amines. (In fact, most studies seem to report levels of total amines.) To date, no one has performed a correlative study of ethanol's effects on NE, DA, 5HT, and ACh, and attempted to alter these observed effects with drugs which affect the levels of specific central neurotransmitters. Such studies, we feel, are vitally necessary before the hypothesis of an intoxicant effect by ethanol through central neurotransmitter action can be supported or rejected. While the present study is not complete enough to fulfill such requirements, it is a necessary beginning step in this direction.

Female albino mice, HA/ICR strain, weighing 20-25 grams were used in the present experiments. An hypnotic dose of ethanol, 4.5 gm/kg given intraperitoneally, was chosen as a dose which could be expected to alter central amines, while producing a behavioral end-point (length of sleep-time) that could be shortened or lengthened by centrally-active drugs which might affect ethanol's activity.

The first experiment was a time-course study of the effects of ethanol on NE, DA, 5HT, and ACh. NE and DA were measured by a spectrofluorometric method utilizing the extraction procedure of Bertler _et al._ (1958) and the separation and analysis procedures of Laverty and Taylor (1968) and Welch and Welch (1969). Determination of 5HT in these same mouse brains in some experiments was made possible by combining the method of Wiegand and Scherfling (1962) with the catecholamine analysis method, along the lines of Atack and Magnusson (1970). Measurement of ACh was by ion-pair extraction and gas-phase analysis through the gas chromatographic-mass spectrometric method of Karlén _et al._ (1974), using deuterated internal standards. Since choline is also readily determined by this method, it was measured as a suggestion of ACh turnover. In all experiments, the mice were killed by decapitation and the brains were rapidly removed, weighed, and homogenized in cold buffer and perchloric acid. In some mice, blood was collected from the severed head region into heparinized tubes and used for ethanol determination using the gas chromatographic method of Le Blanc (1968). In these experiments, n-butanol was used as an internal standard.

Figure 1 shows a composite of results obtained with these various methods. (Individual results will be discussed later.) It can be seen from this graph that the whole brain levels of DA, 5HT, ACh and choline all increased over control values, while the level of NE decreased. The greatest increase over saline-treated control was seen in the case of 5HT, which also peaked the earliest of all the amines. Seventy-five minutes after ethanol administration, DA

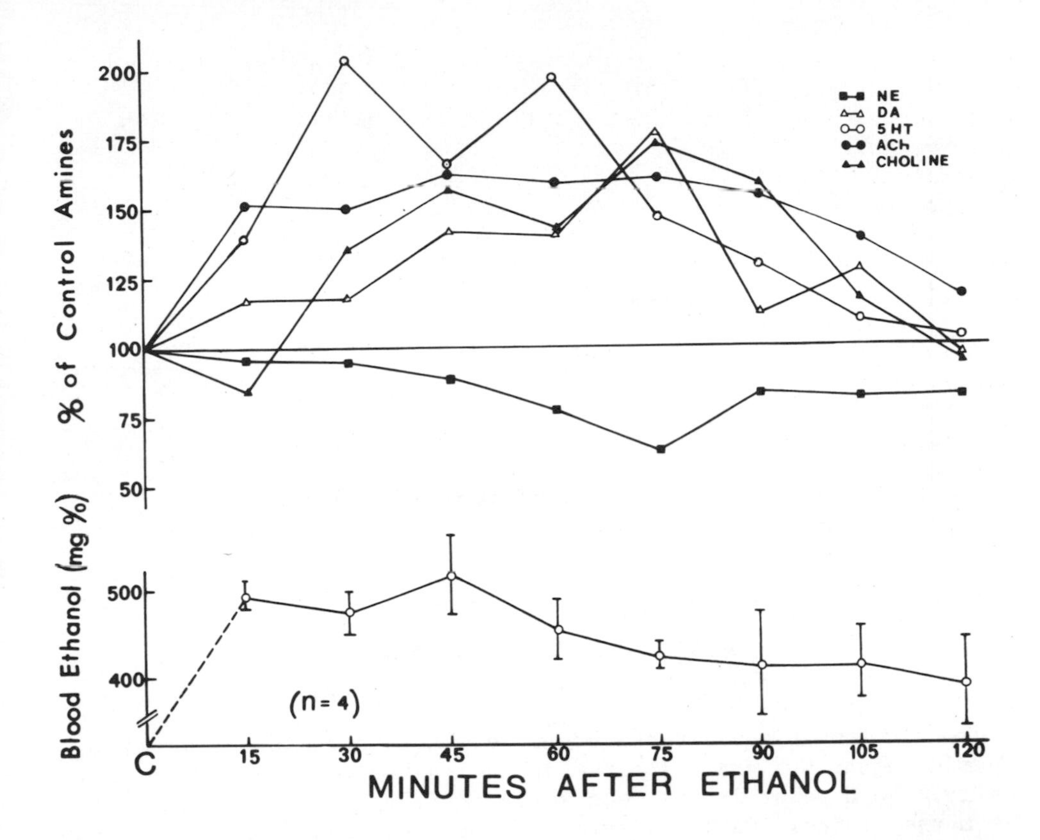

Figure 1. Summary of brain amine changes over time after 4.5 gram/kilogram of ethanol given to mice intraperitoneally. Number of subjects for the blood ethanol curve is 4 for each point, and the number of subjects for amine determinations is 3 or 4. Further detail on the amine curves is presented in later figures.

and choline reached their peak values, while NE reached its lowest value. Blood ethanol values (with standard errors indicated) correlate well with the literature (Wallgren and Barry, 1970), rising quickly to around 500 mg% and very slowly decreasing toward zero, with the curve remaining significantly elevated after 2 hours. Blood ethanol changes correlate most closely with 5HT changes in the brain, but the significance of this observation is unclear.

Looking at Figure 2, we see a comparison between NE and DA, with numbers of subjects, standard errors, and statistical significance included. (Darkened symbols represent points that are significantly different from control, at least at the 0.05 level,

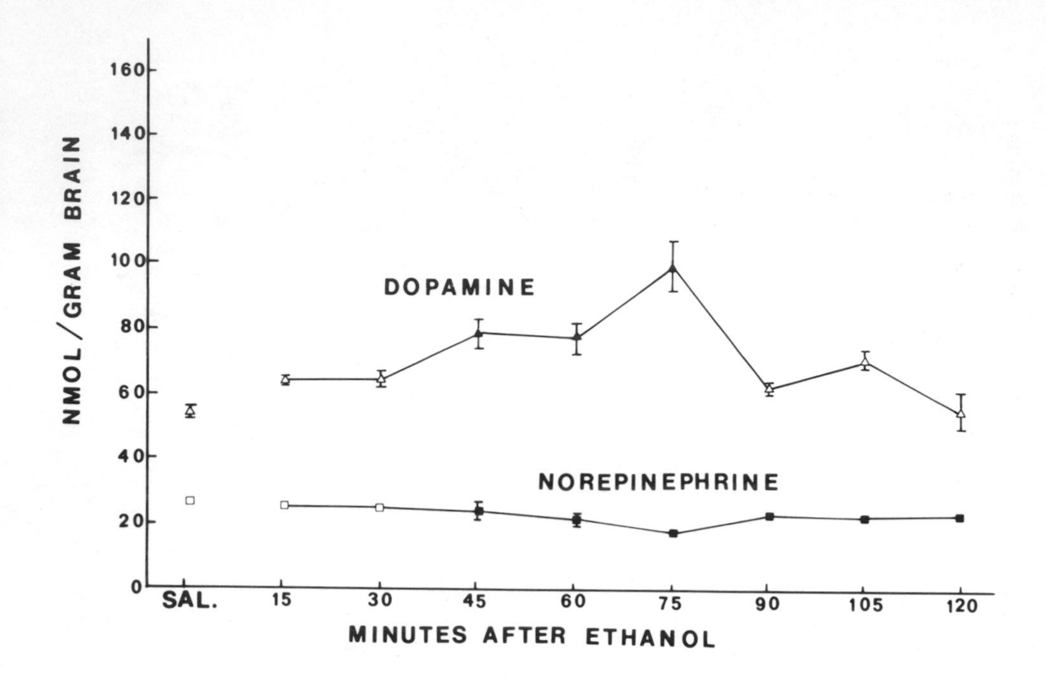

Figure 2. Comparison between brain levels of norepinephrine and dopamine after 4.5 gram/kilogram ethanol given to mice intraperitoneally. Each point is the mean value (± S.E.) of the analyses of 3-4 mouse brains. Darkened symbols represent significant differences from control at least at the .05 level.

with a Student-Newman-Keuls *a posteriori* test after a one-way analysis of variance had shown significant differences for each amine experiment.) The slight but significant decrement in whole brain NE is similar to that seen by Carlsson *et al.* in 1973 (also in mice), but the large increase in whole brain dopamine levels was not seen by these workers. Other studies on the effect of ethanol on DA are similarly negative (e.g. Corrodi *et al.*, 1966). As a cause of the decrease in NE, Israel *et al.* (1973) have suggested that large doses of ethanol may inhibit uptake but not release of the amine.

In Figure 3, we see a comparison between NE and 5HT. As noted earlier, 5HT levels rose higher in relation to control values than any other amine we studied, and the rise occurred sooner and correlated most closely with ethanol blood levels. Palaić *et al.* (1971) have demonstrated an increased whole brain level of 5HT after small acute doses in rats, but a decreased level with chronic doses. Kuriyama *et al.* (1971) have shown an exactly opposite effect in mice

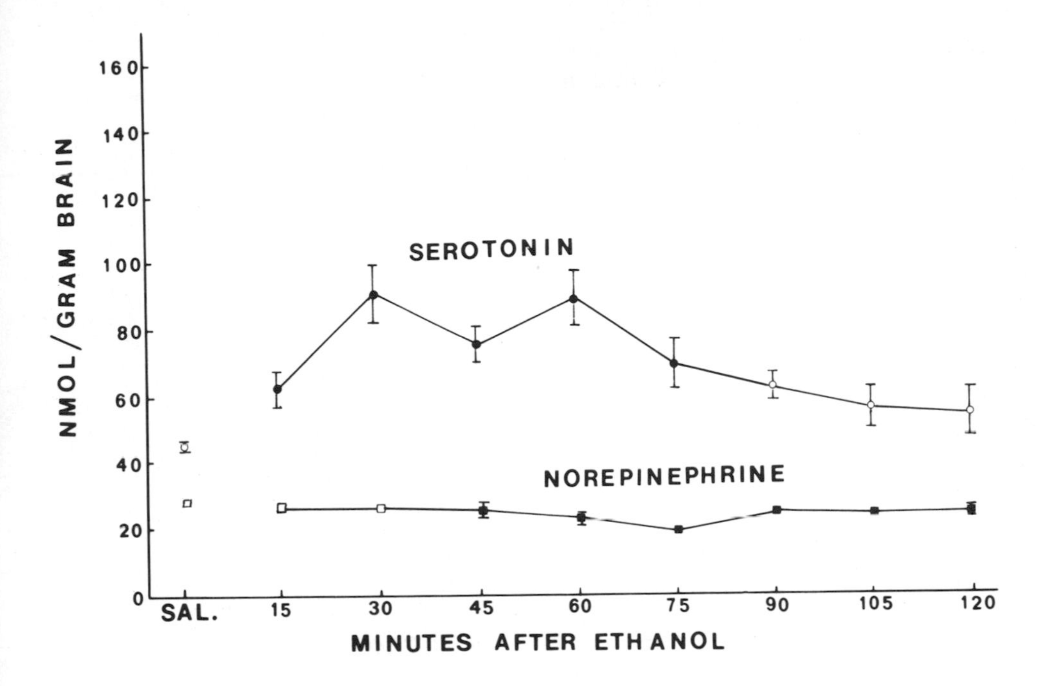

Figure 3. Comparison between brain levels of norepinephrine and serotonin after 4.5 gram/kilogram ethanol given to mice intraperitoneally. Each point is the mean value (± S.E.) of the analyses of 3-4 mouse brains. Darkened symbols represent significant differences from control at least at the .05 level.

with acute doses similar in size to those in our study. Thus it seems that unrecognized procedural or strain differences must play a role in the different results seen by different workers.

The effects of ethanol on ACh are, according to the literature and the present studies, somewhat less controversial. Figure 4 shows that choline increases significantly in the whole mouse brain following the hypnotic dose of ethanol, while ACh increases also, but not significantly. Such results are consistent with earlier studies in our laboratories which have shown an increase in whole brain ACh in rats after a similar dose of ethanol given orally, and we have speculated that this rise is apparently due to decreased release of ACh into the synapse, resulting in more bound ACh in the presynaptic nerve endings (Erickson and Graham, 1973). The increase in whole brain choline after ethanol is unexpected because increased choline levels suggest increased ACh turnover in the brain. However, we and other workers have consistently shown a decreased release of ACh in parts of the cortex (Erickson and Graham, 1973; Phillis and

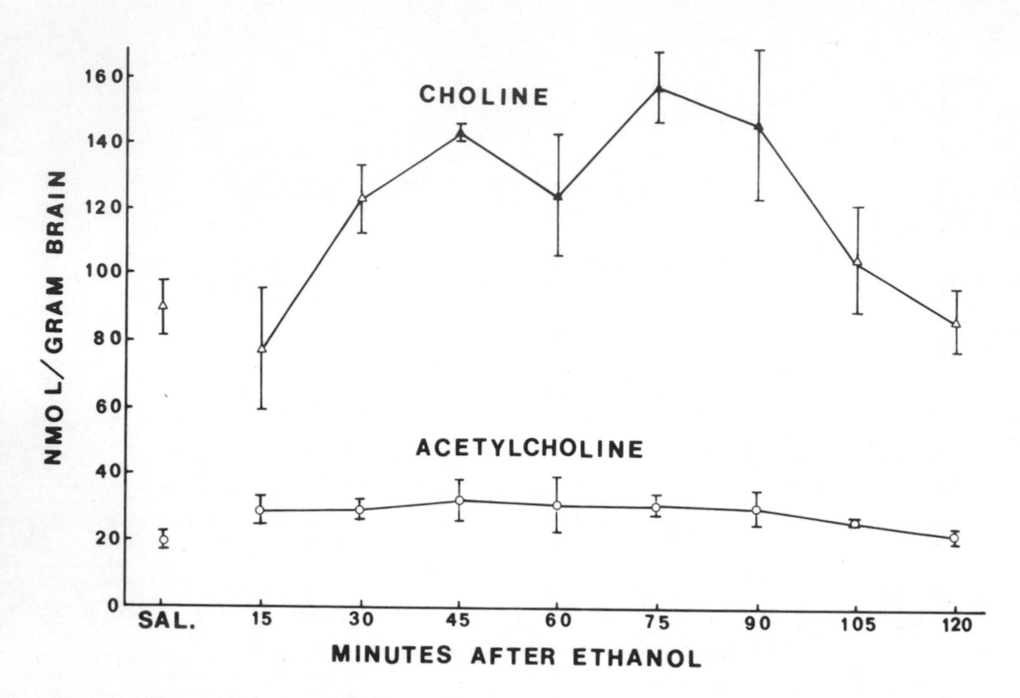

Figure 4. Comparison between brain levels of acetylcholine and choline after 4.5 gram/kilogram ethanol given to mice intraperitoneally. Each point is the mean value (± S.E.) of the analyses of 3-4 mouse brains. Darkened symbols represent significant differences from control at least at the .05 level.

Jhamandas, 1971). There are many theoretical possibilities for such an increase in whole brain choline, and more studies must be done to answer this question. Specifically, measurement of the effect of ethanol on the enzymes choline acetyltransferase and acetylcholinesterase would be indicated. In this regard, other drugs (e.g., physostigmine and oxotremorine; Consolo _et al_., 1972) also increases whole brain choline while causing an increase in whole brain ACh.

It is possible that further insight into the effects of ethanol on central neurotransmitters might be gained by administering drugs which selectively alter one or more neurotransmitters, before treating the mice with an hypnotic dose of ethanol. In our second experiment, we have used p-chlorophenylalanine (pCPA), a selective depletor of 5HT (Koe and Weissman, 1966), alpha-methyl-para-tyrosine (AMT), which inhibits synthesis of NE and DA, but not 5HT (Dominic and Moore, 1969), and d-amphetamine, which increases the whole brain levels of both catecholamines plus 5HT (Smith, 1965). In Figure 5A we see that 3 different doses of pCPA given once daily for 3 days

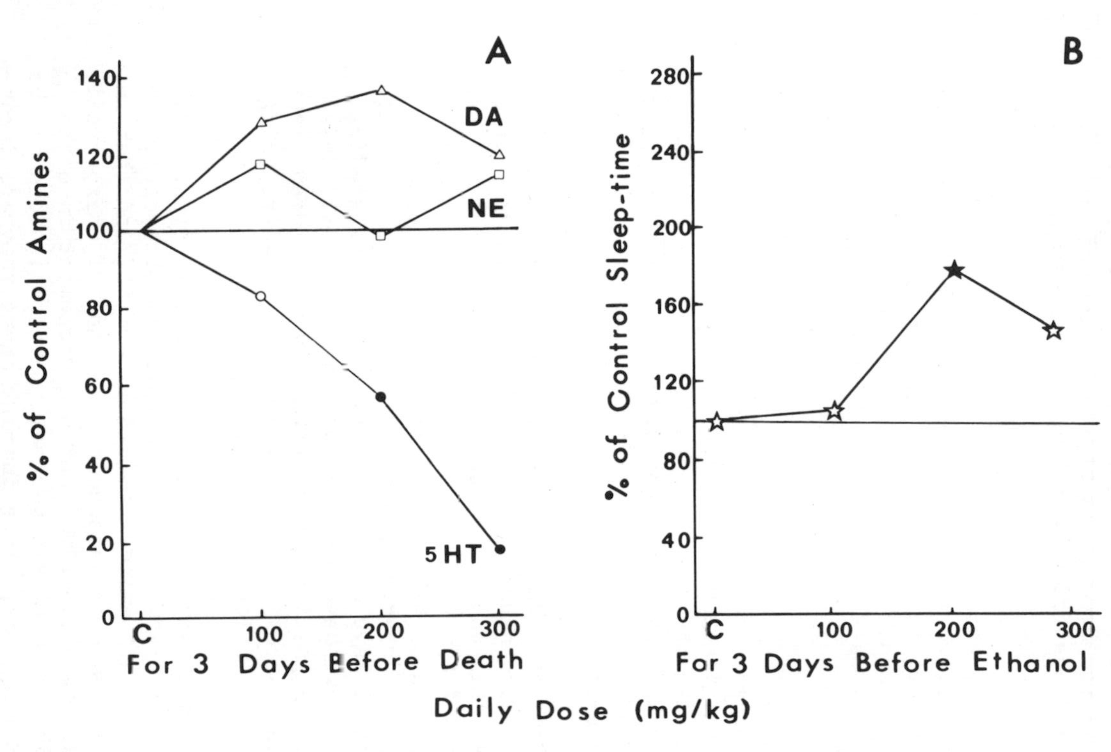

Figure 5. Effects of various doses of p-chlorophenylalanine on brain levels of norepinephrine (NE), dopamine (DA) and serotonin (5HT) in non-ethanol treated mice (A), and on sleep-time induced by a 4.5 gram/kilogram dose of ethanol given intraperitoneally in mice (B). Darkened symbols represent significant differences from control at least at the .05 level. Each point in A represents the mean value of the analyses of 5 mouse brains. Each point in B represents the mean sleep-time of 10 mice.

do indeed decrease whole brain 5HT, while leaving NE and DA intact. A prolongation of sleep time in mice receiving these same doses of pCPA before ethanol is shown in Figure 5B, with the 200 mg/kg dose producing a significant change from ethanol alone. (It should be emphasized that the amine changes shown in Figs. 5A, 6A, and 7A are in mice that have not been given ethanol. The effect of a combination of these drugs and ethanol on the amines has not yet been studied by us.)

Figure 6A shows the expected decrease in catecholamines produced by 3 doses of AMT, while 5HT is relatively unaffected. These 3 doses, even though they reduced catecholamines, prolonged sleep-time (Figure 6B) in a manner similar to the prolongation caused by pCPA. However, in this case, the prolongation was significantly dose-dependent.

Finally, the only drug we have used in this study which shortens ethanol-induced sleep-time is shown in Figure 7. d-Amphetamine in 3 doses significantly increased the levels of all 3 non-cholingeric amines, with the level of 5HT being increased the greatest by the 3 mg/kg dose of d-amphetamine. As seen in Figure 7B, the two highest doses of d-amphetamine significantly decreased ethanol-induced sleep-time.

The above studies are admittedly incomplete. Ostensibly missing are studies using other doses of ethanol; measurements of the effects of pCPA, AMT, and d-amphetamine on ACh; the administration of other drugs which are also known to preferentially affect central catechol- and indolealkylamine levels (e.g., pipradol, reserpine, and 6-hydroxydopamine); and the use of drugs which are known to affect central levels of ACh in sleep-altering studies. In this latter regard, we have previously shown that small doses of physostigmine will significantly shorten ethanol-induced sleep-time in mice, and that atropine will prevent this sleep-reducing effect of physostigmine (Erickson and Burnam, 1971).

All of the preceding studies, then, show that an hypnotic dose of 4.5 gm/kg of ethanol given intraperitoneally in mice causes an increase in all total brain amines studied except NE, and that in general, a further increase in catecholamines and 5HT unexpectedly shortens ethanol-induced sleep-time, while a decrement in either catecholamines or 5HT enhanced sleep-time. Few conclusions can be drawn from these observations, since in many cases they are inconsistent with the previously reported literature. It is apparent that more work needs to be done in standardizing experimental designs and in doing complete correlative studies on neurotransmitter release and turnover in each species and in specific areas of the brain where ethanol purportedly has its greatest activity, such as the cortex, reticular formation, and cerebellum. It is hoped that

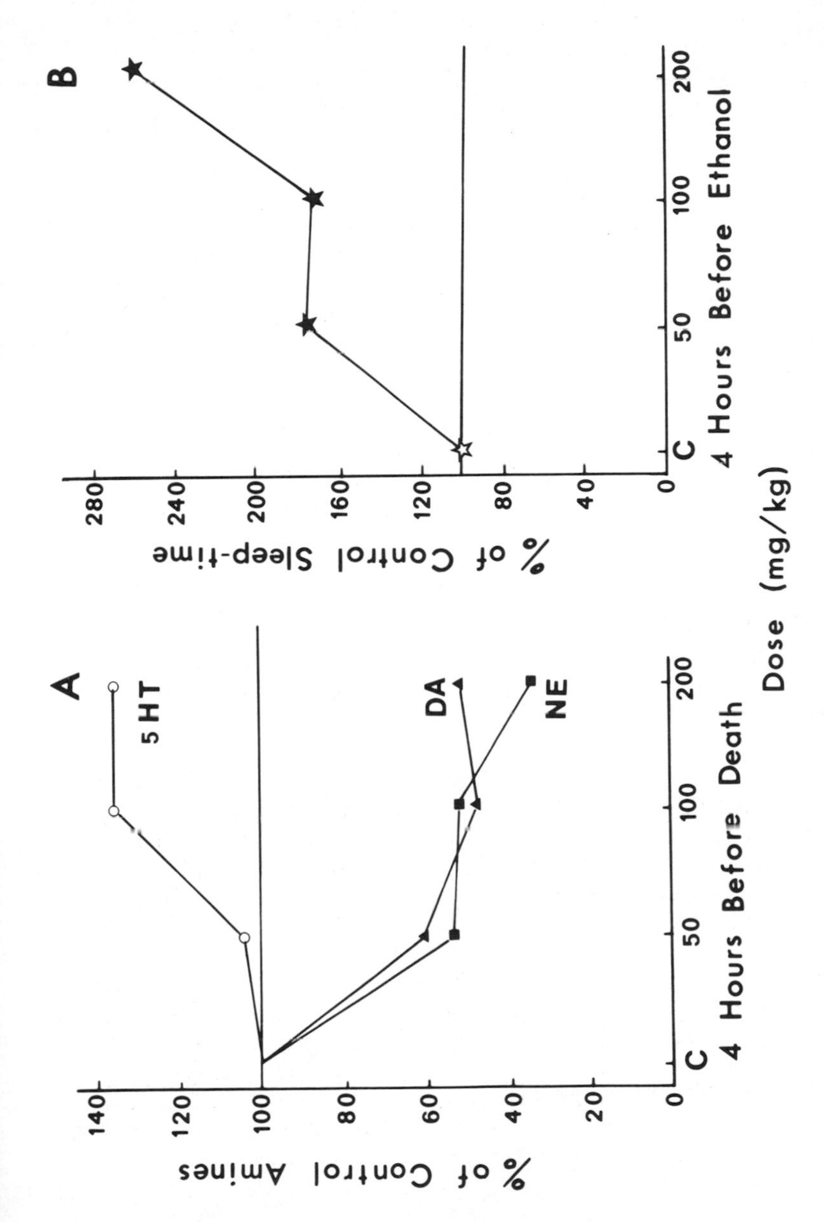

Figure 6. Effects of various doses of alpha-methyl-para-tyrosine on brain levels of norepinephrine (NE), dopamine (DA) and serotonin (5HT) in non-ethanol treated mice (A), and on sleep time induced by a 4.5 gram/kilogram dose of ethanol given intraperitoneally in mice (B). Symbol significance and point representation is the same as in Figure 5.

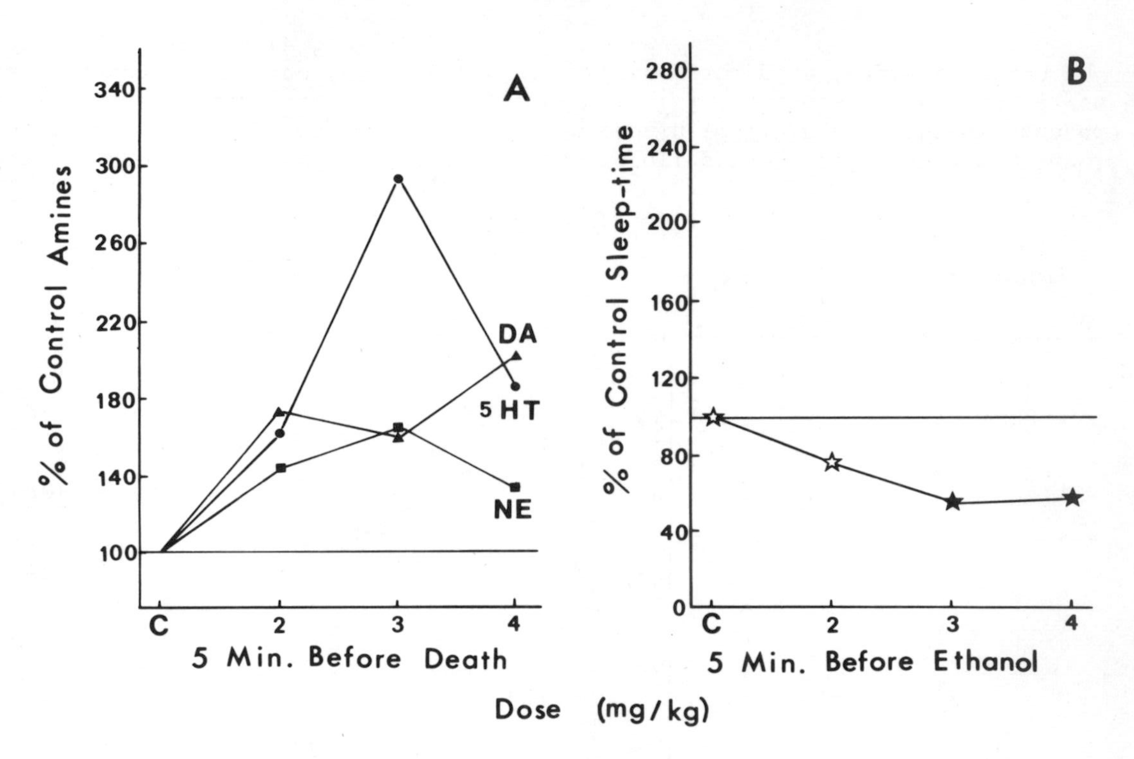

Figure 7. Effects of various doses of d-amphetamine on brain levels of norepinephrine (NE), dopamine (DA) and serotonin (5HT) in non-ethanol treated mice (A), and on sleep time induced by a 4.5 gram/kilogram dose of ethanol given intraperitoneally in mice (B). Symbol significance and point representation is the same as in Figure 5.

such future studies will reveal preferential effects of ethanol on single neurotransmitters and on nerve function in particular areas of the brain so that we can begin to more accurately correlate neurochemical changes with alcohol intoxication.

ACKNOWLEDGMENTS

We are indebted to the following sources for their support of this project: USPHS grant MH-14485 and grants from the Swedish Medical Research Council, B74-40X-199-10C, the Bank of Sweden Tercentenary Fund, 68/53, and the Wallenberg Foundation.

REFERENCES

Atack, C.V. and Magnusson, T. 1970. Individual elution of noradrenaline (together with adrenaline), dopamine, 5-hydroxytryptamine and histamine from a single, strong cation exchange column, by means of mineral acid-organic solvent mixtures. J. Pharm. Pharmacol. 22:625.

Bertler, A.; Carlsson, A.; and Rosengren, E. 1958. A method for the fluorimetric determination of adrenaline and noradrenaline in tissues. Acta Physiol. Scand. 44:273.

Bonnycastle, D.D.; Bonnycastle, M.F.; and Anderson, E.G. 1962. The effect of a number of central depressant drugs upon brain 5-hydroxytryptamine levels in the rat. J. Pharmacol. 135:17.

Carlsson, A.: Magnusson, T.; Svensson, T.H.; and Waldeck, B. 1973. Effect of ethanol on the metabolism of brain catecholamines. Psychopharmacologia 30:27.

Consolo, S.; Ladinsky, H.; Peri, G.; and Garattini, S. 1972. Effect of central stimulants and depressants on mouse brain acetylcholine and choline levels. Europ. J. Pharmacol. 18:251.

Corrodi, H.; Fuxe, K.; and Hökfelt, T. 1966. The effect of ethanol on the activity of central catecholamine neurones in rat brain. J. Pharm. Pharmacol. 18:821.

Dominic, J.A. and Moore, K.E. 1969. Acute effects of α-methyltyrosine on brain catecholamine levels and on spontaneous and amphetamine-stimulated motor activity in mice. Arch. Int. Pharmacodyn. 178:166.

Erickson, C.K. and Burnam, W.L. 1971. Cholinergic alteration of ethanol-induced sleep and death in mice. Agents and Actions 2:8.

Erickson, C.K. and Graham, D.T. 1973. Alteration of cortical and reticular acetylcholine release by ethanol in vivo. J. Pharmacol. Exp. Ther. 185:583.

Graham, D.T. and Erickson, C.K. 1974. Alteration of ethanol-induced CNS depression: Ineffectiveness of drugs that modify cholinergic transmission. Psychopharmacologia 34:173.

Gursey, D. and Olson, R.E. 1960. Depression of serotonin and norepinephrine levels in brain stem of rabbit by ethanol. Proc. Soc. Exptl. Biol. Med. 104:280.

Häggendal, J. and Lindqvist, M. 1961. Ineffectiveness of ethanol on noradrenaline, dopamine or 5-hydroxytryptamine levels in brain. Acta Pharmacol. Toxicol. 18:278.

Israel, Y.; Carmichael, F.J.; and Macdonald, J.A. 1973. Effects of ethanol on norepinephrine uptake and electrically stimulated release in brain tissue. Ann. N.Y. Acad. Sci. 215:38.

Kalant, H.: Israel, Y.; and Mahon, M.A. 1967. The effect of ethanol on acetylcholine synthesis, release, and degredation in brain. Can. J. Physiol. Pharmacol. 45:172.

Karlén, B.; Lundgren, G.; Nordgren, I.; and Holmstedt, B. 1974. Ion-pair extraction and gas-phase analysis of acetylcholine and choline, In: Choline and Acetylcholine: Handbook of Chemical Assay Methods, ed. I. Hanin, Raven Press, New York, 163.

Koe, B.K. and Weissman, A. 1966. p-Chlorophenylalanine: a specific depletor of brain serotonin. J. Pharmacol. Exp. Ther. 154:499.

Kuriyama, K.; Rauscher, G.E.; and Sze, P.Y. 1971. Effect of acute and chronic administration of ethanol on the 5-hydroxytryptamine turnover and tryptophan hydroxylase activity of the mouse brain. Brain Res. 26:450.

Laverty, R. and Taylor, K.M. 1968. The flurometric assay of catecholamines and related compounds. Analyt. Biochem. 22:269.

LeBlanc, A.E. 1968. Microdetermination of alcohol in blood by gas-liquid chromatography. Can. J. Physiol. Pharmacol. 46:665.

Palaić, D.; Desaty, J.; Albert, J.M.; and Panisset, J.C. 1971. Effect of ethanol on metabolism and subcellular distribution of serotonin in rat brain. Brain Res. 25:381.

Phillis, J.W. and Jhamandas, K. 1971. The effects of chlorpromazine and ethanol on in vivo release of acetylcholine from the cerebral cortex. Comp. Gen. Pharmacol. 2:306.

Smith, C.B. 1965. Effects of d-amphetamine upon brain amine content and locomotor activity in mice. J. Pharmacol. Exp. Ther. 147:96.

Wallgren, H. and Barry, H. 1970. In: Actions of Alcohol, Vol. 1, Elsevier, New York.

Welch, A.S. and Welch. B.L. 1969. Solvent extraction method for simultaneous determination of norepinephrine, dopamine, serotonin, and 5-hydroxyindoleacetic acid in a single mouse brain. Analyt. Biochem. 30:161.

Wiegand, R.G. and Scherfling, E. 1962. Determination of 5-hydroxytryptophan and serotonin. J. Neurochem. 9:113.

EXPERIMENTAL ETHANOL INGESTION: SLEEP VARIABLES AND METABOLITES OF DOPAMINE AND SEROTONIN IN THE CEREBROSPINAL FLUID

Vincent Zarcone,* Jack Barchas, Eric Hoddes, Jacques Montplaisir, Robert Sack, Richard Wilson

Dept. of Psychiatry and Behavioral Science, Stanford University School of Medicine, Palo Alto, California, *and Palo Alto Veterans Administration Hospital

Many previous investigators have reported the effects of ethanol on sleep. Yules, et al. (1966, 1967), Knowles and Laverty (1968), and Rundell, et al. (1972) have demonstrated that acute doses of ethanol given to normal subjects cause a decrease in rapid eye movement sleep followed by a compensatory increase, or rebound, during withdrawal. Ethanol also increases the amount of beta activity (16-18 Hz) in the sleep EEG, and causes an increase in both the heart and respiration rates. In clinical observations made on alcoholics following heavy drinking periods by Gross et al. (1966), and Greenberg and Pearlman (1967), and in experimental ethanol ingestion studies done by Johnson (1971) and Gross et al. (1971), profound effects on both rapid eye movement sleep and slow wave sleep were noted. In addition, Allen et al. (1971) have noted that REM sleep is fragmented and stage three and four sleep decreased for many weeks after abstinence began.

These observations have naturally led to some speculations concerning the effect of ethanol on the mechanisms which regulate sleep. The likelihood that at least part of this regulation is monoaminergic seems great (Jouvet, 1972). Kissin et al. (1973) data probably bears most directly on this. They have shown that experimental ethanol ingestion using a dosage of approximately a quart of beverage ethanol in a ten hour period for four or six days produces profound reduction in rapid eye movement sleep followed by a rebound, and a large increase in slow wave sleep followed during the withdrawal period by a decrease and then return toward baseline. These changes are accompanied by an increase in excretion of tryptamine, a metabolite in the serotonin pathway. This increase and return to baseline roughly parallels that of slow wave sleep in the subjects. Urine levels of norepinephrine metabolite parallel the decrement of REM sleep during the ingestion period, and the subsequent large increase during recovery. Williams and Salamy (1972) have speculated that, initially, ethanol changes serotonin and catecholamine metabolism in a way similar to reserpine. In other words, there is a release of these neurotransmitters. This is followed by a "MAOI-like" phase in which REM sleep decreases to very low levels or to zero. When alcohol ingestion stops there is a large rebound of REM sleep as occurs after MAOI is discontinued.

These kinds of observations and speculations led us to study simultaneously a number of sleep variables and brain amine-metabolites in the cerebrospinal fluid, using an experimental ethanol ingestion design. We used doses of ethanol lower than were commonly used in the studies mentioned above. We expected, and have in fact found that this lower dose of ethanol produces considerable variability of response in both the sleep and biochemical parameters.

METHODOLOGY

The subjects' ages, duration of drinking problem, the number of weeks dry before the study began, and the duration of the last drinking period are given in Table 1. Table 2 shows the MMPI and ALCAAD scores for all the subjects. Figure 1 gives the mean MMPI profile for all the subjects, showing that the eight subjects, when pooled, resemble Byron's "chronic alcoholic profile" (1950).

The subjects were recruited from the in-patient, 24-hour residential alcoholic treatment program at the Palo Alto Veterans Admin-

Table 1

Subjects	Z	D	W	E	A	R	F	J
Ages	36	51	48	38	46	49	46	48
Previous Drinking (Yrs.)	5	25	17	7	10	15	31	3
Dry Before Study (Wks.)	3	6	3	2	20	3	5	18
Duration of Last Drinking Period (Days)	6	2	1	60	60	2	42	2

Table 2

MMPI and ALCAAD Scores

	Subjects	Z	W	D	E	F	R	A	J
MMPI	L	3	3	4	20	4	4	6	3
	F	22	5	10	21	14	4	2	11
	K	11	15	13	4	6	17	19	10
	Hs	35	11	19	12	21	10	16	17
	D	35	27	29	28	28	16	32	17
	Hy	36	23	26	20	33	20	28	23
	Pd	32	33	30	31	34	29	37	28
	Mf	18	25	36	25	30	21	26	27
	Pa	17	16	16	15	14	10	15	12
	Pt	45	30	39	36	44	24	36	24
	Sc	49	21	47	45	47	25	29	28
	Ma	31	18	22	36	28	23	22	29
ALCAAD	A	8	2	3	11	11	8	11	8
	B	8	4	5	10	8	10	10	8
	C	13	7	9	17	18	11	14	11
	D	16	5	5	14	16	8	15	11
	E	17	6	5	15	19	12	9	10
	Total	62	24	27	67	72	49	59	48

istration Hospital. They were recruited with the understanding that upon completion of the approximately one month of sleep and biochemical studies they would return at the same point in time that they left the treatment program in order to complete it. All of them did so. All eight of the subjects had been dried out for at least three weeks prior to the beginning of the study, and had been thoroughly evaluated medically. Prospective subjects were eliminated from this study if they had a past history of severe delirium tremens, convulsions, hepatic disease severe enough to cause BSP retention of greater than 5%, and cardiac or pulmonary disease. They also had to be free of any history suggestive of bleeding gastritis or esophageal varices. The large majority of the subjects were middle-class men who had not lost their occupations. Half of

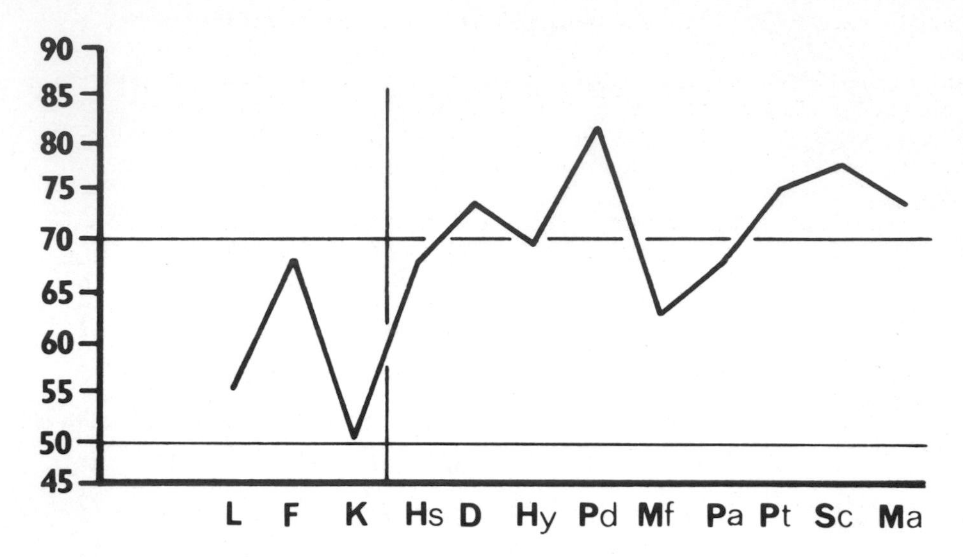

MEAN MMPI PROFILE FOR ALL SUBJECTS

Figure 1

them still maintained their marriages intact as well. Only one subject (F) could be described as bordering on the skid-row type of drinking pattern. The subjects were housed on the alcoholic treatment ward or, later in the study, in a clinical research ward at the Stanford Medical Center. They were under continuous 24 hour-a-day observation and any extra drinking was noted by the nursing personnel. This occurred in only one subject (F) who chose to increase his intake on the last day of ingestion.

The study began with a seven-day baseline period, followed by a seven-day ingestion period, and ending with a fourteen-day withdrawal period. All-night polygraphic monitoring of the central electroencephalogram, electro-oculogram, and electromyogram was done on each subject except for the sixth day of the baseline period, the sixth day of the ingestion, and the third and twelfth days of the withdrawal period, when lumbar punctures were performed to obtain the cerebrospinal fluid for assay of homovanillic acid (HVA) and 5-hydroxyindole acetic acid (5HIAA).

The subjects, by and large, cooperated only moderately well during the study. They were all volunteers and were all paid for participation in the study. Some of the subjects did not tolerate all the lumbar punctures, and one subject refused further lumbar punctures after the initial baseline tap. No medications were given for a period of at least a week prior to the beginning of the study except for one subject (F) who had thioridazine 50 mg., three times a day up to one day before the baseline observations began. Nearly 40% of the subjects originally recruited dropped out of the study before the end of the baseline period, or after they had completed the first spinal tap, or before they had completed the first day of the experimental ingestion period.

During the experimental ingestion period the subjects were given 1cc/kg of body weight of ethanol four times a day (at 8 a.m., 12 p.m., 4 p.m., and 8 p.m.). At each of these times the four to six ounces of eighty-proof beverage was given with either water or a lemon flavored soft drink. Each subject then had a Breathalyzer test (Smith and Wesson Electronics) between 8:30 p.m. and 9 p.m. The mean blood alcohol concentration for the ingestion period was 91 mg.%, with a range of 80 mg.% to 120 mg.%. All of the subjects reported the dosage of ethanol to be less than what they had been accustomed to drinking prior to hospitalization, in their own heavy drinking periods. They all reported that the ethanol consumption resulted in the feeling of a need for more alcohol, particularly after three or four days of ingestion. None of the subjects reported any particular physical discomfort associated with drinking, and only one subject (Z) reported some increase in anxiety. All of the subjects, at one time or another during the ingestion period, reported a decrease in feelings of depression or, conversely, an increase of alertness and stimulation. None of the subjects became noticeably aggressive or violent, except for one subject (F) who did become verbally abusive during the last two days of the ingestion period. (This was the same subject who surreptitiously increased his alcohol intake on the last day of the ingestion period.) The only withdrawal syndrome noted in any of the subjects was a slight tremulousness. There were no hallucinations or delerium tremens in any of the subjects.

RESULTS

Table 3 gives the sleep variables for all the subjects. The baseline days five through seven (the last three days of baseline) were compared to the first three days of ingestion, the last three days of ingestion, the first three days of withdrawal, and the last three days of withdrawal.

As can be seen from inspection of the table, several significant effects were observed when the entire group of subjects was considered as a whole. As expected, the REM sleep percentage decreased a significant amount, dropping from 24% during baseline to 18% in late ingestion (matched pairs t-test, $p<.05$) (Figure 2). REM efficiency also decreased during the same period, from .83 to .79 (Figure 3). Though the number of wakes

Table 3

Mean Sleep Variables for All Subjects

Variables	Baseline Day 5–7	Early Ingestion Day 8–10	Late Ingestion Day 12–14	Early Withdraw Day 15–17	Late Withdraw Day 18–20
Total Sleep Time	382	390	387	384	375
Sleep Latency	29	34	22	34	37
Number of Wakes	16	17	18	18	18
Wake Time After Sleep Onset	47	35	46	52	48
Stage 1	49	53	54	54	52
REM Latency	94	77	106	78	79
REM Percent	24	21	18	22	24
REM in 1st Third of Night	18	16	14	20	19
REM Efficiency	.83	.81	.79	.81	.79
REM Distribution Index	.60	.59	.61	.57	.60
Stage 3	15	18	20	17	15
Stage 4	1	5	7	2	1
Slow Wave Sleep Distrib. Index	.26	.20	.22	.22	.30
Correlation of Variance	.732	.643	.643	.645	.712
HVA	41		37	183	53
5HIAA	44		46	54	64

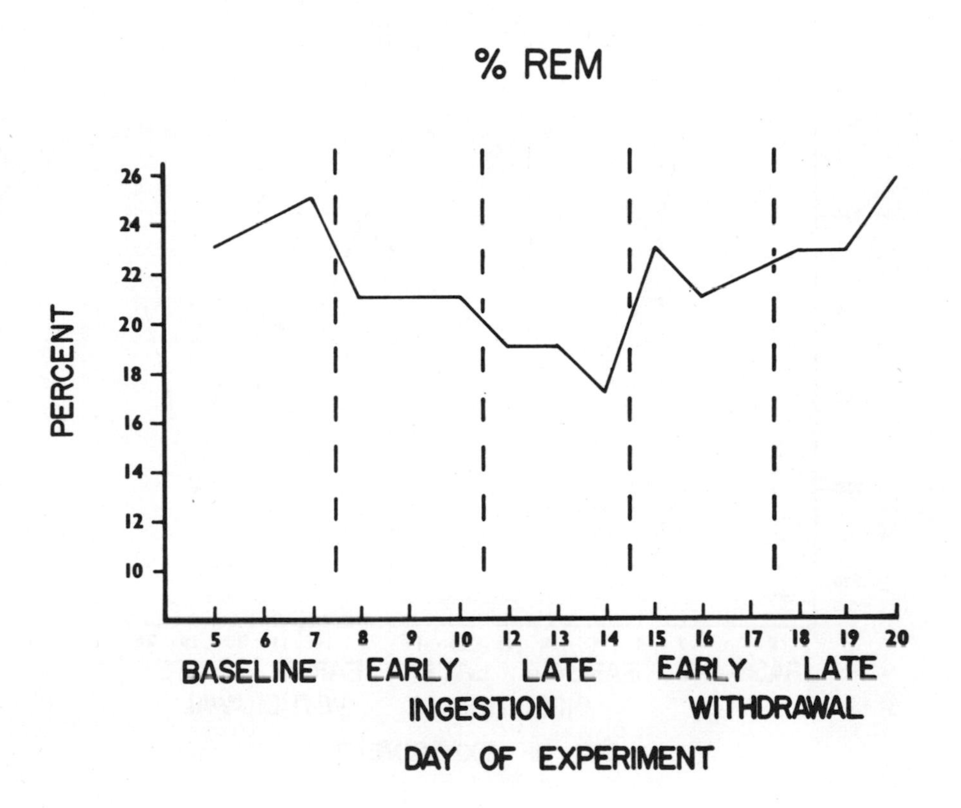
% REM
PERCENT
26
24
22
20
18
16
14
12
10
5 6 7 8 9 10 12 13 14 15 16 17 18 19 20
BASELINE
EARLY
LATE
INGESTION
EARLY
LATE
WITHDRAWAL
DAY OF EXPERIMENT

Figure 2

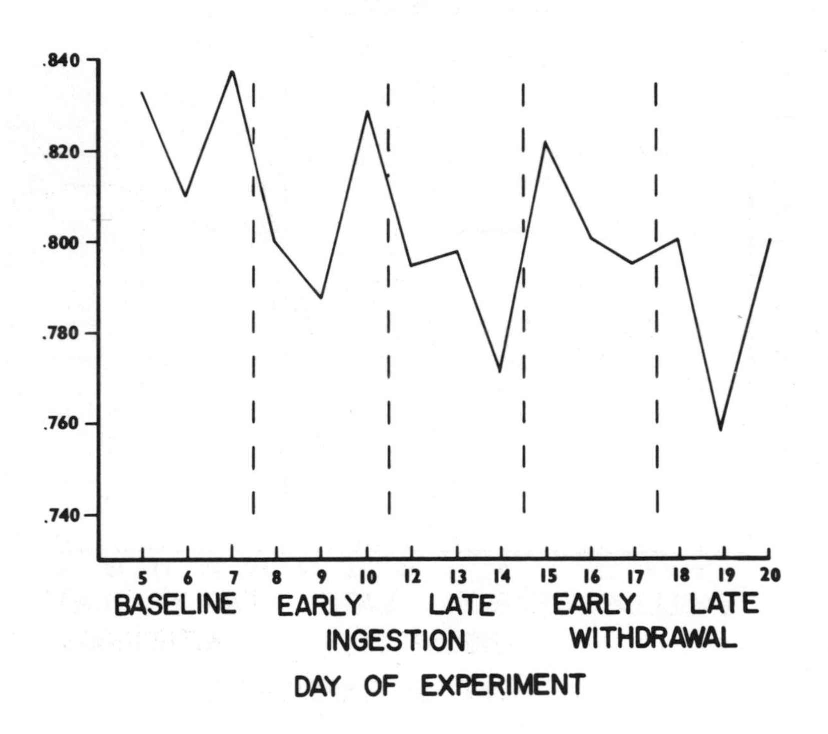
REM EFFICIENCY
.840
.820
.800
.780
.760
.740
5 6 7 8 9 10 12 13 14 15 16 17 18 19 20
BASELINE EARLY LATE EARLY LATE
INGESTION WITHDRAWAL
DAY OF EXPERIMENT

Figure 3

after sleep onset did not change significantly, wake time after sleep onset decreased during early ingestion and then increased during early withdrawal ($p<.05$). There was a slight increase in stages 3 and 4 during ingestion, as well as an increase in stage 1 for the same periods.

The intra-sleep REM-NREM cycle is a periodic phenomenon that was studied by means of a binary autocorrelation (Globus et al, 1972). Each REM epoch was scored as 1, and NREM as 0. Figure 4 shows the results of this autocorrelation for one baseline night and one ingestion night on the same subject (A = baseline, B = ingestion). The distance between successive peaks of high correlation is a measure of the period. The amplitude of the curve (maximum correlation to minimum correlation) has been called the correlation of variance. It is apparent in the figure that the correlation of variance decreases during ingestion without any obvious changes in the REM-NREM period. The data collected for all the subjects in Figure 5 showed that the correlation of variance declined from a baseline of .732 to .643 during early ingestion and remained low during late ingestion and early withdrawal ($p<.05$).

As can be seen from Table 3, total sleep time for the entire group of subjects did not noticeably change during the experimental periods. However, if individual subject's total sleep time is examined we can again see the wide variability in the three basic patterns that emerge. In Figure 6 (subject R) ethanol can be seen to act as a sedative with a significant increase in total sleep time during ingestion followed by a return to baseline during withdrawal ($p<.05$). Figure 7 (subject A) shows ethanol as a sleep depriving substance. Finally, Figure 8 (subject D) shows a subject whose total sleep time was essentially unchanged over the ingestion and withdrawal periods.

Slow wave sleep changed rather dramatically in five subjects over the course of the experiment. For their ages these subjects had low normal levels of slow wave sleep time at the beginning of the experiment, a mean of twenty-two minutes for baseline. Stages 3 and 4 increased to forty minutes during the late ingestion period. In addition, these same five subjects had low normal values of 5HIAA during baseline (mean of 36 ng/ml.), which increased during withdrawal (72 ng/ml.). The three subjects whose slow wave sleep time was below five minutes during baseline, or markedly abnormal, had high values of serotonin during baseline, which remained essentially unchanged throughout the course of the experiment. Figure 9 shows the two groups.

Figure 10 shows the values for the cerebrospinal fluid homovanillic acid for the entire group of subjects. The dramatic increase in HVA during early withdrawal is the result of marked

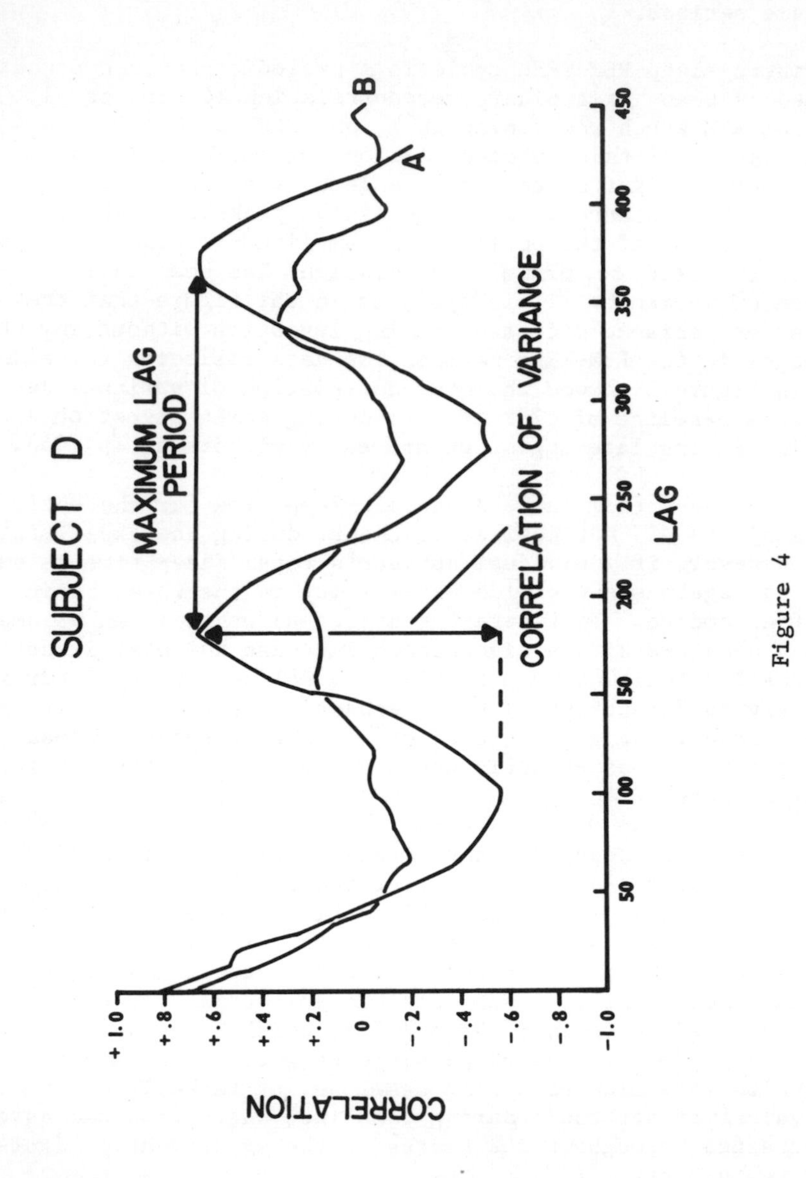
SUBJECT D
MAXIMUM LAG
PERIOD
CORRELATION OF VARIANCE
A
B
CORRELATION
+ 1.0
+.8
+.6
+.4
+.2
0
-.2
-.4
-.6
-.8
-1.0
50
100
150
200
250
300
350
400
450
LAG

Figure 4

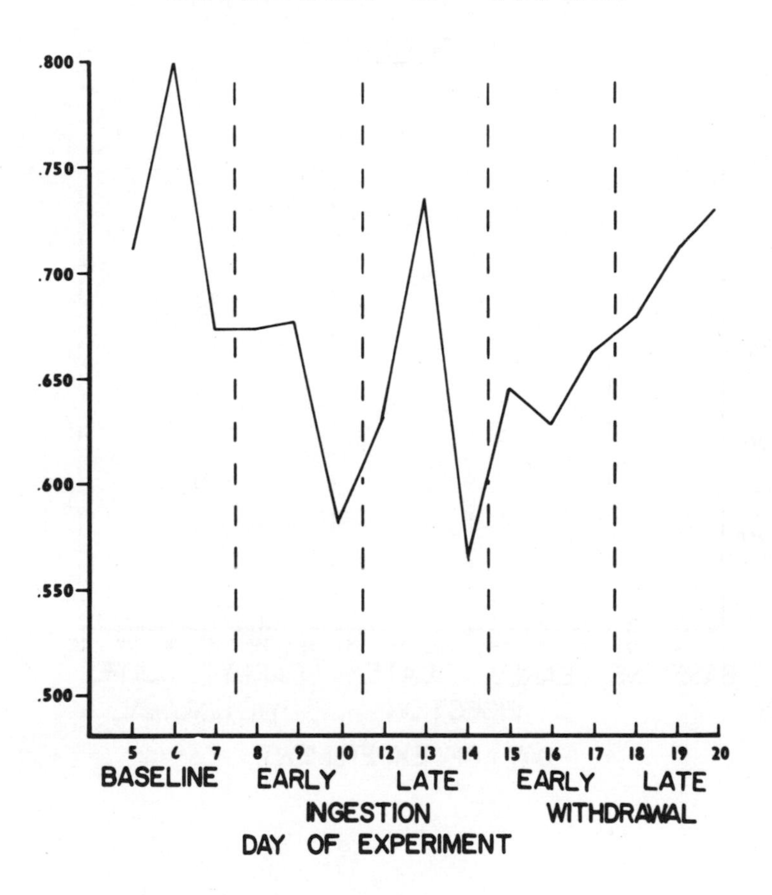
CORRELATION OF VARIANCE
.800
.750
.700
.650
.600
.550
.500
5 6 7 8 9 10 12 13 14 15 16 17 18 19 20
BASELINE EARLY LATE EARLY LATE
INGESTION WITHDRAWAL
DAY OF EXPERIMENT

Figure 5

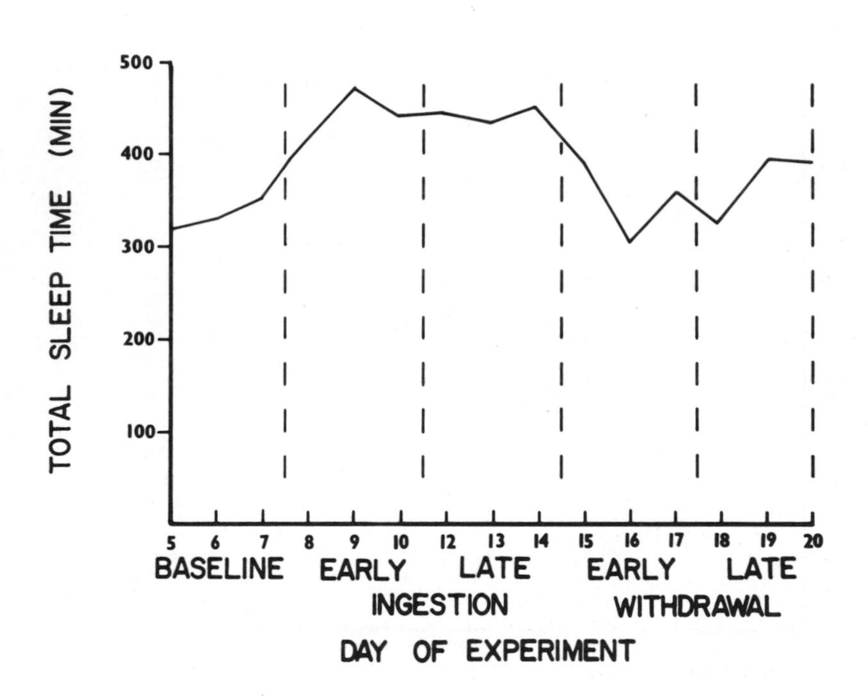
SUBJECT R
TOTAL SLEEP TIME (MIN)
500
400
300
200
100
5 6 7 8 9 10 12 13 14 15 16 17 18 19 20
BASELINE EARLY LATE EARLY LATE
INGESTION WITHDRAWAL
DAY OF EXPERIMENT

Figure 6

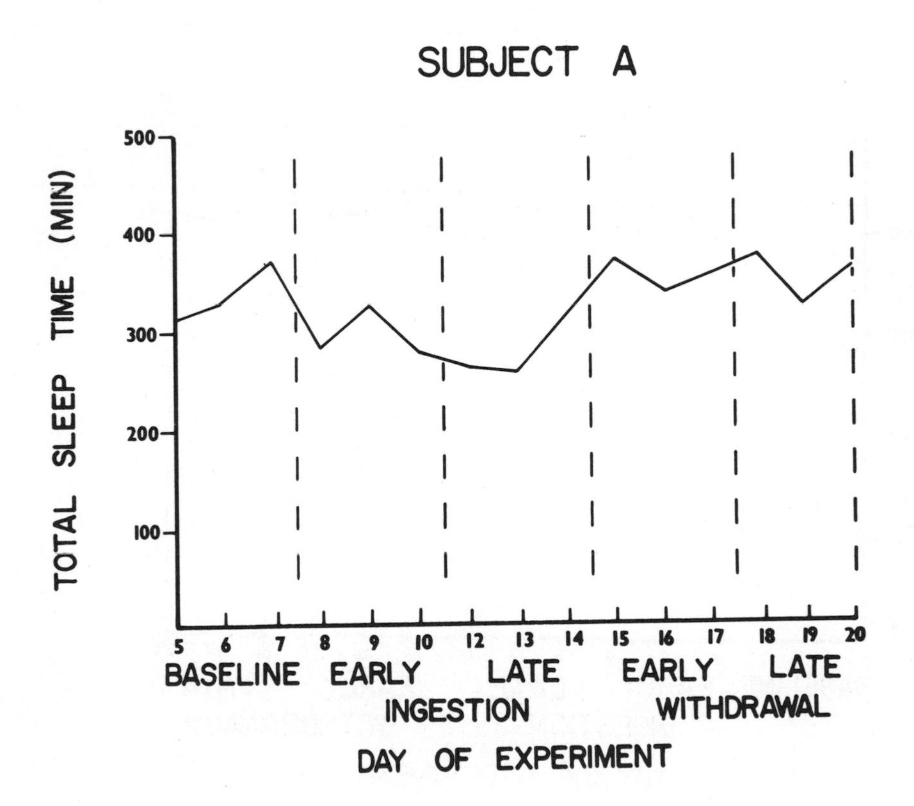
SUBJECT A
500
400
300
200
100
TOTAL SLEEP TIME (MIN)
5 6 7 8 9 10 12 13 14 15 16 17 18 19 20
BASELINE EARLY LATE EARLY LATE
INGESTION WITHDRAWAL
DAY OF EXPERIMENT

Figure 7

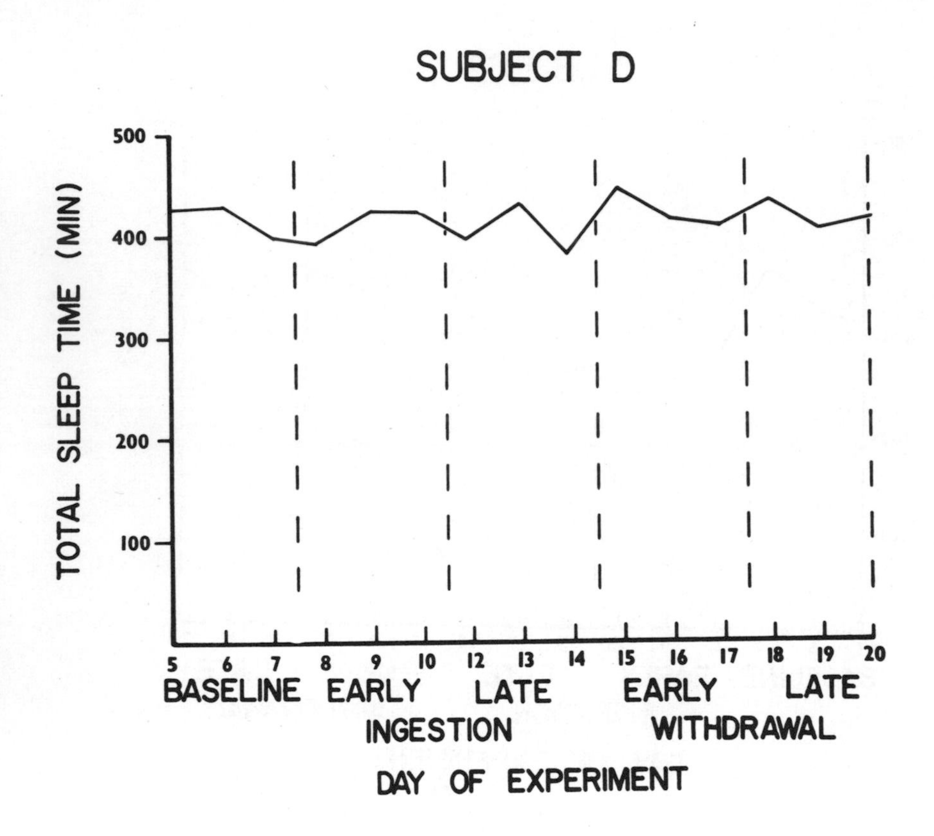
SUBJECT D
TOTAL SLEEP TIME (MIN)
500
400
300
200
100
5
6
7
8
9
10
12
13
14
15
16
17
18
19
20
BASELINE
EARLY
LATE
INGESTION
EARLY
LATE
WITHDRAWAL
DAY OF EXPERIMENT

Figure 8

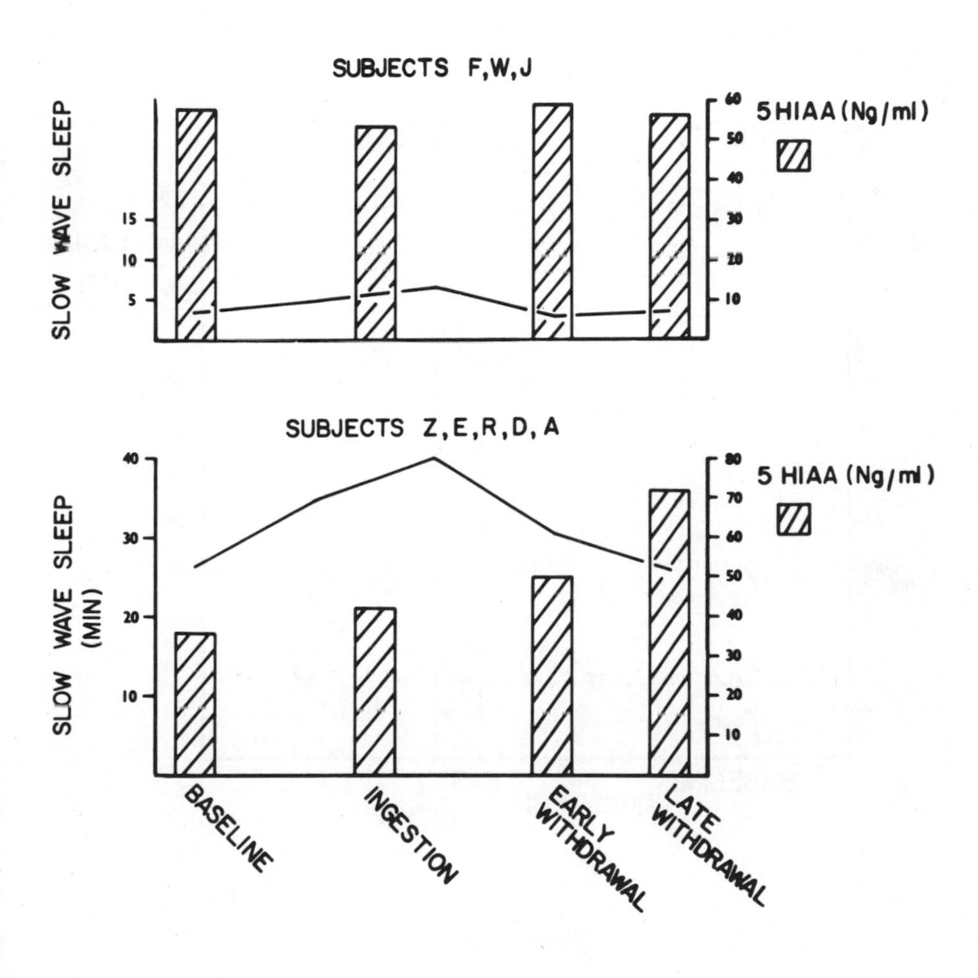
SUBJECTS F,W,J
SLOW WAVE SLEEP
15
10
5
60
50
40
30
20
10
5HIAA (Ng/ml)
SUBJECTS Z,E,R,D,A
SLOW WAVE SLEEP (MIN)
40
30
20
10
80
70
60
50
40
30
20
10
5 HIAA (Ng/ml)
BASELINE
INGESTION
EARLY WITHDRAWAL
LATE WITHDRAWAL

Figure 9

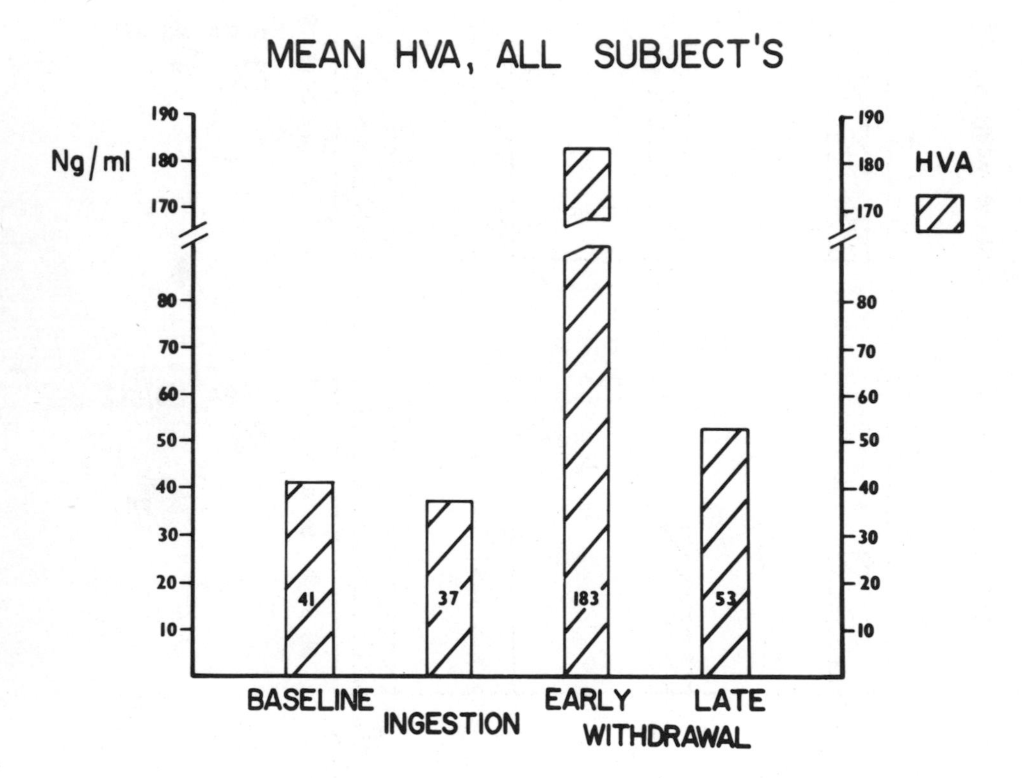
MEAN HVA, ALL SUBJECT'S
Ng/ml
190
180
170
80
70
60
50
40
30
20
10
HVA
41
37
183
53
BASELINE
INGESTION
EARLY
LATE
WITHDRAWAL

Figure 10

increases in three subjects (from baseline values of 26, 64, and 24 nanograms per milliliter to 820, 220, and 84 respectively). Table 4 shows the wake time after sleep onset, sleep latency, stage 1 sleep, and total sleep time for these three subjects with dramatic increases in HVA. In subject J the increased wakefulness in withdrawal seems to be delayed.

Table 4

Wakefulness Parameters in Three Subjects With High HVA Responses

	Baseline	Early Ingestion	Late Ingestion	Early Withdrawal	Late Withdrawal
F					
Total Sleep Time	361	356	391	342	327
Wake After Sleep Onset	55	35	70	81	73
Sleep Latency	38	42	8	38	34
Stage 1	108	96	109	115	125
HVA ng./ml	64	*	62	220	*
J					
Total Sleep Time	366	403	427	437	362
Wake After Sleep Onset	61	29	43	26	58
Sleep Latency	38	23	6	22	38
Stage 1	20	21	20	24	28
HVA ng./ml	24	*	24	84	38
R					
Total Sleep Time	333	441	442	350	368
Wake After Sleep Time	55	25	16	83	32
Sleep Latency	23	6	11	30	39
Stage 1	34	35	48	61	44
HVA ng./ml	26	*	46	820	62

*No LP was done.

DISCUSSION

As previously reported by other investigators REM sleep is particularly sensitive to ethanol ingestion.

The changes in REM percent and REM efficiency in our study were noted even with the lower dose employed. Also the initiation and maintenance of REM sleep (the stability of the REM-NREM cycle) was particularly sensitive to the effects of a moderate dose of ethanol.

Slow wave sleep was less responsive to the level of intoxication produced in our study. The effect of ethanol seemed to be dependent on the baseline level of slow wave sleep. Ethanol increases slow wave sleep only when baseline levels are in the normal range. This type of response has also been reported by other investigators. (See the papers in this volume by Wagman, A. and Allen, R.P., "Effects of alcohol ingestion and abstinence on Slow Wave Sleep of alcoholics" and by Gross, M.M. and Hastey, J.M., "Baseline Slow Wave Sleep level as a predictor of the Slow Wave Sleep response to alcohol in alcoholics"). Further study is indicated to see what prognostic value this finding might have. Slow wave sleep changes, if they are irreversible or very slowly reversible, may indicate significant brain damage.

The observation that slow wave sleep increments with intoxication were accompanied by increases in the CSF serotonin metabolite, 5HIAA, may indicate that in some patients the system which regulates slow wave sleep is at least partly serotonergic and able to respond to ethanol with an increase in turnover. In those patients with low slow wave sleep during baseline the serotonergic control of slow wave sleep may be less flexible. Perhaps there is already a high rate of turnover as reflected in the high baseline levels of 5HIAA and with ethanol a ceiling effect may occur so that slow wave sleep does not increase.

We began our studies before learning of Jouvet's (1972) hypothesis concerning the regulation of behavioral wakefulness by dopamine neurons in the nigro strial pathway. We were of course interested in the possibility that increased wakefulness during withdrawal from ethanolization might be correlated with changes in the CSF dopamine metabolite HVA. As a group all eight subjects did show decreases in wake time after sleep onset while drinking, followed by increases during withdrawal. In the three subjects who showed marked changes in HVA levels these occurred in early withdrawal, coincident with sleep disturbance as evidenced by increased wake time, sleep latency, or stage 1 time. In subject J, who had the least marked change in HVA, ethanol had a marked sedative effect. By late withdrawal however, his total sleep time decreased to baseline levels of about six hours and his wake time after sleep onset went back up to about one hour. We can speculate that these changes in HVA are indicative of increased turnover of dopamine neurotransmitter in a system which partly determines the level of behavior activation. A postalcoholization denervation supersensitivity in this system plus increased neurotransmitter turnover might be the mechanism for the increased wakefulness during withdrawal.

The observations have to be interpreted with extreme caution however, and the above can only be considered preliminary hypotheses. First of all, the response of the subjects was quite variable and there are only small numbers in each group. Second, we were only able to study levels of two metabolites and could not do indirect probenecid turnover studies. Mass fragmentography may make assay of norepinephrine metabolites and alcoholic metabolites of serotonin, dopamine and norepinephrine possible. These procedures might uncover the presence of false neurotransmitters during alcoholization and withdrawal which would more exactly correlate with changes in sleep. Third, even the metabolite levels we have must be viewed cautiously since we are not sure of the relationship of lumbar spinal fluid levels to brain levels. We do not know the effect of activity or lack of it and we had no independent measure such as cortisol levels to control for the stress effects of drinking and withdrawal which may of course operate independently of any effects of alcohol on the amine neurotransmitters important in sleep regulation. Finally, it should be noted that we had the opportunity to follow up one of these subjects (J) and found that three days after a ten day drinking period (when approximately twelve ounces of 80-proof beverage was consumed daily) both the 5HIAA and HVA were slightly below normal. Therefore we cannot be sure that the response to ethanol is consistent across time in the same subject.

In spite of these reservations it seems already evident that continued study of the relationship between sleep disturbance and prognosis or the reversibility of brain abnormalities during drying out periods should be useful. For instance we would predict that there is a continuum of defects in sleep regulation in alcoholic patients. After a month of drying out some patients may show an intolerance to moderate doses of ethanol at bedtime. They are unable to increase REM time or percent in the second two-thirds of the night so that they can maintain normal REM percent for the night.

In a second type of withdrawal process the patient may show REM fragmentation and a SWS deficit even after several months of drying out.

Finally, a third group may never have a return to normal sleep regulation and their records will show REM fragmentation and SWS deficit for many, many months after drying out. This latter group has a chronic brain syndrome by definition, even if they do not show deficits on psychometric testing--although if they did not, we would ask the psychologist for further testing or the development of more sensitive tests. This latter group may need some type of replacement therapy to restore normal regulation of the basic rest-activity cycle.

We have studies underway to correlate these syndromes as defined by polysomnography with sobriety and psycho-social functioning after treatment in a 90 day residential program. It may be possible that a patient should be treated more intensively if he has continuing REM or SWS abnormalities and that his prognosis is more guarded.

ACKNOWLEDGEMENTS

The authors gratefully extend a very sincere appreciation to Dr. Helena Kraemer, Stanford University, for the statistical analysis in this paper.

Dr. Edgar Radcliffe and the staff of the alcohol treatment service of the Palo Alto Veterans Administration Hospital provided invaluable assistance in selection and clinical management of the patient subjects.

The authors wish to express their gratitude for the assistance of the Clinical Research Center at Stanford University.

REFERENCES

Allen, R., Wagman, A., Faillace, L. and McIntosh, M. Electroencephalographic (EEG) sleep recovery following prolonged alcohol intoxication in alcoholics. J. Nerv. Ment. Dis., 153:424-433, 1971.

Byron, M.A. Alcoholic profiles in the MMPI. J. Clin. Psych. 6: 266-269, 1950.

Globus, G., Phoebue, E., Boyd, R., Drury, R. and Leventhal, T. The effect of a tranquilizer on the temporal organization of sleep. Psychobiology, 9:94, 1972.

Greenberg, R. and Pearlman, C. Delirium tremens and dreaming. Am. J. Psychiat. 124:37-46, 1967.

Gross, M.M., Goodenough, D.R., Tobin, M., Halpert, E., Lepore, D., Perlstein, A., Sirota, M., Dibianco, J., Fuller, R. and Kishner, I. Sleep disturbances and hallucinations in the acute alcoholic psychoses. J. Nerv. Ment. Dis. 142, 6:493-516, 1966.

Gross, M.M., Goodenough, D.R., Hastey, J., Rosenblatt, S. and Lewis, E. Sleep disturbances in alcoholic intoxication and withdrawal. In Recent Advances in Studies of Alcoholism: An Interdisciplinary Symposium. N.K. Mello and J.H. Mendelson (Eds.). U.S. Dept. of H.E.W. publication HSM 71-9045, pp. 317-397, Wash. D.C., 1971.

Johnson, L.C. Sleep patterns in chronic alcoholics. In Recent Advances in Studies of Alcoholism: An Interdisciplinary Symposium. N.K. Mello and J.H. Mendelson (Eds.). U.S. Dept. of H.E.W. publication HSM 71-9045, pp. 285-316, Wash. D.C., 1971.

Jouvet, M. The role of monoamines and acetylcholine containing neurons in the regulation of the sleep-waking cycle. Rev. Psychiol. 64:166-307, 1972.

Kissin, B., Gross, M.M., Schutz, I. Correlation of urinary biogenic amines with sleep stages in chronic alcoholization and withdrawal. In Alcohol Intoxication and Withdrawal: Experimental Studies I, Advances in Experimental Medicine and Biology, M.M. Gross, (Ed.). Plenum Publishing Corp., New York, 1973.

Knowles, J., Laverty, S. and Kuechler, H. Effects of alcohol on REM sleep. Quart. Journ. Stud. Alc. 29, 2:342-349, 1968.

Rundell, O.H., Lester, B.K., Griffiths, W.J., Williams, H.L. Alcohol and sleep in young adults. Psychopharmacologia 26: 201-218, 1972.

Yules, R.B., Friedman, D.X., Chandler, K.A. The effect of ethyl alcohol on man's electroencephalographic sleep cycle. Electroenceph. Clin. Neurophysiol. 20:109-111, 1966.

Yules, R.B., Lippman, M.E., Friedman, D.X. Alcohol administration prior to sleep: The effect on EEG sleep stages. Arch. Gen. Psychiat. 16:94-97, 1967.

Williams, H.L., Salamy, A. Alcohol and sleep. In Biology of Alcoholism. Kissin, B., Begleiter, H. (Eds.). Vol. II, pp 435-483, Plenum Press, New York, 1972.

EFFECTS OF ALCOHOL INGESTION AND ABSTINENCE ON SLOW WAVE SLEEP OF ALCOHOLICS

Althea M. I. Wagman and Richard P. Allen*

Maryland Psychiatric Research Center and
Baltimore City Hospital
Baltimore, Maryland, U.S.A.

Disturbance of sleep is a frequent component of the clinical pathology associated with depression (Hartmann, 1965), schizophrenia (Stern, et. al., 1969) and drug addiction (Watson, et. al., 1972). Reduction in total sleep time, increased awakening, long sleep latency, and REM sleep suppression occur frequently as a function of short-term stresses and are responsive to symptomatic treatment. Slow Wave Sleep (SWS) impairments are more often associated with chronic conditions which respond poorly to situational therapy and may be the result of long-term stress. Short-term as well as long-term sleep disturbances are often associated with alcoholism. Experimental studies which have been concerned with the effect of alcohol on sleep were designed to assess the role of alcohol in producing both kinds of sleep dysfunction in normal subjects and chronic alcoholics.

The effects of alcohol on normal subjects were studied by Yules, et. al. (1967) and Knowles, et. al. (1968). Their subjects were studied for several baseline nights and several drinking nights. REM percentage decreased on the first drinking night and gradually increased to baseline levels across subsequent drinking nights. SWS was unaffected under these conditions. Both studies suggested that the REM decrease was a direct

*Supported by NIH Grant No. AA 00311-03, NIAAA

function of the effect of alcohol on the central nervous system.

Johnson, et. al. (1970) similarly studied the effects of alcohol on sleep; however, chronic alcoholics were used and two drinking nights were followed by ten withdrawal nights. These subjects averaged 17 years of excessive drinking and, therefore, probably differ from normal subjects in their ability to tolerate alcohol. On the second drinking night, alcohol produced REM and SWS suppression occurred as compared to the final withdrawal night. REM sleep disturbance continued throughout the withdrawal period characterized by fragmentation of the REM episodes such that the normal periodicity was impaired. SWS remained significantly below normal values even on the tenth withdrawal night.

Gross and Goodenough, et. al. (1973) extended the drinking period to four and six days to determine whether progressively severe sleep impairment occurred with prolonged drinking. REM latency and the amount of REM and SWS, significantly reduced after the four day period. After six days, total sleep time was significantly lower, and number of awakenings increased as well as significant suppression of REM and SWS. Withdrawal produced further sleeplessness, longer sleep latencies, some REM rebound and very little SWS.

In the present study, in order to partition out the confounding effect of number of days of drinking from amount of alcohol consumed per day, several alcohol doses were administered, and each dose was studied for at least five consecutive days. This design made it possible to compare SWS amount after 18 oz., 26 oz. and 32 oz. of 95° alcohol to SWS values obtained from two abstinent periods for chronic alcoholics. These parameters of alcohol ingestion were chosen because of their similarity to drinking patterns of alcoholics in the non-hospitalized situation. Mean energy content of the EEG bands for delta, theta, alpha and $beta_1$ was also obtained for each comparison day. This measure provided information to evaluate alcohol effects on the synchronous slow wave systems associated with SWS as well as the high frequency energy associated with arousal. In addition, the design assessed the EEG tolerance effects of five consecutive days of ingestion of the two lowest doses. Tolerance was observed as a tendency to recover baseline values with prolonged drinking. Finally the recovery of SWS function during prolonged abstinence was assessed with a second group of abstinent alcoholics.

METHOD

Two groups of alcoholic subjects were studied. The first group consisted of six male alcoholics, whose mean age was 35.5. The mean number of years of problem drinking was eight. The subjects were admitted to this protocol from the hospital emergency ward within 24 hours of their last drink. Subjects were evaluated by a physician at the time of admission to determine whether the subject was in good health, other than acute inebriation. Subjects were maintained on either alcohol (18 oz.) 95 proof or Librium for three days in order to stabilize their adjustment to the laboratory after which the first withdrawal period began. After at least seven days of withdrawal, alcoholization began. The general protocol required subjects to receive 18 oz. 95 proof alcohol for five consecutive days, followed by 26 oz. for the next five days, and 32 oz. for at least one day. Alcohol dose was then gradually reduced to 18 oz. prior to withdrawal.

Sleep records were obtained each night from electrode placements recommended by the Rechtschaffen and Kales (1968) Sleep Manual. Subjects were permitted ad libitum sleep; however, they were required to be in bed by midnight.

Sleep records were scored according to the criteria of the Sleep Manual by raters in the Baltimore City Sleep Laboratory whose inter-rater reliability is .90.

FM recordings from these patients were analyzed using spectral analysis techniques to ascertain information about the specific energy content within the Delta, Theta, Alpha, and Beta frequency bands. These assessments were obtained using four active Butterworth filters* designed to measure frequencies between .5-4Hz for Delta, 4-8Hz for Theta, 8-12Hz for Alpha, and 12-20Hz for $Beta_1$. These data in volt/seconds were then integrated and averaged for each five minute period throughout the night using a PDP-12 computer.

The second group of subjects consisted of 20 male alcoholics with a mean age of 35.5 obtained through A.A.

---

*Attenuation characteristics of 24 db/octave

These subjects were verified abstinent alcoholics. Sleep records were obtained from the subjects on three consecutive nights and only the third night was assessed for Sleep Stages. This group provided information on the recovery process of SWS which could not be obtained within the other protocol.

RESULTS

Figure 1 shows the progressive increase in average SWS percentage that occurred with continued drinking. The initial abstinent values of 3% increased to 15% after

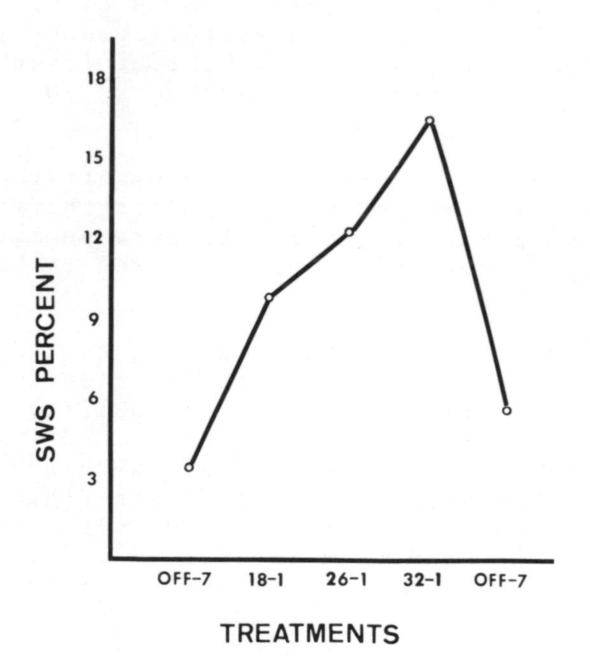

Figure 1

The effect of prolonged drinking on SWS percent. Each OFF-7 refers to seven days of abstinence, 18-1 refers to the first day of 18 oz. consumption and so forth. The first OFF-7 period occurred prior to the experimental procedure and the second OFF-7 occurred subsequent to drinking.

32 oz. and then returned to baseline level after seven abstinent days. Repeated measures analysis of variance indicate that this increase was significant.

The mean EEG energy content in each frequency band was also evaluated by four separate analysis of variance for repeated measures.

Subjects were studied as a main effect to determine whether consistent idiosyncratic EEG phenomena could be obtained for all frequencies. Measures were averaged for each half hour throughout the night and compared to determine whether systematic energy cycles occurred during the night. Table 1 presents the F ratios and degrees of freedom associated with all of these comparisons.

A subject main effect was obtained for each analysis. Only delta energy showed consistent patterns across subjects. The two subjects with the highest delta energy in the abstinent situation produced the most energy on the first night of drinking 18 oz.. In each case delta energy remained at a very high level until the dose was increased at which time delta energy was dramatically reduced. The subject with the lowest abstinent delta energy remained low across all conditions. Variable patterns were obtained for the other three subjects. Figure 2 shows the effect of the alcohol conditions on the energy of each EEG band. The data are represented as difference scores from baseline abstinent values. Baseline was obtained from the EEG analysis of the first seven days abstinent following admission to the ward and is shown as a horizontal reference line for each band. Significant augmentation of delta occurred for each measurement period except the fifth day of 26 oz.. The maximum increase occurred after the first day of drinking 18 oz.. Theta energy was significantly enhanced after the first day of drinking but returned toward baseline following five days at this dose. Increase of dose again significantly increased theta and the level remained high for the next five days. Alpha showed changes similar to theta. Beta energy did not significantly increase until the subject had been drinking for ten days. The Scheffe Multiple Comparisons found on Table 2 shows that these comparisons were significant. All values returned to baseline condition after seven days of abstinence.

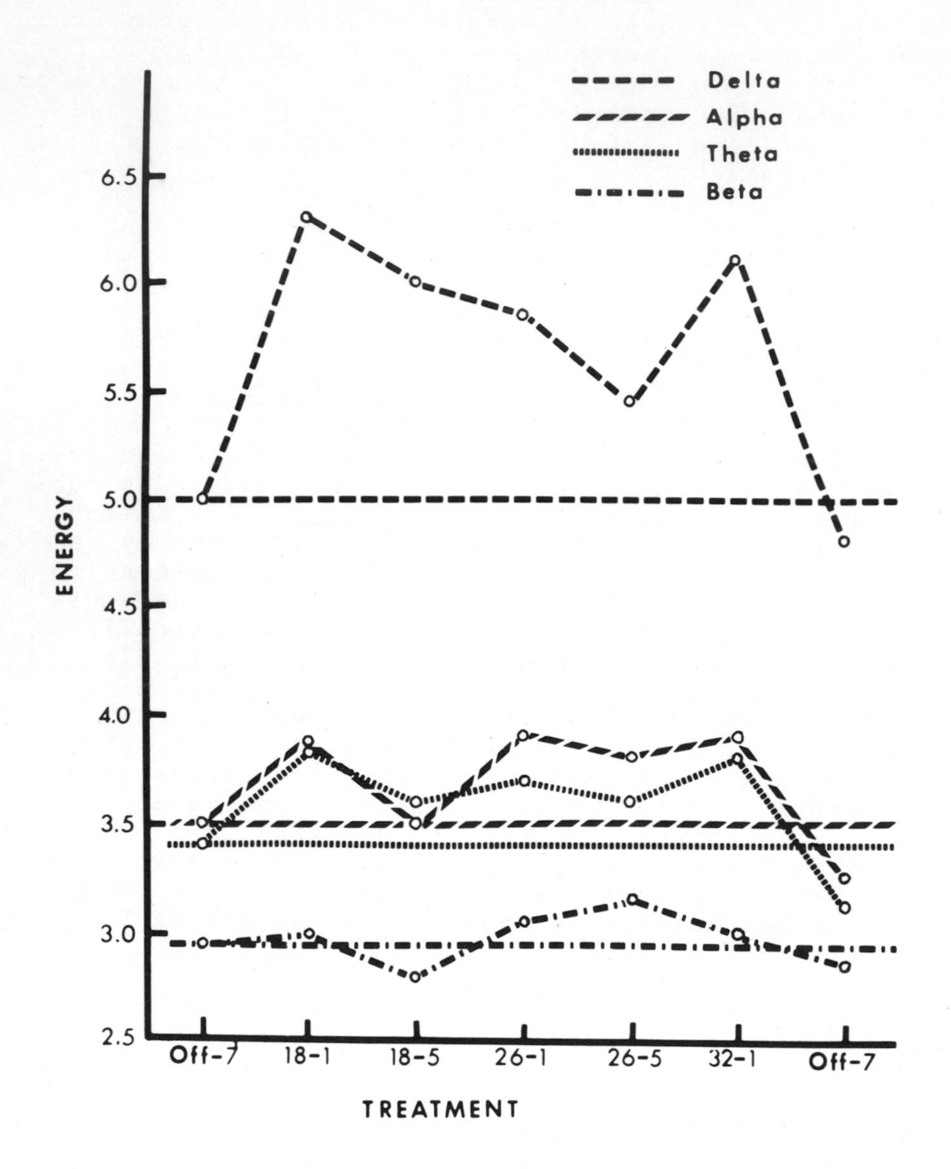

Figure 2

Spectral analysis of the EEG in volt-seconds for each dose. Horizontal lines represent baseline (OFF-7) levels for each frequency band.

Table 1

F ratios derived by Analysis of Variance for the comparison of subjects by treatment by hour of the night, across the four frequency bands.

		Energy Band			
Comparison	df	Delta	Theta	Alpha	Beta
Subjects	5/120	6.998**	94.657**	203.592**	64.876**
Treatment	4/120	5.999**	19.088**	17.827**	10.545**
Hour	3/120	4.371**	8.003**	29.248**	5.383**
Subject x Treatment	20/120	4.413**	19.939**	13.231**	8.641**
Subject x Hour	15/120	.577	2.875**	4.175**	4.404**
Hour x Treatment	12/120	.974	1.531	1.802	.111
Subject x Treatment x Hour	60/120	.829	5.559**	4.924**	4.157**

* Significant .05
** Significant .01

Table 2

Alcohol Treatment Groups, Scheffe Multiple Comparisons

Comparison (oz.-day)	Δ	Θ	α	β
18-1 vs. OFF-7	F=24.443*	F=52.762*	F=28.644*	F= 1.313
18-5 vs. OFF-7	14.391*	8.366	.169	2.224
18-5 vs. 18-1	1.310	19.102*	24.346*	6.987
26-1 vs. OFF-7	9.807*	28.296*	37.328*	6.147
26-1 vs. 18-1	3.284	3.780	.568	1.768
26-1 vs. 18-5	.445	5.887	32.402*	14.280*
26-5 vs. OFF-7	2.915	12.125*	16.550*	17.478*
26-5 vs. 18-1	10.475*	13.286*	1.645	9.194
26-5 vs. 18-5	4.375	.534	13.320*	32.241*
26-5 vs. 26-1	2.028	2.872	4.167	2.889

*Significant at $p < .05$ (F crit. 9.64, df 4/236)

The relationship of SWS increase to EEG energy was explored by Pearson r. The correlation of delta and theta energy with SWS attained significance. The relationship between these measures and SWS during abstinence was r = .84 and r = .73 respectively. As SWS increased with progressive drinking, these correlations decreased. After ten days of drinking no relationship was found (r = .02), (r = .01).

Clinical inspection of the EEG records during the last five days of drinking indicated that alpha-delta patterns occurred for some subjects. Hauri and Hawkins (1973) first reported on this phenomena and found that it was rare in normal subjects and was almost always associated with neurological dysfunction. After the tenth day of drinking, a large amount of alpha-delta pattern was found in the sleep records. The EEG energy of the alpha and delta bands was compared to determine whether similar relationships occurred between the bands during abstinence as after ten days of drinking. Figure 3 shows that alpha energy progressively declines throughout the night while delta waxes and wanes for the

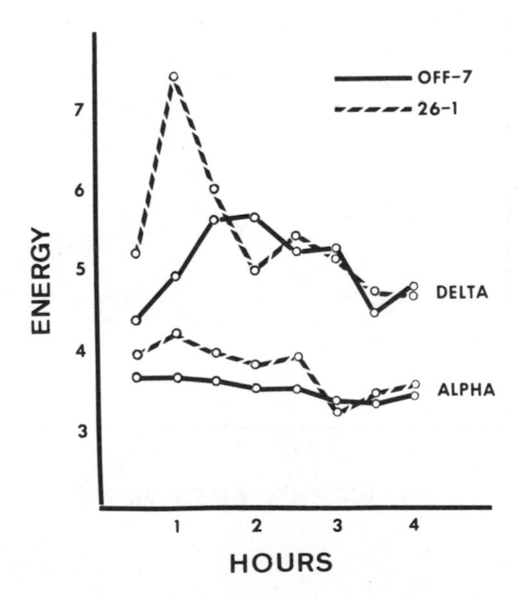

Figure 3

Alpha and delta energy in microvolt-seconds for baseline and after prolonged drinking.

abstinent state, suggesting that alpha energy is not related to delta values. After five days of ethanol ingestion, alpha energy was significantly augmented and remained at a high level for several hours. This period corresponded to the period of maximum delta activity. The correlation between the bands during baseline for the first four hours of the night is $r = .004$ while it increased to $r = .573$ after ten days of drinking.

The recovery of SWS function across time for the abstinent alcoholics is shown on Figure 4. Abstinent duration in weeks was converted to log units in order to describe a linear relationship to SWS percent. SWS suppression is seen for at least 3 1/2 weeks subsequent to drinking. Recovery was complete (18 percent) after 200 weeks. Multiple regression was used to evaluate the contribution of additional variables to SWS recovery. Significant relationships were obtained for the variables of subject age, log duration of abstinence, and Stage 4 percentage.

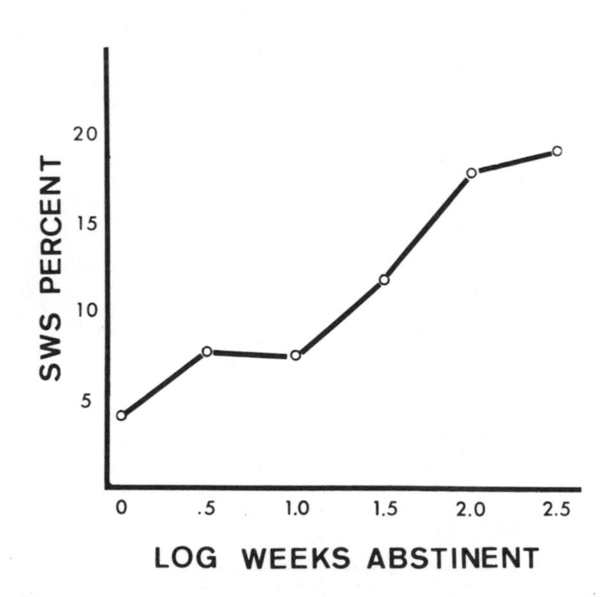

Figure 4

SWS recovery after prolonged abstinence

DISCUSSION

SWS percentage was found to be near zero during the baseline seven days prior to the experimental procedure. Dose-related increase in SWS was observed after each dosage increase. Gross and Goodenough, et. al. also found augmented SWS following four days of drinking a fixed amount of alcohol. The amount of SWS progressively returned toward baseline level with continued drinking for two more days. These data together with the present findings suggest that tolerance to a single heavy dose of alcohol begins after four days of consistent drinking and that increase in dosage counteracts the tolerance effect.

SWS recovery was found to occur slowly as a function of duration of abstinence and subject age. The findings of Johnson, et. al. and Gross and Goodenough, et. al. confirm the relative lack of SWS during the first ten and seven days of withdrawal, respectively. SWS suppression was observed in all three of these studies regardless of the duration or amount of drinking.

Long-term recovery of SWS was also studied by Adamson and Burdick (1973). Ten alcoholics abstinent between 1-2 years, (mean age 42) were studied for three consecutive nights. The range of SWS percentage was 0-13% with a mean of 3.8%. Figure 4 indicates that an average of 13% SWS would be expected, based on the present sample after 52 weeks abstinence (log = 1.7) and with a mean age of 35 years. The SWS percentage discrepancy between the two studies is probably due to age differences between the two samples as well as to individual differences produced by the inherent sampling bias of drawing from the alcoholic population. Adamson and Burdick as well as Allen, et. al. (1971) have suggested that these individual differences may be accounted for by the number of years of excessive drinking. Subjects with shorter histories tend to show more SWS. Quantity consumed per day, age of drinking onset, as well as current age and duration of abstinence may all be important factors.

Spectral analysis of the EEG subsequent to alcohol ingestion during sleep has not received prior attention. EEG samples obtained in the waking state from eight young men subsequent to ingestion of a heavy dose of alcohol were studied by Rosadini, et. al. (1974). Power spectrum analysis revealed slowing of alpha by 1.4Hz.

Integration of the spectrum divided into the four main bands revealed a large increase in delta power, a similar trend for theta and a marked reduction for alpha. No changes were observed for beta. These results were not statistically analyzed. Data obtained in the present study during the sleeping state also revealed augmentation of delta, theta and alpha after the first day of drinking. No consistent changes in beta activity were observed until subjects had been drinking for a long while. The discrepancy between the two studies in alpha amount may be due to the suppression of alpha in the waking, eyes open state used by Rosadini, et. al. and release during drowsy, eyes closed condition (present study). In addition, energy obtained from fixed band filters tends to overestimate the actual energy due to an overlap of similar frequency components at the edge of each filter. Thus energy obtained for an 8Hz signal may be detected by both alpha and theta filters to some degree. Power analysis does not have this shortcoming.

The spectral analysis demonstrates that the three low frequency bands respond to acute alcohol intoxication while the high frequency beta band does not. Delta frequency shows the greatest increase followed by a progressive decrease in energy toward baseline levels with continued drinking regardless of dose. Theta and alpha bands decrease towards baseline with prolonged drinking at each fixed dose but respond with increased energy when the dose is increased. These findings suggest that there are two different low frequency response mechanisms to alcohol: the delta system which is immediately and maximally affected by the alcohol per se and the theta-alpha system which responds to alcohol dose. The beta band shows no systematic changes until prolonged drinking has taken place. This indicates that arousal mechanisms are unaffected by early alcoholization but become significantly involved in long-term drinking. SWS systems on the other hand seem to be immediately affected by alcohol.

The correlation between SWS and spectral EEG analysis proved of interest. During the abstinent periods the amount of SWS correlated significantly with delta ($r = .84$) and theta ($r = .73$) energy. During the alcoholization period these correlations tended towards zero as duration of drinking increased. The breakdown of this correlation may be due to the presence of alpha-delta patterns in the records. Persistent alpha-delta or delta waves interspersed with fast activity would tend to augment SWS scoring by the Sleep Manual but would tend to decrease delta energy towards baseline while alpha and/or beta energy might increase. These observations suggest that sleep patterning and EEG energy cycling should be studied further.

REFERENCES

Adamson, J. and Burdick, J. A. Sleep of dry alcoholics. Archives of General Psychiatry, 1973, 28, 146-149.

Allen, R. P., Wagman, A., Faillace, L. and McIntosh, M. Electroencephalographic (EEG) sleep recovery following prolonged alcohol intoxication in alcoholics. Journal of Nervous and Mental Diseases, 1971, 153, 424-433.

Gross, M. M., Goodenough, D. R., Nagarajan, M. and Hastey, J. M. Sleep changes induced by 4 and 6 days of experimental alcoholization and withdrawal in humans. In (M.M. Gross, editor) Alcohol Intoxication and Withdrawal: Experimental Studies I, Advances in Experimental Medicine and Biology, Vol. 35, pp. 291-304, Plenum Press, New York, 1973.

Hartmann, E. Longitudinal studies of sleep and dream patterns in manic-depressive patients. Archives of General Psychiatry, 1968, 19, 312-329.

Hauri, P. and Hawkins, D. R. Alpha-Delta sleep. Electroencephalography and Clinical Neurophysiology, 1973, 34, 233-237.

Johnson, L. C., Burdick, J. A. and Smith, J. Sleep during alcohol intake and withdrawal in the chronic alcoholic. Archives of General Psychiatry, 1970, 22, 406-418.

Knowles, J. B., Laverty, S. G. and Kuechler, H. A. Effects of alcohol on REM sleep. Quarterly Journal of Studies of Alcohol, 1968, 29, 342.

Rechtschaffen, A. and Kales, A. A Manual of Standardized Terminology, Techniques and Scoring Systems for Sleep Stages of Human Subjects. U. S. Department HEW, National Institute of Health, Neurology Information Network, Bethesda, Maryland.

Rosadini, G., Rodriguez, G. and Siani, C. Acute alcohol poisoning in man: An experimental electrophysiological study. Psychopharmacologia, 1974, 35, 273-285.

Stern, M., From, D. H., Wyatt, R., Grinspoon, L. and Tursky, B. All night sleep studies of acute schizophrenics. Archives of General Psychiatry, 1969, 20, 470-477.

Watson, R., Hartmann, E. and Schildkraut, J. J. Amphetamine withdrawal: Affective state, sleep patterns, and MHPG excretion. American Journal of Psychiatry, 1972, 129, 263-269.

Yules, R. B., Lippmann, M. E. and Freedman, D. X. Alcohol administration prior to sleep. Archives of General Psychiatry, 1967, 16, 94-97.

THE RELATION BETWEEN BASELINE SLOW WAVE SLEEP AND THE SLOW WAVE SLEEP RESPONSE TO ALCOHOL IN ALCOHOLICS

Milton M. Gross and John M. Hastey
Division of Alcoholism & Drug Dependence
Dept. of Psychiatry
Downstate Medical Center
Brooklyn, New York, U.S.A.*

INTRODUCTION

Important advances in the psychophysiological studies of sleep, triggered by the breakthroughs of Aserinsky and Kleitman (1953) and Dement and Kleitman (1957), led to investigations of the possible relevance of such studies to psychopathology. In alcoholics, striking departures from normal sleep psychophysiology were observed during and following acute intoxication and withdrawal (see reviews by Johnson, 1971, Gross et al., 1971, Williams and Salamy, 1972 and Gross et al., 1974). The disturbances observed in alcoholics involved the rhythmicity and composition of sleep. Prominent among the disturbances of sleep composition were those involving Slow Wave Sleep (SWS).

There was general agreement in the findings that during acute alcohol withdrawal the % SWS tended to be decreased to absent, and subsequently tended to increase over increasing time of abstinence (Gross et al., 1966, Gross and Goodenough, 1968, Johnson et al., 1970, Allen et al., 1971, Gross et al., 1971, Johnson, 1971, Lester et al., 1973, Gross et al., 1973, Gross et al., 1974). However, these same studies did not agree on the post-acute withdrawal SWS levels, or, on the effects of alcohol intake on SWS. Furthermore, while comparisons between studies suggested a relation between baseline % SWS (which is, in essence, post-acute withdrawal from the preceding drinking episode) and the effect of alcohol on the SWS, this suggested relationship was based upon differences in the mean baseline % SWS of each sample. Within-sample studies of this relationship had not been done.

*Supported by NIAAA Grant AA01236

The purpose of this study was: (1) to examine the range of baseline % SWS in a sample of alcoholics, (2) to determine the relation between baseline SWS and the effects on SWS of four days of regulated heavy drinking, (3) to compare this relationship on heavy drinking days 1 and 2 with heavy drinking days 3 and 4, and (4) to compare the relationship between baseline SWS and SWS during heavy drinking, for the group of alcoholics who had reduced REM during heavy drinking and those who had no REM during heavy drinking. (The alcoholics who had reduced REM and those who had no REM during heavy drinking were compared elsewhere (Gross, 1973) and appeared to differ in their alcoholization characteristics primarily on the basis of higher blood alcohol concentrations for those who had no REM. However, it is somewhat different in that it represents the interaction of the blood alcohol concentrations and a CNS response).

METHODOLOGY

All participants were paid male alcoholic volunteers whose participation was sought and determined after they were recovered from the acute alcohol withdrawal which had brought them to hospital. Half of them were white, half were black. Their ages ranged from 26-48 with an average age of 33.7 (S.D. 6.7). They were well nourished and free of medical and psychiatric complications prior to the onset of the study. All were gamma alcoholics with histories of more than five years of heavy drinking. All ingested larger daily quantities of alcohol during heavy drinking outside the hospital than they received during the study. One volunteer was studied at a time.

The studies were begun 3-5 weeks after hospitalization with an average interval between admission and study onset of approximately 3.5 weeks. Medication was limited exclusively to the first 6 days in hospital.

Alcohol was administered on a fixed schedule and dosage. Four volunteers received alcohol for 5 days, six received alcohol for 7 days, one received alcohol for 8 days and one received alcohol for 9 days. The first day of drinking was half dose and the remaining drinking days were full dose (approximately 1.6 gm/kilo and 3.2 gm/kilo for half and full dose respectively). Equal doses were administered hourly between 2 PM and midnight with the exception of 3 PM when none was given. Four volunteers also received a dose at 10 AM. During the alcohol administration period, breathalyzer determinations were obtained at 6 AM, 2 PM (prior to the first drink) and midnight (prior to the last drink). A Stephenson Breathalyzer was used.

Sleep was regulated and permitted only from midnight to 6 AM. A member of the research nursing team was always present. Sleep was monitored each study night on an 8 channel Grass EEG console. Electrode placements and scoring were done in accordance with the Rechtschaffen Kales manual (1968). Sleep was recorded and discarded the night prior to the first baseline day.

All participants had 3 consecutive baseline days immediately prior to the period of alcohol intake and 6 consecutive days of withdrawal immediately after alcohol intake. Baseline, alcohol intake and withdrawal periods were continuously monitored.

For this communication the baseline period will be examined in relation to the first four days of heavy drinking.

FINDINGS

The baseline average daily % SWS among the participants ranged from 0.7-44%. Three of the volunteers had average daily levels below 20%, six between 20 and 30%, and three above 30% (Figure 1).

The daily average midnight blood alcohol levels ranged from 220-300 mg/100 ml.

The average daily % SWS during baseline was examined in relation to the average daily % SWS during the first four days of heavy drinking (Figure 1). A significant correlation was found ($r=.79$, $p< .02$; $rho=.77$, $p< .01$).

The data indicated that those with between 20 and 40% SWS baseline levels had an increase of SWS in response to the heavy alcohol intake, and that this increase in SWS was positively related to the baseline level. The data suggested that at baseline levels below 20% SWS the effect of heavy alcohol intake tended to be a reduction of SWS.

Examination of the SWS response in relation to baseline suggested the possibility of a composite pattern of two responses (Figure 1). The sample was partitioned on the basis of those who had reduced REM (RR) and those who had no REM (NR) during heavy drinking. There were six alcoholics in each subgroup. There was no significant difference in the baseline % REM between the groups. The no REM group had significantly higher blood alcohol concentrations and this was assumed to be the primary difference between the two groups.

Figures 2 and 3 demonstrate that, indeed, the total group response was a composite of the two subgroups. The prominent dif-

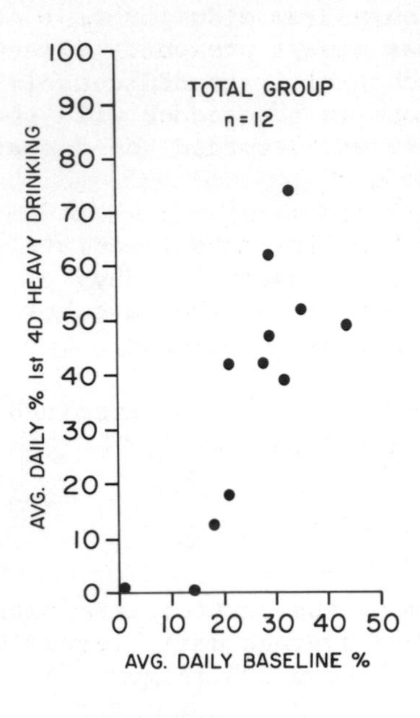

Figure 1

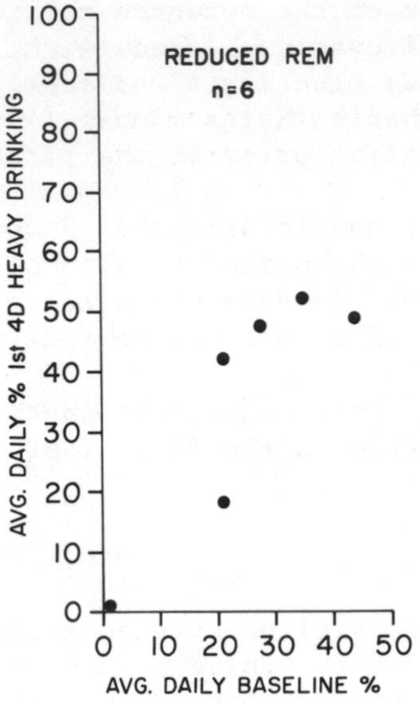

Figure 2

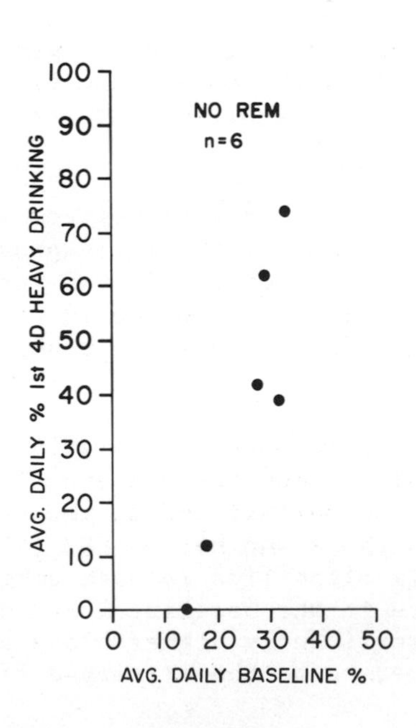

Figure 3

ference between them was in the 30-40% baseline SWS range where the no REM group continued to show an increase in the SWS response while the reduced REM group plateaued.

The correlation between baseline and heavy drinking % SWS for the reduced REM group was significant ($r=.78$, $p<.05$; $rho=.83$, $p<.05$. The correlation between baseline and heavy drinking % SWS for the no REM group was $r=.84$ ($p<.05$) and $rho=.60$ (ns).

The relation between baseline SWS and the SWS during alcohol intake was present for the entire group on heavy drinking Days 1 and 2 and also on heavy drinking Days 3 and 4 (Figure 4). The correlations were significant for both pairs of days (Days 1 and 2, $r=.72$, $p<.01$, $rho=.63$, $p<.05$; Days 3 and 4, $r=.75$, $p<.01$, $rho=.79$, $p<.01$).

The relation between baseline SWS and SWS during drinking for the pairs of heavy drinking days was examined in the reduced REM and no REM subgroups (Figure 5).

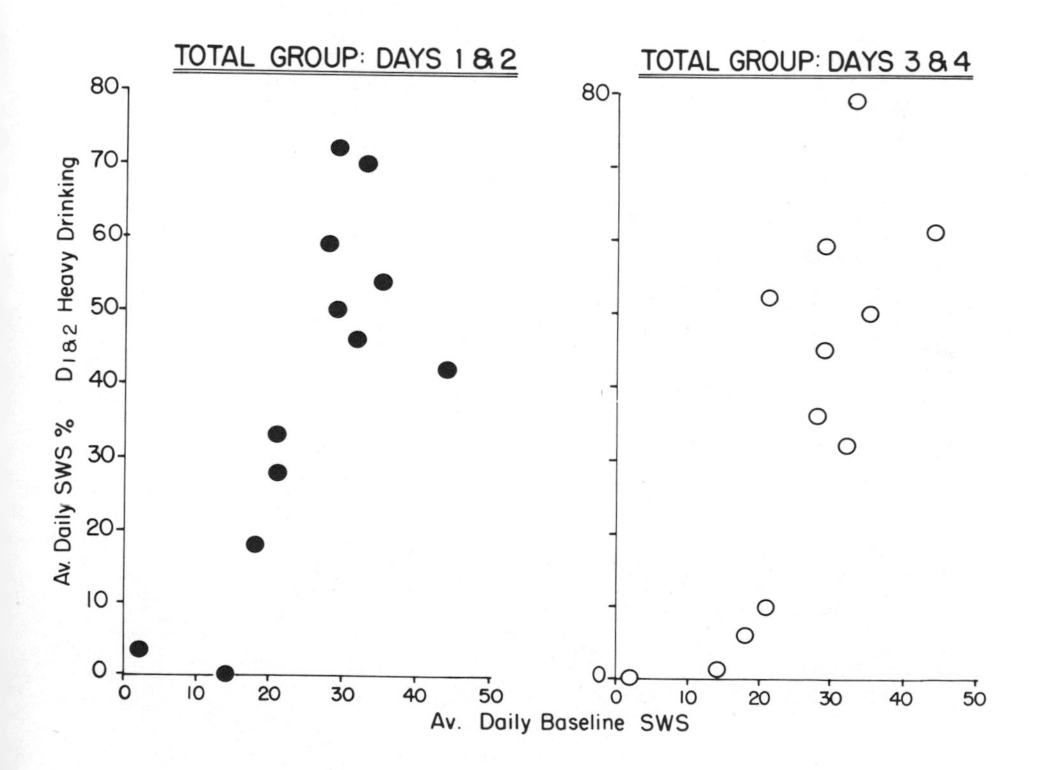

Figure 4

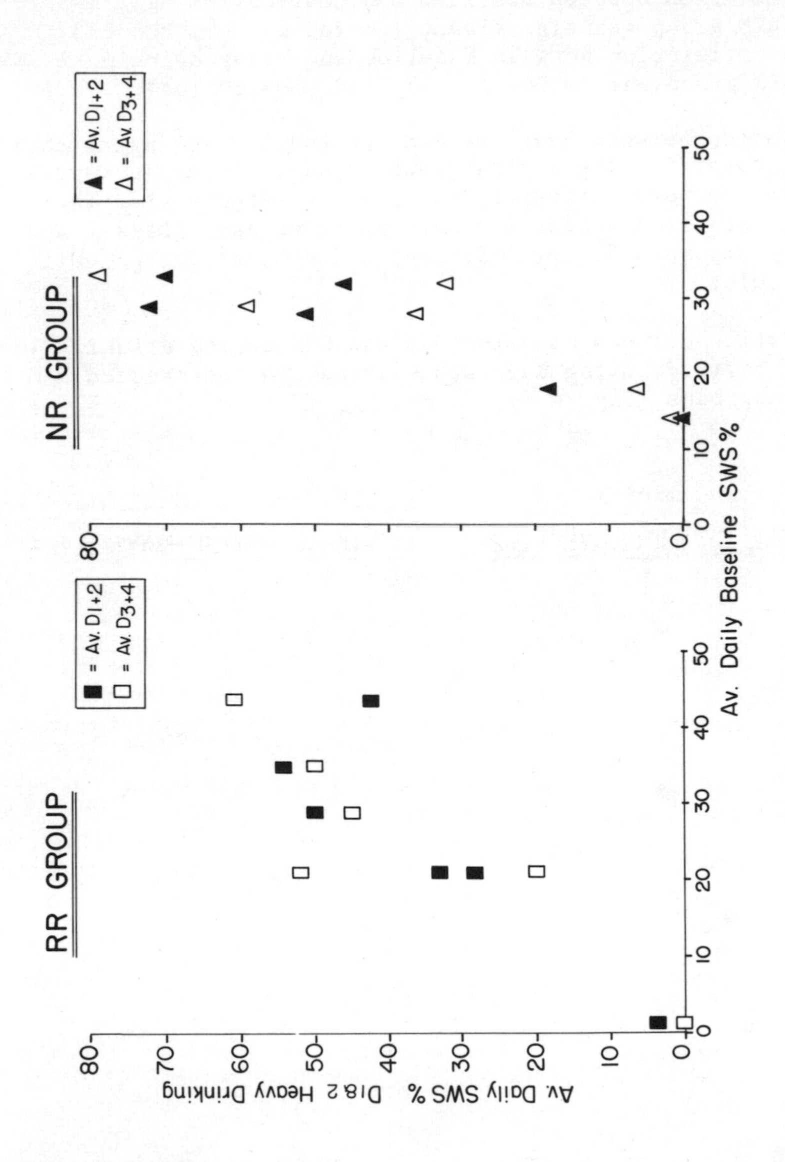
RR GROUP
■ = Av. D1+2
□ = Av. D3+4
NR GROUP
▲ = Av. D1+2
△ = Av. D3+4
Av. Daily SWS % D1 & 2 Heavy Drinking
Av. Daily Baseline SWS %
0
10
20
30
40
50
60
70
80

Figure 5

In the reduced REM group for Days 1 and 2, r=.47, ns; rho=.34, ns; Days 3 and 4, r=.70, ns; rho=.75, ns.

In the no REM group for Days 1 and 2, r=.86, $p < .05$, rho=.54, ns. For Days 3 and 4, r=.80, ns, rho=.60, ns.

DISCUSSION

The findings of this study demonstrated that within a sample of alcoholics the % Slow Wave Sleep during baseline may range from very low to very high. This was observed following treatment for acute withdrawal and a prolonged interval of abstinence from alcohol and psychoactive medication.

The study also demonstrated that the effects of alcohol on the Slow Wave Sleep of alcoholics was related to the baseline level of Slow Wave Sleep. During the first 4 days of heavy drinking: those with baseline levels of less than 20% appeared to have a decrease or no change; those with baseline levels between 20 and 40% tended to have an increase in Slow Wave Sleep. The rate of increase of Slow Wave Sleep in the 20-40% group appeared related to the baseline Slow Wave Sleep level and the blood alcohol concentrations during drinking. Although not reported in this study, the SWS tended to decrease with continued drinking beyond the 4 days.

It appears likely that the differences in the literature are related to several factors among which are sampling differences (including age and pre-study preparation), differences of duration of abstinence and differences in duration and quantity of alcohol administration.

The possible clinical implications of the differences in baseline Slow Wave Sleep and the Slow Wave Sleep response to alcohol are examined elsewhere in this volume (Gross et al., Gross and Best, Wagman and Allen, and Zarcone et al.).

REFERENCES

Allen, R.P., Wagman, A., Faillace, L.A. and McIntosh, M., 1971. Electroencephalographic (EEG) sleep recovery following prolonged alcohol intoxication in alcoholics. J. Nerv. and Ment. Dis. 153:424-433.

Aserinsky, E. and Kleitman, N., 1953. Regularly occurring periods of eye motility and concomitant phenomena during sleep. Science 118:273-274.

Dement, W.C. and Kleitman, N., 1957 Cyclic variations in EEG during sleep and their relation to eye movements, body motility, and dreaming. Electroenceph. Clin. Neurophysiol. 9: 673-690.

Gross, M.M., Goodenough, D.R., Tobin, M., Halpert, E., Lepore, D., Perlstein, A., Sirota, M., DiBianco, J., Fuller, R. and Kishner, I., 1966. Sleep disturbances and hallucinations in the acute alcoholic psychoses. J. Nerv. and Ment. Dis. 142: 493-514.

Gross, M.M. and Goodenough, D.R., 1968. Sleep disturbances in the acute alcoholic psychoses. In: Psychiatric Research Report 24, pp. 132-147, American Psychiatric Association, New York.

Gross, M.M., Goodenough, D.R., Hastey, J.M., Rosenblatt, S.M. and Lewis, E., 1971. Sleep disturbances in alcohol intoxication and withdrawal. In: (Eds.) N.K. Mello and J.H. Mendelson, Recent Advances in Studies of Alcoholism, pp. 317-397, U.S. Government Printing Office, Washington, D.C.

Gross, M.M., 1973. Sensory Superactivity: A preliminary report on an hypothetical model for an hallucinogenic mechanism in alcohol withdrawal. In: (Ed.) M.M. Gross, Alcohol Intoxication and Withdrawal: Experimental Studies, Advances in Experimental Medicine and Biology, Vol. 35, pp. 321-330, Plenum Press, New York.

Gross, M.M., Goodenough, D.R., Nagarajan, M. and Hastey, J.M., 1973. Sleep changes induced by 4 and 6 days of experimental alcoholization and withdrawal in humans. In: (Ed.) M.M. Gross, Alcohol Intoxication and Withdrawal: Experimental Studies, Advances in Experimental Medicine and Biology, Vol. 35, pp. 291-304, Plenum Press, New York.

Gross, M.M., Lewis, E. and Hastey, J., 1974. Acute alcohol withdrawal syndrome. In: (Eds.) B. Kissin and H. Begleiter, The Biology of Alcoholism, Vol. 3, pp. 191-264, Plenum Press, New York.

Gross, M.M., Hastey, J.M., Lewis, E. and Young, N. (This volume) Slow Wave Sleep and carry-over of functional tolerance and physical dependence in alcoholics.

Gross, M.M. and Best, S. (This volume) Behavioral concomitants of the relationship between baseline Slow Wave Sleep and carry-over of tolerance and dependence in alcoholics.

Johnson, L.C., 1971. Sleep patterns in chronic alcoholics. In: (Eds.) N.K. Mello and J.H. Mendelson, Recent Advances in Studies of Alcoholism, pp. 288-316, U.S. Government Printing Office, Washington, D.C.

Johnson, L.C., Burdick, J.A. and Smith, J., 1970. Sleep during alcohol intake and withdrawal in the chronic alcoholic. Arch. Gen. Psychiat. 22:406-418.

Lester, B.K., Rundell, O.H., Cowden, L.C. and Williams, H.L., 1973. Chronic alcoholism, alcohol and sleep. In: (Ed.) M.M. Gross, Alcohol Intoxication and Withdrawal: Experimental Studies, Advances in Experimental Medicine and Biology, Vol. 35, pp. 261-279, Plenum Press, New York.

Rechtschaffen, A. and Kales, A., 1968. A Manual of Standardized Terminology, Techniques and Scoring System for Sleep Stages of Human Subjects. U.S. Government Printing Office, Washington, D.C.

Wagman, A. and Allen, R.P. (This volume) Effects of alcohol ingestion and abstinence on Slow Wave Sleep of alcoholics.

Williams, H.L. and Salamy, A., 1972. Alcohol and sleep. In: (Eds.) B. Kissin and H. Begleiter, The Biology of Alcoholism, Vol. 2, pp. 435-483, Plenum Press, New York.

Zarcone, V., Barchas, J., Hoddes, E., Montplaisir, J., Sack, R. and Wilson, R. (This volume) Experimental ethanol ingestion: Sleep variables and metabolites of Dopamine and Serotonin in the cerebrospinal fluid.

SLOW WAVE SLEEP AND CARRY-OVER OF FUNCTIONAL TOLERANCE AND PHYSICAL DEPENDENCE IN ALCOHOLICS

Milton M. Gross, John M. Hastey,
Eastlyn Lewis and Norma Young
Division of Alcoholism & Drug Dependence
Dept. of Psychiatry
Downstate Medical Center, Brooklyn, New York, U.S.A.*

INTRODUCTION

Alcohol affects sleep. In alcoholics, profound sleep changes have been observed clinically and experimentally before, during and after a period of heavy alcohol intake. Elucidation of the alcohol-sleep interactions may illuminate mechanisms involved in alcoholism. This communication will focus on the post-drinking Slow Wave Sleep (SWS) which tends to be decreased initially and gradually recovers. More specifically the post-drinking SWS will be examined in relation to the rate of reacquisition of functional tolerance and physical dependence in a subsequent drinking episode.

The initial SWS reduction of the post-drinking phase has been observed to be of variable magnitude and duration. The duration of SWS reduction ranges across a continuum from days to months and possibly, even years (Gross et al., 1966, Gross and Goodenough, 1968, Johnson et al., 1970, Allen et al., 1971, Gross et al., 1971, Johnson, 1971, Lester et al., 1973, Gross et al., 1974, Wagman and Allen, this volume, Zarcone et al., this volume). Viewed in relation to the preceding drinking episode, the reduction of SWS appeared to be a sleep correlate of the acute withdrawal syndrome during the first few days (Gross and Goodenough, 1968); when the SWS reduction persisted for extended periods of time, the possibility was suggested that it was a correlate of a subacute, even in some chronic, subclinical withdrawal syndrome (Gross et al., 1971, Gross et al., 1973, Gross et al., 1974).

*Supported by NIAAA Grant AA01236

What of the relation of the post-drinking SWS to the subsequent drinking episode? The fact that the SWS reduction appeared related to a component of the addictive mechanism (the acute withdrawal syndrome), and that the SWS reduction could persist "silently" for extended periods of time thereafter, suggested that the SWS might be related to another persistent "silent" addictive mechanism which Kalant (1973) designated "carry-over."

Carry-over refers to the residual alcohol effects which result in functional tolerance and physical dependence once acquired, being reacquired more rapidly during a subsequent episode of alcohol administration.

The existence of carry-over is consistent with the experimental observations in humans (Mendelson et al., 1966) and in animals (Branchey et al., 1971 and Kalant, 1973). Since the data to be reported were obtained from alcoholic volunteers treated for withdrawal approximately three weeks prior to study, the baseline SWS level was hypothesized as negatively related to the level of carry-over from the preceding episode(s) of drinking and withdrawal. The baseline SWS was then examined in relation to the rate of reacquisition of functional tolerance and physical dependence during experimental alcoholization and withdrawal.

METHODOLOGY

All subjects were paid male alcoholic volunteers. They were approached following recovery from the acute alcohol withdrawal which had brought them to hospital. The experimental studies were begun approximately 3-5 weeks after hospitalization with an average interval between admission and study onset of approximately 3.5 weeks. Medication was used only for the first 6 days in hospital with no further medication administered prior to nor during the experimental study.

The volunteers were well nourished, in good general health and free of functional or organic psychoses. Half of them were white, half were black. All were gamma alcoholics who had histories of more than 5 years of heavy drinking. All ingested larger daily quantities of alcohol when drinking heavily outside the hospital than the quantities they received during the experimental study. Four volunteers received alcohol for 5 days and six volunteers received alcohol for 7 days. The first day of drinking was half dose and the remaining 4 and 6 days respectively were at full dose. Their ages ranged from 26 to 48 with an average age of 34.6 (S.D. 8.1).

All volunteers had 3 consecutive baseline days immediately

prior to the drinking period and 7 consecutive withdrawal and recovery days immediately following the drinking period. On the first drinking day, approximately 1.6 gm alcohol/kilo were given which was followed by 4 or 6 days of approximately 3.2 gm alcohol/kilo/day. Care was taken to maintain good nutrition throughout the study. One volunteer was studied at a time.

Alcohol (Canadian Club) was administered on a fixed schedule and dosage. Equal doses were administered hourly between 2 PM and midnight with the exception of 3 PM when none was given. Two volunteers also received a dose at 10 AM. During the alcohol administration period breathalyzer determinations were made at 6 AM, 2 PM and midnight (prior to the last drink). A Stephenson Breathalyzer was used.

Sleep was permitted on a scheduled basis only, from midnight to 6 AM. The volunteers were under continuous surveillance and napping was not permitted. Sleep was monitored each night on an 8 channel Grass EEG console. Electrode placements and scoring were done in accordance with the Rechtschaffen and Kales Manual (1968). Sleep was recorded and discarded the night prior to the first baseline day.

A member of the research nursing team was with the volunteer at all times. The nurse systematically evaluated him at 10 PM, 6 AM, and 1 PM using a 30 item quantitative clinical instrument, the Total Severity Assessment (TSA) (Gross et al., 1973a; Gross et al., 1974).

Serial daily psychometric testing was done at 1 PM and 3 PM. This coincided with the period prior to and after the first drink of the 10 hour drinking sequence during the alcoholization phase of the study (see Gross and Best, this volume).

Based upon the mean baseline % SWS, the sample was divided into two groups, the higher SWS and lower SWS groups. The higher SWS group had a mean daily baseline of 36% SWS (32-44%); the lower SWS group had a mean daily baseline of 25.3% SWS (14-29%). The average age of the higher SWS group was 32.3 (26-48); the average age of the lower SWS group was 36.5 (28-46).

Of those volunteers who drank for 5 days, 2 were in the higher SWS group and 2 were in the lower SWS group; of those who drank for 7 days, 2 were in the higher SWS group and 4 were in the lower SWS group. Since there was an unequal distribution of volunteers for the 5 and 7 days of drinking in the two SWS groups, the experimental findings in the lower SWS group will be reported on the basis of the mean of the means of the 5 day and the 7 day volunteers. In this way, any distortion of the findings in the lower SWS group resulting from more of the volunteers having a longer period of drinking was avoided. Furthermore, comparisons during the drinking period between the two SWS groups were limited to the first 5 days of drinking only.

FINDINGS

The maximum daily blood alcohol levels obtained were prior to the last drink at midnight. A comparison of the higher and lower baseline SWS groups during the drinking days indicated that the daily average blood alcohol levels in both groups were similar (Figure 1).

A sensitive indication of functional tolerance to alcohol is a decrease of % Total Sleep Time (TST) (Kalant et al., 1971). The average TST during baseline and drinking were compared (Figure 2). The average % TST for both SWS groups were less than 100% during baseline. The lower baseline SWS group appeared to have a lower average % TST during baseline than the higher SWS group. During the drinking days both groups tended to show a parallel decrease in daily average % TST in which the lower SWS group maintained the lower level throughout.

The effects of alcohol on % SWS was compared for the two baseline SWS groups (Figure 3). Both SWS groups behaved similarly

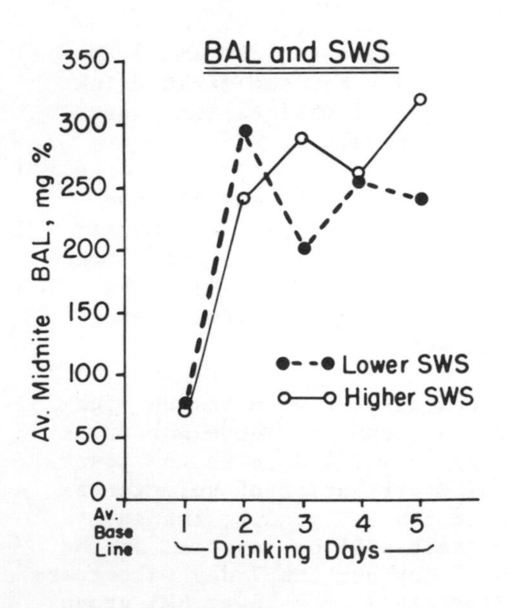

Figure 1

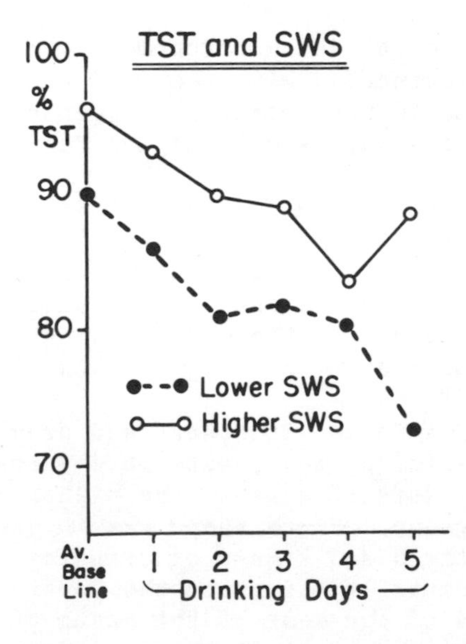

Figure 2

on the first 2 days of drinking. A difference was observed thereafter when the lower baseline SWS group showed a somewhat irregular, progressive decrease of daily average % SWS, while the higher baseline SWS group continued to have an increase of % SWS through the fourth day with a decrease on Day 5.

Behavioral evidence of intoxication during drinking was compared for the two baseline SWS groups using those variables of the quantitative clinical assessment (TSA)-impaired consciousness, clouding of the sensorium, impaired contact, impaired gait and nystagmus-which were related to behavioral intoxication. (These variables were factorially related (Factor III) and have a bimodal distribution, increasing with increasing intoxication and with increasing withdrawal (Gross et al., 1971b, Rosenblatt et al., 1972, Gross et al., 1974). During drinking days the TSA administered at 10 PM was the maximal alcohol effect evaluation.

Comparison of the 10 PM levels of behavioral intoxication (Factor III) between the two baseline SWS groups indicated that

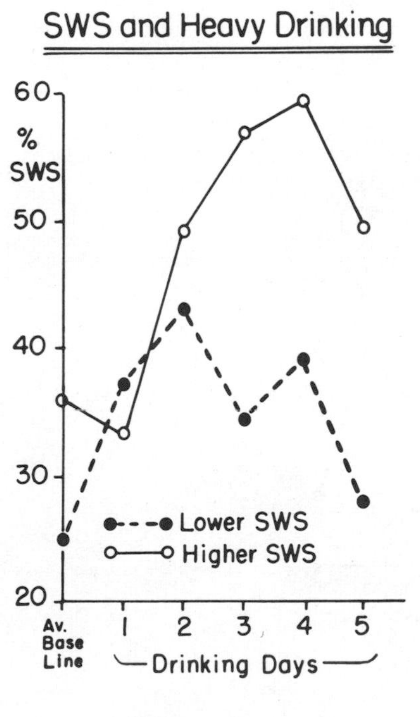

Figure 3

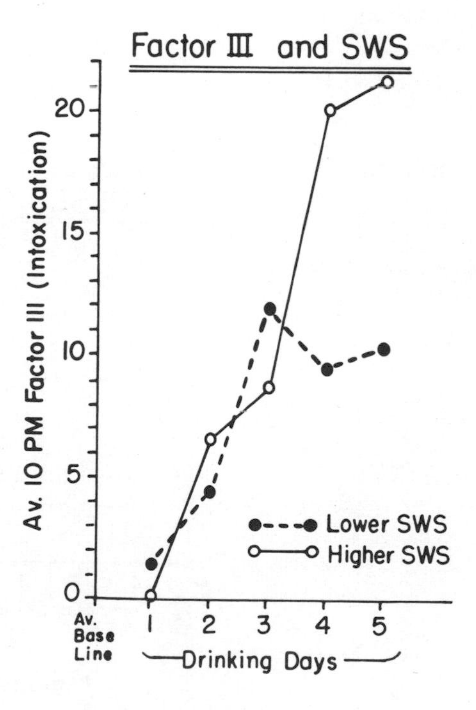

Figure 4

for the first 3 drinking days the groups behaved similarly, while on drinking Days 4 and 5 they behaved differently (Figure 4). There was minimal evidence of intoxication on the first drinking day at the half dose level of alcohol intake. On drinking Days 2 and 3 the average 10 PM levels of intoxication of both groups increased sharply and similarly. The lower baseline SWS group had its peak average daily intoxication level on drinking Day 3. In contrast, the higher baseline SWS group showed a continued increase so that by drinking Day 5, the average level of intoxication was almost twice that which was reached at the peak for the lower baseline SWS group.

The pulse rates at 10 PM were compared for both SWS groups (Figure 5). Similar to the manner of the Factor III variables, pulse rate also tends to have a bimodal distribution, increasing in relation to increasing intoxication and then, to increasing withdrawal. The average daily baseline pulse rates at 10 PM were essentially the same for both SWS groups. On drinking Days 2 and 5 both groups were similar in their average daily pulse although the lower baseline SWS group had a slightly lower average pulse

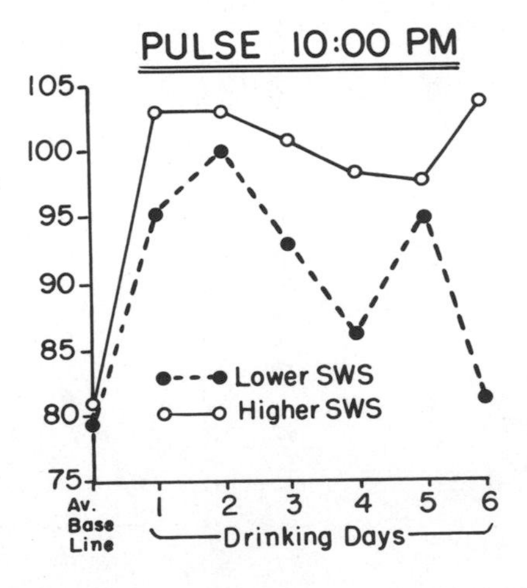

Figure 5

rate on those days. However, on drinking Days 1, 3 and 4 the lower baseline SWS group had considerably lower daily average 10 PM pulse rates. (This difference was even more accentuated on drinking Day 6).

Differences in partial withdrawal (presumably a consequence of differences in physical dependence) were compared for both SWS groups during the periods of non-drinking of the drinking days. The experimental design was such that during drinking days the TSA at 6 AM and 1 PM assessed clinical manifestations of partial withdrawal. (Daily TSA scores organized by experimental days with the sequence of 10 PM, 6 AM and 1 PM).

The 6 AM nausea during drinking days was compared between the two SWS groups (Figure 6). Both groups showed essentially no nausea on the morning following the first drinking day. On the second drinking day the two groups showed a comparable increase. On drinking Days 3 and 4 the average severity of nausea was greater for the lower baseline SWS group. On the fifth drinking day the average 6 AM nausea rose sharply in the higher baseline SWS group so that both SWS groups were again similar.

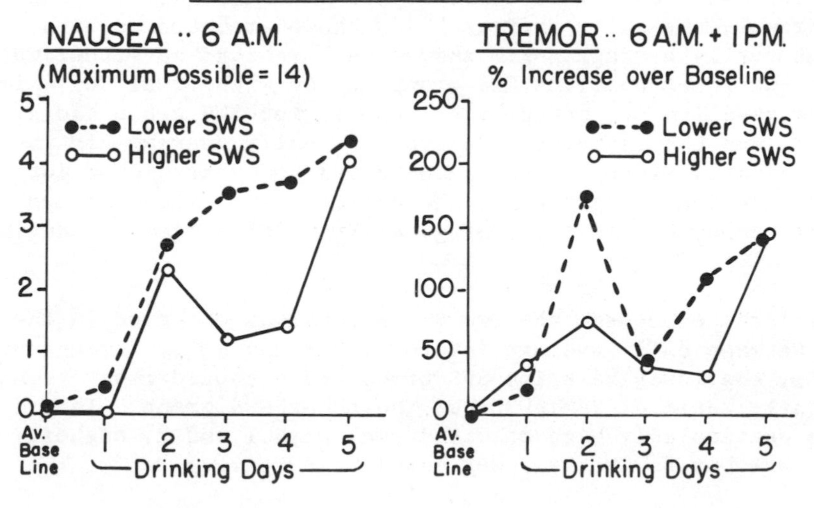

Figure 6

Figure 7

Comparison of the sum of the average % increase of tremor (compared to baseline) at 6 AM and at 1 PM indicated differences between the two SWS groups (Figure 7). The higher baseline SWS group showed relatively low levels of average % increase in tremor for the first 4 drinking days and rose sharply to a 150% increase on drinking Day 5. The lower baseline SWS group showed a tendency for an earlier appearance of greater average increases in tremor severity. On drinking Days 2 and 4 the lower baseline SWS group had an increased % tremor more than twice the level of the other SWS group. On drinking Days 1 and 3 both groups showed similar small increases of tremor and on drinking Day 5 both groups had a comparable considerable average % increase in tremor.

The other clinical variables showed no overall indications of differences between the two SWS groups during partial withdrawal.

During withdrawal, changes of % SWS and % REM were compared for the two baseline SWS groups. Both groups had an average % SWS of approximately 10% on withdrawal Day 1, approximately 20% on withdrawal Days 3 and 5 and 25% on withdrawal Day 6 (Figure 8). The SWS groups differed on withdrawal Days 2 and 4, when the lower baseline SWS group had an average SWS of 18% on each day, while the higher baseline SWS group had 30% and 27% respectively. The lower baseline SWS group had a pattern of gradual SWS recovery with a return to baseline level on withdrawal Day 6, when this group had its peak average SWS level of the withdrawal phase. The level of average % SWS in the higher baseline SWS group peaked on withdrawal Day 2 and oscillated thereafter. By withdrawal Day 6, this higher SWS group was still considerably below baseline levels.

The two SWS groups were compared for % REM (Figure 9). During withdrawal, the daily average % REM showed rebound on withdrawal, the daily average % REM showed peak rebound on withdrawal Day 2 for the lower baseline SWS group and on withdrawal Day 4 for the higher baseline SWS group, i.e., the former SWS group had a more rapid, and the latter a more gradual, daily average REM rebound. The daily average % REM rebound was somewhat higher for the lower baseline SWS group but the difference between the two groups was comparable to the average daily % REM difference during baseline.

A difference between the two SWS groups was observed in the relation between daily average levels of REM and SWS. Throughout withdrawal, the lower baseline SWS group had a considerably higher average daily ratio of REM:SWS than the other SWS group. This ratio was particularly high on withdrawal Days 1 and 2, highest on withdrawal Day 2, in the lower baseline SWS group.

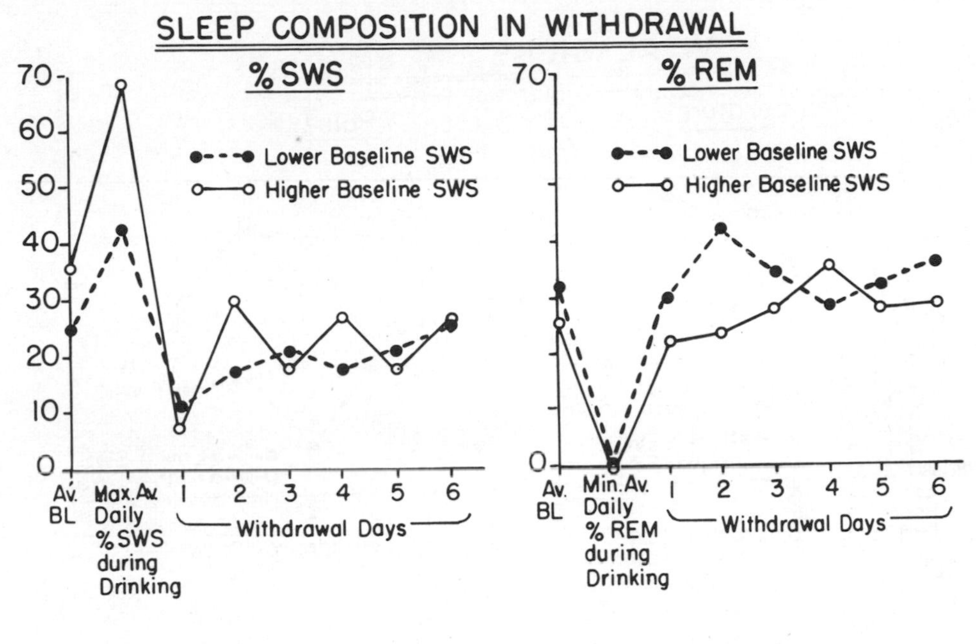

Figure 8 Figure 9

Temperature and pulse at 10 PM were compared for both SWS groups during withdrawal. (The average lowest daily 10 PM temperature for each group during drinking revealed that while this temperature was comparable for both SWS groups, it represented a greater decrease from average baseline temperatures for the higher baseline SWS group). In the higher baseline SWS group the average daily 10 PM temperature returned to average baseline SWS level on withdrawal Day 1 and remained there throughout with a slight increase on withdrawal Day 4 and a slight decrease on withdrawal Days 5 and 6 (Figure 10). The lower baseline SWS group demonstrated a rise in average daily 10 PM temperature to considerably above the average daily baseline level on withdrawal Day 1 and dropped progressively so that it crossed baseline level on withdrawal Day 3 and dropped to levels comparable to maximum decrease during drinking on withdrawal Day 4. Thereafter the temperature rose toward baseline.

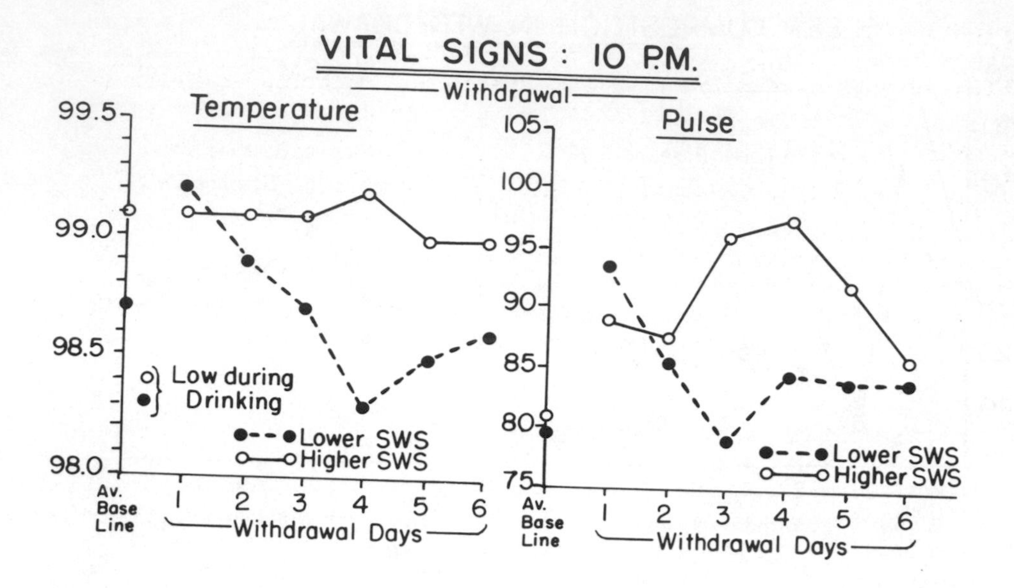

Figure 10

Figure 11

The two SWS groups demonstrated different patterns of change of the 10 PM pulse during withdrawal (Figure 11). For the higher baseline SWS group, on withdrawal Day 1 the 10 PM daily average pulse rate dropped (see Figure 5), fell slightly on withdrawal Day 2, rose for the next 2 days with a peak on withdrawal Day 4 and then declined toward baseline levels. In the lower baseline SWS group, the 10 PM daily average pulse rate increased on withdrawal Day 1, dropped for the next 2 days, returning to baseline on withdrawal Day 3 and rose to a slight plateaued increase for the last 3 days.

Clinically, the most striking difference observed in the quantitative assessment (TSA) of the two SWS groups during the experimental withdrawal was the fact that the only two volunteers who hallucinated (auditory and/or visual) occurred in the group with the lower baseline SWS who also demonstrated no REM during the heavy drinking period (Gross, 1973). Both volunteers who hallucinated had high levels of those TSA variables included in the hallucinogenic factor, Factor I (pruritus, tinnitus, visual disturbances, paresthesia, nausea, myalgia, agitation, sleep disturbances, tactile hallucinations and auditory and/or visual hallucinations) (Gross et al., 1974). Several of the volunteers in the higher baseline SWS group also had high levels of Factor I variables during the experimental withdrawal but reported no hallucinations. It is of interest that the average daily Factor I activity during withdrawal was similar for both SWS groups. Nor was there a clear indication of overall clinical differences between the two SWS groups in the remaining variables of the TSA.

Evidence will be reported elsewhere in this volume that the delayed peak of REM rebound in the higher baseline SWS group tends to be associated with a milder withdrawal syndrome of increased dysphoria (Gross and Best, this volume).

DISCUSSION

For several reasons, the observations reported must be considered preliminary. First, the sample size is small and the problem is complex. Second, the higher baseline SWS group was in fact a high SWS group ranging from 32-44%. However, high SWS levels have also been observed in alcoholics under conditions similar to our own baseline conditions by Zarcone (private communication). Third, the lower SWS group actually consisted of 5 volunteers with SWS within the normal range and only one with a low level of SWS. It will ultimately be necessary to examine the carry-over correlates of high, normal and low baseline SWS in alcoholics. Given these limitations, what indications do these data provide of a possible relation between baseline SWS and carry-over of functional tolerance and physical dependence?

Carry-over is the residual effect(s) of functional tolerance and physical dependence which results in the more rapid reacquisition of the tolerance and dependence on recurrence of alcohol intake (Kalant, 1973). <u>Carry-over is primarily a rate regulating factor</u>. <u>Secondarily, carry-over may have magnitude regulating effects</u>. For example, the more rapid reacquisition of tolerance may diminish the magnitude of behavioral intoxication.

In examining the data on the relation between the SWS groups and functional tolerance it is essential to keep in mind the observation by Kalant (1973) that functional tolerance is not a uniform state but actually a sequence of tolerances with a series of increasing thresholds. Although the classical pharmacologic demonstration of carry-over would involve greater tolerance across increasing alcohol doseage, it is equally reasonable to expect that a comparable demonstration of carrier-over would be the more rapid appearance of tolerance over days of similar dosage. This would be a rate related effect rather than a dose related effect.

The data suggested that the baseline level of SWS was a sleep correlate of carry-over of functional tolerance in which the SWS and carry-over were negatively correlated. The differences in the rates of reacquisition of tolerance of the two SWS groups appeared at different exposure times for different functions although for all the functions examined, the lower baseline SWS group appeared to demonstrate a higher level of carry-over.

While both SWS groups demonstrated a decrease in average % TST from the first drinking day (Figure 2), a lower level of average % TST was observed across drinking days in the lower SWS group. (The fact that the two groups were similarly different at baseline will be discussed in connection with carry-over of physical dependence below).

There was also a suggestion of a difference of the two SWS groups in the tolerance to the alcohol effect of increasing 10 PM pulse rate, which was observed from the first day of drinking (although the differences were most marked on drinking Days 4 and 6) in spite of the average baseline levels being almost identical (Figure 5).

There appeared to be a difference between the two SWS groups in their tolerance to the alcohol increasing the % SWS from drinking Days 3-5 (Figure 3). This effect is described at greater length elsewhere in this volume (Zarcone et al., Wagman and Allen, Gross and Hastey).

The differences in tolerance of the two SWS groups to the behavioral intoxicating effects of alcohol was observed last in the series of functions described, and appeared to occur on drinking Days 4 and 5 (Figure 4).

Psychometric data were consistent with these sleep and clinical observations (Gross and Best, this volume).

The indications of a possible relation between the baseline SWS and carry-over of physical dependence were less clear than was the case with functional tolerance.

During baseline, the lower baseline SWS group had a slightly lower % TST. This may indicate a higher level of carry-over of physical dependence than in the other SWS group. It has been demonstrated that withdrawal reduces % TST (Gross et al., 1973) and it has also been demonstrated that alcoholics under similar baseline conditions have a lower % TST than matched non-alcoholic controls (Lester et al., 1973).

During the hours of partial withdrawal on the drinking days, the data indicated that the lower baseline SWS group appeared to have a greater severity of nausea at 6 AM and tended to have a greater severity of tremor. These findings were consistent with the group with the higher carry-over of functional tolerance also having a greater carry-over of physical dependence. The other signs and symptoms of withdrawal were not clearly different between the two SWS groups during partial withdrawal.

In the sleep during total withdrawal from alcohol, the higher baseline SWS group tended to have a somewhat higher average % SWS during withdrawal. By withdrawal Day 6, both SWS groups were at essentially the same average % SWS level.

A rate regulating function related to baseline SWS was more clearly suggested by the REM during withdrawal. The average % REM of the lower SWS group peaked on withdrawal Day 2 while the average % REM of the higher SWS group peaked on Day 4. The more rapid rebound and the greater the rate of change from the extremely low (to absent) levels of REM during drinking, are likely to have clinical implications. It had been proposed that the withdrawal syndrome is the manifestation of CNS hyperexcitation (Weiss et al., 1964). This clinical implication is further emphasized when one considers that the daily average % REM:SWS is a much higher ratio in the lower baseline SWS group, particularly on the peak average rebound day. In the lower SWS group the ratio on peak rebound day was almost 3:1 while the ratio of the higher SWS group was approximately 3.5:2.5. It has been suggested that the SWS is a manifestation of inhibitory system activity and REM of excitatory system activity. (Feinberg et al., 1969). Within this theoretical framework, the disproportionate REM:SWS on the peak rebound day would be associated with an even greater net increase in excitation than the % REM alone would suggest. The only two volunteers who hallucinated in the course of the study did have the lower levels of baseline SWS. (It is possible that the rationale of treatment of the withdrawal syndrome with current therapies involves the controlled conversion of the more rapid discharge of excitation to a slower discharge by the tapering doses of drugs which are CNS depressants, cross tolerant to alcohol or to greatest advantage, both).

The lower baseline SWS group demonstrated a tendency for the 10 PM temperature and pulse to rise during the initial withdrawal day and then fall rapidly (Figures 10 and 11). It is of interest to note that, concurrent with the later REM rebound of the higher SWS group, there was also a rise in the 10 PM temperature and pulse.

The lack of clear indications of a relation between SWS and carry-over of physical dependence comparable to the evidence consistent with a correlation between baseline SWS and carry-over of functional tolerance, suggests that the physical dependence may involve more complex interactions with additional factors.

For both the functional tolerance and physical dependence, the baseline SWS appears to be a correlate of a basal, rate regulating factor which interacts with the alcohol intake. Primary magnitude effects are probably more critically regulated by the stimulus intensity, i.e., the quantity and duration of alcohol intake and by the blood alcohol concentrations interacting with other factors including the basal state of the organism.

The evidence of this preliminary study suggest that the baseline SWS level is a negative correlate of carry-over of functional tolerance. The evidence is consistent with baseline SWS being a correlate of physical dependence as well, but is not as clear as it is for the tolerance. Data in preparation suggest that differences of carry-over of physical dependence for the two SWS groups are critically related to the interaction with dimensions of alcohol intake.

It is vital to emphasize, that the possibility of the baseline SWS level being a correlate of carry-over, refers specifically to the SWS changes resulting from alcoholism. There are other causes of SWS abnormalities which may very well involve different neurophysical, neurochemical, neurophysiological and psychophysiological mechanisms. Such non-alcohol related abnormalities of SWS may in some, or perhaps no, instances relate to, or interact with carry-over mechanisms.

REFERENCES

Allen, R.P., Wagman, A., Faillace, L.A. and McIntosh, M., 1971. Electroencephalographic (EEG) sleep recovery following prolonged alcohol intoxication in alcoholics. J. Nerv. and Ment. Dis. 153:424-433.

Branchey, M., Rauscher, G. and Kissin, B., 1971. Modifications in the response to alcohol following the establishment of physical dependence. Psychopharmacologia 22:314-322.

Feinberg, I., Wender, P.H., Koresko, R.L., Gottlieb, F. and Piehuta, J.A., 1969. Differential effects of chlorpromazine and phenobarbital on EEG sleep patterns. J. Psychiat. Res. 7:101-109.

Gross, M.M., Goodenough, D.R., Tobin, M., Halpert, E., Lepore, D., Perlstein, A., Sirota, M., DiBianco, J., Fuller, R. and Kishner, I., 1966. Sleep disturbances and hallucinations in the acute alcoholic psychoses. J. Nerv. and Ment. Dis. 142: 493-514.

Gross, M.M. and Goodenough, D.R., 1968. Sleep disturbances in the acute alcoholic psychoses. In: Psychiatric Research Report 24, pp. 132-147, American Psychiatric Association, New York.

Gross, M.M., Goodenough, D.R., Hastey, J.M., Rosenblatt, S.M. and Lewis, E., 1971a. Sleep disturbances in alcohol intoxication and withdrawal. In: (Eds.) N.K. Mello and J.H. Mendelson, Recent Advances in Studies of Alcoholism, pp. 317-397, U.S. Government Printing Office, Washington, D.C.

Gross, M.M., Rosenblatt, S.M., Malenowski, B., Broman, M. and Lewis, E., 1971b. A factor analytic study of the clinical phenomena in the acute alcohol withdrawal syndromes. Alkohologia, 2:1-7.

Gross, M.M., Lewis, E. and Nagarajan, M., 1973a. An improved quantitative system for assessing the acute alcoholic psychoses and related states (TSA and SSA). In: (Ed.) M.M. Gross, Alcohol Intoxication and Withdrawal: Experimental Studies, Advances in Experimental Medicine and Biology, Vol. 35, pp. 365-376, Plenum Press, New York.

Gross, M.M., Goodenough, D.R., Nagarajan, M. and Hastey, J.M., 1973b. Sleep changes induced by 4 and 6 days of experimental alcoholization and withdrawal in humans. In: (Ed.) M.M. Gross, Alcohol Intoxication and Withdrawal: Experimental Studies, Advances in Experimental Medicine and Biology, Vol. 35, pp. 291-304, Plenum Press, New York.

Gross, M.M., 1973. Sensory Superactivity: A preliminary report on an hypothetical model for an hallucinogenic mechanism in alcohol withdrawal. In: (Ed.) M.M. Gross, Alcohol Intoxication and Withdrawal: Experimental Studies, Advances in Experimental Medicine and Biology, Vol. 35, pp. 321-330, Plenum Press, New York.

Gross, M.M., Lewis, E. and Hastey, J., 1974. Acute alcohol withdrawal syndrome. In: (Eds.) B. Kissin and H. Begleiter, The Biology of Alcoholism, Vol. 3, pp. 191-264, Plenum Press, New York.

Gross, M.M. and Best, S. (This volume) Behavioral concomitants of the relationship between baseline Slow Wave Sleep and carry-over of tolerance and dependence in alcoholics.

Gross, M.M. and Hastey, J.M. (This volume) The relation between baseline Slow Wave Sleep and the Slow Wave Sleep response to alcohol in alcoholics.

Johnson, L.C., 1971. Sleep patterns in chronic alcoholics. In: (Eds.) N.K. Mello and J.H. Mendelson, Recent Advances in Studies of Alcoholism, pp. 288-316, U.S. Government Printing Office, Washington, D.C.

Kalant, H., LeBlanc, A.E. and Gibbins, R.J., 1971. Tolerance to, and dependence on some non-opiate psychotropic drugs. Pharmacological Reviews, 23:135-191.

Kalant, H., 1973. Biological models of alcohol tolerance and physical dependence. In: (Ed.) M.M. Gross, Alcohol Intoxication and Withdrawal: Experimental Studies, Advances in Experimental Medicine and Biology, Vol. 35, pp. 3-13, Plenum Press, New York.

Lester, B.K., Rundell, O.H., Cowden, L.C. and Williams, H.L., 1973. Chronic alcoholism, alcohol and sleep. In: (Ed.) M.M. Gross, Alcohol Intoxication and Withdrawal: Experimental Studies, Advances in Experimental Medicine and Biology, Vol. 35, pp. 261-279, Plenum Press, New York.

Mendelson, J.H., Stein, S. and McGuire, M.T., 1966. Comparative psychophysiological studies of alcoholic and nonalcoholic subjects undergoing experimentally induced ethanol intoxication. Psychosom. Med. 28:1.

Rechtschaffen, A. and Kales, A., 1968. A Manual of Standardized Terminology, Techniques and Scoring System for Sleep Stages of Human Subjects. U.S. Government Printing Office, Washington, D.C.

Rosenblatt, S.M., Gross, M.M., Malenowski, B., Broman, M. and Lewis, E., 1972. Factor analysis of the daily clinical course rating scale of the acute alcoholic psychoses. Quart. J. Stud. Alc. 33:1060-1064.

Wagman, A. and Allen, R.P. (This volume) Effects of alcohol ingestion and abstinence on Slow Wave Sleep of alcoholics.

Weiss, A.D., Victor, M., Mendelson, J.H. and La Dou, J., 1964. Electroencephalographic findings. In: (Ed.) J.H. Mendelson, Experimentally Induced Chronic Intoxication and Withdrawal in Alcoholics, Quart. J. Stud. Alc. Suppl. 2, pp. 96-99.

Zarcone, V., Barchas, J., Hoddes, E., Montplaisir, J., Sack, R. and Wilson, R. (This volume) Experimental ethanol ingestion: Sleep variables and metabolites of Dopamine and Serotonin in the cerebrospinal fluid.

DO SLEEP PATTERNS RELATE TO THE DESIRE FOR ALCOHOL?

Richard P. Allen and Althea M. Wagman

Department of Psychiatry, Baltimore City Hospital,* and

Maryland Psychiatric Research Center, Baltimore,

Maryland

Sleep research on drugs of abuse has often been at least implicitly justified by a presumed relation between sleep disturbances and the development of addiction, particularly for hypnotics. Some relation between sleep changes and addiction seems compelling when it is noted that most short-acting hypnotics are drugs of abuse. Smith and Wesson (1974), for example, recently advanced the not unusual opinion that all short-acting hypnotics have abuse potential. Alcohol sleep research in particular has been, on occasion, explicitly justified by this apparent relation between sleep and addiction. Gross et al (1966) and Gross and Goodenough (1968) noted a possible causal relation between excessive drinking and sleep onset insomnia. Lester et al (1973) similarly indicated that alcoholics may start or continue drinking in order to reduce their sleep disturbance. These arguments suggest that an alcoholic's sleep disturbances may be one of the significant factors contributing to his disposition to drink. But aside from general observations there has been no attempt to directly test for this relation. Indeed, there has been relatively little experimental work on factors affecting an alcoholic's disposition to drink.

The experimental assessment of an alcoholic's disposition or probability for starting or continuing alcohol is itself inchoate in the literature on man. Two general types of approaches have been used: namely, behavioral and self-report procedures. Behaviorists have looked at disposition to drink for man in terms of

* supported by Grant No. AA00311-03.

amount of alcohol consumed as a function of such things as social factors (e.g., Nathan and O'Brien, 1971), and drinking related cues (Miller et al 1974). Other investigators have addressed the problem in terms of the alcoholic's self-report of subjective feelings of desire or "craving" for alcohol (e.g., Hore, 1972). Both of these approaches have serious drawbacks for the evaluation of sleep or other organism factors. On one hand, behavioral assessments involve consumption of fairly large amounts of alcohol which will usually more than suffice to significantly alter the organism variables, particularly sleep variables. On the other hand, self-report data is often unreliable, particularly for addictive behaviors where the discrepancy between actions and words is marked. A chronic alcoholic's self-report of disposition to drink and even of craving must be considered to be of uncertain validity. The problems of self-report validity seem inherent in the measure, while the problems of the behavioral assessment seem to be only technical. Thus, as a first approach attention was given to developing a suitable behavioral assessment of disposition to drink.

DEVELOPMENT AND VALIDATION OF A MEASURE OF DISPOSITION TO DRINK

In developing a measure of disposition to drink, consideration was given to four requirements: 1) consumption of no more than two ounces of 95 proof ethanol, which should have minimal effects on metabolic and psychological processes; 2) obtaining a quantitative value in a short testing time of less than one hour; 3) requiring a simple response which even an intoxicated and illiterate subject could learn and perform; 4) permit early morning testing so that drinking disposition is assessed at a time close to the recent sleep period and prior to any alcohol consumption for the day. Models for such a measure were taken from the operant animal literature. Behavioral assessments in animals of differential reinforcement values of drug stimuli have usually involved three general types of approaches: 1) relative rates of responding; 2) progressive ratio schedules; and 3) choice schedules (Griffiths, Brady and Winger, 1974). Only the first two approaches were considered in this study. Application of response rates to evaluate reinforcing stimulus strength has been difficult (Iglauer, 1972), but concurrent variable interval (VI) schedules have proven to be satisfactory (Catania, 1963). For this procedure two response levers are available with reinforcements available for each lever delivered on independent VI schedules. The progressive ratio procedure has been successfully used by Hodos and Kalman (1963) and involves progressively increasing the number of responses required for each reinforcement until the subject gives little or no responding for the reinforcement. In a pilot study a concurrent VI schedule on two levers for small increments of alcohol versus number of cigarettes was compared with a

TABLE I

PROGRESSIVE RATIO FOR ALCOHOL DELAY-REDUCTION RESPONSES (ADR)

Total		Increment	
ADR (No.)	Delay (Mins.)	ADR (No.)	Delay (Mins.)
300	210	300	30
700	180	400	30
1200	150	500	30
1800	120	600	30
2500	90	700	30
3300	60	800	30
3750	45	450	15
4200	30	450	15
4700	15	500	15
5200	0	500	15

progressive ratio schedule for reduction of delay in receiving an alcohol drink. For this progressive ratio reduction of delay can be considered the reinforcer with each successive delay reduction requiring more responses (see Table I). Pilot data indicated that the progressive ratio with a drink of two ounces of 95 proof ethanol provided adequate measures and was simpler than the concurrent schedules.

Measurement Methods for Alcohol Delay-Reduction Responses (ADR)

At 07:45 each day the subject was taken to a special testing area and told he could, if he wished, work on a counter button to obtain a two-ounce drink at about 3-1/2 hours after the test or earlier, depending upon how many responses he made on the event counter. He was shown the schedule for the number of responses required to reduce the drinking delay (Table I) and instructed to push the counter button as many times as he wished. The experimenter emphasized that this procedure was established to determine whether or not the subject wanted the drink and, if so, how soon he wanted the drink. The subject was permitted to respond on the counter until he indicated he was through. The instruction to the subjects and comparison of schedules are presented elsewhere (Funderburk and Allen, 1974). The number of ADR (alcohol delay-reduction responses) on the counter was recorded for each test session as a presumed direct measure of disposition to drink. The maximum number of ADR (5200) required approximately 10 to 15 minutes of continuous responding.

Validation of ADR as a Measure of Disposition to Drink

For validation purposes data were available for this measure applied to three specific conditions which are generally assumed to strongly affect disposition to drink, namely: alcohol withdrawal, experimental intoxication and low level alcohol consumption. It was specifically assumed that each of these conditions would increase disposition to drink and accordingly it was hypothesized that each condition would increase the number of ADR. Subjects were 20 to 45-year-old male alcoholics with more than four years of heavy drinking. They were admitted to a research ward intoxicated and were treated for three days with either moderate alcohol doses (9 to 18 ounces of 95 proof ethanol daily) or with chlordiazepoxide (100 to 200 mgs. daily). No other psychoactive medication was used aside from an occasional single routine dose of tylenol for headaches. Alcohol abstinence was considered to begin after the last treatment dose of either alcohol or chlordiazepoxide.

Alcohol Withdrawal. Eight subjects were tested, five were arbitrarily assigned to an early abstinence (EA) group and ADR were recorded starting immediately on their first day of alcohol abstinence. Three subjects were assigned to the late abstinence (LA) group with ADR recorded starting in the second week of abstinence. For each group the days in Figure 1 represent the days since starting the ADR during abstinence. An analysis of variance indicates

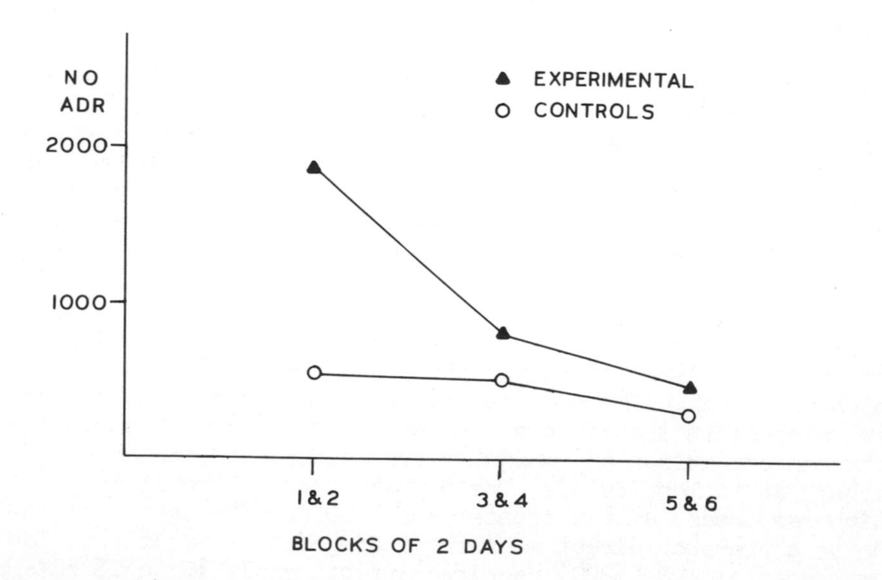

Figure 1. Mean ADR for Experimentals (EA) and Controls (LA).

TABLE II

ANALYSIS OF VARIANCE FOR NUMBER OF ADR FOR ALCOHOL ABSTINENCE CONDITIONS

Source Between Subject	MS	df	F	P<
Group (EA vs. LA)	2.12×10^6	1	14.8	.01
Error	1.43×10^5	6		
Within Subject				
Days	1.22×10^6	2	6.16	.02
Group x Days	6.29×10^5	2	3.17	.08
Error	1.98×10^5	12		

highly significant group and days effect with a borderline significant interaction effect (Table II). Since the EA group gave significantly more ADR than the LA group the effect was attributed to withdrawal rather than possible novelty of the test situation. (Duration of stay in the hospital could, of course, also have accounted for this effect, but it seems unlikely that an extra seven days in the hospital would by itself, without alcohol withdrawal, produce such a large effect.)

Experimental Intoxication. To evaluate the effects of experimental intoxication, three progressive dose levels were selected, namely: 18, 24 and 32 ounces daily of 95 proof ethanol mixed in juice or soda. Drinking was distributed from 8 a.m. to 11 p.m. and continued at each dose level for five consecutive days, followed immediately by the next dose level. Seven subjects who had been off alcohol and all other drugs for more than two weeks were placed on this schedule; two subjects failed to complete the protocol due to medical complications. Baseline ADR were obtained for five days prior to starting alcohol consumption.

Six of the seven subjects showed increases in ADR when on alcohol compared to baseline (sign test, $p < 0.10$). Figure 2 presents the average number of ADR for the last three days at each dose level for the five subjects completing the protocol. Baseline abstinence condition represents three days immediately prior to the day before alcohol consumption began and the final abstinence condition represents data for three days immediately prior to the day before terminating the research procedure after at least seven days of abstinence. One subject withdrew from the experiment after only three days in the final abstinence period and his data are not included in the averages for this period.

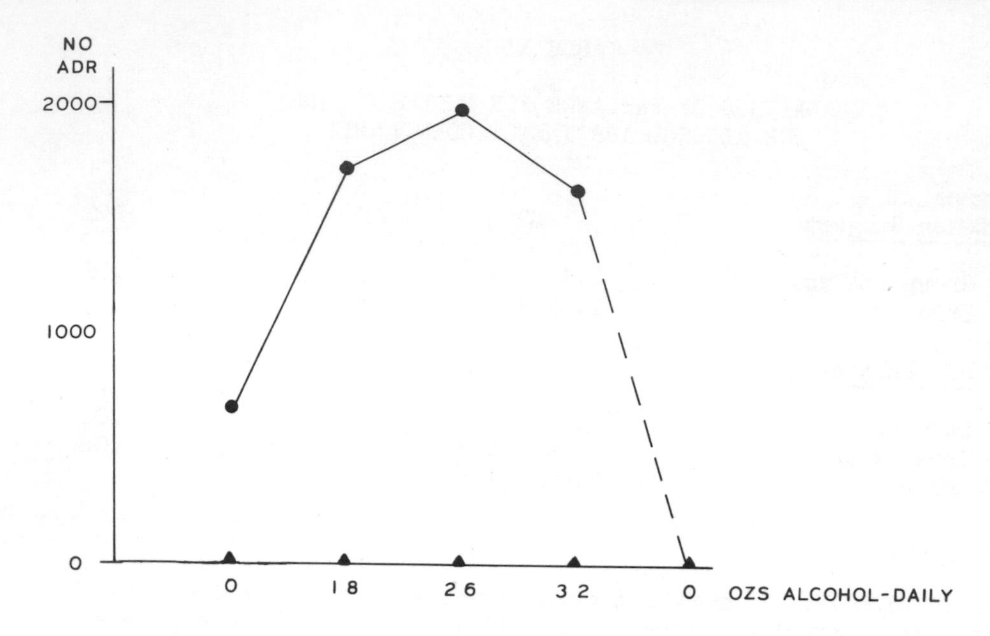

Figure 2. Mean ADR of five alcoholics during various amounts of alcohol consumption.

As can be seen in Figure 2 the number of ADR increases after 18 and 24 ounces but decreases at the highest dose. The final abstinence period produced values even lower than the baseline abstinence period.

The last two subjects in the experimental intoxication procedure were also asked to rate their attitude toward their first scheduled morning drink. They were asked to rate their subjective feelings about the drink on a seven point scale ranging from: "1 - I can't wait to get it" to "7 - I can't stand the idea of another drink." This scale was developed by Gross and Lewis (1974). The correlations for the 15 days of alcohol consumption of the ADR and the self-report attitudes for these two subjects were: 0.07 and -0.15.

Low Dose Alcohol Consumption (Priming): One subject who had been off ethanol for two weeks was given a morning drink of three ounces of 95 proof ethanol (primer dose) one hour before the subject was permitted to give ADR. The number of ADR for five baseline days immediately before starting the primer drink was zero; for three days during which the subject received the primer the ADR exceeded 4200 daily (Mann-Whitney $U = 0$, $p < .05$). The ADR subsequently decreased to baseline on the third day after the primer doses.

DISCUSSION

Each of the three procedures—withdrawal, experimental intoxication and a low dose alcohol drink produced significant increases in the number of ADR. These changes indicate a concordance of this behavioral measure with changes expected from clinical judgments regarding disposition to drink.

The changes in ADR are also in accord with the existing experimental literature on disposition to drink. Thus for self-administration of addictive drugs in animals, Schuster and Thompson (1969) have shown that the number of responses for the drugs increase with dose up to some high dose after which they decrease. This parallels the "U" shaped response curve for the number of ADR during experimental intoxication. All but one of the subjects showed a decrease in ADR at the highest dose and all subjects showed lower ADR after experimental intoxication than before. The change in ADR after a primer dose is similarly consistent with existing literature. Cohen *et al* (1971) showed that the amount of money required to stop an alcoholic from continuing to drink was increased by increasing doses of a primer drink. It seems hardly surprising that alcohol should serve as a stimulus for continued alcohol consumption. It may be useful, nonetheless, to evaluate both dose factors and the relative significance of this effect compared to setting or withdrawal conditions. It is reassuring that the ADR measure seems applicable to this study since the low level of alcohol consumption reduces the problems of order effects.

The results of the comparison of self-report of attitude toward drinking and the number of ADR is particularly interesting. Not only are the correlations near zero, but at times the subjects would respond with a 6 or 7 on the self-report scale indicating "I don't like (or can't stand) the idea" of the next drink, yet they had worked for several minutes giving 5200 ADR in order to have the drink immediately. Perhaps the lack of concordance between behavior and self-reports of drinking attitude should be expected for alcoholics, but the degree of disparity seemed surprising. It should be noted, though, that ADR were obtained in the morning before any alcohol consumption and after eight hours of alcohol deprivation. The ADR, therefore, reflect in part the effects of partial alcohol withdrawal. It should also be noted that the questionnaire was phrased in terms of attitude and not intended behavior. Thus the "I shouldn't" but "I will" dichotomy may have developed. It seems quite reasonable to expect that an alcoholic really feels badly about continuing to drink, especially in a hospital setting, yet he intends to do it anyway.

SLEEP AND DISPOSITION TO DRINK

Alcohol Withdrawal. The period following alcohol withdrawal was selected for study since during this time both increased disposition to drink and sleep disturbances are common. It was particularly hypothesized that the EEG sleep parameters would relate to the number of ADR during this period and that this relation would be significant even when the effect of the number of days abstinent was included in the analyses.

Four male alcoholics who met the general criteria described earlier were admitted intoxicated and detoxified for two to three days using a daily maintenance dose of 10 to 18 ounces of ethanol. No other psychoactive medications were used. The withdrawal period was defined as starting on the night following the last maintenance alcohol dose. EEG sleep was recorded and the data read using 20-second epochs following the procedures described in the Rechtschaffen and Kales Manual (1968). Sleep onset and awakenings were defined as nine consecutive epochs (three minutes) of a recognized sleep stage or stage W, respectively. All epochs between awakenings and the next sleep onset were removed from the EEG sleep stage analyses. The number of ADR each morning were paired with the EEG sleep parameters for the immediately preceding night's sleep period. Thus morning ADR were correlated with the prior EEG sleep. EEG sleep was recorded for every night for at least seven days of abstinence; ADR were recorded for each subsequent morning. EEG sleep records were generally read for each of the first ten days of the withdrawal period and for every third day thereafter.

TABLE III

MULTIPLE REGRESSION FOR ADR AS A FUNCTION OF DAYS AND EEG SLEEP PARAMETERS

Factor	Partial-R	Comp. t	Prob <
Days	-0.66	-5.13	.001
W%	-0.50	-3.46	.002
R%	-0.47	-3.12	.004
3%	0.40	2.66	.013

Mult. R = 0.84

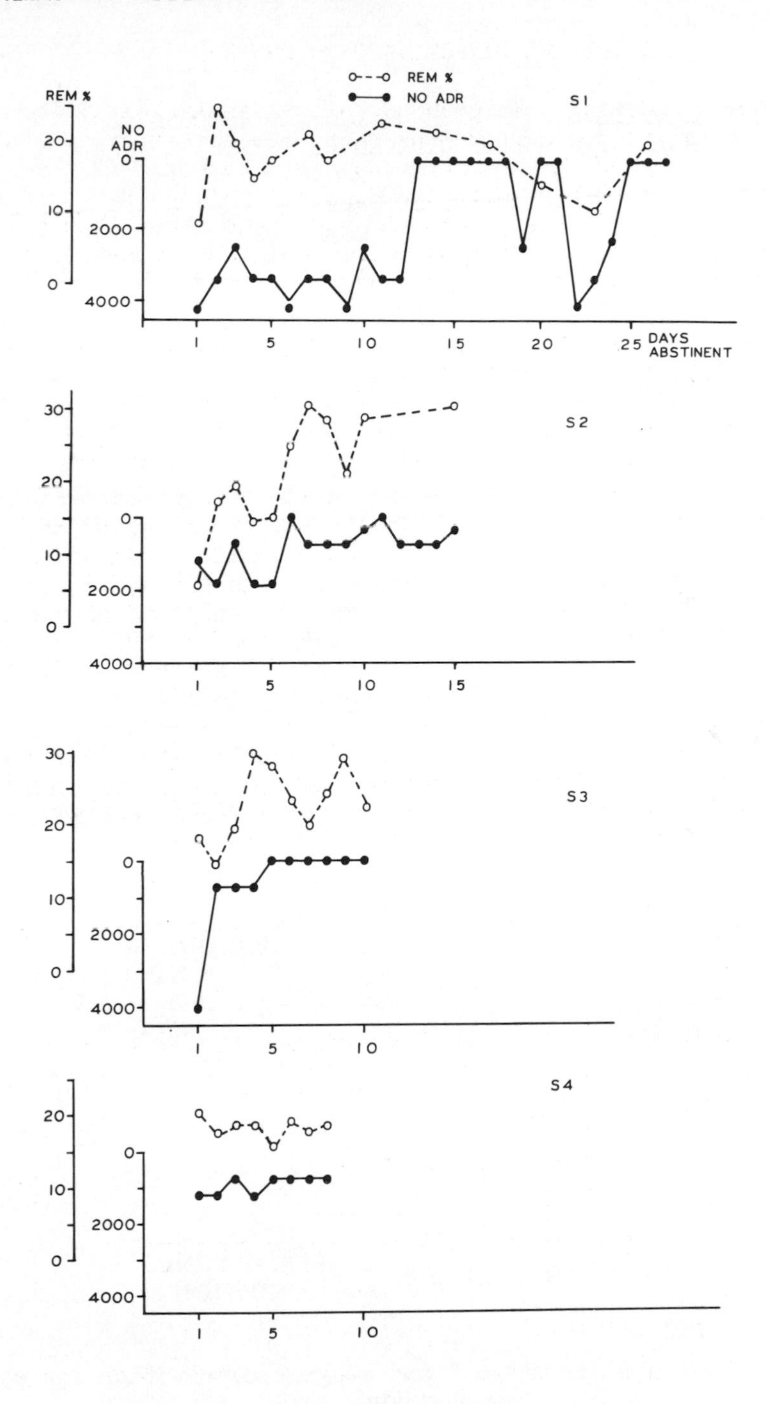

Figure 3. REM% and ADR for days abstinent.

TABLE IV

NUMBER OF DAYS WITH HIGH OR LOW ADR ASSOCIATED WITH REM% VALUES DIVIDED BY THIRDS

REM%	No. of Days Low ADR (≤ 700)	No. of Days High ADR (> 700)	Total
Low Third (≤ 18%)	5	9	14
Middle Third	8	5	13
Upper Third (>21.1%)	10	3	13

Table III presents the step-wise multiple regression analysis for ADR as a function of days abstinent and percentage of EEG sleep stages after removal of subject effects. Results indicate that days, REM% and W% are inversely correlated with ADR while stage 3% is directly correlated with ADR. Analysis by subjects showed reasonable variation for three of the four subjects in REM%, but for only one of the four subjects in both stage W% and stage 3%. In fact, three of the four subjects had very little stage 3 sleep. Figure 3 presents the REM% and ADR data for each subject. As can be seen, there is a striking relation between low stage REM% and the number of ADR. To further evaluate the apparent relation between high ADR and low REM a set of categories was established.

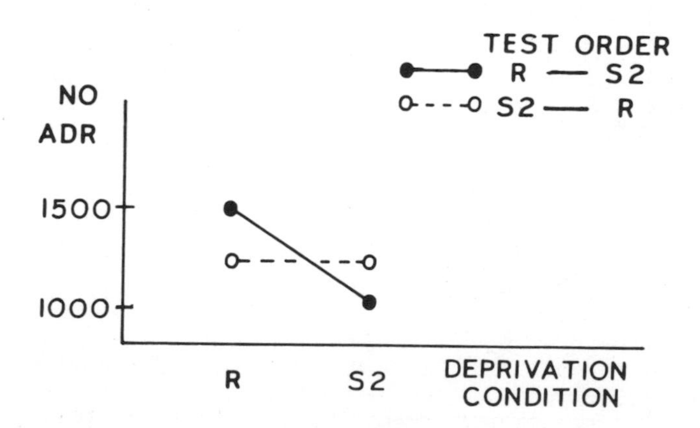

Figure 4. Mean ADR for stage R and stage 2 deprivation for each test order.

Since the subjects were giving 700 ADR or less after seven days of abstinence, 700 or lower ADR was considered to be a low number of responses. To assess the significance of low REM%, the REM values were divided into thirds (low $\leq$ 16%, n = 14; high $\geq$ 21.1%, n = 13; medium 16.1% to 21%, n = 13). Table IV presents the occurrence of high and low ADR for each REM% level. High ADR followed nine out of 14 nights with low REM% and only three out of 13 nights with high REM%. (Fisher exact, $p < .05$). Indeed, the three nights with abnormally low REM% (< 11%) are each associated with high ADR even though one of these nights occurred at 23 days of abstinence.

Stage REM Deviation. To test whether or not REM% deprivation itself leads to an increase in ADR, six male alcoholics off alcohol and other psychoactive drugs for at least seven days were experimentally deprived of stage REM and stage 2 sleep. Deprivation was continued for three consecutive days for each sleep stage with three days of undisturbed sleep between deprivation conditions. Order effects were counterbalanced with arbitrary assignment of three patients to each order condition. The average number of ADR on the days following each deprivation condition is presented in Figure 4 for each order group. For the group with REM deprivation first, the number of ADR was slightly higher during the REM deprivation than during stage 2 deprivation period. This effect was consistent for all three subjects in this group. For the group with stage 2 deprivation first the number of ADR was exactly the same for each deprivation condition. There was, overall, no significant effects of REM deprivation although there was a suggestion of an order-interaction effect.

DISCUSSION

The multiple regression data above indicate that both the number of days abstinent and the EEG sleep parameters relate to the number of ADR during short term abstinence. The day effect was, of course, previously confirmed for the validation of the ADR reported earlier in this paper. The significant EEG effects support the general hypothesis that sleep factors relate to disposition to drink. Since the stage 3 and stage W factors are attributable principally to one subject each, the generalities of the conclusion for these sleep stages seem doubtful. The stage REM effects, in contrast, were consistent for three of the four subjects with low REM% being associated in each case with high ADR. Only subject four failed to show this effect. He showed, in fact, little change in either REM% or the number of ADR; thus his data are not inconsistent with the relation between low REM% and high ADR. During the early period of alcohol abstinence REM% values are characteristically low for at least one or two days and then may fluctuate widely (Gross et al, 1973; Allen et al, 1971). The recovery of REM% to at least near normal levels appears to signal the end of the abstinence period during which the alcoholic shows a much greater disposition to drink than is characteristic of later abstinence. Thus for the alcoholic the REM% instabilities of early abstinence appear related to a period of increased risk for returning to alcohol consumption, as had been previously conjectured (Allen et al, 1971).

The one episode of low REM% sleep in later abstinence (after 20 days) deserves note. This period of low REM% was associated with a high number of ADR, although for several days before there were zero ADR. These data suggest that the relation between low REM% and high ADR may apply generally and not simply to the early abstinence period. The data on experimental REM deprivation indicate, however, that the REM% changes themselves do not suffice to produce changes in disposition to drink. REM% decreases could occur naturally for a large variety of reasons and presumably an observed relation between disposition to drink and low REM% results from these effects. It should be noted, however, that three days of REM% deprivation may not have been long enough to produce changes in disposition to drink.

In view of other existing literature it is tempting to hypothesize a general relation between decreased REM% and increased disposition to drink, particularly in the extreme cases. Alcohol consumption itself reduces stage REM% (Gross et al, 1973) and as noted in this paper generally increases disposition to drink. Conversely delirium tremens have been associated with elevated stage REM (Gross et al, 1966; Greenburg and Pearlman, 1967) and have been clinically reported for one case to relate to a refusal to drink alcohol (Allen et al, 1974). Clearly an hypothesized relation between low REM% and disposition to drink needs to be considered cautiously, particularly in view of the limited number of subjects in this study. Further data are needed covering a wider range of subjects and experimental conditions; but it is encouraging that a behavioral assessment shows a suggestive relation between psychophysiological parameters and the alcoholics' disposition to drink.

SUMMARY

A measure of disposition to drink in alcoholics was developed using a progressive ratio schedule for reduction of delay in receiving a drink of two ounces of 95 proof ethanol. This measure showed increased disposition to drink during early abstinence compared to lab abstinence (more than seven days) during experimental intoxication compared to lab abstinence and shortly after a low dose drink during late abstinence. A multiple regression analysis of this measure for four subjects after alcohol withdrawal, showed significant effects for days abstinent and EEG sleep variables, particularly stage REM%. Low REM% was associated with high values for the disposition to drink. REM sleep deprivation, however, failed to significantly alter the measure of disposition to drink. Results are interpreted as supporting the hypotheses that sleep disturbances relate indirectly to the disposition to drink.

REFERENCES

Allen, R., Faillace, L., Wagman, A. (in prep.). EEG sleep and psychological recovery after alcoholic delerium tremens.

Allen, R., Wagman, A., Faillace, L., McIntosh, M., 1971. Electro-encephalographic (EEG) sleep recovery following prolonged alcohol intoxication in alcoholics, J. nerv. ment. Dis. 153: 424-433.

Catania, A., 1963. Concurrent performance: a baseline for the study of reinforcement magnitude, J. Exp. Anal. Behav. 6: 299.

Cohen, M., Liebson, I., Faillace, L., Speers, W., 1971. Alcoholism: controlled drinking and incentives for abstinence. Psychol. Reports 28: 575-580.

Funderburk, F., Allen, R. (in prep.). A simple progressive ratio technique for assessing disposition to drink in alcoholics.

Greenburg, R., Pearlman, C., 1967. Delerium tremens and dreaming. Amer. J. Psychiat. 124: 133-142.

Griffiths, R., Brady, J., Winger, G., 1974. Drug-maintained performance procedures and the evaluation of sedative-hypnotic dependency potential. Upjohn Conference on hypnotics, Kalamazoo, Michigan (in press).

Gross, M. and Goodenough, D., 1968. Sleep disturbance in the acute alcoholic psychoses. Psychiat. Res. Report 24, American Psychiatric Association, New York.

Gross, M., Goodenough, D., Nagarazin, M., Hastey, J., 1973. Sleep changes induced by 4 and 6 days of experimental alcoholization and withdrawal in humans. In: Gross, M. (ed.) Alcohol Intoxication and Withdrawal, Advances in Experimental Medicine and Biology 35: 291-304, Plenum Company, New York.

Gross, M., Goodenough, D., Tobin, M., Halpert, E., Lepore, D., Perlstein, A., Sirota, M., Dibianco, J., Fuller, R., Kishner, I., 1966. Sleep disturbances and hallucinations in the acute alcoholic psychoses, J. nerv. ment. Dis. 142: 493-514.

Gross, M., Lewis, E. (in prep.). The bender ender.

Hodos, W., Kalman, J., 1963. Effects of increment size and reinforcement volume on progressive ratio performance, J. Exp. Anal. Behav. 6: 387.

Hore, B., 1972. Craving for alcohol. Presented at 30th International Congress on Alcoholism and Drug Dependence, September 1972.

Iglauer, C., 1972. Response-strength effects of different magnitude of reinforcement assessed in individual organisms. (unpublished manuscript).

Lester, B., Rundell, O., Cowden, L., Williams, H., 1973. Chronic alcoholism, alcohol and sleep. In: Gross, M. (ed.) Alcohol Intoxication and Withdrawal, Advances in Experimental Medicine and Biology 35: 261-279, Plenum Company, New York.

Miller, P., Hergin, M., Eisler, R., Epstein, L., Wooten, L., 1974. Relationship of alcohol cues to the drinking behavior of alcoholics and social drinkers: an analogue study. Psychol. Record 24: 61-66.

Nathan, P., O'Brien, J., 1971. An experimental analysis of the drinking behavior of alcoholics and non alcoholics during prolonged experimental drinking: a necessary precursor of behavior therapy? Behav. Therapy 2: 455-476.

Rechtschatten, A., Kales, A. (eds.), 1968. A Manual of Standard Terminology, Techniques and Scoring System in Sleep Stages of Human Subjects, U. S. Government Printing Office, Washington, D.C.

Schuster, C., Thompson, T., 1969. Self administration of and behavioral dependence on drugs. Annual Review of Pharmacology 9: 483-502.

Smith, D., Wesson, D., 1974. Drugs of abuse 1973: trends and developments. Annual Review of Pharmacology 14: 513-520.

A NOTE ON REM REBOUND DURING EXPERIMENTAL ALCOHOL WITHDRAWAL IN ALCOHOLICS

Milton M. Gross and John M. Hastey
Division of Alcoholism & Drug Dependence
Dept. of Psychiatry
Downstate Medical Center
Brooklyn, New York, U.S.A.*

INTRODUCTION

Until recently the effect of alcohol on rapid eye movement sleep (REM) appeared to be a relatively simple and clear aspect of the complicated relationship between alcohol and sleep. In normals, alcohol, even in moderate amounts, reduced REM (Gresham et al., 1963, Yules et al., 1966, Knowles et al., 1968). This was followed by a transient increase in REM, a "REM rebound," and return to baseline. When increased REM was observed in patients in acute alcohol withdrawal, this was assumed to be a manifestation of the REM rebound following the reduction of REM by the alcohol (Gross et al., 1966, Greenberg and Pearlman, 1967, Gross and Goodenough, 1968, Johnson et al., 1970). Patients in delirium tremens were often observed to have very considerable increases of REM and a level of 100% REM was reported in several instances (Gross et al., 1966, Greenberg and Pearlman, 1967, Johnson et al., 1970). Under experimental conditions, alcoholics appeared to have the REM reduction during alcohol intake and transient REM rebound when alcohol intake was stopped (Greenberg and Pearlman, 1967, Gross et al., 1971, Gross et al., 1973a, Gross et al., 1973b). The high levels of REM observed clinically during alcohol withdrawal in alcoholics were assumed to be quantitative effects of more alcohol intake for longer periods of time than the alcohol administered to normals or experimentally given to alcoholics. There was no evidence that there might be anything more complex in the REM differences between normals and alcoholics than the alcoholics' ability to tolerate greater alcohol intake, and thereby, produce greater REM reduction and subsequent

*Supported by NIAAA Grant AA01236

rebound. The REM reduction and rebound effects appeared to be shared by the other sedative-hypnotics.

However, Feinberg et al. (1974) drew attention to the fact that while barbiturates produced REM reduction, this was not consistently followed by REM rebound. They pointed out the need to determine what factors determined whether or not rebound occurred.

In an earlier descriptive report Gross (1973) noted that alcoholics who were experimentally given heavy alcohol intake and withdrawn could be partitioned into 2 groups. One group had a complete absence of REM (no REM) during heavy alcohol intake while the other group had traces of REM (reduced REM) during the same experimental conditions. That communication focused on these two REM groups and a possible hallucinogenic mechanism during withdrawal, and noted that the "no REM group" appeared to have REM rebound while the "reduced REM group" didn't. In seeking the possible factors involved in determining whether the alcohol volunteer had traces of REM or no REM during drinking, it was noted that the "no REM group" appeared to have a lower baseline REM level and a higher blood alcohol concentration than the "reduced REM group." These differences were not examined statistically.

In the light of the questions posed by Feinberg and his coworkers regarding barbiturates, (1974) it was apparent that our data on experimentally alcoholized alcoholics similarly indicated that alcohol consistently reduced REM, but did not consistently result in withdrawal rebound. But, our data indicated an additional complexity, namely, that under the conditions of our studies, it was only the group of alcoholics who had <u>total</u> REM suppression during drinking that demonstrated withdrawal REM rebound. It is essential to note that the alcoholic volunteers in the "reduced REM group" had only traces of REM which amounted to less than an average of 1% REM/night during the heavy drinking period. This suggested that in alcoholics there may be a difference which would make them more resistant than normals to REM rebound following alcohol intake.

In this note, the previously published data were examined to determine if the "non REM group" had a significant REM rebound, and to determine if there were significant differences in baseline REM and blood alcohol concentrations between the non REM and reduced REM groups.

METHODOLOGY

Approximately 2-3 weeks after completion of treatment for acute alcohol withdrawal, a group of male gamma alcoholic volunteers were experimentally alcoholized and withdrawn. At the time of study onset they had been abstinent of alcohol for 3-4 weeks and off psychoactive medication for 2-3 weeks. After 3 consecutive baseline days 4 volunteers received approximately 1.6 gm alcohol/kilo for 1 day and approximately 3.2 gm alcohol/kilo for 4 days while 6 volunteers received approximately 1.6 gm alcohol/kilo for 1 day and approximately 3.2 gm alcohol/kilo for 6 days. This was followed by 7 days of withdrawal.

Sleep and alcohol intake were regulated. Sleep EEG was obtained from midnight to 6 AM. Electrode placements and scoring criteria were according to Rechtschaffen and Kales (1968). Sleep recordings were obtained the night before study onset and discarded to eliminate first night effect. Alcohol was administered in equal doses from 2 PM to midnight with the omission of the 3 PM dose. Blood alcohol concentrations were obtained at 6 AM, 2 PM and midnight, using the Stephenson Breathalyzer. (For complete details of procedure see Gross et al., this volume).

The reduced REM and no REM groups each had 2 volunteers with 4 days of heavy drinking and 3 volunteers with 6 days of heavy drinking.

FINDINGS

The reduced REM group had a daily mean baseline of 33.2% REM (S.D. 06.4); the no REM group had an average daily baseline of 23.2% REM (S.D. 10.9). The student's t test (two-tailed) revealed that the difference was not significant ($t=1.78$).

The daily mean blood alcohol concentrations (BAC's) were consistently higher in the no REM group than the reduced REM group at midnight, 6 AM and 2 PM. However the BAC's were fairly close to each other (with considerable overlap of standard deviations) at 6 AM and 2 PM. At midnight the no REM group had a daily mean BAC of 302 mg/100 ml (S.D. 59.4) while the reduced REM group had a daily mean BAC of 221 mg/100 ml (S.D. 35.7). The student's test (two-tailed) yielded a $t=2.62$, significant at $p < .05$.

The daily average % REM during withdrawal revealed no REM rebound in the reduced REM group. A significant rebound was demonstrated in the no REM group during withdrawal. In paired t tests (two-tailed) comparing the mean differences between baseline and the second and third withdrawal nights (% REM was still low on the first withdrawal night), the no REM group had a mean difference of 7.50, a mean S.D. of differences of 4.5689, $t=3.671$ ($df=4$), $p < .02$. Significant REM rebound was not demonstrated during the later withdrawal period.

DISCUSSION

Is there evidence that alcoholics may have an altered response to alcohol which makes them more resistant to REM rebound during alcohol withdrawal? The data obtained from this study of heavy intake of alcohol by gamma alcoholics would suggest that there is such an alteration. Only the group of alcoholics who had complete absence of REM during the 4-6 days of heavy drinking demonstrated REM rebound during withdrawal. It was striking that the group of alcoholics who only had traces of REM during heavy drinking did not demonstrate rebound. On the basis of the available evidence, one would anticipate that in normals, the blood alcohol concentrations and the marked REM reductions observed in the reduced REM group would result in REM rebound during withdrawal.

The data indicated that the blood alcohol concentration at bedtime was a factor in the absence vs. traces of REM and the respective presence or absence of withdrawal rebound. As was previously noted, (Gross, 1973) it was our impression that the higher blood alcohol concentrations appeared to be related to a more rapid ingestion of the doses of alcohol. This impression was based on observations rather than systematic measurements. There was a suggestion that the baseline level of REM might play a role, but in this small sample the difference of baseline REM between the no REM and reduced REM groups was not significant.

It would seem most likely that an altered REM response during withdrawal in the alcoholics would be an acquired characteristic. This and other aspects of the nature of such an altered REM response, the underlying mechanisms and the clinical implications remain to be determined.

REFERENCES

Feinberg, I., 1974. Absence of REM rebound after barbiturate withdrawal. Science, Vol. 185, No. 4150, pp. 534-535.

Greenberg, R. & Pearlman, C., 1967. Delirium tremens and dreaming. Amer. J. Psychiat., 124:133-142.

Gresham, S.C., Webb, W.B. and Williams, R.C., 1963. Alcohol and caffeine: Effect on inferred visual dreaming. Science, 140: 1226-1227.

Gross, M.M., Goodenough, D.R., Tobin, M., Halpert, E., Lepore, D., Perlstein, A., Sirota, M., DiBianco, J., Fuller, R. and Kishner, I., 1966. Sleep disturbances and hallucinations in the acute alcoholic psychoses. J. Nerv. and Ment. Dis., 142: 493-514.

Gross, M.M. & Goodenough, D.R., 1968. Sleep disturbances in the acute alcoholic psychoses. In:(Ed.) J. Cole, Clinical Research in Alcoholism, Psychiatric Research Report, Amer. Psychiat. Assn., Vol. 24, pp. 131-147.

Gross, M.M., Goodenough, D.R., Hastey, J.M., Rosenblatt, S.M. and Lewis, E., 1971. Sleep disturbances in alcohol intoxication and withdrawal. In: (Eds.) N.K. Mello and J.H. Mendelson, Recent Advances in Studies of Alcoholism pp. 317-397, U.S. Government Printing Office, Washington, D.C.

Gross, M.M., 1973. Sensory Superactivity: A preliminary report on an hypothetical model for an hallucinogenic mechanism in alcohol withdrawal. In: (Ed.) M.M. Gross, Alcohol Intoxication and Withdrawal: Experimental Studies, Advances in Experimental Medicine and Biology, Vol. 35, pp. 321-330, Plenum Press, New York.

Gross, M.M., Goodenough, D.R., Nagarajan, M. and Hastey, J.M., 1973a. Sleep changes induced by 4 and 6 days of experimental alcoholization and withdrawal in humans. In: (Ed.) M.M. Gross, Alcohol Intoxication and Withdrawal: Experimental Studies, Advances in Experimental Medicine and Biology, Vol. 35, pp. 291-304, Plenum Press, New York.

Gross, M.M., Goodenough, D.R., Hastey, J.M. and Lewis, E., 1973b. Experimental study of sleep in chronic alcoholics before, during, and after four days of heavy drinking, with a nondrinking comparison. Annals of the New York Academy of Sciences, Vol. 215, pp. 254-265.

Gross, M.M. and Hastey, J.M. (This volume) The relation between baseline slow wave sleep and the slow wave sleep response to alcohol in alcoholics.

Johnson, L.C., Burdick, J.A., Smith, J., 1970. Sleep during alcohol intake and withdrawal in the chronic alcoholic. Arch. Gen. Psychiat., 22:406-418.

Knowles, J.B., Laverty, S.G. and Kuechler, H.A., 1968. Effects of alcohol on REM sleep. Quart. J. of Stud. on Alc. Vol. 29, No. 2, pp. 342-349.

Rechtschaffen, A. and Kales, A., 1968. A Manual of Standardized Terminology, Techniques and Scoring System for Sleep Stages of Human Subjects. U.S. Government Printing Office, Washington, D.C.

Yules, R.B., Freedman, D.X. and Chandler, K.A., 1966. The effect of ethyl alcohol on man's electroencephalographic sleep cycle. Electroenceph. Clin. Neurophysiol., 20:109-111.

"AROUSAL" AND ALCOHOLISM: PSYCHOPHYSIOLOGICAL RESPONSES TO ALCOHOL[1]

A. M. Ludwig and L. H. Stark

University of Kentucky, College of Medicine, Department of Psychiatry, Lexington, Kentucky 40506

It is assumed that any comprehensive theory of alcoholism presupposes an understanding of the biological mechanisms of action subserving or associated with this disorder. It is further assumed that the determinants of excessive drinking behavior pertain not only to environmental events and contingencies but to interoceptive bodily cues induced by alcohol itself (1). Unfortunately, though alcohol represents one of the oldest and most widely used drugs known to man, there is a surprisingly scant and often contradictory literature relevant to its physiological and neurophysiological effects in alcoholics. Almost all prior studies (a) pertain to animals rather than humans, (b) pertain to normal, healthy male volunteers rather than habituated or detoxified alcoholics, thereby ignoring critical factors of metabolic and tissue tolerance (2), (c) employ extremely small, non-representative samples (d) do not take into account the "Mellanby effect" (3) (i.e., differential effects of alcohol pertaining to the ascending and descending limb of the blood alcohol curve) or time-dose-response characteristics and (e) do not control for the extremely important influences of mental set and physical setting.

If the actions of alcohol are to be investigated, then it should prove helpful to organize and interpret observations within some reasonable, theoretical framework. Since the psycho-neurophysiological concepts of "arousal" and "activation" have been adopted to interpret many kinds of psychopathology (4), they may likewise prove useful as explanations of the biological basis for certain kinds of alcoholic behavior. In this regard, perhaps the most intriguing but yet unreplicated study is that of Kissin, et al. (5), which assayed six physiological systems in alcoholics. According to these investigators, alcoholics showed evidence of an impaired sympathetic nervous system, overactive parasympathetic nervous system, underactive 17-hydroxycorticoid and 17-ketosteroid production and overactive muscle tension functioning. With the exception of 17-ketosteroid production, the actions of alcohol appeared to "normalize" the functioning of these systems. While Coopersmith and Woodrow (6) concluded that alcoholics and non-alcoholics did not differ in arousal level, as measured by basal skin conductance, Garfield and McBrearty (7) demonstrated that the ingestion of small to moderate amounts of alcohol raised basal conductance but decreased over-all reactivity in alcoholics. Other scattered reports indicate that low doses of alcohol administered to alcoholics heightened "vigilance" behavior (8) and increased critical flicker fusion discrimination (9), a measure presumed to be associated with "arousal" (10,11). Aside from the acute effects of alcohol, chronic administration characteristically seems associated with greater "emotionality" and anxiety (12,13). With findings such as these, the simplistic classification of alcohol as a depressant drug, especially in alcoholics, must be open to serious question.

The present study derives from a prior one (14), investigating the effects of the "first drink" of alcohol on craving, alcohol acquisition behavior and selected physiological responses, in which certain "discrepant" findings were obtained. In brief, despite the differential effects produced by high and low doses of alcohol on tremor and CNV amplitudes, both doses were associated with increases in heart and respiratory rate, as well as percent time alpha activity. Although increased alpha activity represents a constant reproducible effect of alcohol (8,15), reports of the acute effects of alcohol on heart and respiratory functions in alcoholics are mostly anecdotal and somewhat contradictory (16-18).[/2] More important, there have been no systematic attempts made to reconcile the apparent "dissociation" between these autonomic nervous system (ANS) and central nervous system (CNS) responses. Increases in heart and respiratory rate are consistent with the concept of "sympathetic dominance" or ANS "arousal" (21) while increases in alpha activity have become traditionally associated with the concept of CNS "inhibitory" functioning (22,23).

Since our own theory of relapse to drink (14) presupposes the presence of a "conditioned withdrawal syndrome" or comparable state of "arousal," it seemed necessary to clarify the meaning of this apparent dissociation of ANS and CNS indices. In essence, we wished to determine whether it was possible for alcoholics to respond in a "normal" or "aroused" manner during an "inhibitory" or "depressed" electrocortical state or to respond in a "normal" or "retarded" manner in the presence of peripheral signs of ANS arousal. In addition, we hoped to gather empirical data on the nature of the relationships among subjective, behavioral-perceptual, physiological and neurophysiological indices of arousal-activation associated with different blood alcohol levels.

PROCEDURE

In order to assess the responses of alcoholics to progressive increments in blood alcohol concentration during a single session, testing periods were selected to differentiate the ascending limb, peak and descending limbs of the blood alcohol curve from baseline. Five testing periods were employed, with alcohol (i.e, 100% ethanol plus four ounces of lemonade) administered after completion of the measures in the first three periods. The schedule for the testing periods, which took about 20 minutes to conduct, and the doses of alcohol given were as follows: T(1) (baseline measures) -- then equivalent of 0.6 c.c./kg. absolute ethanol; T(2) (30 minute measures) -- then equivalent of 0.8 c.c./kg. absolute ethanol; T(3) (90 minute measures) -- then equivalent of 0.3 c.c./kg. absolute ethanol; T(4) (150 minute measures) -- no ethanol administered; T(5) (210 minute measures) -- no ethanol administered.

The group means and standard deviations for actual blood alcohol levels (obtained by Breathalyzer) corresponding to each testing period are given in Figure 1 below.

Sample Selection

Forty detoxified alcoholics from the Veterans Administration Hospital constituted the study sample. Selection criteria pertained to an upper age limit of 55 years, no gross evidence of cardiovascular, liver or organic brain damage, positive history of alcohol withdrawal symptoms and voluntary participation. Subjects were informed that the purpose of the experiment was to evaluate their physical responses to alcohol. They were paid at the rate of $2.00 per hour and were expected to remain off all psychotropic medication for at least three days prior to the experimental session. Selected sample characteristics were as follows: mean age = 45.5, S.D. = 5.8; mean number of hospital admissions for alcoholism = 3.6, S.D. = 3.5; and mean educational level (in grade) = 10.6, S.D. = 2.8.

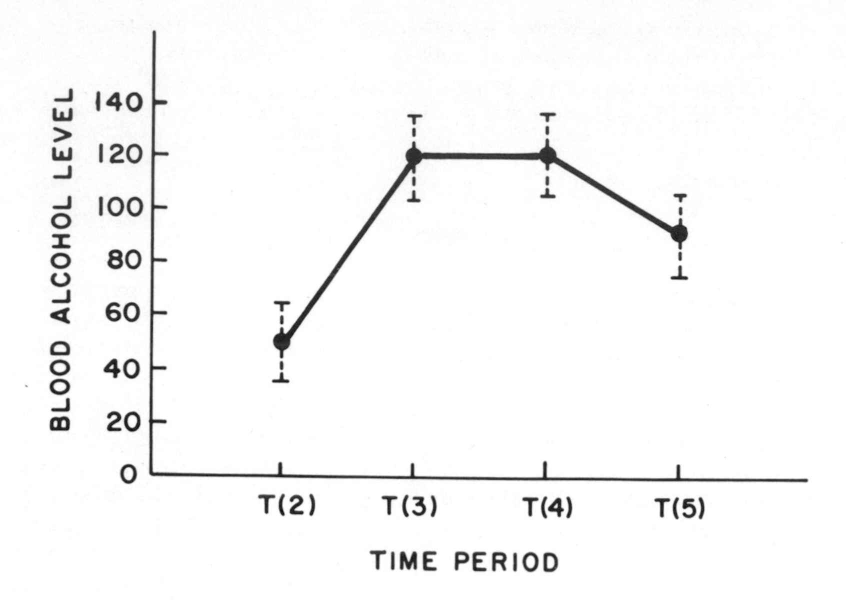

Figure 1. Summary of Blood Alcohol Levels

Measures and Tests

Several days prior to the experimental session, all subjects were required to fill out a detailed Drinking and Craving Questionnaire (DCQ) (24) which contained scales pertaining to prior craving experiences (PCE scale) and prior alcohol withdrawal symptoms (PAW scale). In addition, the Reitan Trail-Making Test (25) was administered to assess relative organicity.

For the actual experimental session, a wide variety of measures were selected as potential indices of arousal and/or activation. All of these measures were systematically obtained during each of the five testing periods. Subjects sat throughout the experimental session in a comfortable recliner chair within a wire-shielded laboratory and were monitored in an adjacent observation-control room. Despite the vast and often controversial literature pertaining to the "validity" of some of these indices, only brief justification for their selection will be provided.

Subjective Assessment

Craving Meter (CRAV). This button-operated scale records intensity of craving (from 0 to 100%). In a prior study (14), responses on this scale correlated highly with actual alcohol acquisition behavior.

Arousal Scale (AROSL). This brief, five item scale (four point continuum for each item) indicates the extent to which subjects feel "alert" and "keyed up." Self-report as an activation indicant has some degree of construct validity (26). In one study (27), a high relationship was found between such a measure and heart rate and basal skin conductance.

Drunkenness Scale (DRUNK). This brief, five-point intensity scale permits subjects to rate their degree of "drunkenness," thereby supplying a measure of their ability to differentiate among different blood alcohol concentrations.

Behavioral-Perceptual Assessment

Simple Reaction Time (RT). This represents a derivative measure from the CNV paradigm (see below) in which subjects are instructed to press a button immediately after the appearance of a tone. The interval between the onset of the tone and the button-press (in msec.) constituted the measure. An average of eight trials was employed.

RT presumably represents a sensitive indicator of attention and arousal (28,29). Though there is sufficient evidence to support an increase in RT following the ingestion of alcohol (30,31), some investigators dispute these findings for alcoholics (31,32).

Continuous Performance Test (CPT). A modification of the CPT (33,34) with auditory rather than visual stimuli was used to assess "sustained vigilance." Three hundred letters were presented at one second intervals and subjects were required to press a button only when an "A" was followed by "X". There were 60 such opportunities for correct responses. The actual responses of the subject yielded measures on errors of omission, errors of commission and total errors. Because of the high correlations found among these measures, only the total error score will be used for the statistical analyses.

According to Mirsky and Kornetsky (34), agents that impair the CPT preferentially act specifically on that portion of the ascending reticular activating system which has the responsibility of producing relatively sustained (tonic) cortical arousal while interfering less with the mechanisms involved in the production of relatively brief (phasic) cortical arousal.

Critical Flicker Fusion Threshold (CFF). This measure assesses the threshold (at a fixed light intensity) at which subjects can no longer distinguish flickering at different light flash frequencies. The higher the threshold for fusion, the greater the perceptual ability to discriminate flashes. The selection of this measure derives from prior research with the Two Flash Threshold (TFT) which seems highly sensitive to reticular activating system stimulation (35), barbiturates and amphetamines (36). While Venables and Wing (10) present compelling evidence for the use of this measure as an indicant of arousal, other investigators (11,37,38) have not confirmed these findings. Alcohol produces a significant reduction in CFF in normals but its effect on this parameter in alcoholics is somewhat debatable (39-41).

For obtaining this measure, a Lafayette Flicker Fusion Apparatus (Model 12020) was used. The method of ascending and descending limits, with three trials each, was employed by the experimenter with the mean value representing the subject's score.

Physiological Assessment

A Grass Model 78B 12 Channel EEG and Polygraph Data Recording System was employed to record the following responses: (a) heart rate (S-HR); (b) respiratory rate (RR) by thermister clipped to nostril; (c) percent time reactivity (i.e., deviation above 5 mm. from baseline) of spontaneous skin potential response (S-SPR). All these measures have been widely used as classical indices of arousal or SNS dominance. In addition, (d) eyeblink rate (EYBLK) was obtained. This particular measure has been advocated as a practical index of generalized muscle tension (42), and, more specifically, forehead muscle activity (43). All these measures, as well as alpha activity, were made simultaneously for a two minute period with eyes open and a two minute period with eyes closed./3 In contrast to these basal measures, two reactive indices were selected. These pertained to (e) evoked heart rate (E-HR) and (f) evoked skin potential response (E-SPR) during a 20 second exposure to Grass Strobe Light flashes at 13 c.p.s., a frequency most effective in eliciting photomyoclonus and/or photoconvulsions in withdrawing alcoholics (M. Victor, personal communication).

Reports of the acute effects of alcohol in alcoholics on all these parameters are both sparse and non-definitive.

Neurophysiological Assessment

Percent Time Alpha Activity (ALFA). Monopolar recordings from scalp electrodes (P_z - A_1) were fed into the Polygraph and written out for direct "eyeball" scoring of alpha activity during a two minute period with eyes open (EO) and eyes closed (EC). This method of scoring has proven highly reliable in our laboratory. As indicated previously, this index has been assumed to be related to states of relaxation or cortical non-activation. Sufficient studies have been conducted to substantiate the induction of alpha activity by alcohol both in non-habituated and habituated individuals (8,15,44).

Integrated EEG Voltage (EEG). The signals from monopolar scalp electrodes (C_z - A_1) were fed into a voltage integrator via a multipurpose filter set to pass a band from 0 - 30 Hz. The accumulated voltage was recorded as number of "bins" filled during comparable two minute time intervals. Although this measure is presumably susceptible to eyeblink and muscle artifact, it was employed as a very crude index of high versus low voltage electrocortical activity with the assumption that arousal would produce a "shift to the right" and inhibition a "shift to the left" on the frequency spectrum.

Contingent Negative Variagion (CNV). The amplitude of the contingent negative variation or "expectancy wave" (45) is purportedly monotonically related to attention and non-monotonically related to arousal (46). Kopell, et al. (47), have reported an increased amplitude following marihuana and a decreased amplitude of the CNV following alcohol administration in normals. In our prior study (14), we found the amplitude enhanced by low dose and depressed by high dose of alcohol compared to placebo. Many methodological problems pertain to the use of this measure (48).

The CNV was obtained from monopolar recordings of scalp electrodes (from C_z - A_1), the signals of which were relayed from the amplifier into a Hewlett-Packard Signal Averager (Model #5480-B). The paradigm employed pertained to the presentation of a warning stimulus (light) followed one second later by an imperative stimulus (Sonalert tone). The subject was instructed to terminate the tone (by a button-press) as soon as it appeared. The electrocortical activity during this one second interval was averaged over eight trials spaced 10 seconds apart. Measures for the CNV included (a) summed amplitudes at 200, 400, 600, 800 and 1000 msec. (CNV-T), (b) peak amplitude (CNV-P), and (c) latency of peak amplitude (CNV-L).

RESULTS

Direct Effects of Alcohol

Analyses of variance were run to assess significant changes over testing periods for the raw scores, change scores from baseline and percentage change from baseline scores for all variables. Because the results were relatively comparable and because of space considerations, only the raw score analyses will be presented. The mean scores for all variables and testing periods, together with F values and multiple comparison of means at the .05 level of significance, are given in Table 1.

Table 1. Summary of Analyses of Variance

VARIABLE	T(1)	T(2)	T(3)	T(4)	T(5)	F	p
S-HR	85.1	83.6	87.5	87.6	89.2	7.6	.0001
E-HR	82.7	84.7	88.2	88.5	91.2	10.8	.0001
S-SPR	6.1	5.6	5.0	2.3	3.9	2.4	.05
E-SPR	17.1	10.1	10.0	5.7	8.9	8.0	.0001
ALFA-EO	14.7	17.9	24.4	27.4	30.4	14.8	.0001
ALFA-EC	50.3	57.9	58.5	59.3	60.5	3.7	.01
EYBLK	30.4	30.3	27.5	26.9	23.3	2.7	.05
EEG	15.9	16.1	16.1	16.3	14.9	0.5	NS
RR	19.0	19.0	19.5	20.3	20.4	4.0	.005

As inspection of the results reveals, except for CFF, integrated EEG and CNV peak latency, the administration of alcohol produced significant changes in all other variables. The magnitude of these changes at the different testing periods conformed roughly to changes in blood alcohol level although only alpha activity (eyes open) and the CPT and RT tasks seemed sensitive enough to discriminate among the levels. Blood alcohol levels at about 50 mg.% (T(2)) could be distinguished from no alcohol (T(1)) only by evoked skin potential, alpha (eyes closed), CNV peak amplitude, reaction time, arousal and craving measures. It should be noted that reaction time was significantly faster at T(2) than at T(1) or subsequent testing periods.

Table 1. (continued)

VARIABLE	T(1)	T(2)	T(3)	T(4)	T(5)	F	p
CFF	37.2	36.8	38.1	36.5	36.0	1.0	NS
CPT	9.3	10.1	15.4	18.0	12.2	12.4	.0001
RT	358.0	327.4	360.4	355.8	338.7	2.8	.05
CNV-T	27.2	21.8	19.1	12.8	16.8	2.5	.05
CNV-P	15.6	13.4	12.6	12.1	12.2	3.6	.01
CNV-L	836.6	845.5	877.6	893.2	880.3	1.7	NS
CRAV	14.2	27.5	34.8	35.8	30.0	11.0	.0001
AROSL	9.9	7.6	7.6	7.3	6.8	10.8	.0001
DRUNK	0.0	1.2	2.4	2.5	2.9	43.8	.0001

Means underscored by the same kind of line are not significantly different from one another at the .05 level of significance.

Factor Analyses

For data reduction purposes, as well as for the detection of patterning among variables, factor analyses, employing a principal factoring with iteration and fairly oblique rotation, were performed. The number of significant components retained for the final rotated solution was based on eigenvalues greater than or equal to 1.0. This criterion insured that only components accounting for at least the average amount of variance of a variable would be treated as significant.

These factor analyses were conducted for all 18 variables for each separate testing period. Each analysis yielded seven inferred factors. The assignment of variables to factors, as well as their actual factor loadings, are given in Table 2. The percent of common variance accounted for by each factor at each testing period is presented in Table 3.

The results indicated some shifting in factor assignment for certain variables over the different testing periods with several relatively stable factors emerging at and after T(3). The more consistent factors pertained to (a) the subjective measures of craving, arousal and drunkenness, (b) the performance measures of CPT and RT, frequently in association with RR or E-SPR, (c) HR and E-HR, (d) alpha, EO and EC, (e) EYBLK and EEG, and (f) CNV total and peak amplitudes.

Inspection of the intercorrelation matrices for all variables revealed that all significant correlations among variables are represented by the factor assignments if a criterion of 10% common variance for correlation values was used (i.e., r = 0.32).

Other Analyses

Scattergram printouts were obtained for all percent change scores for each variable to assess unusual distributions. Observation of responses indicated that all distributions conformed to parametic assumptions.

Multiple chi square analyses were run to determine relationships between percentage change scores of variables for the _direction_ rather than _magnitude_ of response. While some interesting findings emerged, they were not consistent over time periods, thereby indicating the prospects of Type I errors.

In order to determine whether the magnitude of the percentage of change for any given variable, in comparison with all other variables, reflected different hierarchical patterns of response for different individuals, these magnitudes were ranked across all

Table 2. Summary of Factor Loadings

	T(1)		T(2)		T(3)		T(4)		T(5)	
	FTR	LOAD	FTR	LOAD	FTR	LOAD	FTR	LOAD	FTR	LOAD
S-HR	1	949	4	936	2	-976	1	-940	3	-922
E-HR	1	936	4	932	2	-881	1	-985	3	-934
S-SPR	5/3	-278 -266	6	377	7	-687	7	475	7	-729
E-SPR	7	531	7	816	7	-758	2	-319	5	-360
ALFA-EO	4	-765	3	632	5	699	5	-984	2	-703
ALFA-EC	2	-467	3	653	5	609	5	-399	2	-731
EYBLK	3	-810	6	627	3	983	6	-778	6	-772
EEG	3	-842	6/5	480 -414	3	690	6	-594	2	-499
RR	5	623	2	-541	3/6	239 -195	7	-490	3	-322
CFF	4	-622	6	497	5/4	264 225	3/1	226 195	5	-416
CPT	5	659	2	-872	1	504	2	731	5	658
RT	5/7	498 456	2	-695	6	-879	2	776	7	384
CNV-T	2	-782	1	-999	1	-925	4	-906	4	-953
CNV-P	2	-952	1	-846	1	-939	4	-850	4	-968
CNV-L	6	-343	5	-735	1	-572	7	-576	2	-438
CRAV	7	780	3/5	553 490	4	-484	3	-777	1	-806
AROSL	6	-640	6/3	-365 287	4	-510	3	-592	1	-584
DRUNK	1	394	3/4	368 312	4	-836	3	-636	1	-937

Coding of Table: FTR, factor number; LOAD, factor loading; leading zero and decimal point were left out of the table.

selected variables for each subject. For five of the variables, subjects showed bimodal responsivity patterns at T(2) -- i.e., showing both most and least changes compared to other variables.

Table 3. Summary of Common Variance

	FACTOR 1	2	3	4	5	6	7
T(1)	23.6	19.7	15.2	14.5	11.5	9.1	6.6
T(2)	26.5	17.1	15.2	14.0	10.9	10.4	5.9
T(3)	29.8	28.7	15.2	12.6	9.6	7.8	6.9
T(4)	27.2	16.4	15.3	14.0	12.6	8.3	6.2
T(5)	21.7	21.5	18.3	15.0	11.6	6.5	5.3

Since comparable bimodality did not pertain at T(3), it seemed likely that these distributions could be accounted for on the basis of chance.

Demographic and Predictor Variables

The correlational relationships among selected pre-experimental variables are given in Table 4 below. Not surprisingly, it can be noted that the highest correlations are between age and Reitan scores, the PCE and PAW scales, and number of admissions and the PAW scale.

When the above variables were intercorrelated with the 18 variables at T(1), T(2), and T(3) and the criterion of 10% common variance employed, the following significant relationships were found: (a) age was related to EEG at T1 and T3 ($r = -.33$ for both, $p < .05$) and to S-SPR and E-SPR at T(3) ($r = -.35$ and $-.37$, respectively, $p < .05$); (b) number of admissions was related to CNV-P at T(1) ($r = -.36$, $p < .05$); (c) the PCE scale was related to craving and arousal at T(1) ($r = +.43$ and $+.52$, respectively, $p < .01$), craving and drunkenness at T(2) ($r = +.58$ and $+.48$, respectively, $p < .01$) and craving and drunkenness at T(3) ($r = +.73$ and $+.46$, respectively, $p < .01$); (d) the PAW scale was related to arousal at T(1) ($r = +.40$, $p < .01$) and craving at T(3) ($r = +.47$, $p < .01$);

Table 4. Summary of Pre-Experimental Correlations

	ADM	PCE	PAW	REITAN
AGE	.11	-.29*	-.08	.46***
ADM		.27*	-.38***	-.13
PCE			.62***	-.28*
PAW				-.28*

* p less than .10, *** p less than .01.

and (e) the Reitan test was related to RR and CPT at T(1) ($r = +.43$ and $+.55$, respectively, $p < .01$), RR and RT at T(2) ($r = +.39$ and $+.32$, respectively, $p < .05$), and the CPT ($r = +.48$, $p < .01$), CFF ($r = +.44$, $p < .01$), CNV-T ($r = -.47$, $p < .01$), CNV-P ($r = -.44$, $p < .01$) and CNV-L ($r = -.35$, $p < .05$) at T(3).

DISCUSSION

At the outset, it must be emphasized that the study was not designed to compare the differential responses of alcoholics and nonalcoholics to the administration of alcohol. Such a comparison would require the use of an adequately matched control group and the resolution of certain knotty, methodological problems (e.g., the physical discomfort and profound drunkenness produced in normals with the acute administration of alcohol in doses sufficient to achieve blood alcohol levels at about 125 mg.%). Rather, it was our intention to focus only on the responses of alcoholics in an effort to answer certain issues pertaining to arousal, as well as for the generation of new hypotheses. In this regard, while the overall findings are consistent with those of our prior study (14), they still leave many basic issues unresolved.

Initially, we must consider the question of whether alcohol produces generalized depressant or excitatory effects in alcoholics. The data clearly indicate that increasing blood alcohol levels are associated with significant elevations in heart and respiratory rate, whether centrally or peripherally induced, as well as heightened craving, a response previously found directly related to work output for alcohol (14) and subjective symptoms of anxiety and emotional dysphoria (24). Increasing blood alcohol levels are likewise associated with increased alpha activity, depression of the CNV peak amplitude, decreases in evoked skin potential responses, decrements in performance on the CPT and RT tasks and diminished subjective arousal. Certain exceptions to these generalities must be made for blood alcohol concentrations at about 50 mg.%.

If the entirety of these results are to be explained within the classical arousal-activation conceptual framework, then we must conclude that alcohol produces dissociated responses in alcoholics. Such an interpretation is confounded, however, by other considerations. It should be noted that the inhibitory cortical functioning ascribed to alpha activity derives mostly from observations on natural rather than drug-induced changes in electrocortical frequencies. In this regard, several investigators (49, 50) have demonstrated that the administration of certain drugs in animals can produce marked dissociations between electroencephalographic activity and general behavior. It is also important to recognize that to date no investigators have documented significant relationships between alpha activity and ANS measures of "arousal" in groups of subjects (51-53). Conversely, increases in the activity of peripheral indices of arousal need not be associated with heightened vigilance or awareness. In State I-II anesthesia (54) or REM sleep, heart rate, respiratory rate and fluctuations in skin potential are heightened within the context of diminished consciousness (55). Because of these presumed paradoxes, LaVerne Johnson (56) advocates a "psychophysiology for all states," whereby anticipated physiological responses must be interpreted within the context of the observed state of consciousness. In this vein, it may simply be the case that alcohol produces a unique constellation of effects in alcoholics which cannot be evaluated on the same continuum as responses observed during a normal, non-drug waking state.

From these considerations, we would surmise that the alcohol induced alpha activity (within the blood alcohol ranges in this study) may have little to do with diminished arousal or activation and that heightened heart and respiratory rate may have little to do with increased arousal or activation. The increased craving argues against diminished arousal and the decrement in skin potential reactivity and in performance on the vigilance tasks (with the exception of initial dose of alcohol) argues against increased activation. If this diversity of responses cannot be interpreted within the arousal-activation continuum, we must then seek an alternative conceptualization.

If we attribute to alcohol-induced alpha rhythm, instead of an inhibitory cortical functioning, actions capable of producing an inner deployment of attention with a concomitant decrease in external reactivity, such as occurs in certain types of meditative states (57), then many of these presumably discrepant findings can be parsimoniously explained. The decrease in evoked skin potential responsivity, the diminished CNV peak or "expectancy wave," and the decrement in CPT and RT performance (for higher doses) may pertain primarily to difficulty in attending to external events. Moreover, since increases in heart and respiratory rate, along with heightened craving, represent interoceptive cues commonly associated with prior emotional distress and alcohol withdrawal experiences, the alcoholic (because of greater attention directed inward) may interpret these cues as signs of anxiety or somatic distress and respond by increased, "conditioned" behavioral activity focused on acquiring more alcohol for the presumed relief of these feelings.

Aside from the significant responses produced by alcohol, we must also account for the absence of effect on the CFF even at fairly high blood alcohol levels. This intriguing perceptual-behavioral measure has been reported to be highly sensitive to alcohol in non-alcoholics, with higher blood alcohol levels producing corresponding decreases in frequency (58). While Goldberg (39) reported some decrement in frequency with rising blood alcohol levels for all groups of subjects, this decrement in alcoholics was not as great as in non-habituated controls. Our findings, however, are in greater accord with the more recent studies by Weiss, et al. (40) and Hill, et al. (41). The latter of these two groups of investigators propose the intriguing hypothesis that this absence of CFF effect reflects tolerance to alcohol and that this very absence may be used diagnostically to assess chronicity and severity of alcoholism. Though this hypothesis may well prove correct and while we have no alternative hypothesis to offer, we should feel more secure with this explanation if this presumed tolerance effect were not so selective and encompassed some of the other major indices used in our study as well. In any event, the apparent insensitivity of the CFF to fairly high blood alcohol levels in alcoholics seems worthy of further investigation.

Since our study also sought to find commonalities and relationships among the multitude of variables for different blood alcohol levels, factor analyses and correlational analyses were run. In general, the results of these extensive analyses were disappointing. Not only were the magnitudes of correlation among the many physiological-electrocortical measures negligible, as has been reported to be the case for most psychophysiological studies (53,59), although not investigating the effects of alcohol, but the number of derived factors was far too great and far too obvious to aid in later theory construction and hypothesis testing. Aside from the separate constellation of "subjective" verbal responses and performance responses on vigilance tasks (which also related to respiration and evoked skin potential), most of the other factors indicated separate response systems for each major variable. Thus, cardiac variables clustered together, CNV variables clustered together, alpha variables clustered together, and so on. If we still choose to employ the concept of "dissociation," we should have to conclude that there not only is a general dissociation between the CNS and ANS but there are also "multiple dissociations" within each system as well. This dissociation does not imply significant negative relationships opposite to what might be expected but rather a general lack of correlations where some might be expected. At best, these results simply indicate our basic ignorance concerning the laws underlying the relationships among subjective-behavioral-physiological-neurophysiological response systems.

In regard to demographic and pretest variables, several interesting relationships emerge for the three testing periods selected. Age correlates inversely with integrated EEG voltage and skin potential response. The PCE scale is directly related to reported craving and drunkenness following administration of alcohol. The PAW Scale is directly related to reported arousal at baseline and reported craving following alcohol ingestion. The Reitan Trail-Making score is directly related to respiratory rate and performance measures throughout the three testing periods and inversely related to all CNV measures at peak blood alcohol. These relationships, while not astounding, indicate at least the importance of these pretest variables as predictors of response and the need to weigh their influence in any definitive analysis of responsivity to alcohol administration.

With this presentation of results, it is tempting to speculate further on possible reasons for their not conforming to the seductive concepts of arousal and activation. While convincing evidence could be offered and arguments made supporting the validity of these concepts (28), we do not view the prospects of an extensive critique as particularly relevant or productive. In a sense, our use of these concepts as the framework for interpreting our results may be regarded as somewhat of a "strawman," especially since modern psychophysiological and neurophysiological theories have progressed far beyond the assumption that indices of sympathetic innervation need vary together and that these indices, in turn, should be associated with certain types of electrocortical activity. Nevertheless, this simplistic assumption is a valuable one and worthy of testing, mainly because it has never been applied to the effects of alcohol in alcoholics. Obviously, with complex interactions occurring between the reticular activating system and other brain structures, the existence of feedback and feedforward circuits, continuing homeostatic adjustments among a multitude of biological response systems and the differential actions of alcohol on the above, it is perhaps overly naive to expect to uncover lawful relationships among a selected number of variables along the continua of SNS activity and cortical activation.

Though most of our statistical analyses relied on the assumption of linear relationships among variables, we also explored the possibility that the data conformed to certain other psychophysiological concepts, such as an inverted U-shaped distribution, relationships based on direction rather than magnitude of change (28), individual response stereotypy and hierarchical patterning of responses (60) and directional fractionation of response patterns (53). None of these approaches to data analysis and interpretation yielded fruitful insights into the laws governing the relationships among variables.

For the present, therefore, it seems safe only to conclude that the acute administration of alcohol produces a wide variety of subjective, behavioral, psychophysiological and neurophysiological effects (which are to some extent dose related) in alcoholics, the relationships among which cannot be explained completely by any extant psychophysiological theories. Discouraging though this be, it should serve as a spur rather than deterrent for gathering more knowledge of the physiological-neurophysiological mechanisms of action underlying the behavior of alcoholics and how these mechanisms of action are affected after the ingestion of a powerful drug like alcohol.

ACKNOWLEDGEMENTS

We are grateful to Miss Karen Osborne, our polygraph operator, for her important contributions to the study.

FOOTNOTES

/1 Supported by Grant No. 1AA00390-03 ALC, NIAAA.

/2 In normals, intoxicating amounts of alcohol are reported to increase resting cardiac output, heart rate and myocardial oxygen consumption without a change in stroke volume (19). Moderate doses of alcohol purportedly stimulate respiration while higher doses depress it (20).

/3 Only the results for the second minute of the eyes open data will be reported.

REFERENCES

1. Ludwig, A.M., Wikler, A., Craving and relapse to drink, Quart. J. Stud. Alc., Vol. 35, No. 1, pp. 108-130, March 1974.

2. Isbell, H., et al., An experimental study of the etiology of "rum fits" and delirium tremens, Quart. J. Stud. Alc., 16: 1-33, 1955.

3. Kalant, H., et al., Tolerance to, and dependence on ethanol, in Y. Israel and J. Mardones (eds.), Biological Basis of Alcoholism, Wiley, N.Y., 1971.

4. Alexander, A.A., Psychophysiological Concepts of Psychopathology, in N.S. Greenfield & R.A. Sternbach (eds.), Handbook of Psychophysiology, Holt, Rinehart & Winston, N.Y., 1972.

5. Kissin, B., et al., The acute effects of ethyl alcohol and chlorpromazine on certain physiological functions in alcoholics, Quart. J. Stud. Alc., 20:480-492, 1959.

6. Coopersmith, S., Woodrow, K., Basal conductance levels of normals and alcoholics, Quart. J. Stud. Alc., 28:27-32, 1967.

7. Garfield, Z.H., McBrearty, J.F., Arousal level and stimulus response in alcoholics after drinking, Quart. J. Stud. Alc., 31:832-838, 1970.

8. Doctor, R.F., et al., EEG changes and vigilance behavior during experimentally induced intoxication with alcoholic subjects, Psychosom. Med., 28:605-615, 1966.

9. Lewis, E.G., et al., The effect of alcohol on sensory phenomena and cognitive and motor tasks, Quart. J. Stud. Alc., 30: 618-633, 1969.

10. Venables, P.H., Wing, J.K., Level of arousal and the subclassification of schizophrenia, Arch. Gen. Psychiat., 7: 114-119, 1962.

11. Kopell, B.S. et al., Variations in some measures of arousal during the menstrual cycle, J. Nerv. Ment. Dis., 148:180-187, 1969.

12. Tamerin, J.S., Mendelson, J.H., The psychodynamics of chronic inebriation: observations of alcoholics during the process of drinking in an experimental group setting, Am. J. Psychiat., 125:886-899, 1969.

13. McNamee, H.B., et al., Experimental analysis of drinking patterns of alcoholics: concurrent psychiatric observations, Am. J. Psychiat., 124:1063-1069, 1968.

14. Ludwig, A.M., Wikler, A., Stark, L.H., The first drink: psychobiologic aspects of craving, Arch. Gen. Psychiat., Vol. 30, April 1974.

15. Wikler, A., et al., Electroencephalographic changes associated with chronic alcoholic intoxication and the alcohol withdrawal syndrome, Am. J. Psychiat., 113:106-114, 1956.

16. Gould, L., et al., Cardiac effects of alcohol in alcoholics, Quart. J. Stud. Alc., 33:966-978, 1972.

17. Doctor, R.F., Bernal, M.E., Immediate and prolonged psychophysiological effects of sustained alcohol intake in alcoholics, Quart. J. Stud. Alc., 25 (Suppl. 2): 438-450, 1964.

18. Mendelson, J.H., LaDou, J., Experimentally induced chronic intoxication and withdrawal in alcoholics, Pt. 2: psychophysiological findings, Quart. J. Stud. Alc., 25 (Suppl. No. 2) 14-39, 1964.

19. Riff, D.P., et al., Acute hemodynamic effects of alcohol on normal human volunteers, Amer. Heart J., 78:592-597, 1969.

20. Richie, J.M., The aliphatic alcoholics (chap. 2), in L.S. Goodman, A. Gilman (eds.), <u>The pharmacological basis of therapeutics</u>, 4th edition, Macmillan Co., N.Y., 1970.

21. Sternbach, R.A., <u>Principles of Psychophysiology</u>, Academic Press, N.Y., 1966.

22. Lindsley, D.B., Psychological phenomena and the electroencephalogram, EEG Clin. Neurophysiol., 4:443-456, 1952.

23. Heineman, L.G., Emrich, H., Alpha activity during inhibitory brain processes, Psychophysiol., 7:442-450, 1971.

24. Ludwig, A.M., Stark, L.H., Subjective and situational aspects of craving, submitted for publication.

25. Reitan, R., Validity of the trail-making test as an indication of organic brain damage, Percept. Mot. Skills, 8:271-276, 1958.

26. Dermer, M., Berscheid, E., Self-report of arousal as an indicant of activation level, Behav. Sci., 17:420-429, 1972.

27. Thayer, R.E., Activation states as assessed by verbal report and four psychophysiological variables, Psychophysiol., 7: 86-94, 1970.

28. Duffy, E., Activation, in N.S. Greenfield & R.A. Sternbach (eds.), Handbook of Psychophysiology, Holt, Rinehart & Winston, N.Y., 1972.

29. Isaac, W., Arousal and reaction time in cats, J. Compar. and Physiol. Psychol., 53:234-236, 1960.

30. Carpenter, J.A., Effects of alcohol on some psychological processes: a critical review with special reference to automobile driving skill, Quart. J. Stud. Alc., 23:274-314, 1962.

31. Young, J.R., Blood alcohol concentration and reaction time, Quart. J. Stud. Alc., 31:823-831, 1970.

32. Talland, G.A., et al., Experimentally induced chronic intoxication and withdrawal in alcoholics, Pt. 4: Tests of motor skill, Quart. J. Stud. Alc., 25 (Suppl. No. 2): 68-73, 1964.

33. Rosvold, H.E., et al., A continuous performance test of brain damage, J. Consult. Psychol., 20:343-350, 1956.

34. Mirsky, A.F., Kornetsky, C., On the dissimilar effects of drugs on the digit symbol substitution and continuous performance test, Psychopharmacologia, 5:161-177, 1964.

35. Lindsley, D.B., The reticular system and perceptual discrimination, in H.H. Jasper, et al. (eds.), Reticular Formation of the Brain, pp. 513-535, Churchill, London, 1958.

36. Kopell, B.S., et al., The effect of thiamylal and methamphetamine on the two-flash fusion threshold, Life Sci., 4:2211-2214, 1965.

37. Hume, W.I., Claridge, G.S., A comparison of two measures of "arousal" in normal subjects, Life Sci., 4:545-553, 1965.

38. Cautela, J.R., Barlow, D., The relation between anxiety and flicker fusion threshold in a nonstressful situation, Psychology, 2:8-12, 1965.

39. Goldberg, L., Quantitative studies on alcohol tolerance in man: the influence of ethyl alcohol on sensory, motor and psychological functions referred to blood alcohol in normal and habituated individuals, Acta Physiol. Scand., 5 (Suppl. No. 16): 1-128, 1943.

40. Weiss, A., et al., Experimentally induced chronic intoxication and withdrawal in alcoholics, Pt. 6: Critical flicker fusion studies, Quart. J. Stud. Alc., 25 (Suppl. No. 2): 87-95, 1964.

41. Hill, S.Y., et al., Critical flicker fusion: objective measure of alcohol tolerance, J. Nerv. Ment. Dis., 157-46-49, 1973.

42. Meyer, D.R., On the interaction of simultaneous responses, Psychol. Bull., 50:204-220, 1953.

43. Martin, I., Blink rate and muscle tension, J. Ment. Sci., 104:123-132, 1958.

44. Weiss, A.D., et al., Experimentally induced chronic intoxication and withdrawal in alcoholics, Pt. 7: Electroencephalographic findings, Quart. J. Stud. Alc., 25 (Suppl. No. 2): 96-99, 1964.

45. Cohen, J., Very slow potentials relating to expectancy: the CNV, in E. Donchin & D.B. Lindsley (eds.), Average Evoked Potentials: Methods, Results and Evaluations, Gov't. Printing Office, Washington, D.C., 1969.

46. Tecce, J.J., Contingent negative variation and individual differences: a new approach in brain research, Arch. Gen. Psychiat., 24:1-16, 1971.

47. Kopell, B.S., et al., Contingent negative variation amplitudes: marihuana and alcohol, Arch. Gen. Psychiat., 27:809-811, 1972.

48. Straumanis, J.J., Jr., et al., Problems associated with application of the contingent negative variation to psychiatric research, J. Nerv. Ment. Dis., 148:170-179, 1969.

49. Wikler, A., Pharmacologic dissociation of behavior and EEG "Sleep patterns" in dogs: morphine, n-allylnormorphine, and atropine, Proc. Soc. Exp. Biol. Med., 79:261-265, 1952.

50. Bradley, P.B., Elkes, J., Effect of atropine, hyoscyamine physostigmine, and neostigmine on the electrical activity of the brain of the conscious cat, J. Physiol., 120:15-15, 1953.

51. Sternbach, R.A., Two independent indices of activation, Electroencephal. Clin. Neurophysiol., 12:609-611, 1960.

52. Elliott, R., Physiological activity and performance: a comparison of kindergarten children and young adults, Psychol. Monogr., 78: (Whole No. 587, No. 10), 1964.

53. Lacey, J.I., Somatic response patterning and stress: Some revisions of activation theory, in M.H. Appley & R. Trumbull (eds.), Psychological Stress, Appleton-Century-Crofts, N.Y., 1967.

54. Cohen, P.J., Dripps, R.D., Signs and stages of anesthesia (chap. 3), in L.S. Goodman & A. Gilman (eds.), the Pharmacological Bases of Therapeutics, 4th ed., Macmillan, N.Y., 1970.

55. Berger, R.J., Physiological characteristics of sleep, in A. Kales (ed.), Sleep: Physiology & Pathology, J.B. Lippincott, Philadelphia, 1969.

56. Johnson, L., A psychophysiology for all states, Psychophysiol., 6:501-516, 1970.

57. Wallace, R.K., Benson, H., The physiology of meditation, Scientific American, 226:84-91, 1972.

58. Turner, P., Critical flicker frequency and centrally-acting drugs, Brit. J. Ophthal., 52:245-250, 1968.

59. Lucio, W.H., et al., Psychophysiological correlates of female teacher behavior and emotional stability: a seven-year longitudinal investigation, CSE Rep. No. 44, University of California, Los Angeles, 1967.

60. Lacey, J.I., Lacey, B.C., Verification and extension of the principle of autonomic response stereotypy, Amer. J. Psychol., 71:50-73, 1958.

ALCOHOL AND SECOBARBITAL: EFFECTS ON INFORMATION PROCESSING[1]

Van K. Tharp, Jr., O.H. Rundell, Jr., Boyd K. Lester

and Harold L. Williams[2]

University of Oklahoma Health Sciences Center

Oklahoma City, Oklahoma

Most published reports of the effects of drugs on human performance have been empirical and task specific rather than theoretical in conception. For example, alcohol and secobarbital both cause impairment on tasks such as time estimation (Rutschman & Rubenstein, 1966), reaction time (Moskowitz & Roth, 1971; Blum, Stern & Melville, 1964) and verbal retention (Jones, 1973; Evans & Davis, 1969). Yet, alcohol and the barbiturates are not identical, pharmacologically; and it is important to discover whether their similar effects on laboratory tasks result from impairment of different or of the same cognitive processes. One way to address this question is to examine drug effects from the perspective of an information processing model.

Several models of human information processing postulate a sequence of distinct cognitive processes or stages which intervene between presentation of a stimulus and execution of a response (e.g., Smith, 1968; Sternberg, 1969a and 1969b). Thus, as illustrated in Figure 1, for a simple character-recognition task, a typical sequence of events would be (1) stimulus preprocessing and encoding at a sensory-perceptual level, (2) stimulus categorization, wherein the item or information about the item is compared to other items stored in memory, (3) response selection and organization,

---

1 This study was supported in part by Grant No. DADA17-73-C-3157 from the office of the Surgeon General and by Grant No. MH 14702-04 from the National Institute of Mental Health.

2 Dr. Williams is now at the Department of Psychology, University of Minnesota, Minneapolis, Minnesota.

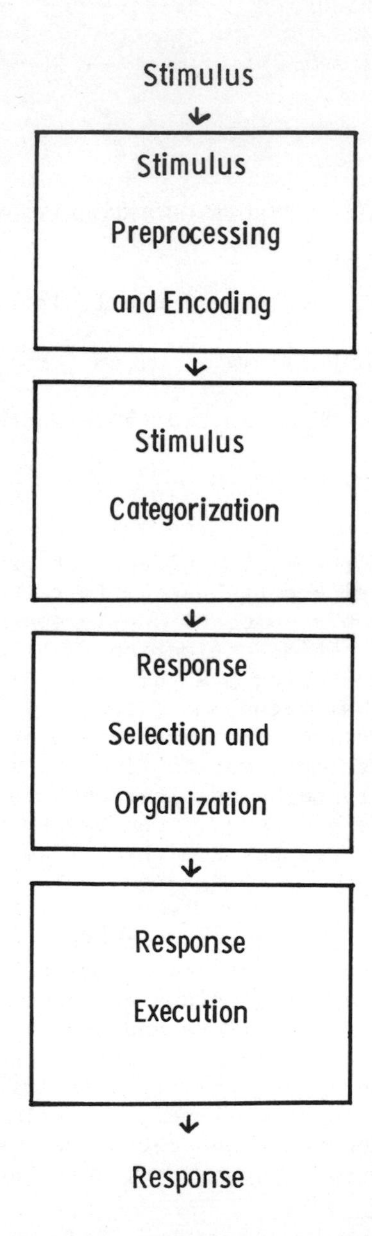

FIGURE 1. The four primary operations in sequence which are postulated by Smith (1968) as being necessary to process information.

and (4) response execution (see Smith, 1968, for a review of this and other information-processing models). Such paradigms usually contain the testable assumption that the successive stages of information processing are additive and non-overlapping.

Suppose that reaction time (RT) is the dependent variable. The additive model assumes that the total time between stimulus and response is divided into discrete, non-overlapping intervals, each of which corresponds to one of the stages shown in Figure 1. Obviously, two different experimental treatments could cause the same quantitative impairment of RT by influencing entirely different cognitive processes. Thus, a drug might cause slowing in any or all of the stages shown in Figure 1; it might cause overlap between normally independent stages; it might alter the normal sequence of cognitive stages; or it might reduce accuracy of performance by impairing the output of one or more stages.

Sternberg's (1969a and 1969b) approach to the validation of additive-stage models of this type is to conduct multiple-treatment experiments, examining patterns of additivity and interaction between the effects of rationally selected treatment variables. If the concept of distinct additive cognitive processes is valid, then the effects of two treatments that influence different stages in the model should be statistically independent. Conversely, rationally selected experimental conditions whose effects show strong positive interactions[3] probably influence the same cognitive stage. Using this "additive factor" method to examine performance on character-recognition tasks, Sternberg (1967, 1969a, 1969b) identified several experimental treatments, each of which apparently influences a specific stage of information processing. For example, a treatment which alters the discriminability of the stimulus probably influences the stimulus-preprocessing and encoding stage, whereas a treatment that alters either the difficulty of mapping the stimulus on the response or the distribution of response probabilities probably influences the stage of response selection and organization. Hereafter in this paper, a treatment whose probable locus of effect has already been identified will be called an "established treatment."

These ideas suggest a line of investigation which may reveal the locus or loci of a drug effect on human performance. Thus, the patterns of additivity and interaction between the effects of a drug treatment and those of established task-related treatments could indicate which cognitive process or processes (i.e., processing stages)

---

3 A _positive interaction_ between two treatments is a relationship in which their joint effect is greater than the sum of their individual effects. The relationship between two treatments may be described as _additive_ when their joint effect is equal to the sum of their individual effects.

are most vulnerable to the drug. For example, a strong positive interaction between the effects of a drug and an established treatment would imply that both experimental variables influence the same stage of information processing. On the other hand, an additive relationship between the effects of a drug and an established treatment would suggest that each variable influences a separate stage.

Using this rationale, we conducted a series of experiments with character-recognition tasks in which we examined the combined effects of either alcohol or secobarbital and several established, task-related treatments. We selected those drugs because both depress the central nervous system, both are representative of a group of compounds which show cross-tolerance, and both are addictive. Nevertheless, their sites of action in the brain probably are not identical (Williams & Salamy, 1972).

Figure 2 shows in outline the major task-related treatments used in the three experiments reported here and indicates which processing stages are probably selectively influenced by these treatments. The response execution stage of the model was not tapped in these experiments. A detailed account of the methodology of the three core experiments can be found in Tharp, Rundell, Lester & Williams (in press), so only a brief outline is presented here.

Experiment 1: Drug Effects on Letter Recognition

Method

Experiment 1 required written transcription of letters of the English alphabet, presented one at a time. With accuracy as the dependent variable, the established task-related treatments were _stimulus discriminability_ (DISC) and _stimulus-response compatibility_ (SRC). The first treatment varied the signal to-noise ratio of the stimuli in that the tape-recorded letters, which were presented at an intensity level of approximately 83 dB, were masked by a white noise of either 70 dB (high DISC) or 80 dB (low DISC). The second treatment varied the difficulty of mapping the stimulus on the response. On half of the trials, subjects (_S_s) wrote down the letter of the alphabet presented (high SRC), whereas on the remaining trials, they wrote down the next successive letter in the alphabet (low SRC). _Stimulus discriminability_ and _SRC_ were chosen in this experiment because they probably influence selectively, different cognitive stages; i.e., stimulus preprocessing and response selection-organization, respectively (Sternberg, 1969a, 1969b; Biederman & Kaplan, 1970).

From a group of 18 young, male volunteers with normal hearing and no history of drug abuse or heavy drinking, nine _S_s were randomly

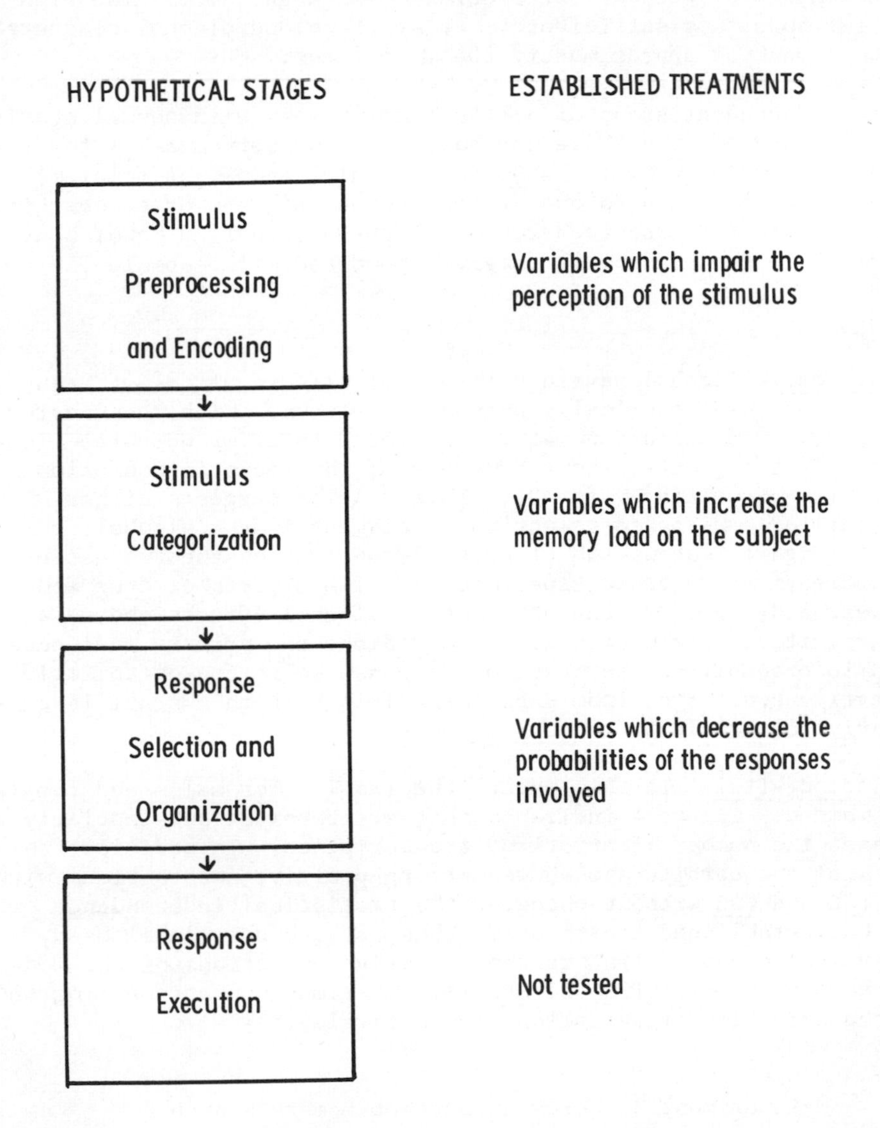

FIGURE 2. Variables (established treatments) used in our core experiments with their associated hypothetical stages.

selected and given a dose of 1.3 ml per kg of 95% ethanol. The remaining Ss were given a placebo consisting of four ml of ethanol floated on top of each of three drinks of orange juice. The high dose of ethanol was sufficient to raise the blood alcohol concentration to a peak of approximately 100 mg percent.

An independent group of 32 young volunteers with normal hearing and no history of drug abuse or heavy drinking served as Ss for the secobarbital study. Half the Ss (eight males and eight females) were randomly assigned to one of two groups, placebo or secobarbital. The secobarbital Ss received 200 mg of the drug per 70 kg of body weight, while the placebo Ss received powdered milk capsules.

Results

Alcohol. Alcohol caused a large increase in errors of transcription of the acoustically presented letters. With these errors distributed in a confusion matrix, we found that for both the alcohol and placebo groups, about 80% were of the acoustic confusion variety (e.g., writing V for B). This finding suggests either a perceptual or a response selection impairment in the alcohol condition. Figure 3 shows the effects of the drug on the two established treatments used in Experiment 1. The effects of drug and DISC were additive. On the other hand, alcohol effects showed a strong positive interaction with SRC. Since the primary influence of SRC is probably on the stage of response selection-organization, a tentative conclusion from Experiment 1 is that this stage is particularly vulnerable to alcohol effects.

Secobarbital. In this paper, the results for males and females are combined. Figure 4 indicates that secobarbital substantially increased the number of errors of transcription. In addition, the effects of the barbiturate showed strong positive interactions with both DISC and SRC without changing the statistical independence of these two established treatments. Thus, at the dose level used, secobarbital probably impairs the cognitive operations of the model associated with both DISC and SRC (i.e., stimulus preprocessing and response selection-organization, respectively).

Experiment 2: Drug Effects on Memory Search

Method

Experiment 2 combined visually presented digits with a binary, manual response in the well-known memory search paradigm described by Sternberg (1967). Here, S is required to decide whether a visually presented digit is or is not a member of a previously memorized set of digits. The task-related treatments, DISC (superimposing a

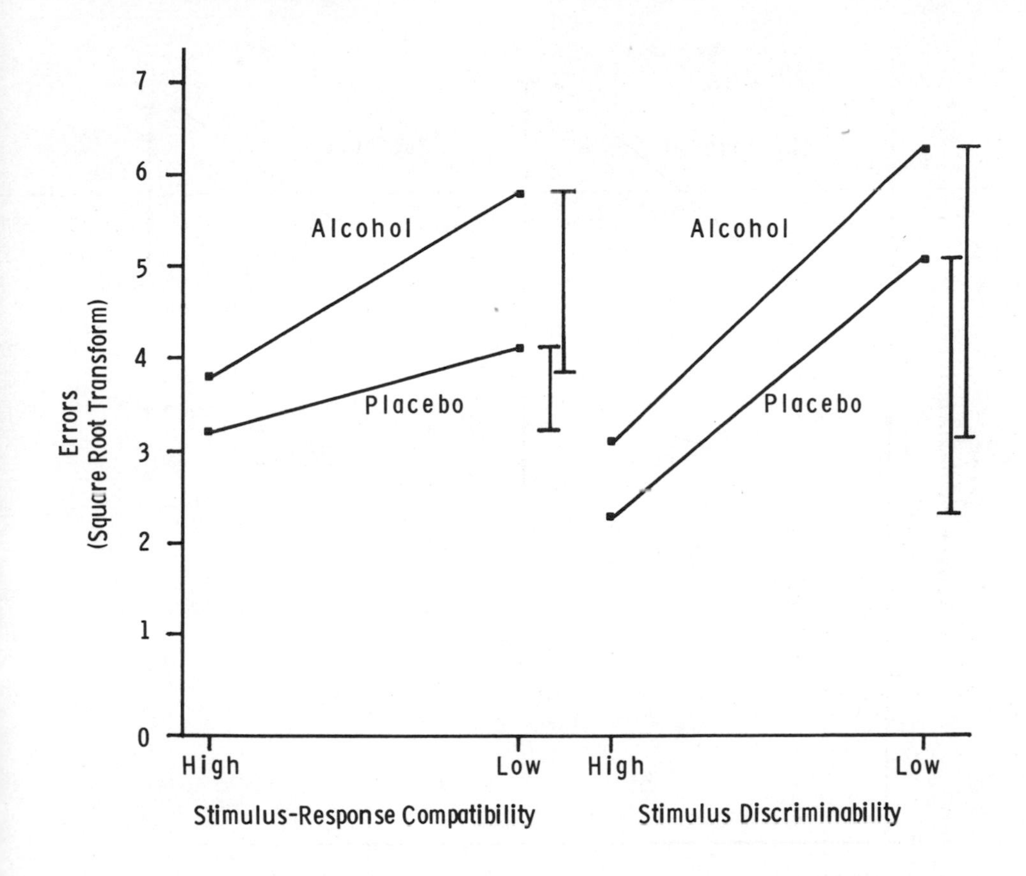

FIGURE 3. Effects of Alcohol, Stimulus-Response Compatibility and Stimulus Discriminability on Errors of Transcription. The difference in the size of the two vertical lines beside each graph indicate the size of the interaction.

checkerboard grid over the digit on half the trials), <u>size of the memory set</u> (in different trial blocks the size of the memorized list varied from one to four) and <u>response type</u> (yes or no), were varied, using speed and accuracy as dependent variables. The probabilities of occurrence of "yes" and "no" items were 4/15 and 11/15, respectively. These three treatments appear to influence selectively, stimulus preprocessing, stimulus categorization, and response selection-organization, respectively.

Twenty-four young, male volunteers with normal vision and no history of drug abuse or heavy drinking, were randomly assigned to alcohol and placebo groups as in Experiment 1. All <u>S</u>s were run for three sessions on three consecutive days, receiving their alcohol and placebo doses on the third day. Another group of 12 normal <u>S</u>s

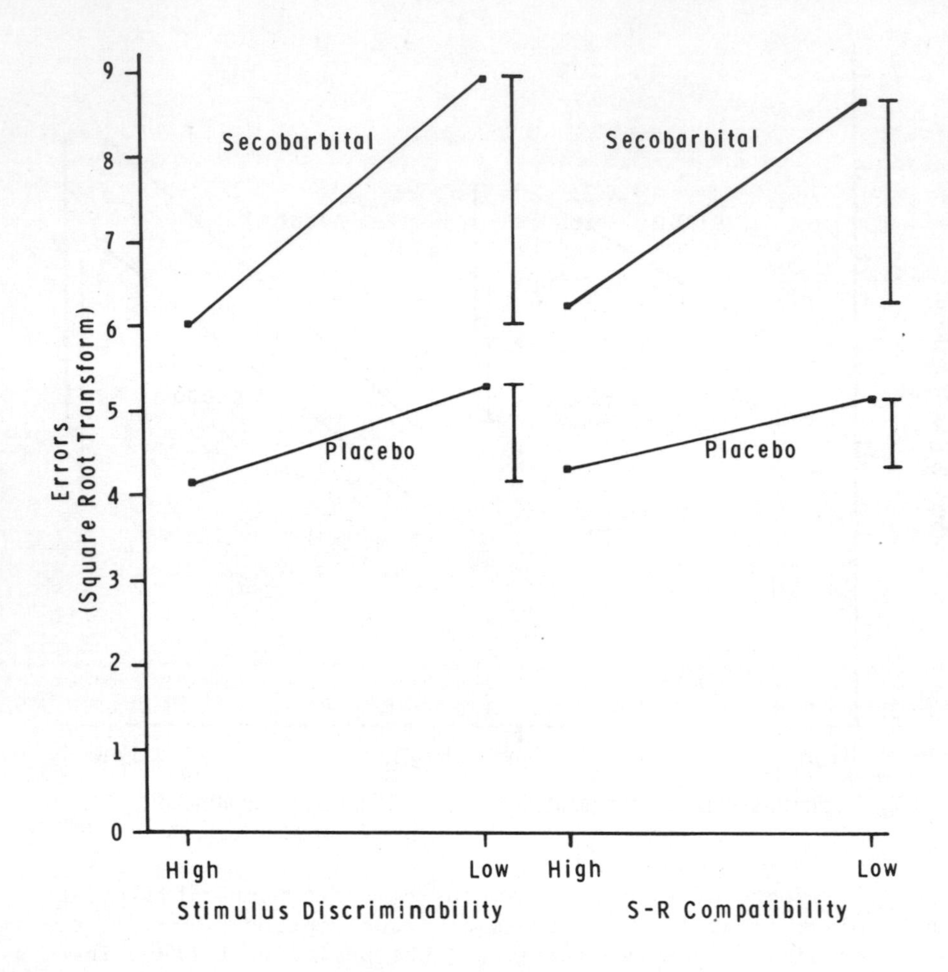

FIGURE 4. Effects of Stimulus-Response Compatibility, Stimulus Discriminability, and Secobarbital on Errors of Transcription. The difference in size of the two vertical lines beside each graph indicates the size of the interaction.

(ten males and two females) served as volunteers for the secobarbital part of the experiment. For this group, scores from sessions two days before and after the drug session were averaged as a baseline.

Subjects were assessed penalty points both for errors and long reaction times (RT). Low penalty scores were then rewarded with bonus pay.

Results

Alcohol. Alcohol caused a very small increase in RT, but this effect did not interact with the effects of any of the task-related treatments. The drug caused a strong, systematic decrease in accuracy, however, in that alcohol *Ss* made almost twice as many errors as placebo *Ss*. Figure 5 shows that this effect consisted almost entirely of an increase in errors to probe stimuli belonging to the memorized list. Thus, with accuracy as the dependent variable, the positive interaction between the effects of alcohol and response type was highly significant. Conversely, the effects of alcohol on accuracy were additive with those of the other two established treatments, *DISC* and *size of the memory set*. Thus, as in the previous experiment, the alcohol impairment may be localized to one particular stage of the model, response selection-organization.

Secobarbital. Unlike alcohol, secobarbital caused a strong, systematic increase in RT, when compared with the mean of the two baseline sessions. As shown in Figure 6, the effects of the drug interacted positively with those of two of the established treatments, *DISC* and *response-type*. The drug effect was additive, however, with the third task-related treatment, *size of the memory set*. Thus, as suggested by Experiment 1, secobarbital appeared to impair two stages of the model, stimulus preprocessing on the one hand, and response selection-organization on the other, without altering the statistical independence of these two stages. The time needed for correct categorization of the probe stimulus as a member or non-member of the memory set did not appear to be affected by the drug.

Experiment 3: Drug Effects on a One-to-One Matching Task

Method

Experiment 3, taken from Sternberg (1969b), required verbal recognition responses to visually presented digits. With verbal RT or accuracy as dependent variables, *DISC*, *SRC*, and *stimulus-response* uncertainty (SRU) were the established treatments. As in Experiment 2, *stimulus discriminability* involved superimposing a checkerboard grid over the digit on half the trials (low DISC). As in Experiment 1, *stimulus-response compatibility* varied the difficulty of mapping the stimulus on the response in that on half the trials *S* responded with the digit presented (high SRC), whereas on the remaining trials *S* responded with the next digit in ordinal progression. *Stimulus-response uncertainty* varied the number of response alternatives by selecting the digits for each block from a population of either two (low SRU) or eight (high SRU) alternatives. The last two treatments both vary response difficulty and their effects show a strong positive interaction (Sternberg, 1969b). Thus, both treatments

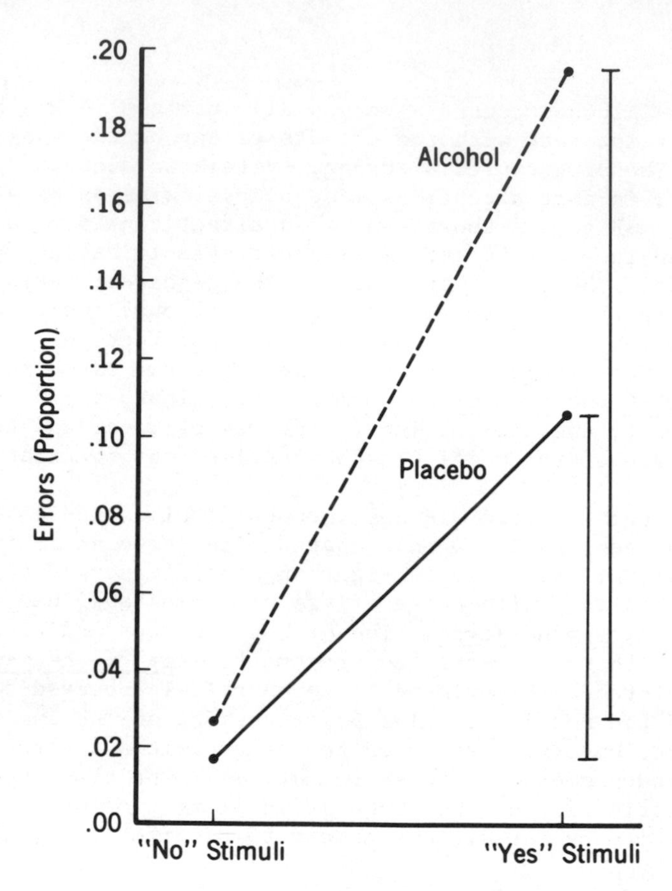

FIGURE 5. Effects of Alcohol and Response Type on Accuracy. The difference in size of the vertical lines beside the graph indicates the magnitude of the interaction.

probably influence the response selection-organization stage of the model.

This was a within-subjects experimental design in which 18 young, normal, male volunteers participated in four experimental sessions. The first three sessions were on consecutive days, while the fourth session followed two days later. On the third day all Ss were given 1.3 ml per kg of 95% ethanol. Scores from the second and

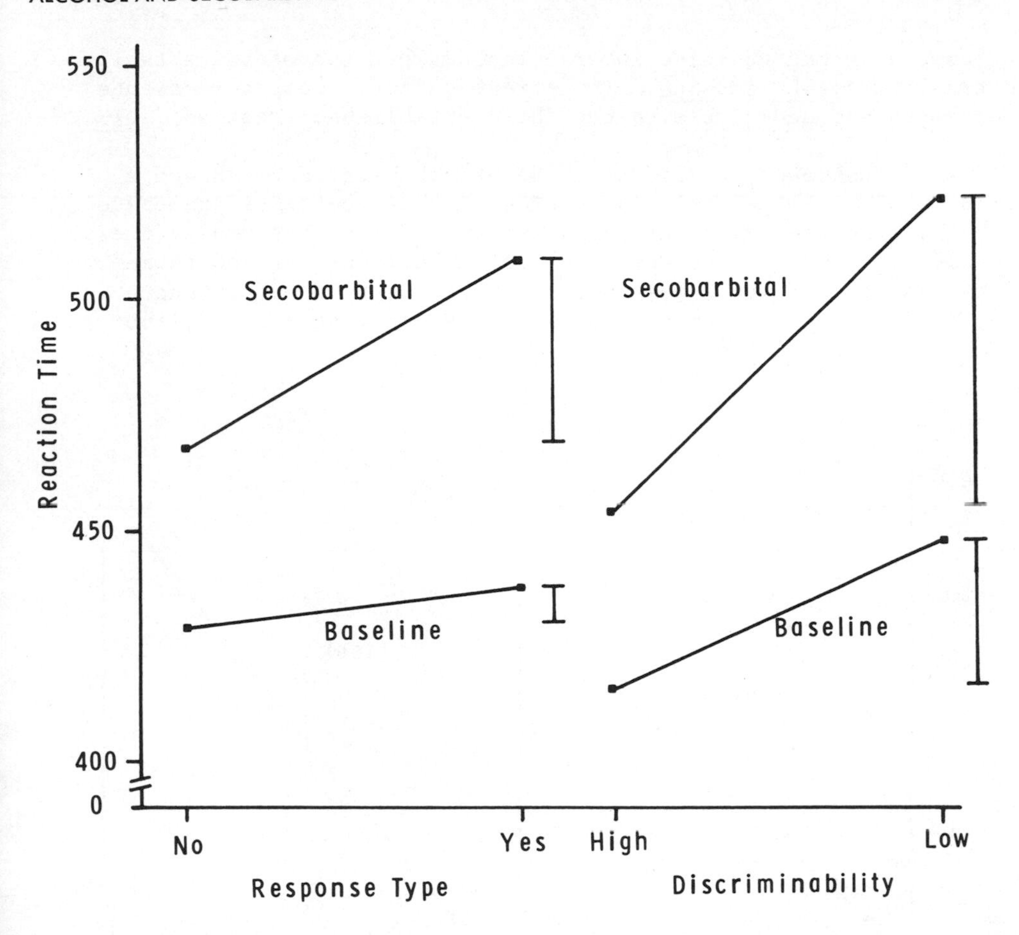

FIGURE 6. Effects of Secobarbital, Response Type, and Stimulus Discriminability on Reaction Time. The difference in the size of the two vertical lines beside each graph indicates the size of the interaction.

fourth sessions were averaged to provide a baseline. We have not yet used this experimental paradigm with secobarbital.

Results

We increased the financial penalty for errors in this experiment. Perhaps, because of this change in payoff, accuracy was nearly 100% in all experimental conditions. For RT, as anticipated, we found a strong positive interaction between two of the established treatments, SRC and SRU. The effect of these response-oriented treatments was additive with the third established treatment, DISC.

As illustrated in Figure 7, the effect of alcohol showed a strong positive interaction with each of the treatments associated with the response selection-organization stage of the model (i.e., SRC and SRU). The three-way interaction between drug and these two task-related treatments was just short of the 0.05 significance level. As expected from the results of Experiment 1 and 2, the effects of alcohol and DISC were additive.

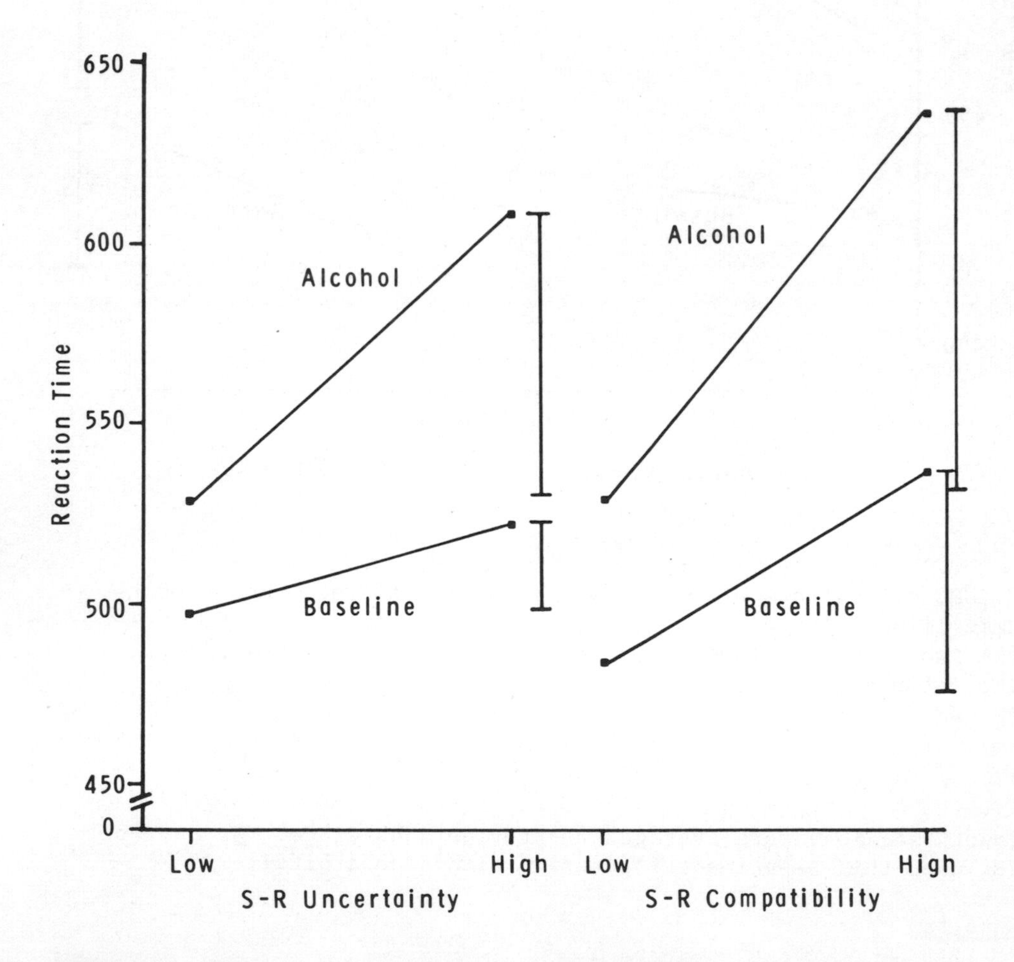

FIGURE 7. Effects of Alcohol, S-R Uncertainty, and S-R Compatibility on Verbal Reaction Time. The difference in the size of the two vertical lines beside each graph indicates the size of the interaction.

DISCUSSION

Figure 8 summarizes the pattern of results for the alcohol treatment condition. The effects of alcohol consistently show strong positive interactions with the effects of established task-related treatments linked with those cognitive operations associated with selecting and organizing a correct response, i.e., stimulus-response compatibility, stimulus-response uncertainty, and response type. The effects of alcohol were additive with all other treatments. Furthermore, in all three experiments the expected patterns of additivity and interaction between the effects of established treatments were not altered by the drug state. Thus, at the dose used, alcohol does not appear to cause overlap between normally independent cognitive operations nor does it alter their normal sequence. Apparently, in simple character recognition tasks, moderate alcohol intoxication impairs the cognitive operations involved in selecting and organizing an appropriate response, but does not impair either of those operations (i.e., the stages) associated with preprocessing and encoding the stimulus or scanning short term memory.

In Experiments 1 and 2, the patterns of additivity and interaction for secobarbital were different than for alcohol. In those experiments, the effects of the drug interacted both with the effects of DISC, and (like alcohol) with those established treatments associated with response selection-organization (see Figure 9). The effects of these two sets of established task-related treatments remained statistically independent, however, suggesting that secobarbital does not alter their normal relationship. Thus, the barbiturate seems to impair both the cognitive operations (i.e., stages) associated with preprocessing and encoding the stimulus, and those associated with selecting and organizing an appropriate response. Since the effects of secobarbital and size of memory set were additive in Experiment 2, the drug does not seem to influence the cognitive operations involved in categorization of the stimulus.

Our results suggest two general conclusions:

1. In simple character-recognition tasks, the normal relations among the effects of established task-related treatments are not disrupted by large doses of either alcohol or secobarbital. Thus, the additive, sequential-stage information processing model employed here, appears to be robust to the effects of these depressant drugs.

2. Although alcohol and secobarbital are similar depressant drugs, they may have different effects on the cognitive operations required for character recognition. With either speed or accuracy as the dependent variable, alcohol seems implicated in only one stage of information processing, response selection-organization.

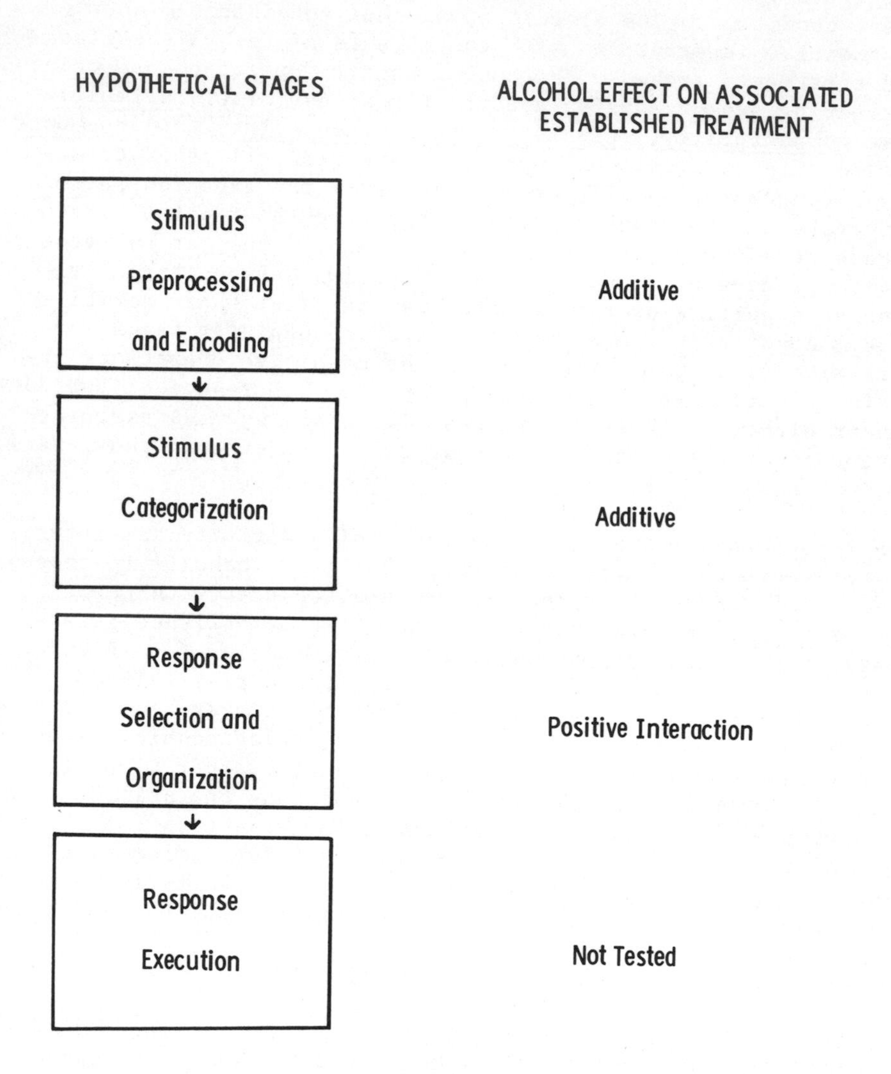

FIGURE 8. Summary of the effects of alcohol on established treatments associated with various stages of the model. The results were consistent in all three core experiments.

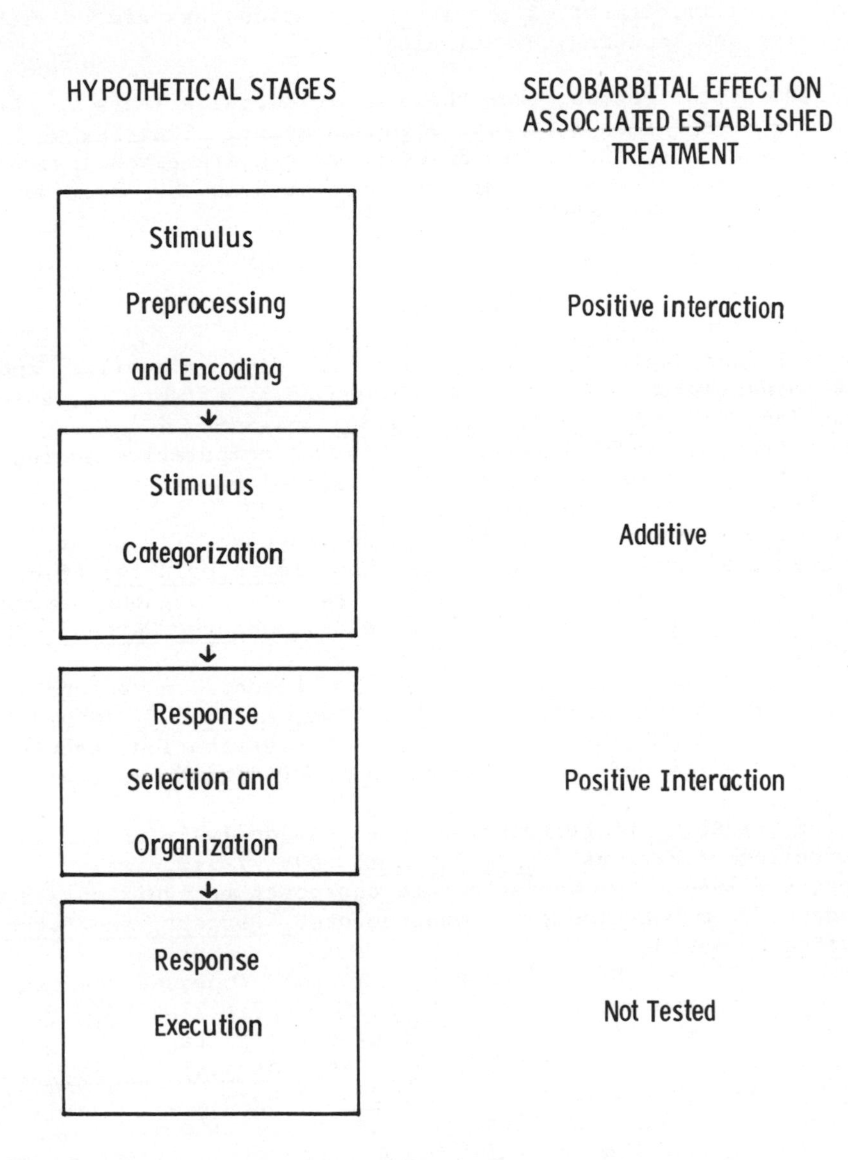

FIGURE 9. Summary of the effects of secobarbital on established treatments associated with various stages of the model. The results were consistent in the first two core experiments. Secobarbital has not yet been administered in the third experiment.

Secobarbital is also implicated in that stage, but in addition probably causes impairment of the input operations associated with preprocessing and encoding the stimulus.

We realize, of course, that these differential effects may be quantitative rather than qualitative in character. That is, we have no evidence that the single doses used here are psychologically or physiologically the same. Dose-response studies now under way in our laboratory may settle that issue.

REFERENCES

Biederman, I. and Kaplan, R., 1970. Stimulus discriminability and S-R compatibility: Evidence for independent effects in choice reaction time. J. Exper. Psychol. 86: 434-439.

Blum, B., Stern, M. and Melville, K., 1964. A comparative evaluation of the action of depressant and stimulant drugs on human performance. Psychopharmacologia 6: 173-177.

Evans, W. O. and Davis, K. F., 1969. Dose response effects of secobarbital on human memory. Psychopharmacologia 14: 46-61.

Jones, B. M., 1973. Memory impairment on the ascending and descending limbs of the blood alcohol curve. J. Abnorm. Psychol. 82: 24-32.

Moskowitz, H. and Roth, S., 1971. Effect of alcohol on response latency in object naming. Quar. J. Stud. Alcohol 32: 969-975.

Rutschmann, J. and Rubinstein, L., 1966. Time estimation, knowledge of results and drug effects. J. of Psychiat. Res. 4: 107-114.

Smith, E. E., 1968. Choice reaction time: An analysis of the major theoretical positrons. Psychol. Bull. 69: 77-110.

Sternberg, S., 1967. Two operations in character recognition: Some evidence from reaction time measurements. Percept. and Psychophysics 2: 45-53.

Sternberg, S., 1969a. Memory scanning: Mental processes revealed by reaction-time experiments. Amer. Sci. 57: 421-457.

Sternberg, S., 1969b. The discovery of processing stages: Extensions of Donders' method. Acta Psychologia: Attention and Performance (W.G. Koster, Ed.) 30: 276-315.

Tharp, V. K., Rundell, O. H., Lester, B. K. and Williams, H. L., (in press). Alcohol and Information Processing. Psychopharmacologia.

Williams, H. and Salamy, A., 1972. Alcohol and sleep. In: Kissin, B. and Begleiter, H. (Eds.), The Biology of Alcoholism, Vol. 2: Physiology and Behavior, pp. 435-474. Plenum Press, New York-London.

ALCOHOL AND BILATERAL EVOKED BRAIN POTENTIALS

Bernice Porjesz and Henri Begleiter, Dept. of Psychiatry

Downstate Medical Center, State University of New York

Brooklyn, New York, U.S.A. *

Man is cerebrally unique in that he is the only primate with marked functional asymmetry, where the two halves of the brain are specialized to serve separate functions. In most individuals (95%), the left hemisphere plays a dominant role in all language function - speech, verbal perception and thinking, while the right hemisphere dominates non-verbal contents, e.g., form perception and feelings.

Responses to non-meaningful blank flashes are mediated by the non-dominant right hemisphere. Experiments performed at Beck's laboratory indicate that right lobe evoked responses are typically larger than left lobe responses to flashes in normal subjects. This finding of interhemispheric asymmetry was significant at central locations (C_3 and C_4) in normal subjects, but was not present in alcoholics, who also displayed lower visual evoked potential amplitudes at all electrode locations, namely occipital, frontal and central (Schenkenberg et al., 1970, 1972).

In another experiment at the same laboratory, Lewis, Dustman and Beck (1970) report that cross-hemispheric asymmetry that was manifested in central areas before alcohol ingestion, disappeared after each alcohol intake. Visual evoked potentials from the right central area were significantly reduced, while those on the left remained unimpaired.

Although visual evoked potential interhemispheric asymmetry has been reported to exist in some subjects, the results of systematic studies of large populations of subjects are equivocal. Harmony

* Supported by NIAAA Grant AA01231

et al. (1973) assessed visual evoked potential asymmetry at central, occipital and temporal locations using both monopolar and bipolar recordings in 139 normal subjects. No significant differences in peak-to-peak amplitudes were obtained between the two hemispheres, and amplitude differences were found to be less than 40%. Ninety percent of the correlations between waveforms obtained simultaneously from the two hemispheres exceeded .85 from all electrode locations. The experimenters therefore concluded that the two hemispheres are strikingly symmetrical in the normal population.

Evoked potentials have been useful in investigating the differential effects of alcohol in the central nervous system, as they are highly susceptible to its suppressive influence by becoming concomitantly reduced in amplitude.

In a study of the effects of 100 cc's of alcohol on the auditory evoked response, Gross et al. (1966) established that alcohol significantly reduces the auditory evoked response. The maximal effects were obtained 15-30 minutes after alcohol ingestion.

Salamy and Williams (1973) investigated the effects of varying concentrations of alcohol on somatosensory evoked potentials recorded from bipolar leads in the post-Rolandic, parasaggital plane, and monopolar vertex recordings. They reported an inverse relationship between the dose of alcohol, the blood alcohol level, and the amplitude of the evoked potential. Early components were found to be more resistant to the effects of alcohol than later components. The major effect was obtained at vertex, particularly amplitude N_1-P_2, which was markedly reduced with increasing blood alcohol levels.

Lewis, Dustman and Beck (1970) studied the effects of two doses of alcohol mixed with grapefruit juice (.41 g/kg and 1.23 g/kg) on the visual and somatosensory evoked responses of moderate drinkers, recorded at occipital and central cortical locations. No effect was obtained with the low dose of alcohol. With the high dose they found a significant decrease in amplitude of both visual and somatosensory evoked responses recorded from central areas, particularly of the later components. Visual evoked response amplitudes were only significantly reduced from the right central electrode placement. No significant changes in amplitude of any component were obtained at occipital locations.

The purpose of the present experiment is to: first, ascertain whether or not bilateral asymmetry is apparent in the normal visual evoked potential and, second, to assess the effects of a moderate dose of alcohol on this interhemispheric asymmetry and, thirdly, to investigate whether there are differential effects of alcohol at different bilateral locations on visual evoked potential amplitude.

METHODS

The subjects were 13 healthy right-handed, adult male graduate students, with a mean age of 24, and a mean weight of 169 pounds. All subjects were occasional drinkers.

They were required to come to two recording sessions on non-consecutive mornings in the same week, and were requested to eat large identical breakfasts on both mornings. On one morning they received 95% ethyl alcohol mixed with orange juice, while on the other morning an equivalent amount of only orange juice was administered.

The amount of alcohol each subject received was dependent on his body weight in the ratio of 1.04 g/kg (1.3 ml/kg) of 95% alcohol. This amount of alcohol comprised 1/4 of his total drink, and to it were added 3 parts of orange juice.

Monopolar recordings were obtained from electrodes secured to the scalp with collodion. Electrodes were placed at central and occipital locations, bilaterally, using the linked ears as reference, and the naseum as ground. The central leads were placed 4 cm on either side of the midline on the inter-aural line, corresponding to C_3 (left) and C_4 (right) of the 10-20 International System. Occipital placements were 2.5 cm anterior to the inion and 2.5 cm on either side of the midline, corresponding to O_1 (left) and O_2 (right) of the 10-20 International System. Resistances were maintained below 5000 ohms.

Each subject was seated in a sound-attenuated, electrically shielded enclosure, with his head resting on an adjustable chin-rest, so that he was looking directly into a viewing hood, fixating in the center of his visual field. A Grass photostimulator set at an intensity of 16, delivered flashes at a random rate of 1-5 seconds apart, for a total of 32 flashes. The visual stimulus was a 5 cm square neutral density filter, that cut down the amount of light being transmitted by 80%.

The drink was administered after the first (baseline) run, and the subject was required to drink it in 10 minutes, spacing it as evenly as possible over the 10 minutes. Each recording session consisted of seven runs, one before the administration of the drink (baseline), and the others at 15, 30, 45, 60, 90, and 120 minutes after the drink had been finished, respectively. Blood alcohol levels were monitored with Breathalyzer readings which were obtained throughout the experiment immediately preceding each run.

Visual evoked potentials for each electrode placement were recorded on a Grass polygraph and summated simultaneously in

4 channels of a Hewlett Packard Signal Analyzer, for a 500 msec epoch.

Four amplitude measures were obtained from the occipital leads as the perpendicular distance in (μV) between successive peaks. Specifically, these are: the negative-going Amplitude A (approximately 60-90 msecs), the positive-going Amplitude B (90-120 msecs), the negative Amplitude C (120-165 msecs), and positive Amplitude D (165-220 msecs after the flash).

Two amplitude measures were also obtained from the central electrodes: namely, P_1-N_1, occurring at approximately 100-140 msecs and N_1-P_2, occurring from 140-200 msecs after the flash.

RESULTS

In order to assess cortical interhemispheric asymmetry, paired t-tests were performed on differences between visual evoked potential amplitudes from the right and left hemispheres. Consistent interhemispheric asymmetry was found in Amplitude B (N_1-P_2) and C (P_2-N_2) at occipital sites, but not from central areas. Neither the early components of the central response (P_1-N_1) nor the late components (N_1-P_2) displayed any significant bilateral asymmetry.

As can be seen in Figure 1 of group means for Amplitude B, the right hemisphere occipital recordings (O_2) were significantly larger than the responses simultaneously evoked by the left hemisphere (O_1) throughout the non-alcohol day.

The occipital asymmetry that was present in Amplitude B throughout the recording session on the control day, disappeared completely with the ingestion of alcohol. (See Figure 2)

Significance tests of the same amplitude under identical conditions but on the alcoholization day, indicate that while interhemispheric asymmetry is present in Run 1, baseline ($t = 3.2746$, $p \angle .01$), before alcohol administration, it disappears immediately after alcohol ingestion. There are no significant amplitude differences between right and left hemispheres, even two hours after the completion of drinking, although the two curves are just beginning to diverge at that time. The right hemisphere response was reduced to a much greater degree than the left hemisphere response, which was relatively unaffected by alcohol.

Figure 3 illustrates typical right and left hemisphere occipital evoked potentials, recorded from the same subject, comparing the same run on the alcohol and control days. The top two traces were obtained from left and right leads respectively, 45 minutes after

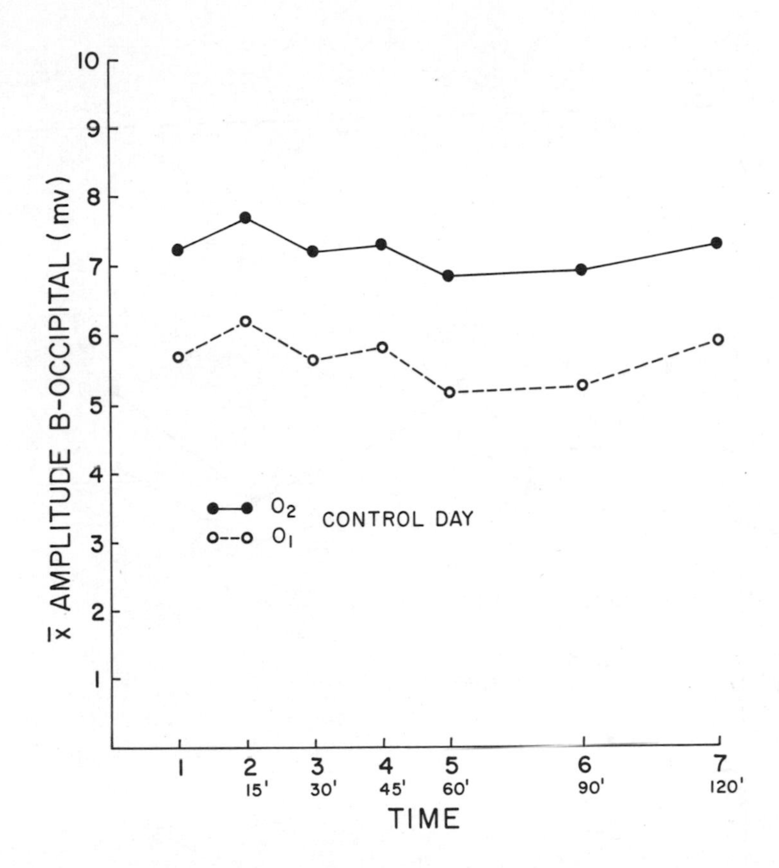

Figure 1. Mean Amplitude B (N_1-P_2) for all subjects (N = 13) recorded simultaneously from the right (O_2) and left (O_1) occipital area on the control day. Amplitudes were significantly larger from O_2 than O_1 as follows:
Run 1 (baseline) t = 2.2968, p $\angle$.05
Run 2 (time 15) t = 3.0783, p $\angle$.01
Run 3 (time 30) t = 2.7452, p $\angle$.02
Run 4 (time 45) t = 2.2869, p $\angle$.05
Run 5 (time 60) t = 3.7159, p $\angle$.01
Run 6 (time 90) t = 3.2653, p $\angle$.01
Run 7 (time 120) t = 2.6636, p $\angle$.05

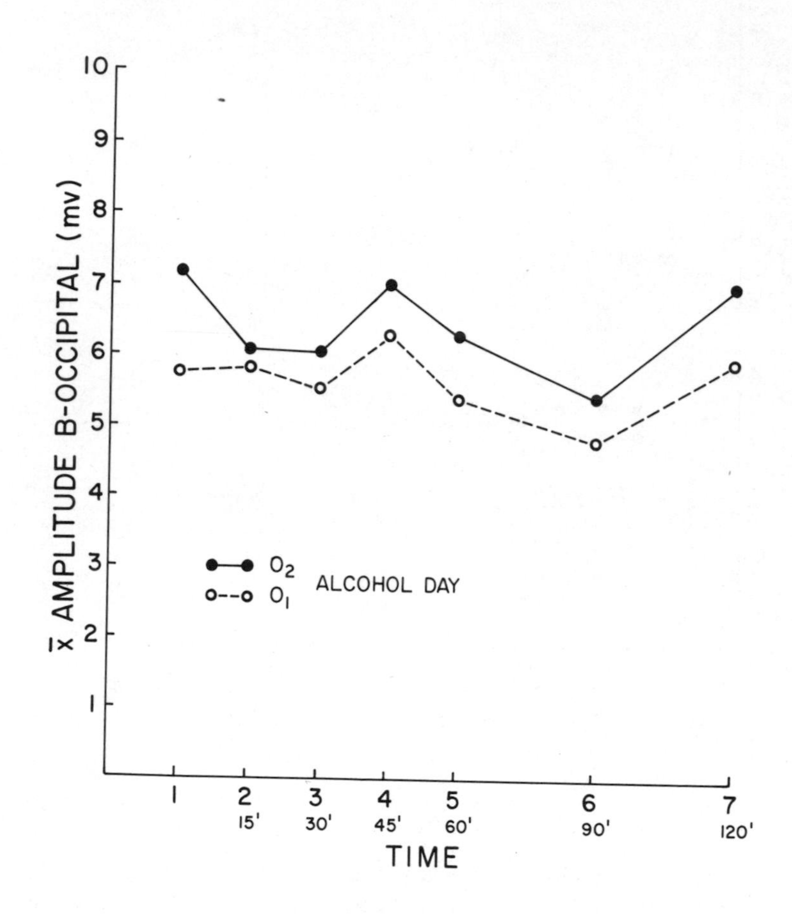
10
9
8
7
6
5
4
3
2
1
x̄ AMPLITUDE B-OCCIPITAL (mv)
O2
O1
ALCOHOL DAY
1
2
15'
3
30'
4
45'
5
60'
6
90'
7
120'
TIME

Figure 2

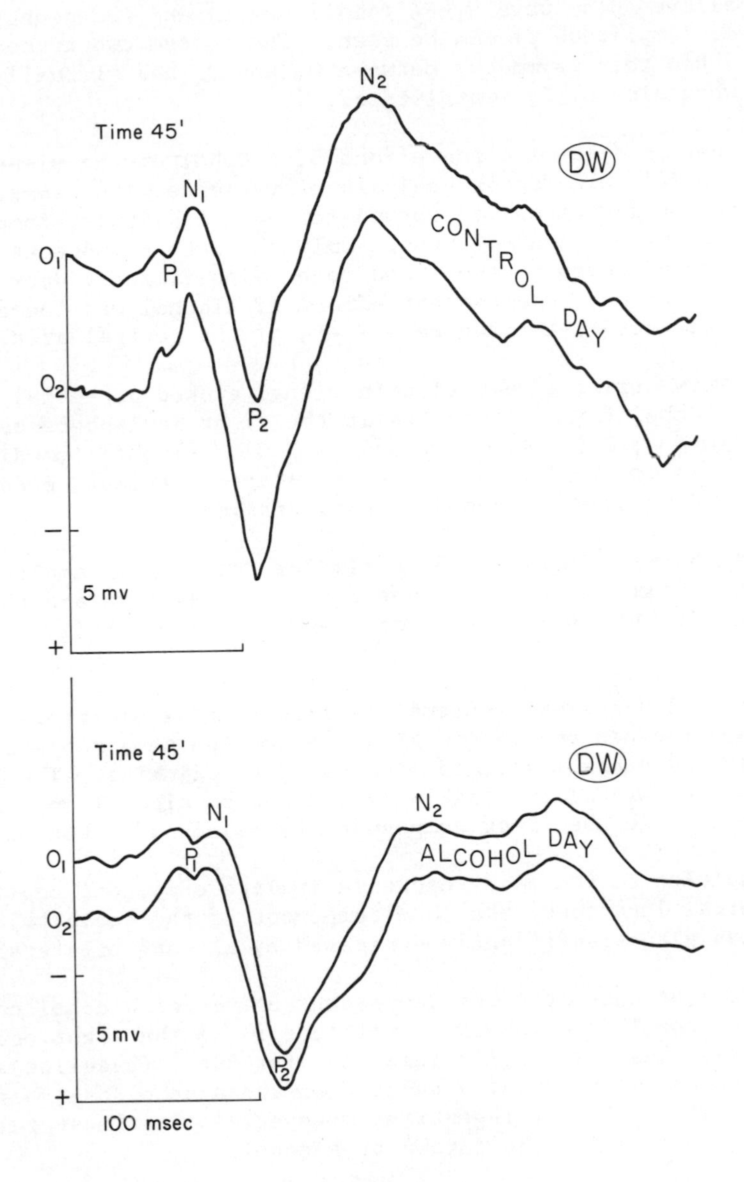
Time 45'
N2
DW
N1
CONTROL DAY
O1
P1
O2
P2
5 mv
Time 45'
DW
N1
N2
ALCOHOL DAY
O1
P1
O2
5mv
P2
100 msec

Figure 3

only orange juice was ingested. The marked asymmetry in amplitude of the positive-going wave N_1-P_2 (Amplitude B) and the negative-going P_2-N_2 (Amplitude C) can be seen. The bottom two traces illustrate how this asymmetry between O_1 and O_2 has virtually vanished when alcohol is administered.

In order to determine the effect of alcohol on the visual evoked potential, a 2-factor analysis of variance with repeated measures on one factor was performed for each amplitude, comparing the 6 alcohol with 6 control runs. Only the late components of all electrode placements were found to be significantly decreased by alcohol. The major depressant effect of alcohol was found in the large late positive-going wave N_1-P_2 of the central area.

This very striking reduction in visual evoked potential amplitude with alcohol for all subjects at the right hemisphere can be seen in Figure 4; $F (1,24) = 19.6554$, $p < .01$. Alcohol immediately depresses this amplitude, and its effects are sustained, even outlasting the 2-hour post-alcohol testing session.

As can be seen in Figure 5, a similar decrease in amplitude was obtained from the left hemisphere central response and the Analysis of Variance was also significant at $p < .01$, $F (1, 24) = 13.7836$.

Typical visual evoked potentials from central electrodes of both hemispheres can be seen in Figure 6 for the same subject on the control day and alcohol day, 15 minutes after drinking. The large late component (N_1-P_2) is strikingly reduced by alcohol in both hemispheres, while the first component (P_1-N_1) is affected much less.

In addition to the major decrease in late component amplitude at the central locations, the late component of the occipital response was also significantly depressed by alcohol bilaterally.

Figure 7 demonstrates the depressant effect of alcohol on mean magnitude of the late component (Amplitude D) of the right occipital. This amplitude was essentially identical for Run 1 (baseline) on both days, but became significantly reduced on the alcohol day, beginning immediately after alcohol ingestion; however, it is almost fully recovered 2 hours after the intake of alcohol.

The same result was obtained at the left occipital (Figure 8), which also almost returned to its original magnitude 2 hours after alcohol ingestion. There was no longer any significant difference between alcohol and control days at this time. An illustration of this occipital late component decrease can be seen in Figure 3, which is a typical record taken 45 minutes after liquid intake; there is a marked difference in amplitude of the large late component between control and alcohol days.

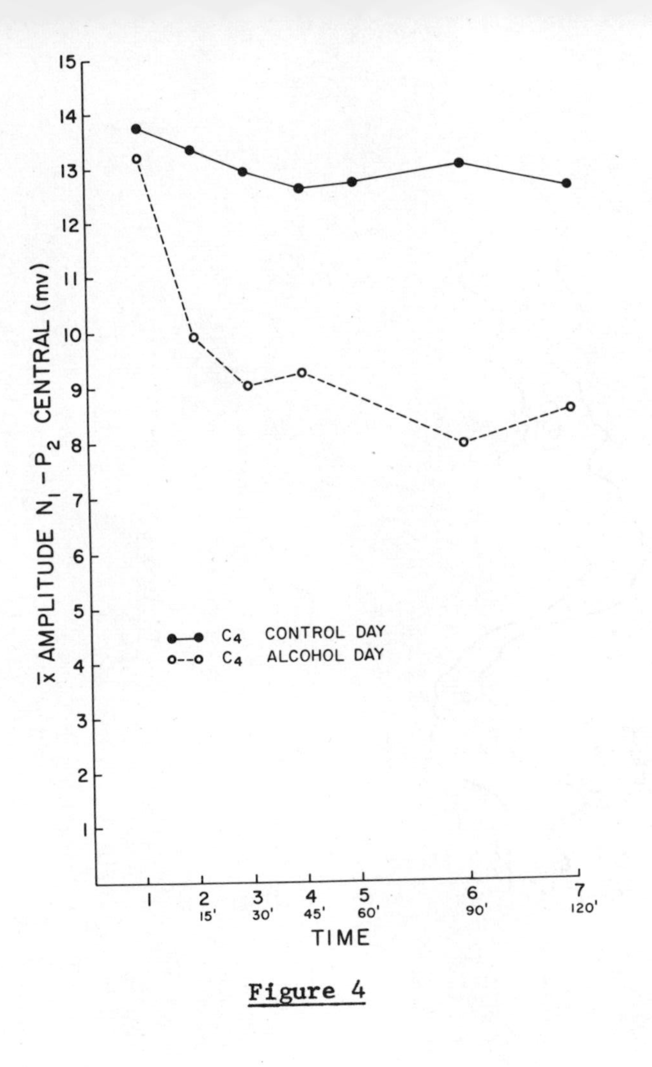

Figure 4

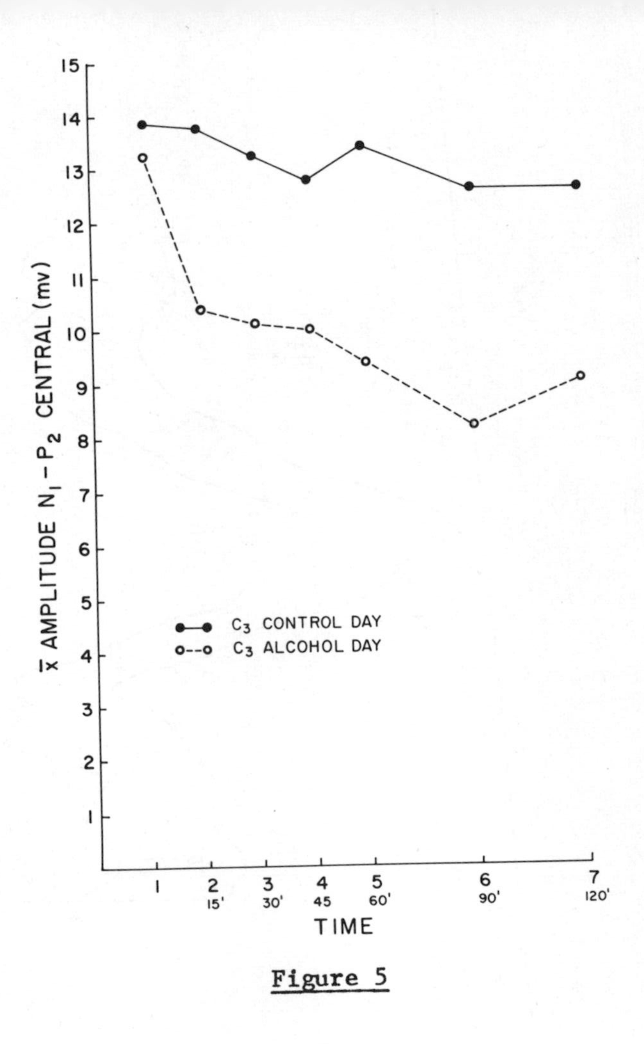

Figure 5

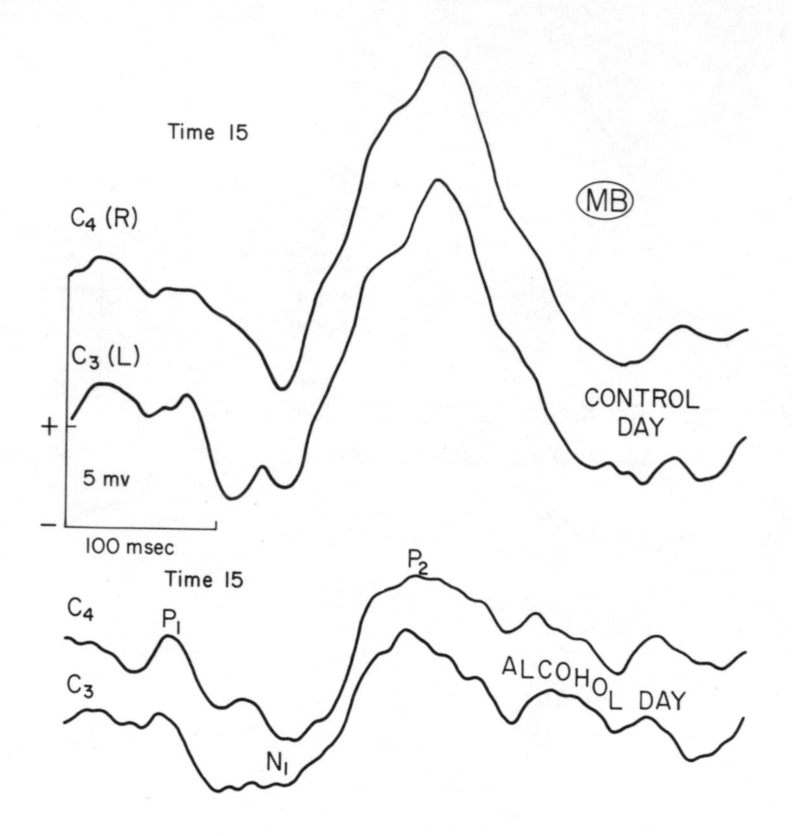

Time 15
C4 (R)
C3 (L)
MB
CONTROL DAY
+
5 mv
−
100 msec
Time 15
P2
C4
P1
C3
ALCOHOL DAY
N1

Figure 6

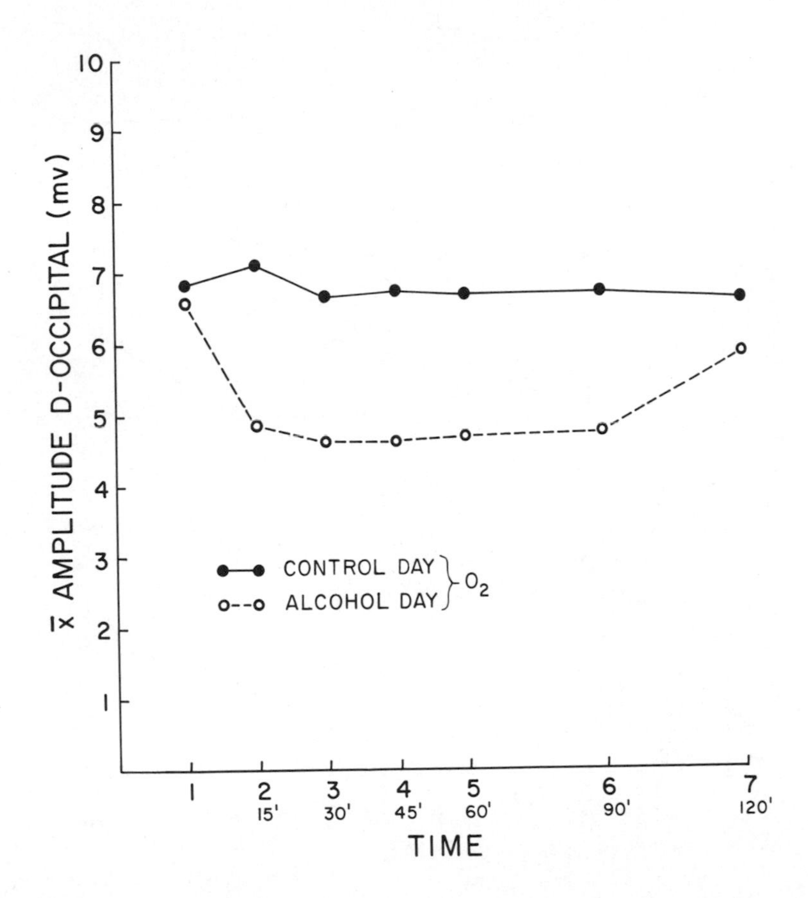

Figure 7. Mean Amplitude D for all subjects recorded from the right occipital (O_2) on alcohol and control days. Alcohol significantly reduces this amplitude as follows:
Run 1 (baseline) $t = 0.4087$, n.s.
Run 2 (time 15) $t = 2.9032$, $p < .02$
Run 3 (time 30) $t = 2.7295$, $p < .02$
Run 4 (time 45) $t = 2.405$, $p < .05$
Run 5 (time 60) $t = 2.5284$, $p < .05$
Run 6 (time 90) $t = 2.6371$, $p < .05$
Run 7 (time 120) $t = 0.8054$, n.s.

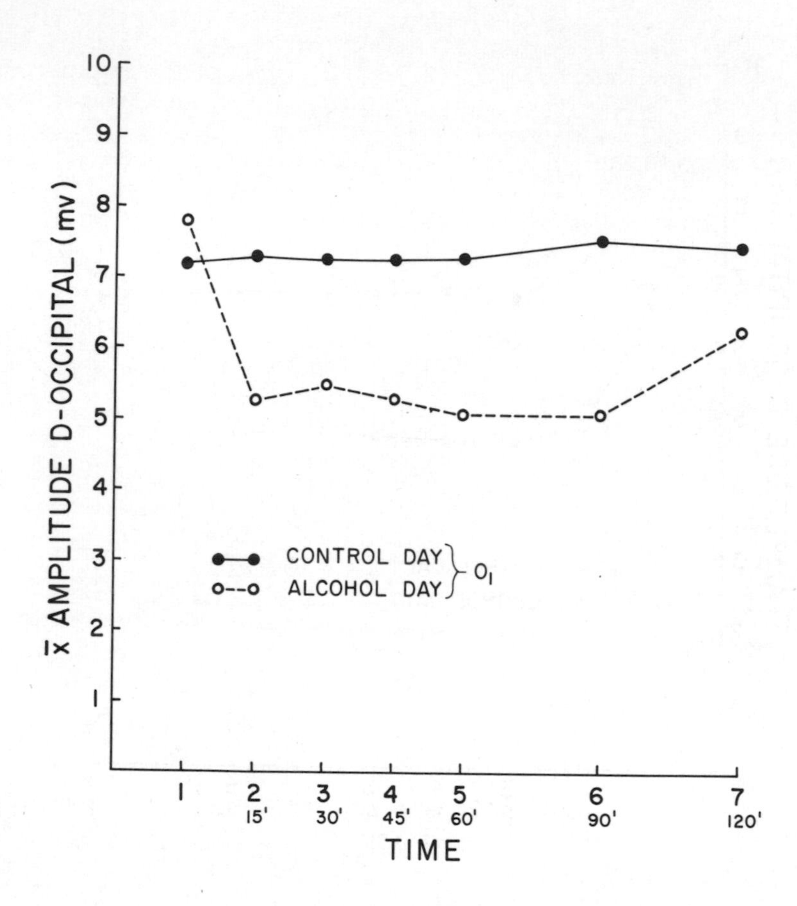

Figure 8. Mean Amplitude D for all subjects recorded at the left occipital (O_1) on alcohol and control days. Alcohol significantly reduces this amplitude as follows:
Run 1 (baseline) t = 0.7017, n.s.
Run 2 (time 15) t = 2.9537, p /.02
Run 3 (time 30) t = 2.9837, p /.02
Run 4 (time 45) t = 2.1797, p /.05
Run 5 (time 60) t = 2.3605, p /.05
Run 6 (time 90) t = 3.3501, p /.01
Run 7 (time 120) t = 1.269, n.s.

DISCUSSION

The results of the present experiment support the contention that responses to non-meaningful blank flashes are mediated by the non-dominant right hemisphere. Bilateral asymmetry was present at occipital sites, where the right lobe responses were significantly larger than left lobe responses. Despite the finding from Beck's laboratory that right central evoked potentials in normals are consistently larger than left central responses, we did not obtain asymmetry at that location. However, the asymmetry studies from Beck's laboratory are based on very small sample sizes. Our finding that neither the early nor late components of the central response demonstrate interhemispheric asymmetry is in agreement with Harmony et al. (1973), who report a remarkable degree of bilateral symmetry in the visual evoked response at identical electrode locations with large samples of subjects.

Therefore, great caution is suggested in evaluating the results of studies comparing interhemispheric amplitude asymmetry between normal and pathological groups. Until normative data can be established for the non-alcoholic population, no frame of reference exists with which to compare the alcoholic condition.

The ingestion of alcohol seems to have the effect of dissipating any existent bilateral asymmetry. In the present experiment, where VEP asymmetry was displayed at occipital locations, the ingestion of alcohol virtually abolished interhemispheric differences, decreasing responses from the right hemisphere, while leaving left hemisphere responses relatively unaffected. A similar dissipation of cross-hemispheric asymmetry with alcohol intake was reported by Lewis, Dustman and Beck (1970) for central areas. However, in contrast to our findings, they report that the left hemisphere was unaltered by alcohol intake at central locations. In the present experiment, although the right central area was found to be most susceptible to alcohol ingestion, the late components of central responses from both hemispheres were found to be particularly sensitive to alcohol.

The finding that the early components are more resistant to the depressant effects of alcohol than the later components supports the results of Salamy and Williams (1973) at a similar electrode location, namely vertex.

In addition to the major decrease in late component amplitude at both central locations, we found that the late component of the occipital response was also significantly depressed by alcohol bilaterally. This is in contrast to the findings by Lewis, Dustman and Beck (1970), who reported that all amplitudes of the occipital response are unaffected by alcohol. However, the results of the present study with visual responses are consistent with those of

Salamy and Williams (1973) with somatosensory responses. They found that the late component of the somatosensory evoked potential is significantly reduced in the primary receiving area, but that the early components are resistant to its effects. Similarly, in the present experiment, the early components of the visual evoked potential in the primary receiving area (occipital) are not susceptible to the depressant effects of alcohol, while the late components are markedly reduced.

The late components of the visual evoked potential are often considered to reflect the more cognitive aspects of perception, while the early components are taken to indicate sensory input. It seems, therefore, from the present findings that alcohol differentially affects cortical functioning, primarily impairing cognitive (or output) processes, while leaving incoming sensory process intact. Furthermore, the major change in visual evoked potential amplitude occurred at the central electrode placements which are adjacent to association cortex, suggesting that higher cortical functioning is more susceptible to alcohol than the primary sensory area (occipital).

There is evidence supporting this contention in information processing reaction-time experiments. In recent studies, Rundell et al. (1973) and Tharp et al. (this volume) report that stimulus pre-processing and encoding are unaffected by alcohol, but that alcohol affects the more central (output) stages of information processing, for example, response selection and organization.

In conclusion, the present experiment seems to indicate that interhemispheric asymmetry is present only at occipital locations, and is dissipated by the ingestion of alcohol. Furthermore, the present findings suggest that alcohol differentially affects central nervous system activity, primarily suppressing higher cortical functioning. However, the exact sites of action of alcohol, and the progression of its effects on the brain remain to be clarified.

REFERENCES

Gross, M. M., Begleiter, H., Tobin, M., and Kissin, B. Changes in auditory evoked response induced by alcohol. Journal of Nervous and Mental Disease, 1966, 143, 152-156.

Harmony, T., Ricardo, J., Otero, G., Fernandez, G., Llorente, S., and Valdes, P. Symmetry of the visual evoked potential in normal subjects. EEG clin. Neurophysiology, 1973, 35, 237-240.

Lewis, E. G., Dustman, R. E., and Beck, E. C. The effects of alcohol on visual and somato-sensory evoked responses. EEG clin. Neurophysiology, 1970, 28, 202-205.

Rundell, O. H., Tharp, V., Lester, B. K., and Williams, H. L. Some effects of acute intoxication on information processing stages. Alcohol Technical Reports, 1973, 1, 25-32.

Salamy, A. and Williams, H. L. The effects of alcohol on sensory evoked and spontaneous cerebral potentials in man. EEG clin. Neurophysiology, 1973, 35, 3-11.

Schenkenberg, T. and Dustman, R. E. Visual, auditory and somato-sensory evoked response changes related to age, hemisphere and sex. Proceedings of the 78th Annual Convention, American Psychological Association, 1970, 183-184.

Schenkenberg, T., Dustman, R. E., and Beck, E. C. Cortical evoked responses of hospitalized geriatrics in three diagnostic categories. Proceedings of the 80th Annual Convention, American Psychological Association, 1972, 671-672.

Tharp, V. K., Rundell, O. H., Jr., Lester, B. K., and Williams, H. L. Alcohol and secobarbital: Effects on information processing (see this volume).

BRAIN DAMAGE IN ALCOHOLICS: ALTERED STATES OF UNCONSCIOUSNESS

Oscar A. Parsons

Oklahoma Center for Alcohol-Related Studies

University of Oklahoma Health Sciences Center*

It has become popular in recent years to employ the concept of "altered states of consciousness" in attempting to understand the behavioral effects of various drugs including alcohol (Jones, 1974; Mello and Mendelson, 1969; Tart, 1969). Such a concept appears to have value in considering the acute effects of drugs in which a transient alteration in the brain has occurred and where it is recognized by the drug taker as a temporary change from an ongoing non-drug state. For example, most people who drink alcohol report altered subjective states or changed consciousness. But what about the effects of chronic drug use in persons who are no longer using drugs? Are there altered states of consciousness? From the occurrence of "flash-backs" such as LSD-25 users have described (i.e., where the former drug-taker during abstinence has an altered state of consciousness similar to that which he experienced while taking the drug) it seems clear that such experiences are possible with certain drugs. The so called "dry-drink" phenomenon has been reported in some alcoholics. However, these are relatively sporadic and transient episodes. Of much greater permanence and consistency are the behavioral changes suggestive of altered brain states, in the presence of a clear consciousness, in chronic alcoholics.

How can such changes be understood? Dr. Ben Jones from our research group suggested somewhat facetiously that we consider changes in the chronic alcoholic as "altered states of unconsciousness".

* This research was supported, in part by USPHS (NIMH) Grant 14702, and is the product of the efforts of the author and colleagues Ben M. Jones, Ph.D., Ralph Tarter, Ph.D., James Callan, M.S., Jim Bertera, M.S. and Diane Klisz, B.A.

When the initial laughter subsided after his remark, it seemed upon reflection that the concept could be pursued with some profit; indeed it has helped to clarify our thinking about some of the puzzling results which we had obtained in our studies of the neuropsychological deficits in chronic alcoholism. It should be made clear at this point that we are not using "unconsciousness" in the psychoanalytic sense or as a behavioral lack of responsivity. Perhaps a more neutral term such as "non-conscious" or "not-conscious" would be more appropriate but less provocative and stimulating.

In a famous chapter on "Vigilance" Sir Henry Head (Head, 1926) has several eloquent paragraphs on the nature of consciousness:

> Any one of the specific mental processes, which enter into the general stream of consciousness, may be eliminated without materially disturbing its other components. Provided the manifestations of the lesion are purely negative any aspect of sensation, perception, images, various forms of symbolic formulation, and even elementary states of ideation may be lost independently of one another. Sensation can be gravely affected, but images remain intact; the visual images of a patient, who has long suffered from hemianopsia, do not differ from those of a normal man. Certain aspects of speech may be disturbed without any loss of general intelligence except in so far as these particular acts are necessary for formal thinking. A lesion of the left pre-frontal region of the brain may affect the elementary processes underlying ideation in such a way that it becomes possible to formulate and to act upon two fundamentally incompatible statements. One of my patients who received a small wound in the left frontal region which injured the brain, appeared to be in every way normal. In daily intercourse he behaved rationally and showed executive ability in the work of the ward; but he wrote a long letter, asking detailed questions about the family, to his mother, although he recognized that she had been dead for three years. He thought that there were two towns of Boulogne, one of which on the homeward journey from the Front, lay near Newcastle; the other one, in France, was reached "after you had crossed the sea." He had lost that fundamental sense of reality which would render two statements of this kind incompatible to the normal person. This lack of mental cohesion lies at the basis of Korsakow's psychosis, so commonly the results of toxic agents such as alcohol.

> These various mental activities, sensation, perception, imagery, symbolic formulation and ideation may be affected independently by an organic lesion, because the physiological processes, which subserve them are bound up with the vitality of different parts of the brain, any one of which can be destroyed without of necessity affecting the others. Each such local disturbance is associated with some specific psychical loss of function. But the activity of the mind as a whole is uninterrupted; removal of one aptitude does not cause collapse of the whole structure like a house of cards. The field of consciousness remains continuous as before; it closes over the gap as the sea leaves no trace of a rock that has crumbled away. (Head, 1963, pp. 493-494)

In these passages and others in his discussion of the "continuity of consciousness" Head is making the point that consciousness as subjectively considered is always "full". The fact that it may not be as "rich" a consciousness or lead to as effective action as the individual was once capable of, does not appear subjectively as diminution of the conscious experience. Consider the patient with visual field scotomata; he typically is unaware of the magnitude of his loss or indeed even the presence of a loss; consider the patient with petit-mal epilepsy where he does not realize that he has had a seizure -- the experience of consciousness is continuous; consider the frontal lobe patients whose judgment and accountability for their behavior is impaired from the observers' points of view but the patients remain unaware of subjective changes.

Recently a new dimension of the conscious-unconscious states has emerged from the "split-brain" studies in humans (Sperry, 1968). If the human brain is divided into two largely separate hemispheres by cutting the connecting commissures (corpus callosum anterior and posterior commissures), and if material is presented to the left hemisphere only, it responds verbally and skillfully to various problems. When the same subject is presented with the same stimuli to the right hemisphere, he makes appropriate behavioral choices but cannot verbally explain his behavior. It is as though the behavior is not in the focus of consciousness or at least it cannot be communicated by the right hemisphere. These findings have led workers such as Eccles (1973) to postulate that consciousness is located primarily in the left hemisphere. While this position clearly does not account for all of the facts, it does seem that consciousness is more commonly associated with the overt or covert activity which we call language whether we are attempting to order our conscious activity subjectively or for the purpose of communication. When language functions are disrupted by brain disturbances, an awareness of difficulty is likely to occur; when language functions remain intact, awareness of altered brain states appears less likely.

From the foregoing discussion there are two main points. First, subjective consciousness represents only a small part of the brain's activity and may be experienced as "full" or unchanged despite the omission or change of significant aspects of brain functioning, i.e., altered state of unconsciousness. Second, such unawareness of change is more likely to be present when language functions are relatively intact. In the language of neuropathology left hemisphere lesions are most likely to give rise to altered states of consciousness, while lesions of the right hemisphere and the frontal lobe are less likely to give rise to altered states of consciousness, or as we stated earlier, more likely to lead to altered states of unconsciousness. And now, how does this discussion relate to alcoholism? It is our contention that the neuropathological and neuropsychological findings in alcoholism strongly suggest frontal lobe damage, possible right hemisphere dysfunction but relatively intact left hemisphere functioning.

There is now ample evidence that chronic ingestion of alcohol over the period of many years leads to neuropathological changes in the brain (see extensive discussions in Wallgren & Barry, 1971; Seixas & Eggleston; 1973). While there is debate as to whether impairment of function or neuropathology can be uniquely attributed to ethanol, there is a growing literature which suggests that alcohol does indeed have direct neurotoxic effects (Freund, 1973). Postmortem studies done by Courville (1955) several decades ago indicate pronounced cortical atrophy, especially in frontal-pariental areas in chronic alcoholics. Pneumoencephalographical studies have demonstrated enlarged ventricles and cerebral atrophy (Feuerlein & Heyse, 1970; Haug, 1968; Shimojyo, Sheinberg & Reinmuth, 1967; Tumarkin, Wilson & Snyder, 1955).

Two recent studies have provided additional evidence of localizing importance. Brewer and Perrett (1971) found that 31 of 33 alcoholic patients had some evidence of cortical or ventricular atrophy. Of the 19 with cortical atrophy, nine had frontal atrophy only, 2 had parietal only and the remainder had both frontal and parietal atrophy. These findings have been confirmed by Ferrer (1970), who studied chronic alcoholics in Chile. The pneumoencephalographic analysis indicated that 44 of 45 patients had discernible cortical atrophy; in 58% of the patients a cortical and paraventricular atrophy was found; subcortical atrophy was present in 23% and solely cortical atrophy in 19%. The results of both studies strongly suggest that chronic alcohol ingestion is associated with frontal lobe and paraventricular atrophy. The neuropathological damage is in areas of the brain traversed by frontal-limbic circuits. Intensive studies of Korsakoff's psychosis have come to similar conclusions with greater emphasis on diencephalic (hypothalamic and thalamic) damage (Victor, Adams & Collins, 1971).

These findings are also in accord with the extensive studies in Russia by Segal and his colleagues (1970). Segal compared a large number of alcoholics with a group of patients with "diencephalic" syndrome from non-alcoholic causes and found many similarities between the patient groups. Segal et. al. concluded that the "withdrawal syndrome" which he found to emerge after 6 to 9 years of hard drinking indicated hypothalamic damage had occurred. The close connection between the frontal lobes and hypothalamic centers has been emphasized by Nauta (1972). In summary, the neuropathological evidence consistently implicates frontal-limbic circuits damage in chronic alcoholics. But before discussing the neuropsychological evidence in accord with these findings, we shall consider the right hemisphere hypothesis.

Neuropsychological Changes: The right hemisphere hypothesis

At the outset it is important to note that few investigators have found deficits in alcoholics on cognitive tasks involving language usage or verbal information processing (Kleinknecht & Goldstein, 1972, Jones & Parsons, 1972). Verbal coding appears more resistant to the many pathological changes than perceptual-cognitive behaviors.

When deficits are found they most frequently appear on tests involving visual-spatial-motor or visual-spatial-conceptual behavior, functions often disrupted by right hemisphere lesions. For example, many investigators Wechsler, 1958; Goldstein et al, 1970; Claeson & Carlson, 1970; and Ferrer, 1970; have reported impaired performance by alcoholics on the Block Design Test of the Wechsler or some variant of the test. In our laboratories we have found impaired performance on the Halstead Category Test, (Jones & Parsons, 1971) Raven's Progressive Matrices, The Wisconsin Card Sorting Test (Tarter & Parsons 1971, Tarter, 1973) and Maze Tracing (Tarter, 1971).

The above findings together with some evidence from our laboratories for left hand performance being less efficient than right hand performance in alcoholics (Parsons, Tarter & Edelberg, 1972; Tarter & Jones, 1971) led us to perform an experiment which was to serve as a test of the right hemisphere hypothesis. The experimental paradigm used was that of Stimulus Response compatibility (Callan, Klisz and Parsons, 1974). The task for the subject is presented in Figure 1; the S is seated with forefingers on two keys one to the left and one to the right of the body axis. When a high tone is presented to either ear he must press the right key if a low tone is presented he responds with pressing the left key, a choice reaction time paradigm (Donder's b). After a series of trials with uncrossed hands he is asked to cross his hands and respond to the same rule. In a second part of the experiment, the subject responds with pressing the key to one tone or not pressing the key to a different tone with the same hand, a "go - no go" or Donder's c reaction time paradigm. It is apparent that a number

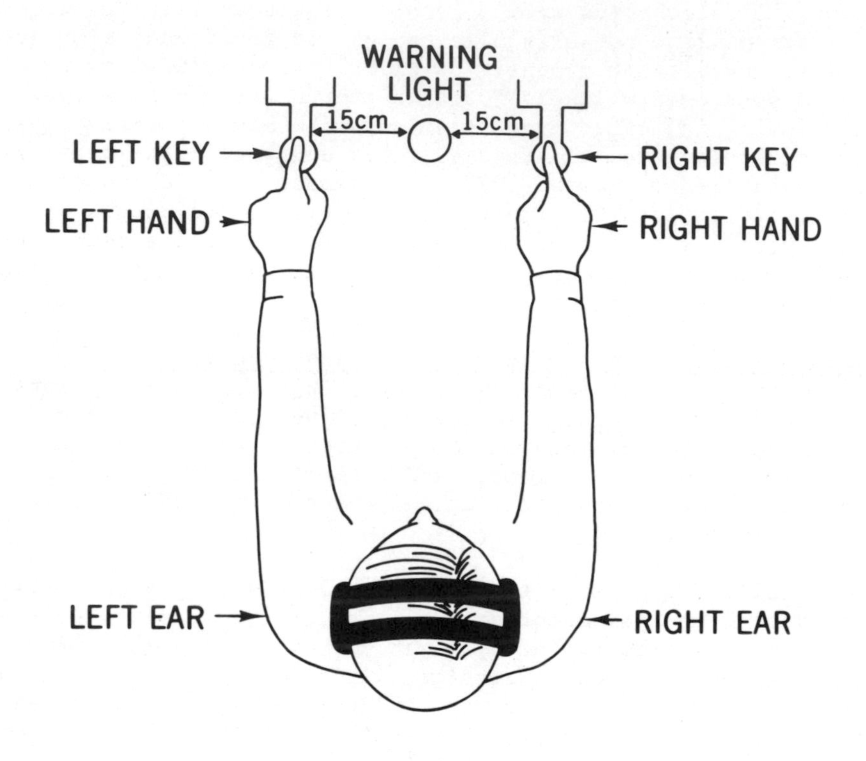

MEASURES

1. Overall ($\bar{x}$) Level	RT
2. LH vs RH	Motor
3. LE vs RE	Sensory
4. LS vs RS	Hemisphere

5. Interactions - Correspondence Effect
   Ear × Hand
   Ear × Side

Figure 1. S-R Correspondence Studies Experimental Set-up

of measures can be derived from this paradigm; hand effects, ear effects and side effects. By inference the differential function of the right and left hemispheres can be studied.

The results are pictured in Figure 2. A young normal group performs on the choice RT exactly as expected from previous studies. (Callan, Klisz & Parsons, 1974) The left side is faster when the tone is presented to the left ear than when the tone is presented to the right ear and vice versa for the right side. These results hold for both crossed and uncrossed hands. Two groups of male, middle-aged, VA alcoholics were run; one group (N=20) were short detoxification patients (two weeks); a second group (N=20) were long detoxification patients (six weeks). The results are clear: in Donder's b there is no difference between the compatibility effects although the general level of RT is elevated in the short detoxification group as might be expected. However, an interesting difference was presented in Donder's c (lower part of figure). In the short detoxification group, the left ear on the left side has a longer RT than the left ear on the right side in 17 of 20 (85%) patients; for the long detoxification group the comparable figures are 4 of 20 (20%). This suggests some altered right hemisphere functioning.

Certainly the right hemisphere hypothesis did not receive hearty support in this experiment. However, the one significant laterality finding is provocative; it occurred in the task where S must inhibit a response and again suggests frontal lobe disturbance as a primary focus in the alcoholic.

Neuropsychological Changes: The frontal lobe hypothesis

The neuropathological evidence offered earlier rather conclusively points to damage of the frontal-limbic circuits of the brain. The psychological effects of frontal lobe damage have been notoriously difficult to document. However, the regulatory role of the frontal cortex with respect to its inhibitory functions has received much attention (Warren & Akert, 1964). We conducted a test of alcoholics' ability to inhibit a response by asking them to turn a knob through 180% as slowly as possible (Parsons, Tarter & Edelberg, 1972). The results are presented in Figure 3. Alcoholics were not able to turn as slowly as controls ($F=17.64$, $1/94$ $p<.01$). Other measures collected in the experiment enabled us to conclude that the results could be attributed to a "central" deficiency in inhibitory control of motor behavior. Interestingly, when we gave the test with eyes closed; the differences between groups were more striking. In Figure 4 it can be seen that controls perform better i.e. slower with their eyes closed; alcoholics perform better with their eyes open. The interaction is significant. The results are reminiscent of the consistent differences between alcoholics and control patients on the Halstead Tactual Form Board Test (Fitzhugh et al, 1960; Smith et al, 1973), also done with the eyes closed.

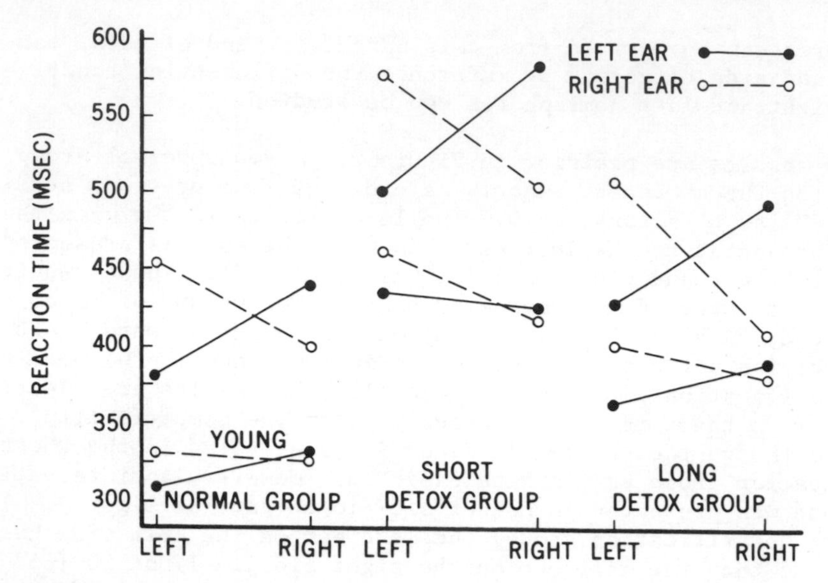

Figure 2. S-R Compatibility Effects for Donders' b (upper figures) and Donders' c (lower figures) conditions. Note the difference in the short detox group on the left side in Donders' c.

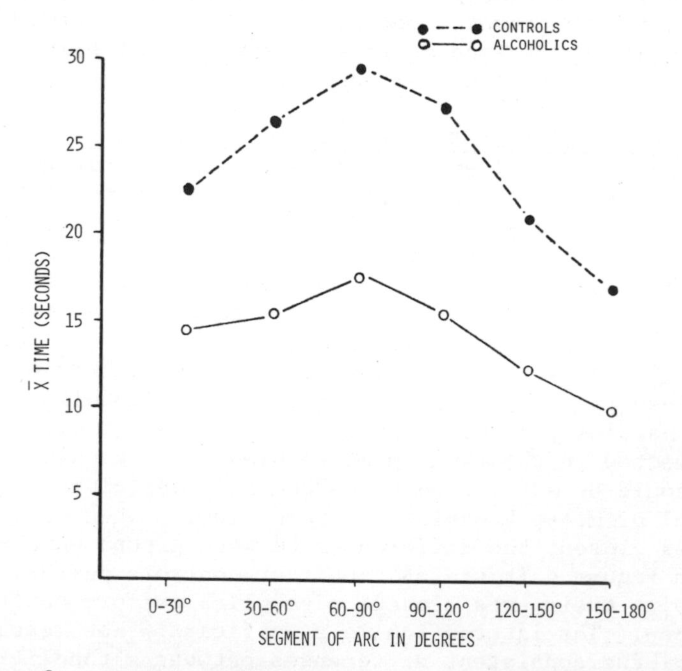

Figure 3. In this experiment, the task for the subject is to turn as slowly as possible therefore a longer time is better performance.

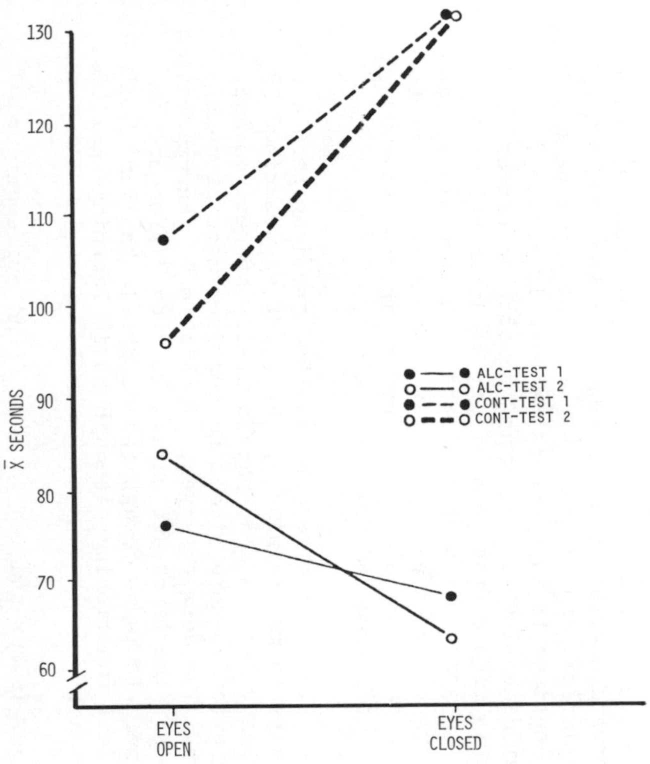

Figure 4. Interaction of groups x conditions (eyes open vs eyes closed). Tests 1 and 2 refer to repeating the task with left and right hand in counterbalanced order.

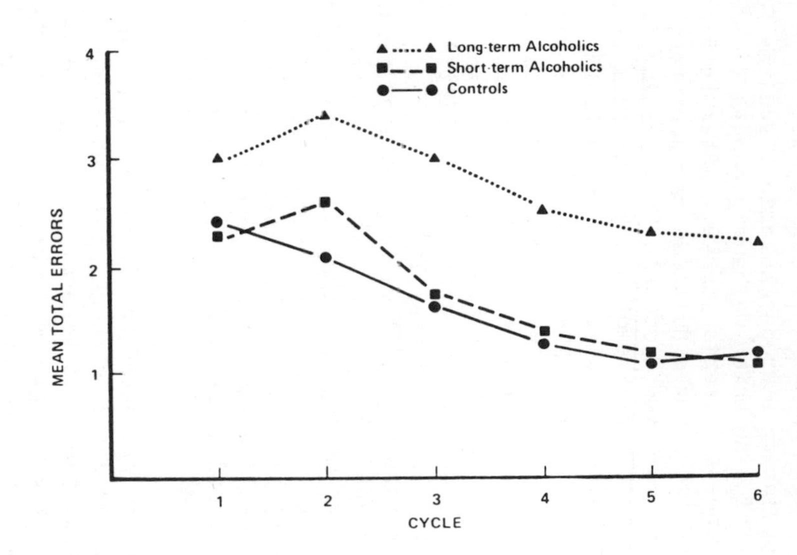

Figure 5. Mean total errors on each cycle for the controls, short- and long-term alcoholics. Each cycle represents sorting to color, form, and number in a constant order.

The most convincing neuropsychological studies for impaired frontal lobe functioning which we have produced in our laboratories are those with concept formation tasks. After confirming the impairment of alcoholics on the Halstead category test (Jones & Parsons, 1971), we used the Wisconsin Card Sorting Test in a series of studies. The test consists of sixty-four stimuli which can be sorted into four piles according to color (N=4) form (N=4) or number (one to four). The S is given feedback as to correctness or incorrectness of his sort; after ten correct sorts, without warning the E shifts the reinforcement to another attribute. The task can be continued for as many shifts from one attribute to another as deemed desirable. Frontal lobe animals have difficulty on such tasks as do frontal lobe patients (Teuber, 1972). In our first study (Tarter and Parsons, 1971) we demonstrated impairment in alcoholics on this task. Alcoholics with ten or more years of excessive drinking did more poorly than those who had been drinking less than ten years, even when age was matched. In a more definitive study (Tarter's 1971 doctoral dissertation) we selected long vs short term alcoholics and administered the WCST for six cycles during which the S had to shift from color to form to number, giving seventeen shifts in all.

The results are presented in Figure 5. The long term alcoholics (again all groups equated on age) were significantly different ($P<.01$) than the short term alcoholics; the latter and the control patients were not significantly different. Most interesting, however, was the error analyses. Perservative errors (continuing inappropriately a previously reinforced sort) are presented in Figure 6. The long term alcoholics have significantly greater numbers of perservative errors than do the other two groups who are again quite similar. Having made an error of any type the long term alcoholics continue to make errors as seen in Figure 7. In Figure 8, another dimension of understanding the performance is presented. In this figure, the probability of interrupting a correct sequence of responses is plotted. The long term alcoholics interrupt significant sequences of correct responses with errors. Needless to say, as far as could be determined in the experiments these Ss remained motivated and tried to do well.

Our interpretation of the results is that the long term alcoholic has difficulty in maintaining his set, but as described by Goldstein (1969) many decades ago, paradoxically perseveres and when he makes an error is more likely to continue more of them. But does he learn? The answer is yes! In Figure 9, the regression lines are pictures. The lines are parallel; they differ only in level not in slope. It is as though the alcoholic has more "noise" in his psychological functioning although basic learning abilities are relatively intact.

In conclusion, in our studies we have found evidence for

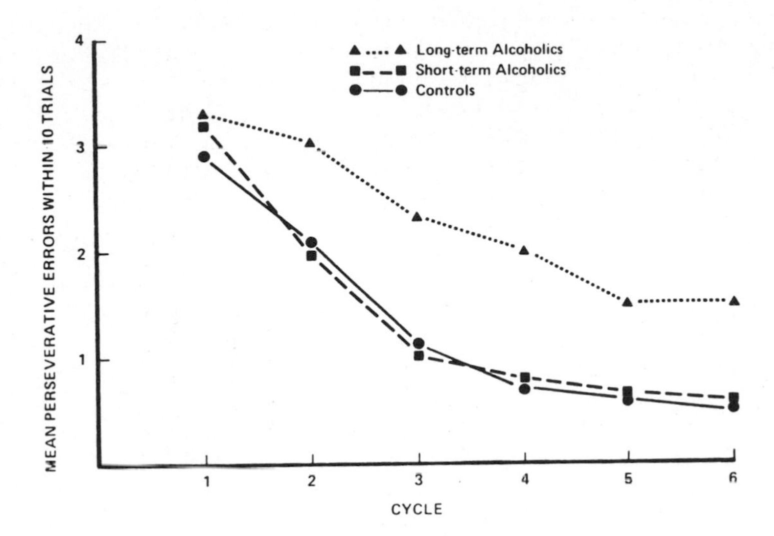

Figure 6. Mean perseverative errors emitted within the first ten trials after a set shift on each cycle for the controls, short- and long-term alcoholics.

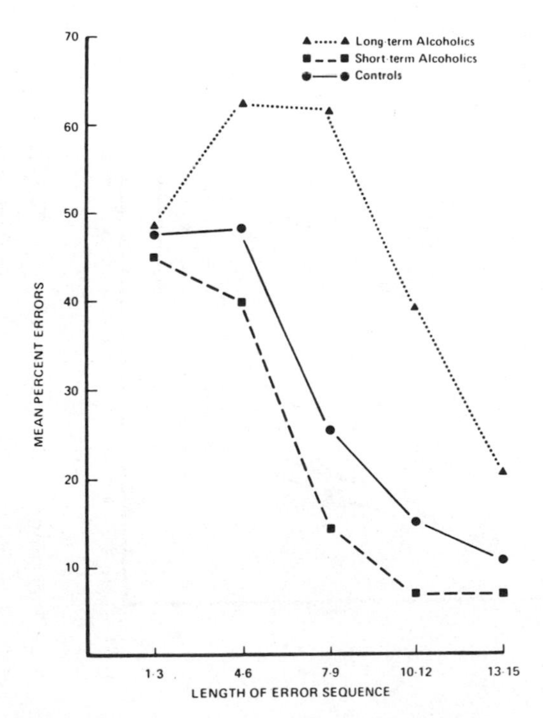

Figure 7. Percentage of times that an error succeeded a sequence of consecutive incorrect responses.

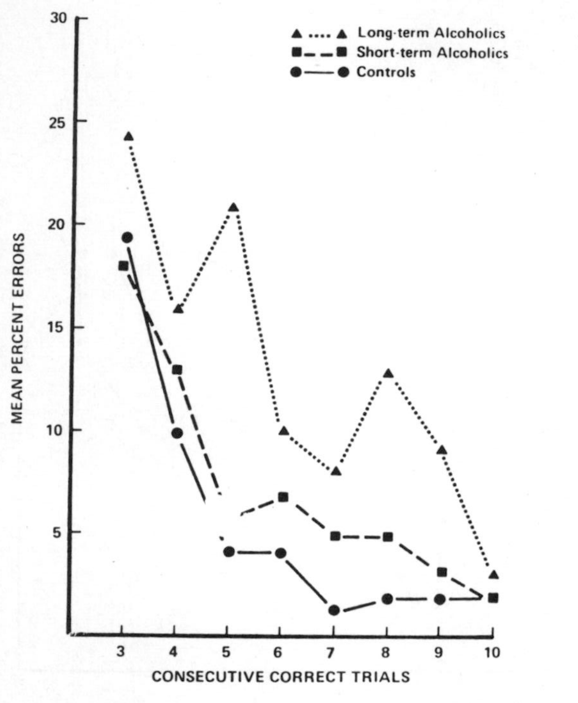

Figure 8. Percentage of times that an error succeeded a sequence of consecutive responses. Consider the points at 5 consecutive correct trials; long-term alcoholics interrupt the correct sequence approximately 21% of the time whereas the short-term and control subjects interrupt only about 5% of the time.

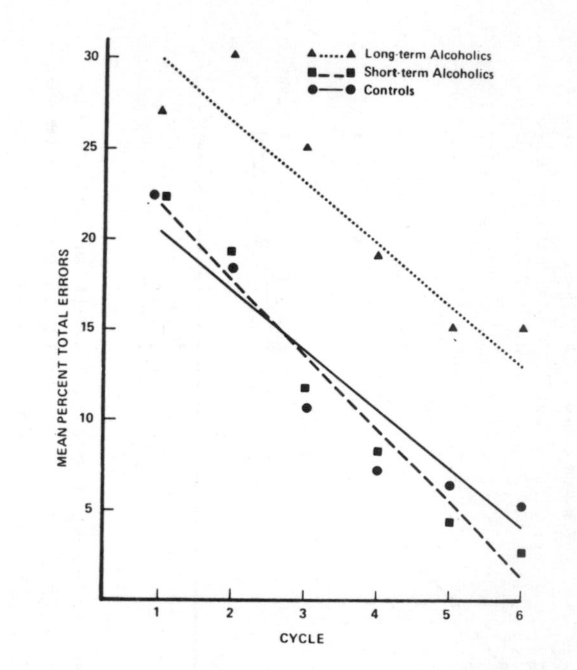

Figure 9. "Learning to learn" curves for the controls, short- and long-term alcoholics.

impaired visual-spatial conceptual and motor behavior, deficits in inhibitory regulation of motor movement, altered conceptual abilities on conceptual shifting tasks, interruption of successful sequences of behavior, perseverative tendencies but intact language abilities. Most of the deficits appear to lie outside of the awareness of the alcoholic leading to what we have termed an altered state of unconsciousness. Currently these data seem best explained by the frontal-limbic hypothesis. Two general circuits of neuropsychological importance in the frontal lobe have been described in primates (Teuber, 1972). One is the dorsolateral frontal cortex: head of the caudate, the lateral part of the globus pallidus and the subthalamic nucleus - lesions here result in impairment in delayed alternation. The second is the orbital-frontal cortex which has connections to the thalamic nucleus such as the ventralis laterales, ventralis anterior and medialis dorsalis. Lesions in the orbital - frontal areas result in impairment on object reversal tasks. The impairment in sequential problem solving and the visual-spatial difficulties could both be explained by frontal lobe dysfunction. More specific tests of these hypotheses are needed and are receiving attention currently in our laboratories.

References

Brewer C. & Perrett, L. Brain Damage Due to Alcohol Consumption: An air-encephalographic, psychometric and electroencephalographic study. Brit. J. of Addic., 1971, 66, 170-182.

Callan, J., Klisz, D. & Parsons, O.A. Strength of Auditory Stimulus-Response Compatibility as a Function of Task Complexity. J. of Exp. Psychol. 1974, 102, 1039-1045.

Carlsson, C., Claeson, L.E. Cerebral Dysfunction in Alcoholics: A psychometric investigation. Quart. J. Stud. Alc., 1970, 31, 317-323.

Courville, C.B. Effects of Alcohol on the Nervous System of Man. Los Angeles; San Lucas Press, 1955.

Eccles, J.C. The Understanding of the Brain. New York: McGraw Hill Book Co., 1973.

Ferrer, S. Complicationes Neurologicas Cronicas Del Alcoholismo. Editorial Universitaria, S.A., 1970.

Feuerlein, W., Heyse, H. Arch. Psychiatrie Nervenkranze. 1970, 213, 78-85.

Fitzhugh, L.C., Kitzhugh, K.B. & Reitan, R.M. Adaptive Abilities and Intellectual Functioning in Hospitalized Alcoholics. Quart. J. of Stud. on Alc. 1960, 21, 414-423.

Freund, G. Chronic Central Nervous System Toxicity of Alcohol. Ann. Rev. of Pharmacol. 1973, 12, 217-227.

Goldstein, K. Functional Disturbances in Brain Damage. In S. Arieti's American Handbook of Psychiatry. New York: Basic Books, 1969, 770-796.

Goldstein, G., Neuringer, C. & Klappersack, B. Cognitive, Perceptual and Motor Aspects of Field Dependency in Alcoholics. J. of Genetic Psychol. 1970, 117, 253-266.

Haug, J.O. Penemoencephalographic Evidence of Brain Damage in Chronic Alcoholics. Acta Psych. Scand. 1968, 203, 135-143.

Jones, B.M. Altered States of Consciousness and Alcohol. Symposium SWPA Meeting, El Paso, Texas, May 1974.

Jones, B. & Parsons, O.A. Impaired Abstracting Ability in Chronic Alcoholics. Arch. of Gen. Psych. 1971, 24, 71-75.

Jones, B. & Parsons, O.A. Specific vs. Generalized Deficit of Abstracting Ability in Chronic Alcoholics. Arch. of Gen. Psych. 1972, 26, 380-384.

Mello, N.K. & Mendelson, J.H. Alterations in States of Consciousness Associated with Chronic Ingestion of Alcohol. (Eds.) Zuben, J. & Shagars, C. Neurobiological Aspects of Psychopathology. New York: Grune & Shalton, 1969, 183-218.

Nauta, W.J.H. Neural Associations of the Frontal Cortex, Acta Neurobiol. Exp. 1972, 32, 125-140.

Parsons, O.A., Tarter, R.E. & Edelberg, R.E. Altered Motor Control in chronic alcoholics. J. of Abnorm. Psych. 1972, 80, 308-314.

Segal, B.M., Kushnarev. V.M., Urakov, I.G., & Misionzhnik, E.U. Alcoholism and Disruptions of Activity of Deep Cerebral Structures. Quart. J. of Stud. on Alc. 1970, 31, 587-601.

Seixas, F.A. & Eggelston, S. (Eds.) Alcoholism and the Central Nervous System. Annals of the New York Academy of Sciences, 1973, 215, 5-389.

Shimojyo, S., Scheinberg, R. & Reinmuth, O. J. Clin. Investigation. 1967, 46, 849-854.

Sperry, R.W. Hemispheric Deconnection and Unity in Conscious Awareness. Amer. Psychol. 1968, 23, 723-733.

Tarter, R. A Neuropsychological Examination of Cognitive and Perceptual Capacities in Chronic Alcoholics. Doctoral Dissertation, University of Oklahoma, 1971.

Tarter, R.E. & Parsons, O.A. Conceptual Shifting in Chronic Alcoholics. J. of Abnorm. Psychol. 1971, 77, 71-75.

Tarter, R.E. & Jones, B.M. Motor Impairment in Chronic Alcoholics. Diseases of the Nervous System. 1971, 32, 632-636.

Teuber, H. L. Unity and Diversity of Frontal Lobe Functions. Acta Neurobiol. Exp. 1972, 2, 615-649.

Tumarkin, B., Wilson, J.D. & Snyder, G. Cerebral Atrophy Due to Alcoholism in Young Adults. U.S. Armed Forces Medical Journal, 1955, 6, 67-74.

Victor, M., Adams, R. & Collins, G. The Wernicke-Korsakoff Syndrome, Contemporary Neurology Series. 1971, F.A. Davis Company.

Wallgren, H. & Barry, H., III Actions of Alcohol, Vol. I: Biochemical, Physiological and Psychological Aspects, Vol. II: Chronic and Clinical Aspects. Amsterdam: Elsevier 1971.

Warren, J.M. & Akert, K. (Eds.) The Frontal Granular Cortex and Behavior. New York: McGraw Hill Book Co., 1964.

Wechsler, D. The Measurement and Appraisal of Adult Intelligence. Baltimore: Williams & Winkins Company. 1958.

ALCOHOLIC BLACKOUTS AND KORSAKOFF'S SYNDROME

Donald W. Goodwin, Shirley Y. Hill and Saul Hopper

Dept. of Psychiatry, Washington University Medical

School, St. Louis; and John O. Viesselman, U.C.L.A.*

Several studies in recent years (Goodwin, 1970; Tamerin, et al., 1971) have indicated that alcohol-induced intoxication is sometimes associated with a specific short-term memory deficit during the period of intoxication followed by amnesia (blackout) for the period during which short-term memory was impaired. Hence, alcoholic blackouts, defined as alcohol-related inability to remember ordinarily memorable events, appears to represent an anterograde form of amnesia.

Several authors have noted the similarity between this amnesia and that which occurs in Korsakoff's syndrome, in which the patient's memory may be fairly intact in most regards except for a striking inability to retain new information (Goodwin, 1971). Korsakoff's syndrome and other organic brain conditions characterized by short-term memory loss have been associated with specific anatomical lesions in the brain stem, commonly involving the dorsomedial nucleus of the thalamus, the mammillary bodies, and the posteromedial portions of the hippocampi (Victor, Adams and Collins, 1971). Conceivably, if alcoholic blackouts and Korsakoff's syndrome have pathophysiological features in common, the acute amnesia of intoxication may also involve these anatomical structures.

Korsakoff's syndrome is classically described as a condition, usually associated with alcoholism, in which the patient is mentally alert, has a normal IQ, but has a striking anterograde amnesia, often coupled with apparent retrograde elements, and a personality

---

* Supported by Grant Nos. AA00209, DA4RG008, MH05804 and a Research Scientist Development Award AA47325 from the National Institute of Mental Health (Dr. Goodwin).

change characterized mainly by apathy (Tallant, 1965). In the past, the amnesia was believed to be total and irreversible, representing a failure to "consolidate" new information into long-term storage form. Recent work, however, mainly by Butters, Cermak and their colleagues at Boston University, has indicated that the amnesia may be selective rather than global and that at least some long-term storage may occur, with retrieval of information possible if the patient is appropriately "cued" (Cermak and Butters, 1973; Weiskrantz and Warrington, 1970). Korsakoff patients are particularly susceptible to "semantic" or "verbal" memory loss, but in memory tasks involving primarily auditory, visual, or tactile "memory," they perform more or less normally. As reported in previous studies, the patients have normal "immediate" memory (e.g., they perform digit span tests normally) but there is a rapid decay of memory for verbal or semantic material within a period of less than one minute.

The goal of the present study was to determine whether subjects particularly susceptible to alcoholic blackouts show in a sober state a memory "profile" similar to that observed in Korsakoff patients. In other words, is their "nonverbal" memory superior to their "verbal" memory? Originally, the study called for the subjects to be tested during the amnesic period produced by alcohol (as detected by impaired short-term memory), but it was found that the extreme degree of intoxication required to produce short-term memory loss was incompatible with ability to perform the Korsakoff battery of tests described below.

METHOD

Twelve volunteer subjects, recruited from the casual labor office of the Missouri Employment Office in St. Louis, were admitted to the hospital for a 24 hour period. Their average age was 39. Eight were Negro and four were white. All had a history of heavy drinking and 10 were diagnosed as alcoholic by specific criteria (Guze, Goodwin, and Crane, 1969), although none was in treatment at the present time. None showed signs of withdrawal from alcohol and all denied heavy drinking in the recent past. All were in good physical health. All gave a history of frequent alcoholic blackouts. They were chosen, in fact, because of this history, since a previous study (Goodwin, et al, 1970) had indicated that a history of frequent blackouts was a reliable predictor of success in producing experimental blackouts under laboratory conditions.

The experiment consisted of two parts. In the first, the patient in a sober state was given a battery of tests developed by the Butters-Cermak group (1973) to detect selective memory loss. The battery consists, in part, of three tests administered as follows:

1. Word Triads

Twenty cards are presented to the subject. Each card has three one-syllable stimulus words (e.g. leg-old-cup). Five of these are presented in a 0 delay condition. Also, five cards are presented in each of three additional conditions--3, 9 and 18 second delays. Each stimulus card is presented for two seconds, and the subject is asked to read the words aloud. Following the appropriate delay, the subject is given 10 seconds in which to recall the three words.

During the delay interval, the Peterson-Peterson (1959) distractor technique is employed to prevent rehearsal. This involves counting backwards from 100 by threes, starting from the time the stimulus is shown until the subject is asked to respond.

2. Visual Nonsense Forms

Computer-generated random shapes with no verbal associative value are used as stimuli. The subject is presented successively with pairs of stimulus cards. The task is to report whether the second stimulus card is the same or different from the first of the pair. The initial stimulus card is presented for 2 seconds with delay intervals of 0, 9, and 18 seconds before presentation of the second card. The subject is allowed 10 seconds to respond to the second card. Eight pairs are presented in each of these delay conditions. During the 9 and 18 second delays the Peterson-Peterson distractor technique is used.

3. Music Study

Sequences of five random piano notes are presented to the subject. In a non-delay condition the subject is immediately presented with a second five-note sequence and asked to determine if the sequences are the same or different. Stimuli are recorded on tape so that presentation intervals are fixed. The five-note sequence lasts 4 seconds and is followed by the second sequence at either 0, 9, 18 seconds. Ten trials are given in the non-delay condition and 10 each in the delay condition. Once again, the Peterson-Peterson distractor technique is used to avoid rehearsal in the delay conditions.

In the second part of the study subjects were administered between 16 and 18 oz. of 86 proof bourbon in a water vehicle (in terms of absolute alcohol, 2.4 grams of absolute alcohol per kilogram of body weight). After one hour of drinking, testing began. This involved exposing the patient at 30 minute intervals to a "memorable" event and testing his memory function throughout the period of drinking and for one hour afterward.

A "memorable events" battery had been pretested in sober hospitalized alcoholics and found to be recognizable with 100% accuracy and capable of free recall with about 80% accuracy 24 hours after presentation. The "memorable events" consisted of the following: (a) A frying pan was presented containing three life-like toy mice; (b) A light bulb contained in a sock was smashed against a wall; (c) A trick candle was lighted that relighted itself after being blown out by the subject; (d) A trick fountain pen containing cap-gun caps exploded when the top was removed; (e) A large trick safety pin was made to appear as if it had been inserted through the investigator's nose; (f) Vanishing ink was squirted on the subject from a water pistol; (g) A balloon filled with water was smashed on the floor; (h) The subject was asked to drink from a cup containing water and two large toy house flies.

At one minute after presentation and again at 30 minutes after presentation, the subject was asked to recall the "event" he had just previously been shown. If he failed spontaneously to recall it, he was "cued" by presenting part of the "event" and asked to recall the remainder (e.g. showing the pan and asking what it contained). The same procedure was followed 24 hours later when the patient was sober. The one minute presentation tested "immediate" memory and the 30 minute presentation tested "short-term" memory. Testing 24 hours later determined whether the patient had experienced a blackout, "blackout" being defined as inability to remember ordinarily memorable events occurring during a prior drinking period.

RESULTS

As shown in Table 1, only one subject experienced an alcoholic blackout, defined as inability to recall or recognize normally memorable events 24 hours after experiencing them. The other 11 subjects, *despite* having histories of frequent blackouts, did not experience a blackout. This was surprising, since a history of frequent blackouts in a previous study invariably predicted success in producing an experimental blackout (Goodwin, et al., 1970).

There was one important difference, however, in this study and the previous one. In the previous study, to test immediate and short-term memory, the patient was asked to spontaneously recall the event but was not "cued" or tested for recognition. On testing 24 hours after drinking, he *was* cued (i.e., given an opportunity to recognize the materials shown him during the intoxication period); if he was unable spontaneously to recall the event, neither could his memory be improved with cues.

As in the previous study, immediate memory was intact in all subjects, with two exceptions which were equivocal. Except in these two instances, no recognition measure was necessary one minute after presentation. Once highly intoxicated, however, all of the subjects required a recognition measure at 30 minutes, which, in every instance but one, restored recollection of the event. Other memory and intellectual testing, identical to that performed in our earlier study (Goodwin, 1970), similarly failed to show evidence of impaired remote memory, "recent" memory (events leading up to the experimental session), or retrograde memory loss. Intellectual functioning seemed generally intact.

As noted, 11 of the 12 subjects were able to recognize the memorable events but all showed some difficulty, towards the later stages of intoxication, in spontaneously recalling events, both at the 30 minute interval and 24 hours later when sober. In Table 2 those who forgot (without cuing) more than three memorable events were compared to those who forgot less than three. As shown in the table, subjects with the "best" memory while intoxicated also had the least deterioration in recall of words in the Word Triad Test when immediate presentation (testing attention and registration) was compared to retention of word triads 18 seconds after presentation. None of the nonverbal tests in the Korsakoff battery, measuring auditory and visual memory, showed a significant difference between the two groups.

Again, although 11 of the 12 subjects failed to experience a blackout, despite a history of frequent blackouts, there were

Table 1

Recall and Recognition of "Memorable Events"

Subject	1	2	3	4	5	6	7	8	9	10	11	12
HX of BOs	+	+	+	+	+	+	+	+	+	+	+	+
1 min Recall	+	+	+	+	+	+	+	+	?	?	+	+
1 min Recog.									+	+		
30 min. Recall	0	0	0	0	0	0	0	0	0	0	0	0
30 min. Recog.	0	+	+	+	+	+	+	+	+	+	+	+
24 h Recall	0	0	0	0	0	0	0	0	0	0	0	0
24 h Recog.	0	+	+	+	+	+	+	+	+	+	+	+
Remote Mem.	+	+	+	+	+	+	+	+	+	+	+	+

differences in the group with respect to ability spontaneously to recall materials at 30 minutes and 24 hours later. When the group was divided into those with the most severe history of blackouts and those with the least severe, it was found there was a trend for those with the most severe histories to have the greatest difficulty in spontaneously recalling the memorable events. However, the difference was not significant and, with one exception, a history of frequent blackouts in general did not predict success in producing experimental blackouts in this study.

DISCUSSION

The original goal of the study was to determine whether alcoholic blackouts resemble Korsakoff's syndrome in manifesting selective rather than global amnesia. It was discovered early in the study that it was impossible to test severely intoxicated subjects, using the tests devised by Butters and Cermak for distinguishing various types of memory impairment (verbal, visual, auditory). It was found, surprisingly, that 11 of our 12 subjects failed to experience blackouts despite having a history of frequent blackouts and despite receiving highly intoxicating amounts of alcohol (between 16 and 18 oz. of bourbon in a four hour period). In a

Table 2

WORD TRIAD TEST

Forgotten Memorable Events	# Correctly Recalled 0 sec.	18 sec.	% drop
Less than 3	14	11	-21
	13	9	-30
	15	8	-46
	15	9	-33
	9	10	+11
			Av: 24%
More than 3	13	5	-61
	15	14	- 7
	15	6	-60
	15	7	-53
	15	7	-53
	15	4	-73
	15	7	-53
			Av: 51%*

* $t = 1.9277$; $p < .05$

previous study, all subjects with a history of frequent blackouts experienced a blackout during an experimental drinking session.

The previous study differed, however, from the present one in perhaps an important regard: the former study did not have a recognition measure during the period of intoxication but only 24 hours later when both recall and recognition measures were used to determine whether the patient had a blackout. In the present study recognition measures also were tested _during_ the intoxication period. It was found that patients in advanced stages of intoxication were unable spontaneously to recall recent "memorable" events but if their memory was jogged by presenting them with a "reminder" of the previously presented material, they were able to remember the event. This in a sense represented a relearning of the material and may have been an important factor in their continued ability to remember the material the next day while sober.

Certain confirmation of this was provided by a personal communication with Drs. Butters and Cermak. Using the same battery of "memorable events," they were able to produce experimental blackouts in seven subjects out of seven subjects tested. However, they did not cue the patient with a partial recognition presentation but, as was done in our previous study, relied solely on the patient's ability spontaneously to recall events while intoxicated. They made the same observations reported in our previous study, namely, that intoxicated individuals with a history of frequent blackouts had a specific short-term memory deficit (where recall, but not recognition, was involved) followed by a failure to remember the material 24 hours later. Apparently, ability to remember, sober, the next day depends on how the testing was conducted during the intoxication period; i.e. whether cuing and therefore presumably some relearning occurred after failure to spontaneously recall the event.

This suggests that failure to transfer memory from short-term to long-term storage may not entirely explain alcoholic blackouts. The fact that our subjects, when cued 30 minutes later about the previously presented material, could then recall the event suggests that a retrieval deficit, rather than impaired consolidation, may be at least part of the explanation. If so, this suggests that alcoholic blackouts do indeed resemble Korsakoff's syndrome, assuming it is true that Korsakoff's syndrome also in part involves a retrieval deficit.

Although we were unable to administer the Korsakoff battery to these highly intoxicated subjects, we did administer it to them while they were sober. There was suggestive evidence, at least, that the subjects who showed the most memory impairment during intoxication also had a memory "profile" most approximating that reported by Butters and Cermak in their Korsakoff patients. In other words, subjects showing the most difficulty recalling memorable events also were most impaired with regard to verbal materials, while performing equally well on "nonverbal" tasks. This suggests that perhaps individuals with a strong susceptibility to blackouts may have a subclinical, very mild form of Korsakoff syndrome which may or may not progress with further drinking and time.

ACKNOWLEDGEMENT

The above report is part of a larger study being conducted by the authors in collaboration with Drs. Nelson Butters and Laird Cermak at Boston University and the Brockton Veterans Administration Hospital in Boston. These preliminary findings will be enlarged upon in subsequent reports with Drs. Butters and Cermak. Meanwhile, however, we wish to thank Drs. Butters and Cermak for permitting us to use their battery of memory tests developed in the study of Korsakoff patients.

REFERENCES

Butters, N., Lewis, R., Cermak, L., and Goodglass, H. Material-Specific Memory Deficits in Alcoholic Korsakoff Patients. Neuropsychologia, Vol. 11, pp. 291-299, 1973.

Cermak, L., and Butters, N. Information Processing Deficits of Alcoholic Korsakoff Patients. Quart. J. Stud. on Alcohol, Vol. 34, No. 4, pp. 1110-1132, December, 1973.

Goodwin, D., Othmer, E., Halikas, J., and Freemon, F. Loss of short-term memory as a predictor of the alcoholic "blackout." Nature, Vol. 227, pp. 201-202, 1970.

Goodwin, D. "Blackouts and alcohol induced memory dysfunction," in Recent Advances in Studies on Alcohol, Proceedings of the 1970 NIMH Interdisciplinary Symposium, publication No. (HSM) 71-9045, 1971.

Guze, S., Goodwin, D., and Crane, J. Criminality and psychiatric disorders. Arch. Gen. Psychiat., Vol. 20, pp. 583-591, 1969.

Peterson, L. and Peterson, M. Short-term retention of individual verbal items. J. exp. Psychol., Vol. 58, pp. 193-198, 1959.

Talland, G. Deranged memory; a psychonomic study of the amnesic syndrome. New York; Academic; 1965.

Tamerin, J., Weiner, S., Poppen, R., Steinglass, P. and Mendelson, J. Alcohol and memory: Amnesia and short-term memory function during experimentally induced intoxication. Amer. J. Psychiat., Vol. 127, pp. 1659-1664, 1971.

Victor, M., Adams, R., Collins, G. The Wernicke-Korsakoff Syndrome; a clinical and pathological study of 245 patients, 82 with post-mortem examinations. Philadelphia; Davis; 1971.

Weiskrantz, L. and Warrington, E. A study of forgetting in amnesic patients. Neuropsychologia, Oxford, Vol. 8, pp. 281-288, 1971.

SOME ANALYSES OF THE INFORMATION PROCESSING AND SENSORY CAPACITIES OF ALCOHOLIC KORSAKOFF PATIENTS*

Nelson Butters, Laird S. Cermak,

Barbara Jones and Guila Glosser

Psychology Service, Boston VA Hospital and

Aphasia Research Center, Boston University School of

Medicine, Boston, Massachusetts, U.S.A.

For the past five years our laboratory has focused upon the factors underlying the amnesic symptoms (i.e., anterograde and retrograde amnesia) displayed by alcoholic Korsakoff patients. Since two comprehensive reviews of these studies have now been published (Cermak and Butters, 1973; Butters and Cermak, 1974), we shall only briefly discuss these findings and then turn to the results of two recent sets of investigations. The first set involves the role of information processing disorders in the memory impairments of alcoholic Korsakoffs and chronic alcoholics. These studies indicate that chronic alcoholics have some of the same cognitive impairments that may play an important role in the Korsakoffs' amnesic symptoms. The second set of investigations deals with the basic sensory capacities of alcoholic Korsakoff patients. Since the neural structures (n. medialis dorsalis and mammillary bodies) damaged in Korsakoff's disease have been implicated in the olfactory system, we used psychophysical methods to study the discriminative capacities of these patients in the olfactory, visual, and auditory modalities. The findings of these studies demonstrate a very severe olfactory sensory impairment in alcoholic Korsakoff patients.

---

* This research was supported in part by N.I.H. grants AA-00187 and NS-06209 to the Boston University School of Medicine. The authors wish to acknowledge the assistance of Dr. Howard Moskowitz with the analysis of the psychophysical data.

THE VERBAL ENCODING IMPAIRMENTS OF ALCOHOLIC KORSAKOFF PATIENTS

In a number of studies it has been clearly demonstrated that alcoholic Korsakoff patients have a severe short-term memory (STM) problem (for review, see Cermak and Butters, 1973). When presented with three consonants (e.g., J Z L) or three words and prevented from rehearsing this material, the alcoholic Korsakoff is unable to recall or to recognize the previously presented information 18 seconds later. A number of studies (Warrington and Weiskrantz, 1970; Cermak and Butters, 1972) have shown that this memory problem is intimately related to interference from previously presented material (i.e., proactive interference), but only recently have we been able to show that this increased sensitivity to interference is itself a manifestation of a deficit in the Korsakoffs' verbal *encoding* strategies (Cermak, Butters and Moreines, 1974).

Encoding refers to the fact that intact humans actively categorize new information at the time of presentation and storage. Verbal information is usually categorized or encoded along a number of dimensions, e.g., acoustic, associative, and semantic. Acoustic encoding refers to categorizations based upon phonetic similarities between the words (e.g., *bare* and *bear* or *map* and *rap*); associative encoding to categorizations based upon arbitrary or natural associations in the individual's experiences (e.g., *cigarette* and *match*); and semantic encoding to categorizations based upon similarities in the abstract meaning of the words (e.g., *ship* and *boat*). While normal individuals spontaneously encode new information along acoustic, associative, and semantic dimensions, these encoding strategies do not equally facilitate memory storage and/or retrieval. In general the more abstract (i.e., semantic) the encoding strategy the individual employs, the greater the probability of storage and retrieval.

Our previously published experiments have indicated that the alcoholic Korsakoff patients' increased sensitivity to proactive interference (and thus his STM deficit) was directly attributable to a failure to employ *semantic* encoding strategies spontaneously. While the alcoholic Korsakoff patient encodes verbal information according to its acoustic and associative attributes, he does not, unless directed, use the more efficient semantic categorizations. In fact, even when instructed to encode new information semantically, the Korsakoff patient performs such categorizations in an impaired fashion.

INFORMATION PROCESSING DISORDERS OF ALCOHOLIC KORSAKOFFS AND CHRONIC ALCOHOLICS

The results of two other recent studies (Oscar-Berman, 1973; Oscar-Berman, and Samuels, 1973) provide evidence that Korsakoff patients may also have a general impairment in their ability to process the relevant dimensions of visual patterned stimuli. Enlarging upon some scattered evidence (Talland, 1965; Samuels, Butters and Goodglass, 1971) that Korsakoff patients perseverate dominant response tendencies, Oscar-Berman (1973) studied the ability of alcoholic Korsakoff patients to adopt and to modify problem solving strategies. The patients (Korsakoffs, aphasics, alcoholics, and normals) were presented with a series of 16-trial two-choice visual discrimination problems. The stimuli varied in color, size, form, and position, and the patients were told to try to choose the stimulus (i.e., dimension) the experimenter had preselected as correct. On two of the 16 trials the experimenter said "correct" and on two trials "wrong" regardless of which stimulus was chosen. On the remaining 12 trials no feedback was provided. By analyzing the patients' performance on the 12 blank trials it was possible to determine what strategy or hypothesis they had adopted (i.e., what dimensions of the stimuli they were focusing upon) and whether or not they changed their hypotheses following a negative reinforcement. The results showed that the Korsakoff patients, like the other patient groups, adopted strategies or hypotheses in solving the discrimination problems but did not shift hypotheses (e.g., from color to form) following a negative reinforcement. Evidently, once the Korsakoff patient adopted a particular strategy he perseverated this hypothesis despite the reinforcement contingencies.

In a second study, Oscar-Berman and Samuels (1973) attempted to determine whether these perseverative tendencies of alcoholic Korsakoff patients might be related to a limited processing capacity. The patients were trained on a two-choice visual discrimination with the stimuli again differing in form, color, size, and position. Following this training, several test trials designed to determine what stimulus dimension (e.g., form) had become relevant to the patients were administered. In comparison to other brain-damaged patients and to normal controls, the Korsakoff patients had utilized fewer stimulus dimensions. For example, while learning the original discrimination the Korsakoff patients often focused upon the color differences between the two stimuli and failed to notice the differences in form, size, and position. Thus, while it seemed that intact individuals spontaneously analyzed many of the characteristics of multi-dimensional stimuli, the Korsakoff patients were restricted to uni- or at best bi-dimensional analyses.

These findings suggested to the present authors that perhaps a general information processing deficit may account for many of the Korsakoff patients' memory difficulties. It could be that the Korsakoff patients' failure to encode spontaneously the semantic dimensions of words is a reflection of their restricted ability to analyze more than one or two dimensions of any given stimulus complex. Verbal materials, like the visual stimuli employed by Oscar-Berman, are multi-dimensional--that is, they vary at least along acoustic, associative, and semantic dimensions. Since the Korsakoff patients can process only a limited number of stimulus dimensions at one time, it may be that they do not note certain characteristics of verbal materials. They may process the more concrete immediate attributes of sound and association but may fail to process the more abstract dimension of meaning.

To examine this hypothesis of restricted processing Glosser, Butters and Samuels (1974) administered a modified version of the dichotic listening technique to nine alcoholic Korsakoff patients, nine hospitalized chronic alcoholics, and nine normal controls. The dichotic technique involved the simultaneous presentation of different stimuli to the subjects' two ears (via stereo headphones) with the instructions to press a response key if a certain stimulus was presented under specified conditions. The experiments were designed to assess the ability of Korsakoff patients (1) to process selected information while filtering out irrelevant information and (2) to process and integrate spontaneously a broad range of stimulus inputs. In all, six basic conditions were employed. For the first four, each consecutive condition required the subject to process more stimuli or attributes of a given stimulus. The fifth and sixth conditions examined whether the Korsakoffs' processing deficits are time-dependent--i.e., can the Korsakoff process more information from a complex stimulus if given additional time?

In the first dichotic condition, the subjects were asked to respond whenever they heard the critical stimulus "10" in _either_ ear. The number "10" occurred on 19.2% (24) of the trials, 12 trials in each ear, within a series of 125 random digits in each ear. The intertrial interval was 1.2 seconds. The results for this condition showed no significant differences among the three groups.

In the second condition, dichotic lists of numbers were presented again at 1.2 second intervals. The critical number "10" occurred in 20% (25) of the trials in each ear within a series of 125 random digits. The subjects were instructed to attend to one ear only and to respond when the critical stimulus, "10", occurred in that ear. This dichotic tape was presented twice, once with instructions to respond to "10" in the right ear and once with instructions for the left ear. The results for this condition showed no significant differences in total errors between the Korsakoffs and normal controls, but the chronic alcoholics did commit significantly more total errors than did the normal controls. No significant ear effects were noted.

In the third condition, the critical stimulus was the digit pair "9"-"10" (with one digit delivered to each ear), and the subjects were told to respond to the presentation of this critical stimulus regardless of the ears in which the individual numbers occurred. Each of the two possible critical pairings (i.e., "10" to the left ear--"9" to the right ear, and vice versa) occurred in 16% (10) of the 125 trials. The numbers "9" and "10" were each paired with a non-critical stimulus in each ear for 4.8% (6) of the trials. The intertrial interval was constant at 1.2 seconds. The results for this condition showed that when the alcoholic Korsakoff patients were required to process two digits instead of one, they made significantly more total errors than did the normal controls. Although the alcoholics made more total errors than did the normal controls, the difference did not attain statistical significance.

The fourth condition made the greatest demands upon the subjects' processing capacities. Not only was the subject required to detect the specific digit pairs ("9" - "10") but also to detect the ear in which each of the critical digits was presented. The subjects were instructed to respond only to the "9" left ear--"10" right ear stimulus pairs or on a second administration of the tape to the "9" right ear--"10" left ear stimulus pairs. The tape consisted of 125 dichotic pairs of digits presented 1.2 second intervals. The digit pair "9" right ear--"10" left ear occurred in 16% (20) of the trials as did the reversed ear combination, "9" left ear--"10" right ear. Each of the two digits comprising the critical stimulus pair was paired with a non-critical stimulus (i.e., "1" to "8") in 9.6% (12) of the trials in each ear.

The results for this fourth and most demanding condition were striking. Both the alcoholic Korsakoffs and the chronic alcoholics made significantly more total errors, errors of omission (i.e., a failure to respond to the presentation of a critical pair), and errors of commission (a response to a non-critical pair). Clearly the Korsakoff patients and the alcoholics were unable to analyze both the specific digits being presented and to detect the ears to which the digits were presented in the time (1.2 sec.) available to them. It should also be noted that both the Korsakoffs and the alcoholics performed more poorly on the fourth than on the third condition while the normals performed almost identically on both conditions.

The fifth and sixth conditions were repeats of the third and fourth conditions except for one change: the interpair interval was increased from 1.2 seconds to 2.0 seconds. This procedure was followed to assess whether, given additional time to analyze the stimulus inputs, the Korsakoffs and the chronic alcoholics could process more information from the stimuli. The results for the fifth and sixth conditions failed to show any significant differences among the three groups of subjects. Both the Korsakoffs and the alcoholics made significantly fewer errors with conditions five and six than with conditions three and four.

These latter findings (third, fourth, fifth, and sixth conditions) are important for three reasons. First, the alcoholic Korsakoffs' impairments on the third and fourth conditions confirm the hypothesis that they have a general deficit in analyzing or processing all of the dimensions of new information. Whether the stimuli be visual patterns (Oscar-Berman and Samuels, 1974), names of common items (Cermak, Butters, and Gerrein, 1973), or digits presented dichotically (the present study), the alcoholic Korsakoff has difficulty in processing all of the characteristics of the information. Thus, it appears that the encoding deficits of the Korsakoff patient, and his increased sensitivity to interference and STM problems, may reflect a general defect in his analysis of all the dimensions of the information. Second, at least in the present situation, the Korsakoffs' information processing deficit is time-dependent. Given additional time to process the information (conditions five and six) the Korsakoffs can extract more information concerning the presented materials. It is important to note that the Korsakoffs' improvement with conditions five and six indicates that the stimulus materials were correctly registered or "heard" with conditions three and four but that additional time was needed for the Korsakoffs to process all of the information contained in the stimuli. Third, the deficits of the chronic alcoholics on these dichotic tasks indicate some continuity between the chronic alcoholics and the alcoholic Korsakoff patients. While most neurological teaching treats the Wernicke-Korsakoff Syndrome as an acute neurological illness related primarily to thiamine deficiency in chronic alcoholics (Victor, Adams, and Collins, 1971), our present findings suggest that chronic alcoholics may manifest some of the identical cognitive defects as the Korsakoff patients. It is conceivable that defects in the alcoholic's processing of information may be the first indication of an impending Wernicke-Korsakoff breakdown. It is interesting to note in retrospect that while Korsakoff patients always perform more poorly than chronic alcoholics on tests of STM, the performance of the alcoholics has often been inferior to that of normal controls (Cermak and Butters, 1973).

AN ANALYSIS OF THE OLFACTORY CAPACITIES OF ALCOHOLIC KORSAKOFF PATIENTS

Neuropsychological interest in alcoholic Korsakoff patients has in the past usually focused upon their striking memory impairment. However, there are neuroanatomical grounds for the investigation of olfaction in this syndrome. Thalamic (n. medialis dorsalis) and hypothalamic (mammillary bodies) structures damaged in the alcoholic Korsakoff syndrome have been shown by comparative neuroanatomical studies to be possible relays for olfactory inputs (e.g., Cajal, 1955; Nauta, 1956). Damage to these structures might conceivably impair olfactory discrimination, and in fact Talland (1965) reported that nine of his Korsakoff patients "seemed to have no capacity for olfactory discrimination." In the first of two studies Jones, Moskowitz, and Butters (1974) investigated simple discrimination of odors by alcoholic Korsakoff patients and by alcoholic and normal controls. In addition to basic discriminatory capacities, we assessed short-term memory for odors since the Korsakoff patients' retentive abilities in this ancient modality had never been tested.

Three groups of male subjects were assessed: 14 alcoholic Korsakoff patients, 14 chronic alcoholic controls, and 14 non-alcoholic controls. The olfactory test used 10 relatively unfamiliar odorants--food essences and chemical compounds--in a sniff-bottle technique. There were two bottles of each odorant, and 20 pairs were used, each odorant matched once with itself and once with another odorant. For each pair the subject judged whether the second odorant was the same as or different from the first. The 20 pairs were presented once with zero seconds and once with 30 seconds between the two members of each pair. For both delay tasks there was a 30-second inter-pair interval.

Several analyses were performed on the resulting data. An analysis of variance and subsequent *t*-tests showed a significant group effect, with Korsakoffs performing significantly worse than both control groups while the latter two did not differ. The mean performance of the Korsakoffs on both delay tasks did not exceed the chance level as assessed by a binomial test, and thus alcoholic Korsakoffs were shown to have a striking impairment of odor quality discrimination. There were no other significant effects. A second analysis used multidimensional scaling techniques and showed that alcoholic Korsakoffs, in contrast to both control groups, showed relatively little structure in their perceptual space for odors.

There are two possible theoretical explanations for this olfactory impairment: First, alcoholic Korsakoff patients may have elevated thresholds for the perception of odors; and second, their ability to discriminate among odor qualities may be selectively impaired although thresholds for odor perception remain normal. The next study (Jones, Moskowitz, Butters and Glosser, 1974) tested the first of these two hypotheses.

Rather than simply measuring olfactory thresholds, however, this second experiment examined the Korsakoff patients' assessment of olfactory stimulus intensity since this latter method provides a broader view of the range of the patients' perceptual deficits. In this second experiment we also assessed stimulus intensity discrimination (scaling) in two other sensory modalities, vision and audition, in order to determine whether alcoholic Korsakoff patients might have a general discriminative impairment rather than one specific to olfaction.

There were three groups of 10 subjects each: Korsakoffs, chronic alcoholic controls, and non-alcoholic controls. Each subject was tested in stimulus intensity assessment in each of the three modalities: vision with shades of gray, audition with loudness, and olfaction with the intensity of butanol odor. The procedure in all three modalities was the same and utilized two scaling methods, _magnitude estimation_ and _category scaling_.

The mode of response of the subject for all scaling tasks was non-verbal in order to minimize the problem of verbal perseveration in the Korsakoff patients. For _category scaling_ the patient was presented with a chart of numbers corresponding to the number of stimuli in the set, from 1 to n, and was asked to point to the appropriate number in response to each stimulus. For magnitude estimation the subjects responded by adjusting the length of an unmarked metal spring-loaded tape measure which had been embedded in the end of a four-foot wooden beam. The subject was instructed to make the length of the tape correspond to his perception of the intensity of the olfactory, visual, or auditory stimulus presented on a given trial. The greater the perceived intensity of the stimulus, the longer the tape.

Materials for the assessment of grayness were eight Munsell neutral gray cards ranging from values 2 through 9 on the Munsell scale. For auditory assessment six levels of white noise ranging from 45 to 95 decibels were used, and olfactory stimuli were six concentrations of butanol from .6% saturation to 20% saturation presented with the use of an air dilution olfactometer.

The results of this scaling study showed that the Korsakoff patients and the chronic alcoholics had normal absolute intensity thresholds for vision the audition and were able to judge the suprathreshold intensities of visual and auditory stimuli in a normal manner. On the olfactory scaling tasks (both magnitude estimations and category scaling) the alcoholic Korsakoff patients were severely impaired in comparison to both the chronic alcoholics and the normal controls. The Korsakoff patients had elevated absolute thresholds for butanol and judged suprathreshold intensities as less intense than did the other two groups. The chronic alcoholics, however, again did not differ from the normal controls.

The implication of these studies is that the olfactory impairment of Korsakoff patients may be due to the thalamic and/or hypothalamic damage in this disease. The suggestion that the olfactory deficit may be due to damage to olfactory nerves, bulbs, or tracts from head trauma does not seem plausible. The alcoholic controls were equally likely to have suffered head trauma, and yet they performed significantly better than Korsakoffs on all olfactory tasks.

Of the central structures damaged in Korsakoff's syndrome the mammillary bodies seem to us less likely than thalamic structures (n. medialis dorsalis) to be important in olfaction. Five synapses are required for olfactory inputs to reach the mammillary bodies. Further, we administered our test of odor quality discrimination to a traumatic amnesic patient with a diagnosis of bilateral hippocampal damage and found him to be normal in his discriminative capacities. Since the mammillary bodies would receive olfactory inputs via the hippocampus, their contribution to olfactory functions may not be significant. On the other hand, the thalamic structures, that is, the dorsal medial nucleus, damaged in around 90% of Korsakoff patients, and the ventral medial nucleus, damaged in around 60%, may play an important role in olfaction. Both receive direct inputs from the paleocortical olfactory area, which may be a relay for olfactory information, and both send projections to neocortex: the medial dorsal nucleus to orbital frontal cortex, and the ventral medial nucleus to frontolateral cortex. It seems possible, then, that these two thalamic structures may play a vital role in the mediation of olfactory perception.

REFERENCES

Butters, N. and Cermak, L. S. The role of cognitive factors in the memory disorders of alcoholic patients with the Korsakoff syndrome. Annals of the New York Academy of Sciences, 1974, 233, 61-75.

Cajal, S. R. Studies on the Cerebral Cortex, 1955, London, Lloyd-Luke.

Cermak, L. S. and Butters, N. The role of interference and encoding in the short-term memory deficits of Korsakoff patients. Neuropsychologia, 1972, 10, 89-95.

Cermak, L. S. and Butters, N. Information processing deficits of alcoholic Korsakoff patients. Quarterly Journal of Studies on Alcohol, 1973, 34, 1110-1132.

Cermak, L. S., Butters, N. and Gerrein, J. The extent of the verbal encoding ability of Korsakoff patients. Neuropsychologia, 1973, 11, 85-94.

Cermak, L. S., Butters, N. and Moreines, J. Some analyses of the verbal encoding deficit of alcoholic Korsakoff patients. Brain and Language, 1974, 1, 141-150.

Glosser, G., Butters, N. and Samuels, I. Evidence for a general information processing deficit in patients with Korsakoff's syndrome. (Manuscript in preparation)

Jones, B., Moskowitz, H. and Butters, N. Olfactory discrimination in alcoholic Korsakoff patients. Neuropsychologia, in press.

Jones, B., Moskowitz, H., Butters, N. and Glosser, G. Psychophysical scaling of olfactory, visual, and auditory stimuli by alcoholic Korsakoff patients. Neuropsychologia, under editorial review.

Nauta, W. J. H. An experimental study of the fornix system in the rat. Journal of Comparative Neurology, 1956, 104, 247-270.

Oscar-Berman, M. Hypothesis testing and focusing behavior during concept formation by amnesic Korsakoff patients. Neuropsychologia, 1973, 11, 191-198.

Oscar-Berman, M. and Samuels, I. Stimulus-preference and memory factors in Korsakoff's syndrome. Paper presented at the American Psychological Association meetings, Montreal, August, 1973.

Samuels, I., Butters, N., and Goodglass, H. Visual memory deficits following cortical and limbic lesions: effect of field of presentation. Physiology and Behavior, 1971, 6, 447-452.

Talland, G. Deranged Memory, 1965, New York, Academic Press.

Victor, M., Adams, R. D. and Collins, G. H. The Wernicke-Korsakoff Syndrome, 1971, Philadelphia, F. A. Davis.

Warrington, E. K. and Weiskrantz, L. Amnesic syndrome: consolidation or retrieval? Nature, 1970, 228, 628-630.

REDUCTION OF SENSORY SHARPENING PROCESSES ASSOCIATED WITH CHRONIC ALCOHOLISM

Murray Alpert and David D. Bogorad

New York University Medical Center, Dept of Psychiatry

550 First Avenue, New York, N. Y. 10016

ABSTRACT

The alcohol withdrawal syndrome can include hallucinations in several modalities. This investigation was based on a multi-factorial model of hallucinogenesis in which hallucinations are thought to be related to a number of central and peripheral factors. Lateral inhibition is a sharpening mechanism in sensory transduction, and it would be expected that its' failure would produce a "noisy" sensory signal. The toxic effects of prolonged exposure to alcohol on this system in combination with personality factors and with other aspects of the withdrawal process might contribute to the development of hallucinations.

We studied auditory threshold shifts in twenty-seven alcoholics, some within a few days of a drinking spree and others after extended periods of sobriety, and in eighteen normal controls. In our normals we were able to replicate a previous demonstration of a 'Mach Band' in hearing. This consists of an alteration in sensory threshold at the frequency boundary of the masking stimulus. The alcoholics, as a group, showed less of this 'Mach Band' effect. The tendency to show 'Mach Bands' was negatively correlated with years of heavy drinking, but not with age, time since last drink or history of hallucinations. These changes were not demonstrable in routine audiometric testing.

INTRODUCTION

A number of biological, experiential, and social correlates

of alcoholism have been identified, but the mechanism through which these factors combine to contribute to the formation of hallucinations is not understood. Victor and Hope (1953) noted that a long history of alcohol abuse precedes the onset of hallucinosis. This observation suggests that some cumulative toxic effect of alcohol may be 'permissive' of the development of hallucinations during the acute withdrawal phase of the illness. Other factors which cumulate with time could also be suggested, such as development of the skill required to imbide the requisite quantity of alcohol, but a neurotoxic action of chronic exposure to alcohol seems likely. This suggestion is supported by the pathoanatomic work of Wolff and Gross (1968). These authors report necrotic changes in post mortem material from alcoholics' auditory structures which they interpret as indicating a toxic disturbance in the receptor organs. The authors suggest that these pathologic changes might be the basis for positive sensory disturbance.

Indirect evidence implicating peripheral neural mechanisms in the pathophysiology of alcoholic hallucinogenesis may be seen in the comments of a number of authors concerning the 'sensory' quality of alcoholic hallucinations. Bromberg (1932), concluded from his study of tactual perception in alcoholism that ".... changed perception is the basis for the hallucinations of the alcoholic patient. Without question, there are psychic factors that shape the hallucinations, but a definite basis exists in the physiopathology of the senses" (p. 45). Bromberg's work in tactile hallucinations was extended to the auditory sphere by Parker and Schilder (1935) who also commented on the sensory quality of alcoholic hallucinations.

In a previous study from our laboratory (Alpert and Silvers, 1970) we contrasted the phenomenology of the hallucinations of alcoholics and schizophrenics, and we also concluded that the hallucinations of the alcoholic are more 'sensory', while those of the schizophrenic are more cognitive. In general, the alcoholic reported experiences which were like noises becoming intelligible, whereas the hallucinations of schizophrenics were like their thoughts becoming audible. Gross (1973) incorporated similar observations into a more comprehensive theory of alcoholic hallucinations. In his treatment he was concerned with the effects of alcohol on the electrophysiology of sleep as well as the phenomena of withdrawal hallucinations. He postulated a sensory superactivity which, in the presence of alterations in consciousness, could be related to the formation of hallucinations.

Thus, study of the phenomena of hallucinatory experiences suggest that 'noisy' sensory channels may contribute to hallucinogenesis. A physiological mechanism that contributes to sensory sharpening, and that might be effected by chronic alcoholism, was demonstrated by Hartline, Ratliff and Miller (1961), although Ernst Mach had speculated on the existence of such mechanisms much earlier (1959). Mach had devised a physical stimulus with which he could demonstrate sensory mechanisms which accentuate contours and borders in the visual field. These mechanisms will even 'create' contours which do not exist in the physical stimulus. Such subjective bright and dark contours are known as 'Mach Bands' and have been shown to be related to lateral inhibitory mechanisms in the retina (Ratliff, 1965).

Von Bekesy (1967) has speculated on the generality of lateral inhibitory sharpening as a characteristic common to sensory transduction in audition and tactile sensation as well as in vision. He demonstrated the existence of lateral inhibition on the skin surface for both pressure and vibration. Concurrent with this work, von Bekesy was developing his traveling wave theory of pitch discrimination. According to this formulation different pitches would be associated with stimulation of different areas of the cochlea. Observations of the basilar membrane consistent with this theory were reported by von Bekesy. However, the portion of the basilar membrane stimulated by a pure tone is far too wide to account for the extraordinary ability of the ear to discriminate closely spaced pitches. von Bekesy postulated lateral inhibitory nerve nets fed by cochlear potentials to sharpen local excitatory differences. Carterette, Friedman, and Lovell (1969) synthesized masking noises on a digital computer which provided very steep gradients of stimulus amplitude as a function of frequency. With these stimuli they were able to demonstrate lateral inhibitory effects in hearing, auditory 'Mach Bands'.

Thus, lateral inhibition has been demonstrated in visual, auditory, and tactile senses, all modalities involved in alcohol withdrawal hallucinations. In addition, we had noted the involvement of arousal mechanisms in hallucinogenesis in alcoholics (Alpert, Silver and Drossman, 1968), and this hyper-arousal would be consistent with a loss of inhibitory control. For these reasons, we attempted to replicate the demonstration of auditory 'Mach Bands' of Carterette, Friedman, and Lovell (1969) in our laboratory and then investigated this phenomena in an alcoholic population, studying the relation between the magnitude of the sharpening phenomena and the duration of previous exposure to alcohol.

METHOD

Age and sex of our control and experimental groups are summarized in Table 1. We first selected controls from our research staff to replicate Carterette et al.'s (1969) population. We attempted to match the age of the alcoholic group with a control group from the general medical wards, but because of a high incidence of alcoholism among these patients, and difficulties in transporting them to our labs in the Psychiatric Pavilion we were unable to obtain an age matched non-alcoholic group. We obtained chronic alcoholic volunteers from the patient population of Bellevue Psychiatric Hospital. Some were inpatients, admitted for detoxification. The remainder were from the outpatient Alcoholism Clinic. Each patient was paid for his participation in the study. Subjects were excluded from the study if they showed atypical pure tone audiometry thresholds or if their diagnosis was complicated by schizophrenia. The *S*'s were tested in a sound-treated quiet room. They were adapted to the quiet condition for at least 30 minutes, during which time a history including a drinking history was taken and the procedures were explained.

Thresholds were determined at ten frequencies between 480 and 675 Hz with four ascending runs at each frequency. The tones were generated by an audio oscillator fed thru an electronic switch which delivered 250 msec. bursts and equal duration silences with 10 msec. rise and decay times delivered monaurally to *S* via calibrated headphones. The *S*'s right ear was used unless he preferred his left. The *S* indicated the just detectable level by pushing a button. The voltage across the headphones was read on a VTVM. The masking noise was generously provided to us by Carterette, and was a copy of the one he synthesized on a digital computer. The rectangular bandpass noise contains 500 independent random sine components logarithmically spaced within the band between 500 and 1000 Hz with theoretically infinite

TABLE 1

Age and Sex of the Groups

SUBJECT GROUPS	TOTALS	MEAN AGE, YEARS
ALCOHOLICS	27 93% male	42.7 S.D.= 10.0
CONTROLS	18 66% male	25.1 S.D.= 7.5

attenuation rates outside the pass band. The ultimate steepness of the gradient is limited, however, by the recording system and transducer employed.

To determine the masked threshold the bandpass noise was played back from a tape deck, the output of which was paralled with the output of the electronic switch and the thresholds again determined as described above. The level of the masking noise was adjusted in preliminary trials to produce threshold shifts of 50 to 55 dB, which was found by Carterette to be optimal for demonstrating the edge effect, and was then held constant for all S's. Controls were run concurrently with experimentals, and some S's were repeated at different points in the course of the experiment to measure the reliability of the effect.

RESULTS

The upward shift in threshold in the noise mask condition was computed for each S. The auditory Mach band effect may be seen as a relative peak in the masking function near the lower edge of the noise band (500 Hz). Inspection of individual S data from Carterette et al. (1969) revealed that the magnitude of the peak shift was about 3 dB with some inter-S variability. Figure 1 shows there was also variability in our data with S 1 exhibiting a peak shift of 3 dB at 540 Hz and S 2 providing no evidence of contour sharpening.

In order to assess the magnitude of contour sharpening in the alcoholic and control groups, we computed the dB difference between the initial masking peak and the point of return to the plateau masking level for each S. Table 2 contains a summary of these results. The mean peak height for the alcoholics was 1.4 dB and for controls 2.5 dB. The difference in the means is statistically reliable (*t* test, $t = 2.48$, $df = 43$, $p < .01$). In Table 3 is reported product-moment correlations between peak height and 1) age, 2) time since last drink, and 3) years heavy drinking, for the alcoholic group. Only the number of years of heavy drinking was significantly (negatively) related to peak height. The non significant relation between peak height and age may be interpreted as suggesting that the difference in age between the alcoholics and controls would not appear sufficient to undermine the primary finding of a difference in peak height between the groups. In point of fact, a partial correlation was computed between peak height and years of heavy drinking, with the variance due to age removed, and the relation became stronger ($r = -.55$, $df = 25$, $p < .005$). This may be interpreted as supporting the hypothesis that the lateral inhibitory networks are a site of the toxic action of prolonged

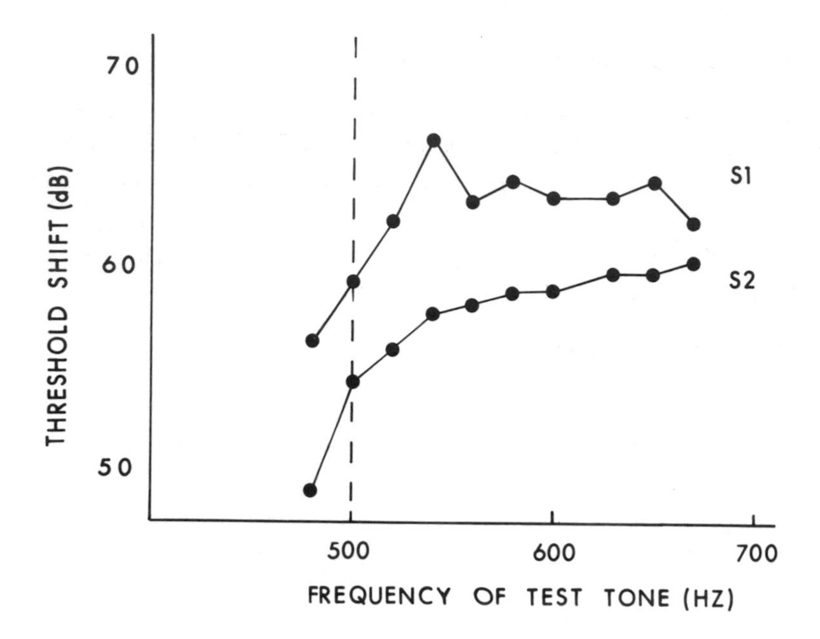

Figure 1. Threshold shift (dB) as a function of test tone frequency (Hz) for two individual subjects.

TABLE 2

Height of Mach Band for each Group

SUBJECT GROUPS	MEAN PEAK HEIGHT, dB
ALCOHOLICS	1.4 S.D.= 1.10
CONTROLS	2.5 S.D.= 1.95

t Test, one tail $t = 2.48$ $p < .01$

TABLE 3

Correlation Analysis for the Alcoholic Group

CORRELATION COEFFICIENTS (PEARSON), ALCOHOLICS (N = 27)	
PEAK HEIGHT VS. AGE	r= -.19 NS
PEAK HEIGHT VS. TIME SINCE LAST DRINK	r= -.16 NS
PEAK HEIGHT VS. YEARS HEAVY DRINKING	r= -.37 p <.05
PARTIAL CORRELATION	
PEAK HEIGHT VS. YRS. DRINKING WITH AGE PARTIALED	r= -.55 p <.005

exposure to alcohol. Several other correlations were examined, and it was found that peak height was not related to the variability of threshold determination or to history of hallucinations.

DISCUSSION

The results that we have presented support the hypothesis that prolonged exposure to alcohol alters auditory transducing mechanisms. The effect was related to years of heavy drinking, rather than to other aspects of the drinking history. However, conclusions as to the significance of the findings should be taken as tentative until replicated against other groups more closely matched to the experimental group in terms of demographic and experiential variables which may affect audiometric functions.

The observed reduction in auditory 'Mach Bands' is of interest because the sharpening effect has been attributed to inhibitory mechanisms. At least some alcohol induced pathology can be interpreted as reflections of hyperaroused states, and our results suggest that this may result from release from inhibitory control. If disinhibition can be established as a site of the toxic action of alcohol, then the measured reduction in auditory 'Mach Bands' could serve as a marker for other disinhibitory processes. In a discussion of sensory sharpening behind the retina, Marshall and Talbot (1942) have speculated that such mechanisms exist at each synapse. There is evidence for excitatory and inhibitory phenomena in other areas of the brain and it is possible that disinhibition of lateral control mechanisms may be associated with the cognitive and the motor manifestations seen in the course of alcoholism as well as in the sensory phenomena.

We studied lateral inhibition in alcoholism for a number of reasons, one being that it appeared that disorder of such a mechanism might contribute to the 'sensory quality' of the hallucinations reported by alcoholics. However, we have not demonstrated any relation between reduction in auditory 'Mach Bands' and other sensory processes. It would be of interest, for example, to determine if patients without 'Mach Band' sharpening were less competent in pitch discrimination. This prediction is directly related to von Bekesy's hypothesis and has not, as far as we can determine, been tested directly. Demonstration of reduced ability to discriminated pitches in alcoholics with reduced 'Mach Bands' would contribute to the confidence that we could place in the interpretation of the present findings.

REFERENCES

Alpert, M. & Silvers, K. N. Perceptual characteristics distinguishing auditory hallucinations in schizophrenia and acute alcoholic psychoses. _American Journal of Psychiatry_, 1970, _127_, 298-302.

Alpert, M., Silvers, K. N., & Drossman, A. K. Sensory and cognitive influences in auditory hallucinations. Paper presented at Eastern Psychological Association, April, 1968.

Bromberg, W. Tactual perception in alcoholism. _Archives of Neurology and Psychiatry_, 1932, 28, 37-51.

Carterette, E. C., Friedman, M. P., & Lovell, J. D. Mach bands in hearing. _Journal of the Acoustical Society of America_, 1969, _45_, 986-998.

Gross, M. M. Sensory super-activity, a preliminary report on an hypothetical model for an hallucinogenic mechanism in alcohol withdrawal. In (M.M. Gross, editor) _Alcohol Intoxication and withdrawal: Experimental Studies I_, Advances in Experimental Medicine and Biology, Vol. 35, pp. 321-330. Plenum Press, New York, 1973.

Hartline, H. K., Ratliff, F., & Miller, W. H. Inhibitory interaction in the retina and its significance in vision. In E. Florey (Ed.), _Nervous inhibition_. New York: Pergamon Press, 1961.

Mach, E. The analysis of sensations (and the relation of the physical to the psychical). Translated by C. M. Williams. New York: Dover, 1959.

Marshall, W. H. & Talbot, S. A. Recent evidence for neural mechanisms in vision leading to a general theory of sensory acuity. Biological Symposium, 1942, 7, 117-164.

Parker, S. & Schilder, P. Acoustic imagination and acoustic hallucination. Archives of Neurology and Psychiatry, 1935, 34, 744-757.

Ratliff, F. Mach bands: Quantitative studies on neural networks in the retina. New York: Holden-Day, 1965.

Victor, M. & Hope, J. M. Auditory hallucinations in alcoholism. Archives of Neurology and Psychiatry, 1953, 70, 659-661.

von Bekesy, G. Sensory inhibition. Princeton: Princeton University Press, 1967.

Wolff, D. & Gross, M. M. Temporal bone findings in alcoholics. Archives of Otolarngology, 1968, 87, 350-358.

QUANTITATIVE CHANGES OF SIGNS AND SYMPTOMS ASSOCIATED WITH ACUTE ALCOHOL WITHDRAWAL: INCIDENCE, SEVERITY AND CIRCADIAN EFFECTS IN EXPERIMENTAL STUDIES OF ALCOHOLICS

Milton M. Gross, Eastlyn Lewis, Suzanne Best,
Norma Young and Leonard Feuer
Division of Alcoholism & Drug Dependence
Dept. of Psychiatry
Downstate Medical Center, Brooklyn, New York, U.S.A.*

INTRODUCTION

A quantitative phenomenological approach to the alcohol withdrawal syndrome in man offers major advantages to the investigator and clinician that are not provided by the current nosology of the withdrawal. These advantages have been discussed at length elsewhere (Gross et al., 1974). A clinical instrument was developed for the purpose of a quantitative phenomenological assessment of the alcohol withdrawal syndrome, the Total Severity Assessment of alcohol withdrawal (TSA). This instrument was based on a review of the literature, observations over thirteen years of approximately fifteen thousand patients in alcohol withdrawal, and extensive clinical trials, several of which have been reported (Gross et al., 1968, Gross et al., 1971a, Gross et al., 1971b, Gross et al., 1973, Gross et al., 1974). The TSA has been applied in experimental studies of intoxication and withdrawal in alcoholic volunteers. In an earlier communication, the prevalence of the signs and symptoms associated with acute alcohol withdrawal during experimental studies were reported (Gross and Lewis, 1973). This communication will examine the incidence and severity of the signs and symptoms during experimental studies of intoxication and withdrawal in an enlarged sample. Circadian effects on temperature changes will also be described.

The TSA is reliable (Gross et al., 1973). Comparison of ratings by nurses and physicians demonstrated that trained nurses' ratings were highly reliable and comparable to physician's ratings (unpublished data). The TSA was sensitive to changing parameters

*Supported by NIAAA Grant AA01236

of alcohol intake (Gross and Lewis, 1973, Gross et al., 1974), differences in drug treatment of withdrawal (Gross et al., 1974), differences in sleep changes during experimental intoxication and withdrawal (Gross et al., 1971c, Gross et al., 1974), changes of cortical excitability during experimental intoxication and withdrawal (Begleiter et al., 1973) and differences in clinical course in relation to initial blood alcohol concentrations (Magrinat et al., 1973).

A factor analysis of the TSA, based upon 100 consecutively rated patients admitted in moderate to severe acute alcohol withdrawal, demonstrated 3 principal factors which accounted for 66% of the common variance (Gross et al., 1971b, Gross et al., 1971c, Rosenblatt et al., 1972). Each factor contained one of the primary triad of the withdrawal syndrome, hallucinations, tremor and clouding of the sensorium. Factor I contained hallucinations (auditory and/or visual), tactile hallucinations, nausea (and vomiting), visual disturbances, tinnitus, paresthesias, pruritus, muscle pain, sleep disturbances and agitation. Factor II contained tremor, sweats, anxiety and depression. Factor III contained clouding of the sensorium, impaired level of consciousness, impaired quality of contact, disturbance of gait and nystagmus. Factor III was noted to be bimodal, a manifestation of marked acute intoxication as well as marked withdrawal (suggesting that it reflected severity rather than etiologic specificity of the acute encephalopathic involvement). The data on the prevalence of the signs and symptoms of the TSA indicated that the individual elements of each factor tended to behave similarly to each other in the course of alcohol intake and withdrawal (Gross and Lewis, 1973). These data also demonstrated that partial withdrawal started early and tended to be progressive during the non-drinking hours of the heavy drinking days. It was felt that the continuous observations of the changes of incidence and severity of the signs and symptoms of the TSA during experimental intoxication and withdrawal might further clarify the relationship of these clinical phenomena to each other, to the effects of alcohol intake and withdrawal, and to circadian effects.

METHODOLOGY

All participants were paid male alcoholic volunteers whose participation was sought and determined after they were recovered from the acute alcohol withdrawal which had brought them to hospital. Half of them were white, half were black. Their ages ranged from 26-48 with an average age of 34.5 (S.D. 8.1). They were well nourished and free of medical and psychiatric complications prior to the onset of the study. All were gamma alcoholics with histories of more than five years of heavy drinking. All ingested larger daily quantities of alcohol during heavy drinking outside the hospital than they received during the study. One volunteer was studied at a time.

The studies were begun 3-5 weeks after hospitalization with an average interval between admission and study onset of approximately 3.5 weeks. Psychoactive medication was limited exclusively to the first 6 days in hospital. The volunteers were in the same unit and were cared for by the same staff from the first hospital day onward.

Alcohol was administered on a fixed schedule and dosage. The basic pattern of alcohol administration involved 10 hours of drinking (intoxication) alternating with 14 hours of non-drinking (partial withdrawal) in each 24 hours. Four volunteers received alcohol for 5 days and six received alcohol for 7 days. The first day of drinking was half dose and the remaining drinking days were full dose (approximately 1.6 gm/kilo and 3.2 gm/kilo for half and full dose respectively). Equal doses were administered hourly between 2 PM and midnight with the exception of 3PM when none was given. Two volunteers also received a single dose at 10 AM. During the alcohol administration period, breathalyzer determinations were obtained at 6 AM, 2 PM (prior to the first drink) and midnight (prior to the last drink). A Stephenson Breathalyzer was used.

All participants had 3 consecutive baseline days immediately prior to the period of alcohol intake and 6 consecutive days of withdrawal immediately after the period of alcohol intake. Baseline, alcohol intake and withdrawal periods were monitored daily with sleep EEG, and psychological testing.

Clinical phenomenology was continuously monitored by the research nursing team. A research nurse was with the participant at all times. Continuous observations were made and entered in the nurses' log. Three times daily (10 PM, 6 AM and 1 PM) there was a systematic quantitative clinical assessment of signs and symptoms associated with alcohol withdrawal (TSA). The assessment involved, for most of the variables, a score for what was observed at the actual time of evaluation (present), and another score for the period since the previous evaluation (overall). (This is described in detail in Gross et al., 1973a). The incidence data will be described on the basis of the present and overall scores. The severity data will be reported for present scores only, for the sake of brevity (the overall scores were similar). The presentation of the TSA data is organized in terms of experimental rather than calendar days, with each sequence of 10 PM, 6 AM and 1 PM being an experimental day. Thus, during drinking, the experimental day began at almost maximum intoxication (10 PM), then went on to early (6 AM) and maximum (1 PM) partial withdrawal. For the drinking period only, the findings of the first 5 days of drinking will be presented.

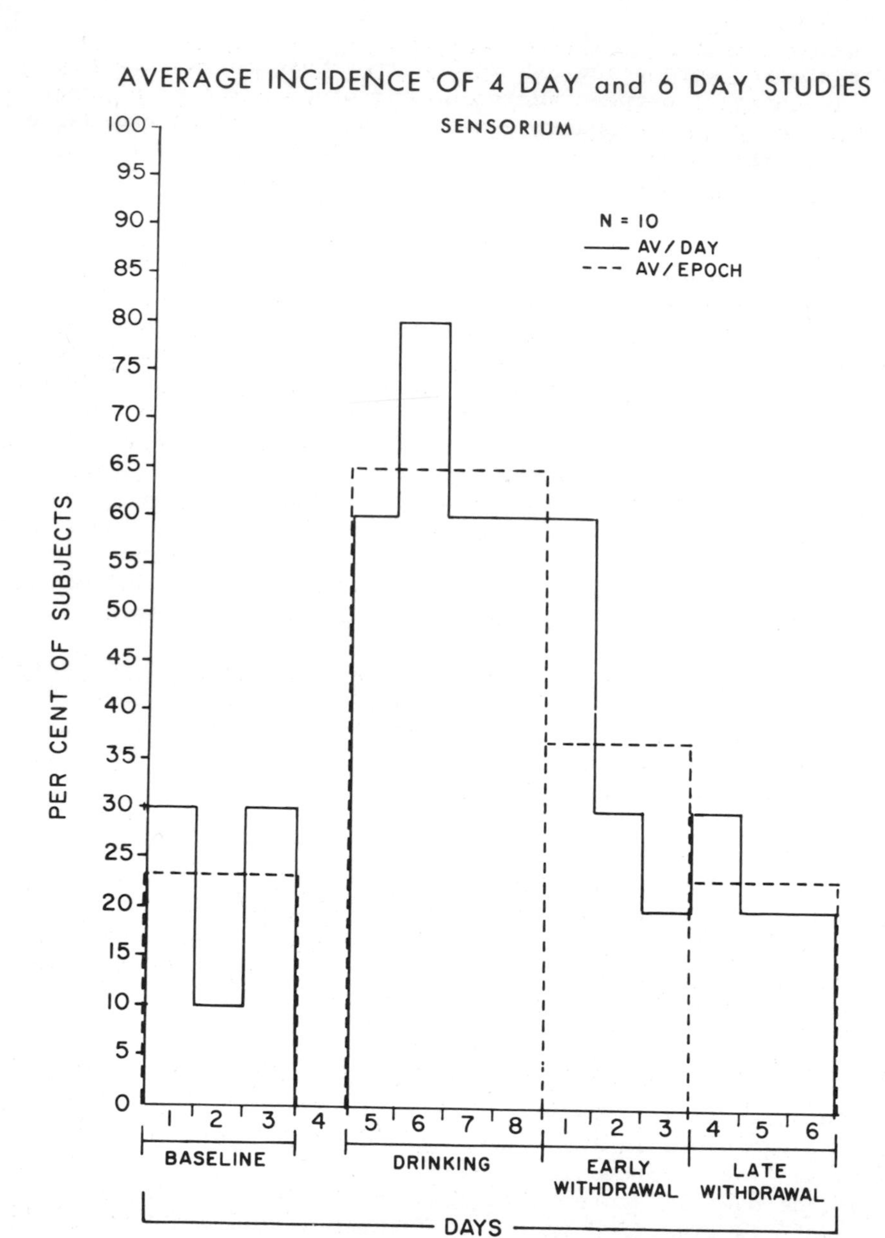
AVERAGE INCIDENCE OF 4 DAY and 6 DAY STUDIES
SENSORIUM
N = 10
AV / DAY
AV / EPOCH
PER CENT OF SUBJECTS
100
95
90
85
80
75
70
65
60
55
50
45
40
35
30
25
20
15
10
5
0
1 2 3 4 5 6 7 8 1 2 3 4 5 6
BASELINE
DRINKING
EARLY WITHDRAWAL
LATE WITHDRAWAL
DAYS

Figure 1

A. Blood Alcohol Concentrations

The average daily group levels at midnight for the first 6 days of heavy drinking ranged from 229 mg/100 ml - 315 mg/100 ml. These were not the maximal levels since they were obtained prior to the last drink of the day.

B. Changes of the TSA

The signs and symptoms fell into 4 major patterns:

1. Maximum activity during intoxication.
2. Considerable activity during baseline; increase during drinking days as a manifestation of partial withdrawal; finally, further increase during total withdrawal.
3. Little or no activity during baseline; some increased activity during the intoxication phase particularly during maximum partial withdrawal; and marked increase of activity during total withdrawal.
4. Change during the intoxication phase and a different pattern of change during total withdrawal.

The first pattern was exemplified by clouding of the sensorium (Figure 1).

The daily incidence revealed that during baseline days 10-30% of the patients demonstrated some clouding of the sensorium with a baseline average incidence of 23.3%. During the first 4 days of heavy drinking the incidence rose sharply. A peak daily incidence of 80% occurred on the second day of heavy drinking, with a daily incidence of 60% on the remaining 3 days. During early total withdrawal, the daily incidence of clouding of sensorium rapidly decreased so that, for the last 3 days of monitored withdrawal the average incidence was back to baseline levels.

When the severity (present) was examined within experimental periods, it was clear that the clouding of the sensorium sharply increased during the heavy drinking period (Figure 2). It was most marked during the 10 PM daily sample which coincided with maximal intoxication and least severe during the 1 PM sampling which was the time of least intoxication and greatest partial withdrawal.

This pattern of severity of clouding of sensorium was further sharpened by the examination of severity by the daily average sampling scores (Figure 3).

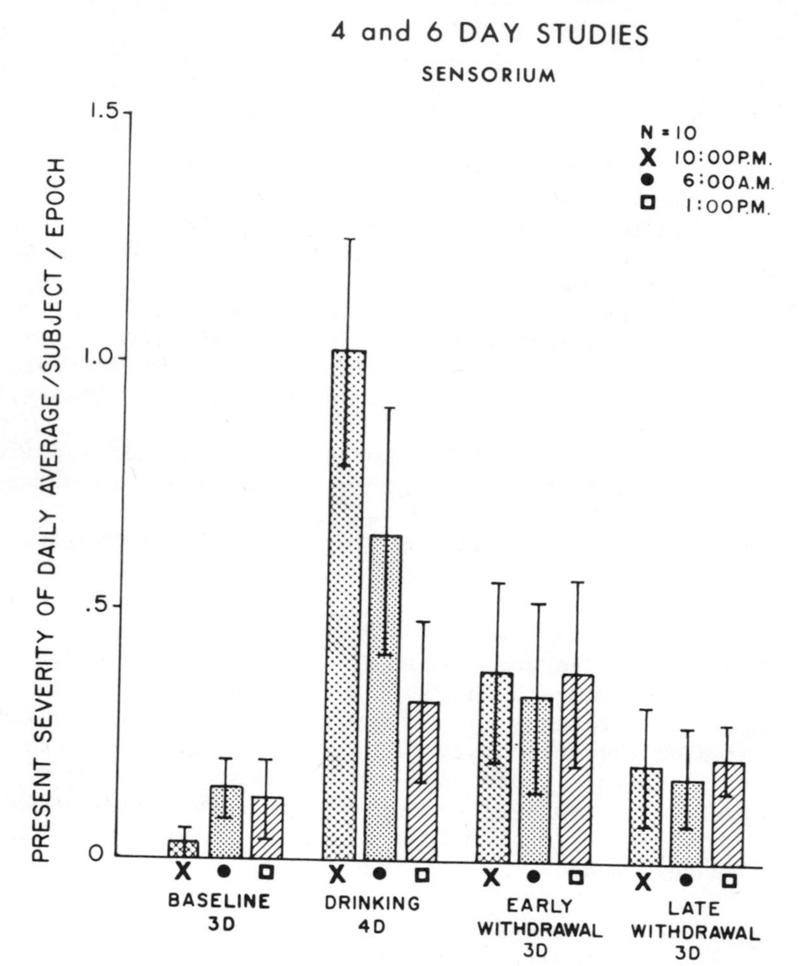
4 and 6 DAY STUDIES
SENSORIUM
PRESENT SEVERITY OF DAILY AVERAGE / SUBJECT / EPOCH
1.5
1.0
.5
0
N = 10
X 10:00 P.M.
● 6:00 A.M.
□ 1:00 P.M.
BASELINE 3D
DRINKING 4D
EARLY WITHDRAWAL 3D
LATE WITHDRAWAL 3D

Figure 2

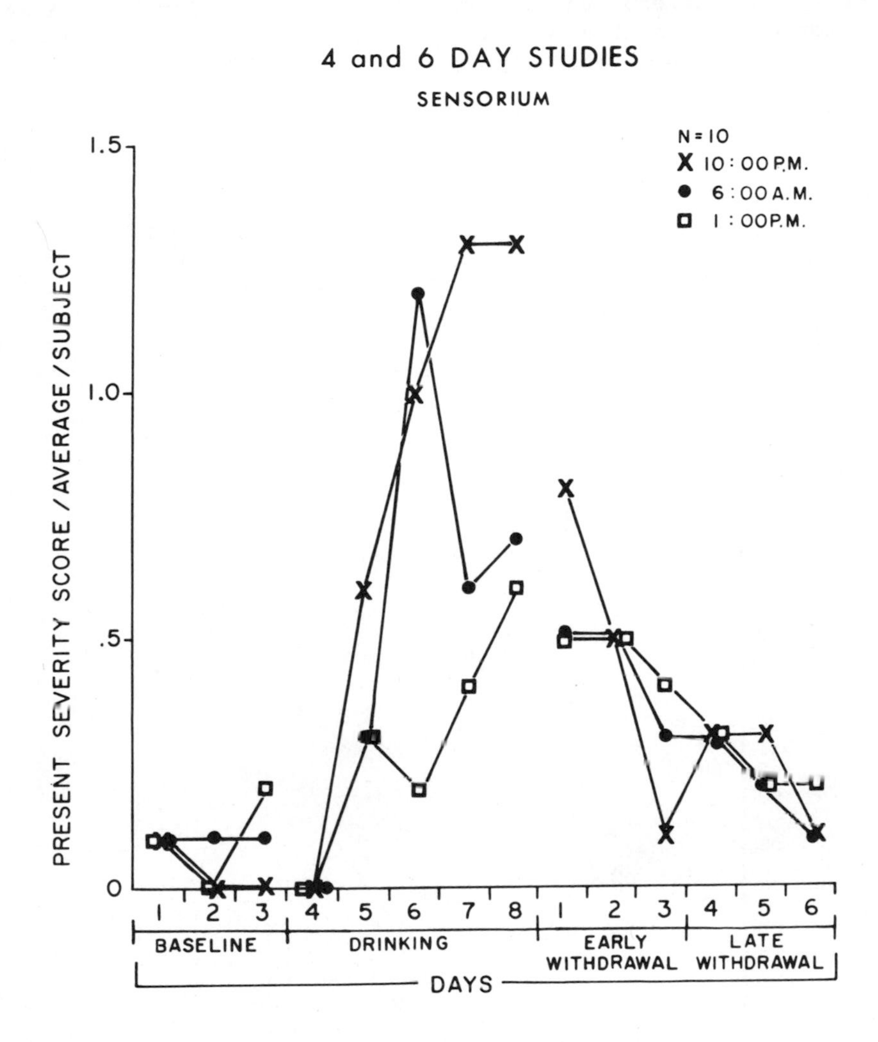
4 and 6 DAY STUDIES
SENSORIUM
N = 10
X 10:00 P.M.
● 6:00 A.M.
□ 1:00 P.M.
PRESENT SEVERITY SCORE / AVERAGE / SUBJECT
1.5
1.0
.5
0
1 2 3 4 5 6 7 8 1 2 3 4 5 6
BASELINE
DRINKING
EARLY WITHDRAWAL
LATE WITHDRAWAL
DAYS

Figure 3

Those variables which were found to be similar to clouding of the sensorium were: disturbances of gait, disturbances of consciousness, quality of contact, nystagmus and increased pulse.

The second pattern was exemplified by tremor. Examination of the incidence revealed that on baseline days the incidence was extremely high, with an average daily incidence of 80% (Figure 4).

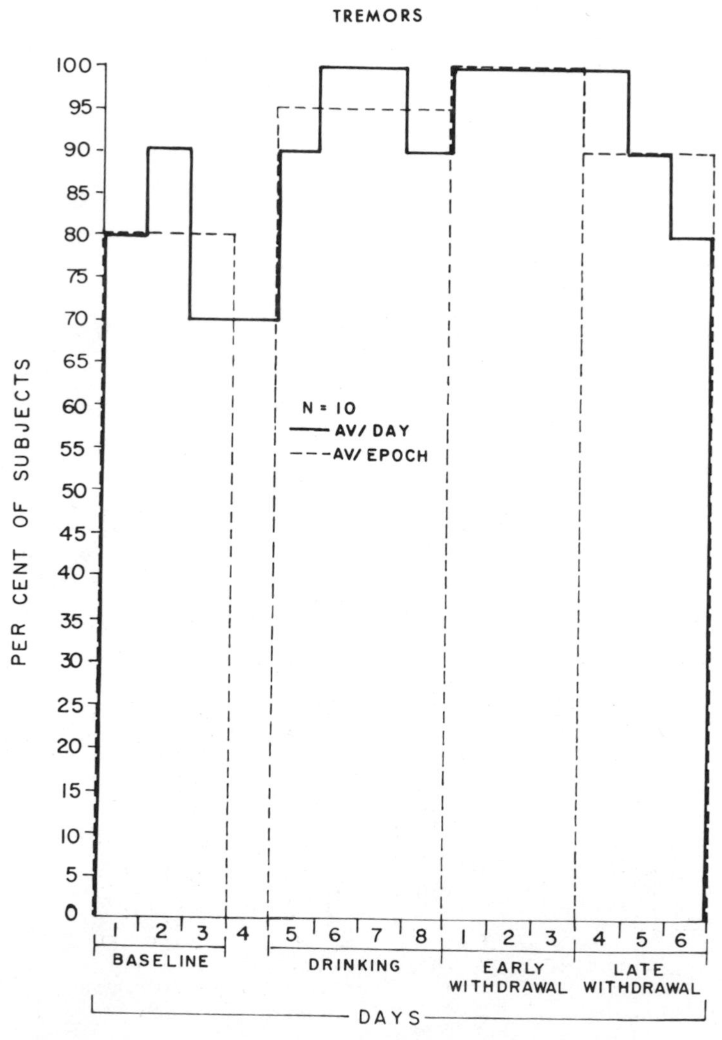

Figure 4

During the heavy drinking period this rose on two days to 100% with an average daily tremor incidence of 95% on the heavy drinking days. Throughout early withdrawal daily incidence remained at 100% and in late monitored withdrawal it gradually decreased to 80%.

The daily average tremor severity (present) by experimental periods was more sharply increased during the heavy drinking and early withdrawal days than was indicated by the incidence. During drinking it was most marked during the 1 PM sample (maximal partial withdrawal) (Figure 5). On the third withdrawal day a substantial

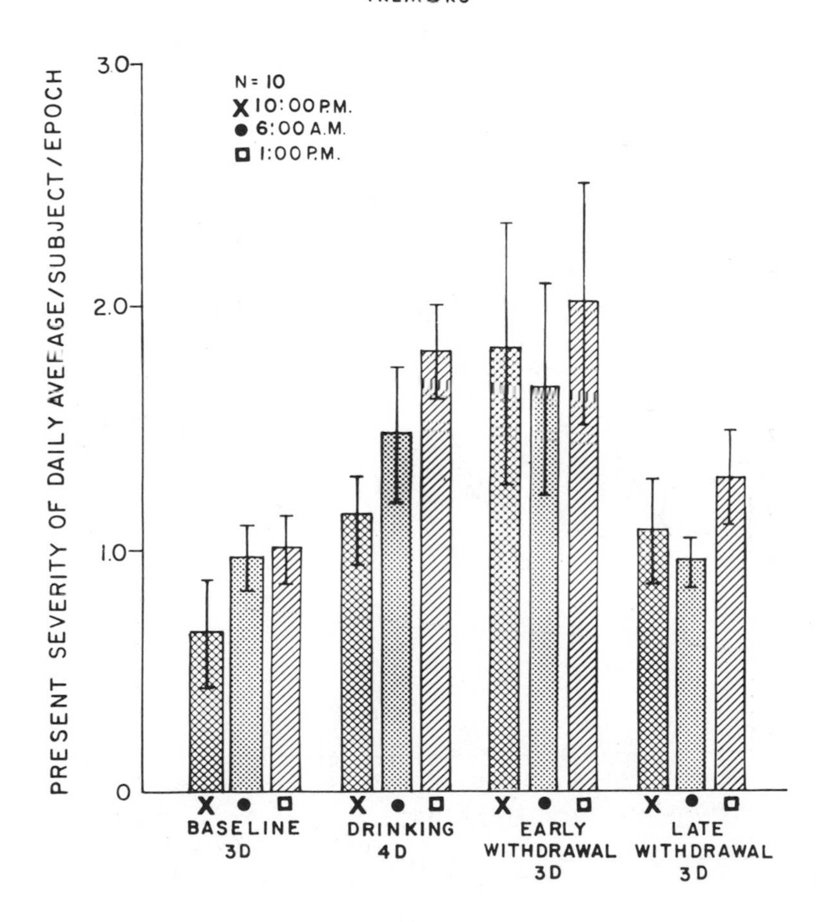

Figure 5

second peak in the severity of tremor was observed at all three sampling times (Figure 6).

The daily average sampling scores emphasized the relation of the tremor severity to partial withdrawal during the first 5 days of drinking and the increased severity, particularly at 10 PM, during the first 3 days of total withdrawal.

Those variables which were found to be similar to tremor were: anxiety, depression, sweats and systolic blood pressure.

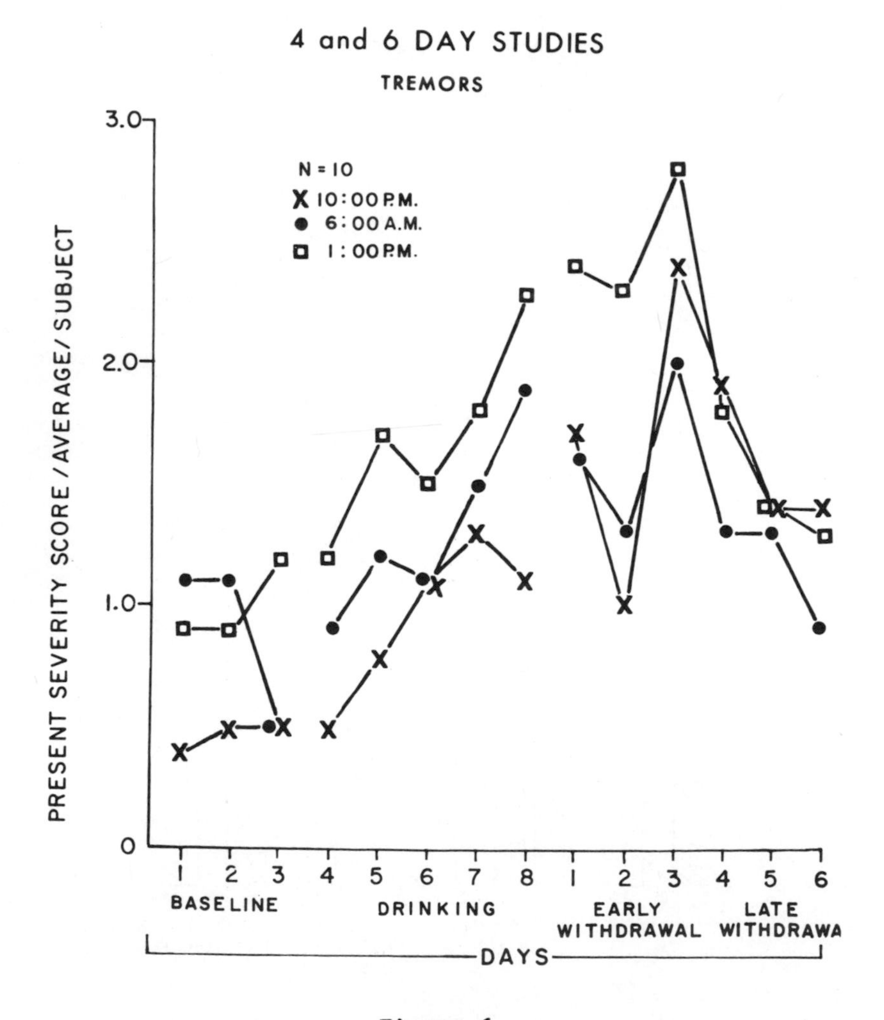

Figure 6

The third pattern was exemplified by muscle pain. The daily incidence revealed a low incidence during baseline, some gradual increase during heavy drinking and a sharp increase to a peak of 80% during total withdrawal which was followed by a gradual reduction (Figure 7).

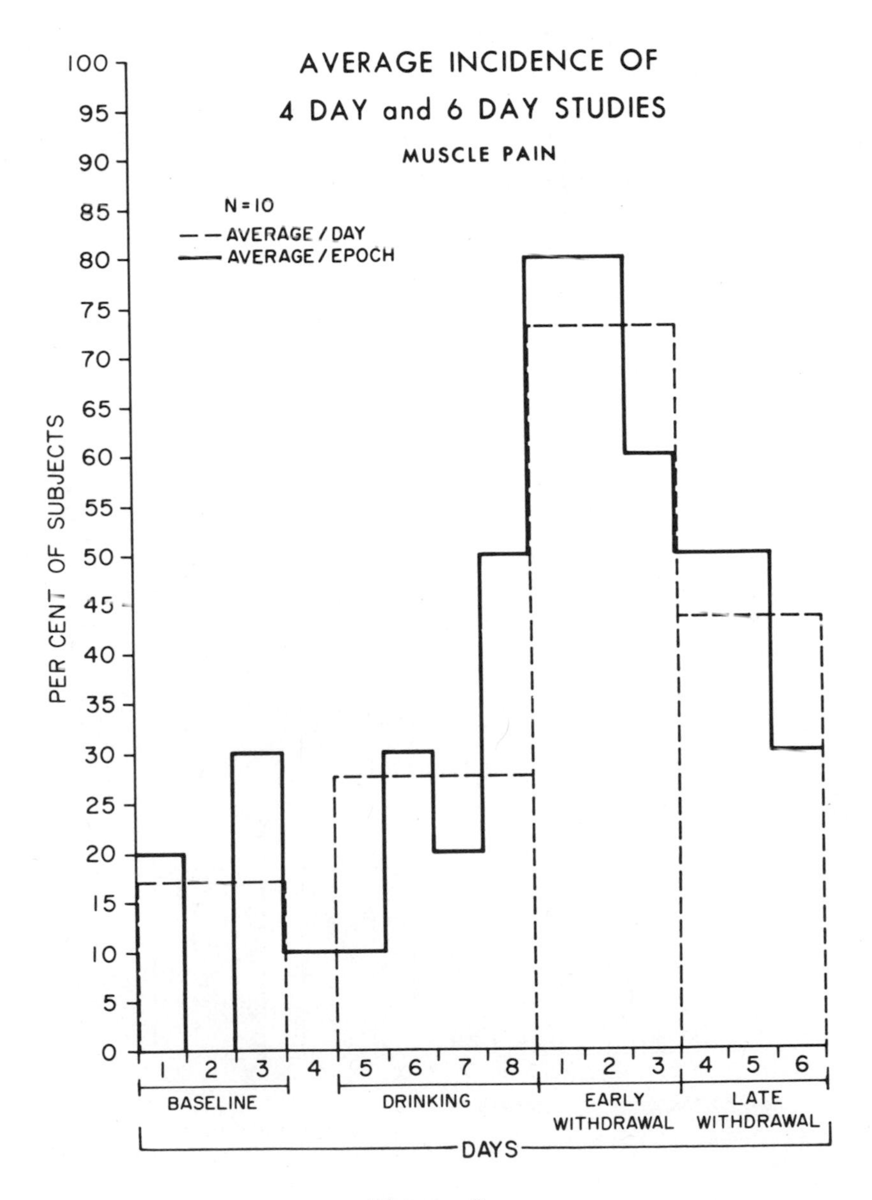

Figure 7

The daily average severity of muscle pain (present) within experimental periods showed some increased severity during partial withdrawal on the drinking days but, compared to the two previous patterns described, showed by far the sharpest increase during total withdrawal (Figure 8).

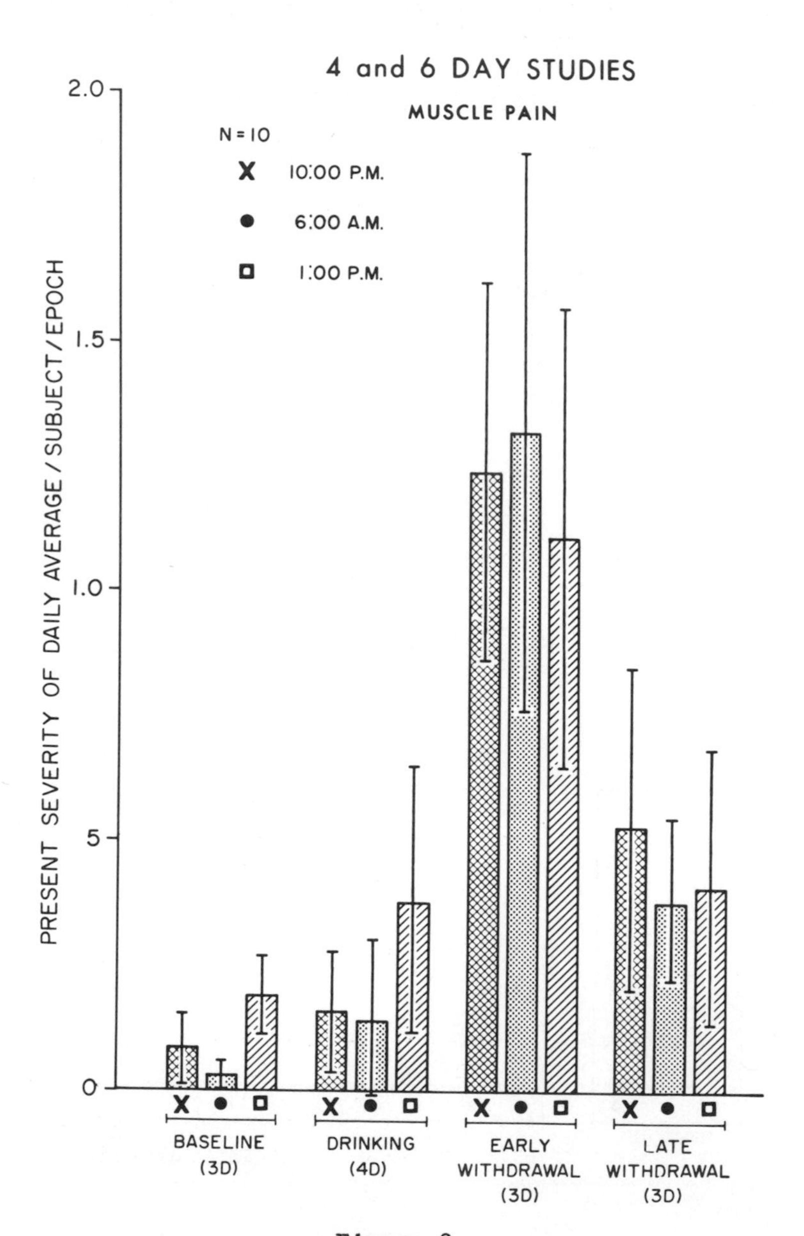

Figure 8

The daily average sampling scores re-emphasized the strikingly increased severity of muscle pain during the initial period of total withdrawal. A similar though lesser secondary peak than seen with tremor was observed on the third day (Figure 9).

Those variables which were found to be similar to muscle pain were: pruritus, visual disturbances, tinnitus, sleep disturbances, nightmares, hallucinations and delusions.

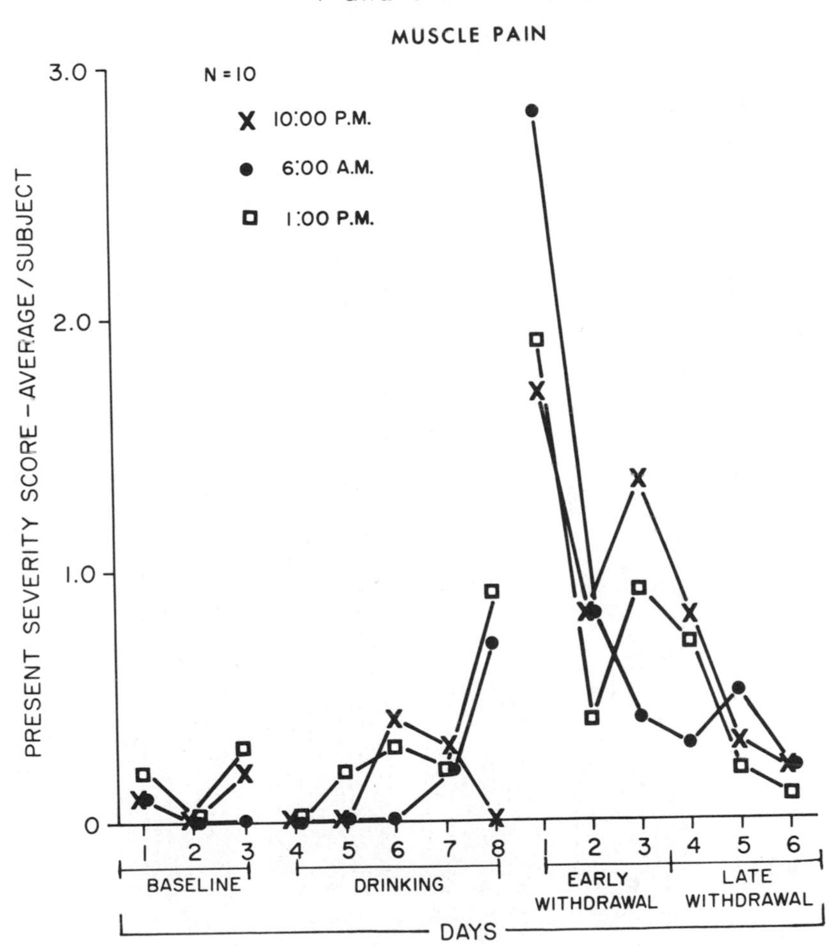

Figure 9

Related though modified activity was observed in the case of nausea and diastolic blood pressure. Both showed no change during maximum intoxication of the drinking phase but showed considerable increase during partial withdrawal. There was comparable activity for nausea during early withdrawal and practically none during monitored late withdrawal. The 10 PM diastolic blood pressure was essentially unchanged across all conditions. However, the 6 AM and 1 PM diastolic blood pressure increased during drinking and this elevation persisted unchanged for the remaining period of the study. The increased average diastolic blood pressure was most

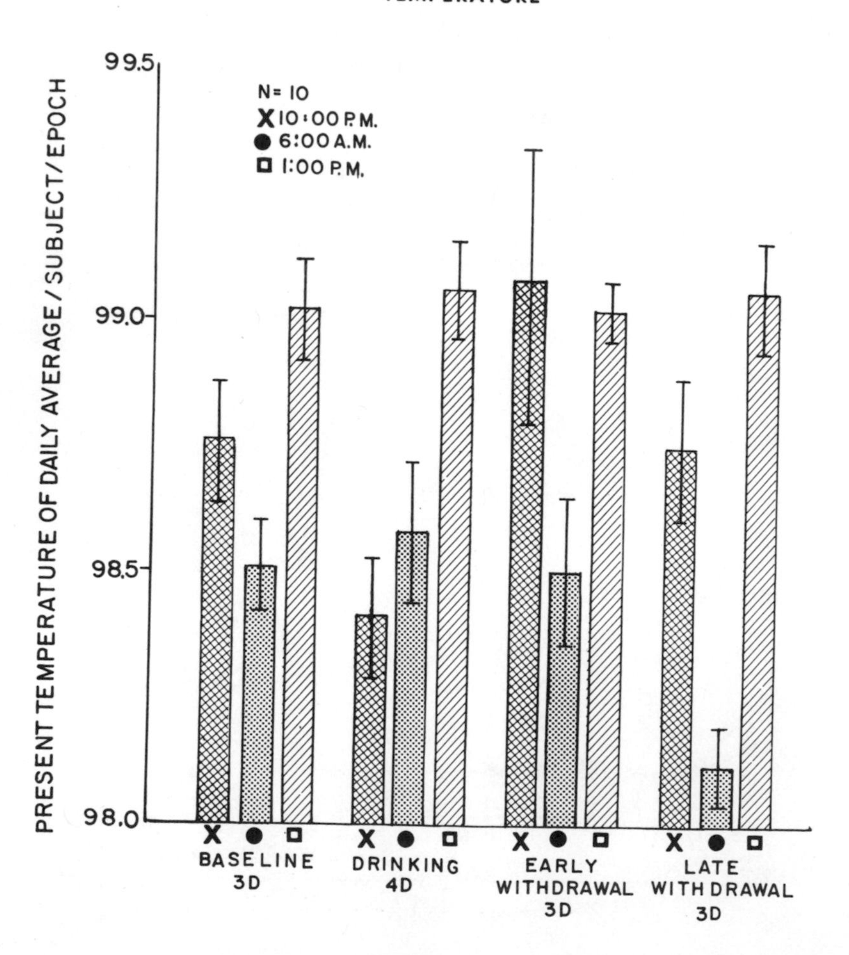

Figure 10

marked at 1 PM during the drinking period as well as early and late withdrawal when it was 79, 78 and 76.5 respectively, compared to 67 at baseline.

The fourth pattern was exemplified by the temperature (rectal). The average daily temperature by experimental periods showed essentially no change across all experimental periods at 1 PM (Figure 10). However, the 10 PM temperature dropped during the drinking period and was elevated during the early withdrawal, following which it returned to baseline. The 6 AM temperature tended to increase during drinking, was somewhat further increased during the first withdrawal day and dropped to below baseline level during the latter period of withdrawal.

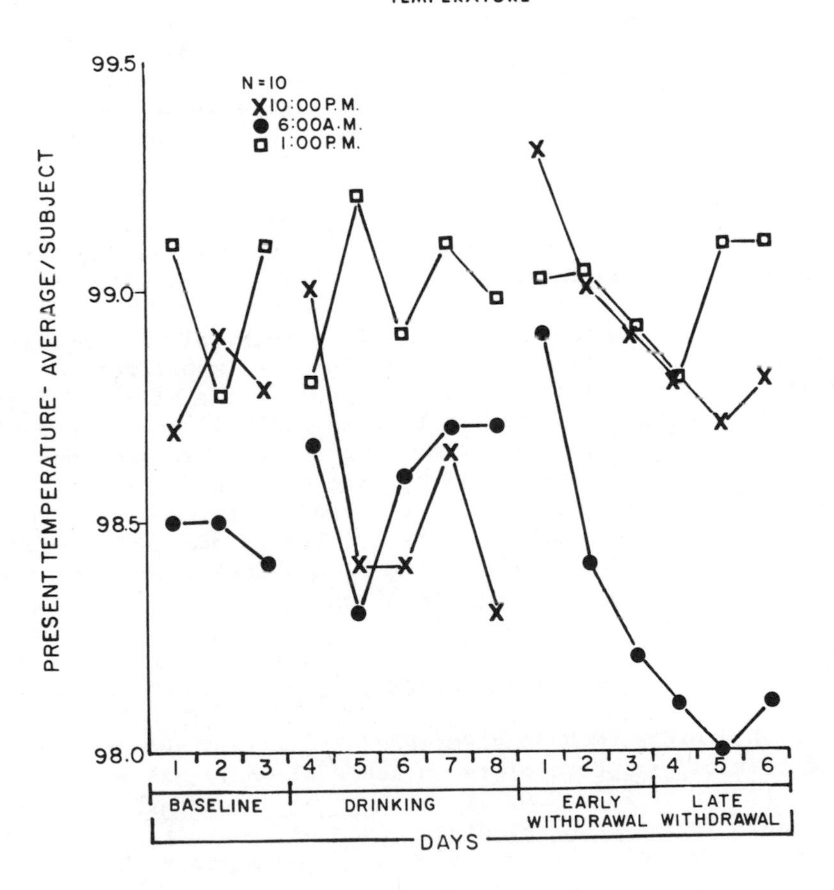

Figure 11

These trends were also observed in the average daily temperatures (Figure 11) namely, the stability of the 1 PM temperature and the tendency for the 10 PM and 6 AM temperatures to change in opposite directions during the drinking phase, and in reverse directions during the withdrawal.

Other variables with considerable activity during drinking and withdrawal, but with different diurnal patterns during drinking and withdrawal were: agitation, paresthesias and eating disturbances.

DISCUSSION

The sample was small so the conclusions must be cautious. Furthermore, even if the sample were larger, the conclusions would also have to be qualified by the limits of the experiment namely, 4 to 6 days of scheduled heavy drinking.

Given these cautions, several observations emerged from the data. As in the earlier report on prevalence (Gross and Lewis, 1973) in examining the incidence and severity scores, the factorially related variables tended to behave similarly to each other during baseline, drinking and withdrawal. The data also suggested that the diastolic blood pressure may be related to Factor I, that the systolic blood pressure may be related to Factor II and that pulse may be related to Factor III, i.e., intoxication under these conditions. Agitation was bimodal with an increase during ratings at the time of maximum intoxication and again during withdrawal.

Once more the importance of partial withdrawal was emphasized. The value of continuous observation and of the importance of diurnal variations were indicated. In particular, the effects of withdrawal on temperature appeared to be critically linked to the time of day. Further elucidation of the fine and complex sequences of change in man, as they are molded by alcohol intoxication, partial withdrawal, total withdrawal and the circadiam rhythms may prove useful in comparative studies of the problem in man and non-human animals, and in the physiological and biochemical studies of the process underlying the addiction to alcohol.

REFERENCES

Begleiter, H., Gross, M.M. and Porjesz, B., 1973. Recovery function and clinical symptomatoloty in acute alcoholization and withdrawal. In: (Ed.) M.M. Gross, Alcohol Intoxication and Withdrawal: Experimental Studies, Advances in Experimental Medicine and Biology, Vol. 35, pp. 407-413. Plenum Press, New York.

Gross, M.M., Halpert, E., Sabot, L. 1968. Toward a revised classification of acute alcoholic psychoses. J. of Nerv. and Ment. Dis. 145:500-508.

Gross, M.M., Rosenblatt, S.M., Chartoff, S., Hermann, A., Schachter, E., Sheinkin, D. and Broman, A., 1971a. A daily clinical course rating scale for the evaluation of the acute alcoholic psychoses and related states. Quart. J. of Stud. on Alc. 32: 611-619.

Gross, M.M., Rosenblatt, S.M., Malenowski, B., Broman, M and Lewis, E., 1971b. A factor analytic study of the clinical phenomena in the acute alcohol withdrawal syndromes. Alkohologia 2:1-7.

Gross, M.M., Goodenough, D.R., Hastey, J.M., Rosenblatt, S.M and Lewis, E., 1971c. Sleep disturbances in alcohol intoxication and withdrawal. In: (Eds.) N.K. Mello and J.H. Mendelson, Recent Advances in Studies of Alcoholism, pp. 317-397, U.S. Government Printing Office, Washington, D.C.

Gross, M.M., Lewis, E. and Nagarajan, M., 1973. An improved quantitative system for assessing the acute alcoholic psychoses and related states (TSA and SSA). In: (Ed.) M.M. Gross, Alcohol Intoxication and Withdrawal: Experimental Studies, Advances in Experimental Medicine and Biology, Vol. 35, pp. 365-376, Plenum Press, New York.

Gross, M.M., Lewis, E. and Hastey, J., 1974. Acute alcohol withdrawal syndrome. In: (Eds.) B. Kissin and H. Begleiter, The Biology of Alcoholism, Vol. 3, pp. 191-264, Plenum Press, New York.

Magrinat, G., Dolan, J.P., Biddy, R.L., Miller, L.D. and Koral, B., 1973. Ethanol and methanol metabolites in alcohol withdrawal. Nature, 244:234-235.

Rosenblatt, S.M., Gross, M.M., Malenowski, B., Broman, M. and Lewis, E., 1972. Factor analysis of the daily clinical course rating scale of the acute alcoholic psychoses. Quart. J. Stud. Alc. 33:1060-1064.

BEHAVIORAL CONCOMITANTS OF THE RELATIONSHIP BETWEEN BASELINE SLOW WAVE SLEEP AND CARRY-OVER OF TOLERANCE AND DEPENDENCE IN ALCOHOLICS

Milton M. Gross and Suzanne Best
Division of Alcoholism & Drug Dependence
Dept. of Psychiatry
Downstate Medical Center,
Brooklyn, New York, U.S.A.*

INTRODUCTION

Once functional tolerance to and physical dependence on alcohol have been acquired, they are more readily reacquired (Mendelson et al., 1966, Branchey et al., 1971 and Kalant, 1973). Kalant (1973) designated this effect "carry-over" indicating a residual effect of the preceding alcohol tolerance and dependence.

In a pilot study elsewhere in this volume, evidence was presented which suggested that the Slow Wave Sleep level (during the interval between drinking episodes) may be negatively related to the level of carry-over (Gross et al.). On the basis of the average baseline % Slow Wave Sleep, a sample of alcoholic volunteers was hypothetically partitioned into a greater and lesser carry-over group. The greater carry-over group was hypothesized to be those with the lower baseline Slow Wave Sleep (lower SWS); the lesser carry-over group was hypothesized to be those with the higher baseline Slow Wave Sleep (higher SWS). In the above-cited communication, the findings of the sleep and clinical comparisons of the two Slow Wave Sleep groups were presented. This communication will compare the two groups on repeated measures of anxiety, depression, and Purdue Pegboard performance.

METHODOLOGY

All subjects were paid male alcoholic volunteers. Their participation was obtained following recovery from the acute alcohol

*Supported by NIAAA Grant AA01236

withdrawal which had brought them to hospital. The experimental studies were begun approximately 3-5 weeks after hospitalization with an average interval between admission and study onset of approximately 3.5 weeks. Medication was used only for the first 6 days in hospital with no further medication administered prior to nor during the experimental study.

The volunteers were well nourished, in good general health and free of functional or organic psychoses. Half of them were white, half were black. All were gamma alcoholics who had histories of more than 5 years of heavy drinking. All ingested larger daily quantities of alcohol when drinking heavily outside the hospital than the quantities they received during the experimental study. Four volunteers received alcohol for 5 days and six volunteers received alcohol for 7 days. The first day of drinking was half dose and the remaining 4 and 6 days respectively were at full dose. Their ages ranged from 26 to 48 with an average age of 34.6 (S.D. 8.1).

All volunteers had 3 consecutive baseline days immediately prior to the drinking period and 7 consecutive withdrawal and recovery days immediately following the drinking period. On the first drinking day, approximately 1.6 gm alcohol/kilo were given which was followed by 4 or 6 days of approximately 3.2 gm alcohol/kilo/day. Care was taken to maintain good nutrition throughout the study. One volunteer was studied at a time.

Alcohol (Canadian Club) was administered on a fixed schedule and dosage. Equal doses were administered hourly between 2 PM and midnight with the exception of 3 PM when none was given. Two volunteers also received a dose at 10 AM. During the alcohol administration period breathalyzer determinations were made at 6 AM, 2 PM and midnight (prior to the last drink). A Stephenson Breathalyzer was used.

Sleep was permitted on a scheduled basis only, from midnight to 6 AM. The volunteers were under continuous surveillance and napping was not permitted. Sleep was monitored each night on an 8 channel Grass EEG console. Electrode placements and scoring were done in accordance with the Rechtschaffen and Kales Manual (1968). Sleep was recorded and discarded the night prior to the first baseline day.

A member of the research nursing team was with the volunteer at all times. The nurse systematically evaluated him at 10 PM, 6 AM, and 1 PM using a 30 item quantitative clinical instrument, the Total Severity Assessment (TSA) (Gross et al., 1973; Gross et al., 1974).

Serial daily psychometric testing, which included the "today" form of the Zuckerman Multiple Affect Adjective Checklist (MAAL) and the Purdue Pegboard (all volunteers were right-handed), was done at 1 PM and 3 PM. This coincided with the period prior to and after the first drink of the 10 hour drinking sequence during the alcoholization phase of the study.

Based upon the mean baseline % SWS, the sample was divided into two groups, the higher SWS and lower SWS groups. The higher SWS group had a mean daily baseline of 36% SWS (32-44%); the lower SWS group had a mean daily baseline of 25.3% SWS (14-29%). The average age of the higher SWS group was 32.3 (26-48); the average age of the lower SWS group was 36.5 (28-46).

Of those volunteers who drank for 5 days, 2 were in the higher SWS group and 2 were in the lower SWS group; of those who drank for 7 days, 2 were in the higher SWS group and 4 were in the lower SWS group. Since there was an unequal distribution of volunteers for the 5 and 7 days of drinking in the two SWS groups, the experimental findings in the lower SWS group will be reported on the basis of the mean of the means of the 5 day and the 7 day volunteers. In this way, any distortion of the findings in the lower SWS group resulting from more of the volunteers having a longer period of drinking was avoided. Furthermore, comparisons during the drinking period between the two SWS groups were limited to the first 5 days of drinking only.

FINDINGS

A. Drinking Period

On all 4 heavy drinking days (Drinking Days 2-5 the average Zuckerman Anxiety score was lowered by the 2 PM drink in the higher SWS (lesser carry-over) group (Figure 1). This was not the case for the lower SWS (greater carry-over) group. This was consistent with the lower SWS group having a greater carry-over of tolerance than the higher SWS group.

There was no apparent difference between the two groups in the comparison of Zuckerman Depression scores.

Before comparing the two groups in their performance on the Purdue Pegboard during drinking, it is useful to examine the non-drinking performance during a "dry" run which was obtained for the four volunteers who participated in the study of 5 days of heavy drinking (Figure 2). The average performance during the first 3 days was better for the right hand than the left. Over the next 5

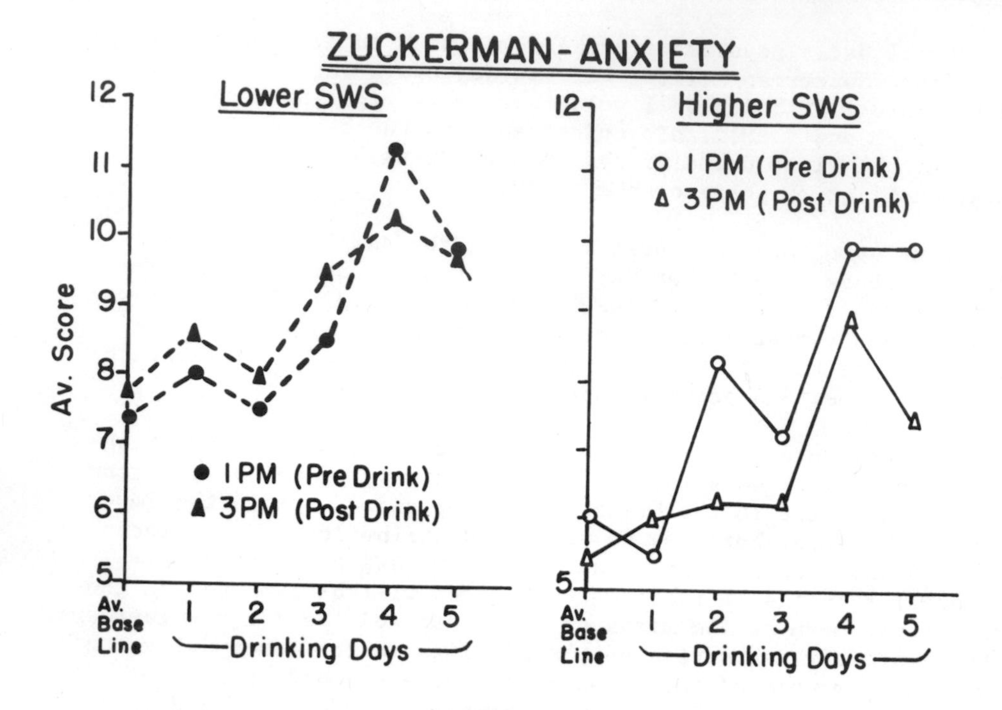

Figure 1

days (equivalent to the days when drinking occurred on the "wet run") at 1 PM and 3 PM both right and left-handed performance tended to increase. During those same 5 days the average right-handed performance at 3 PM was consistently higher than the average right-handed performance at 1 PM. This was not true of the left-handed performance. Thus, for the right hand there was indication of within day and between day learning, while, for the left hand there were only indications of between day learning.

During drinking days the learning effect was markedly reduced. In fact, where performance improved over time during the dry run, it tended to decrease over time during the drinking period (Figures 3 and 4).

In right-handed performance, if the learning effect is set aside for the moment, the comparison of pre-drink (1 PM) and

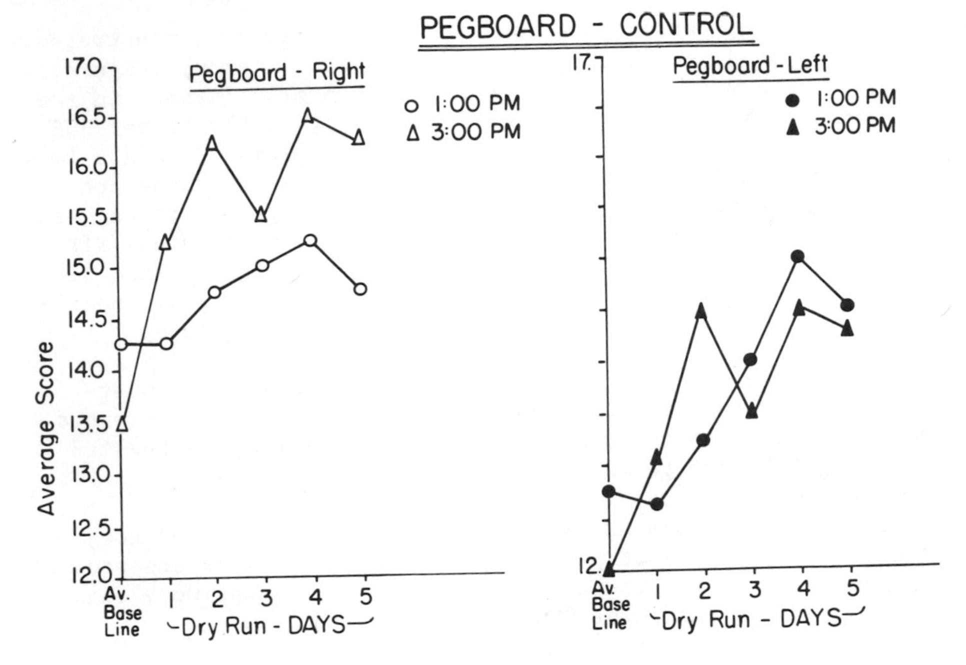

Figure 2

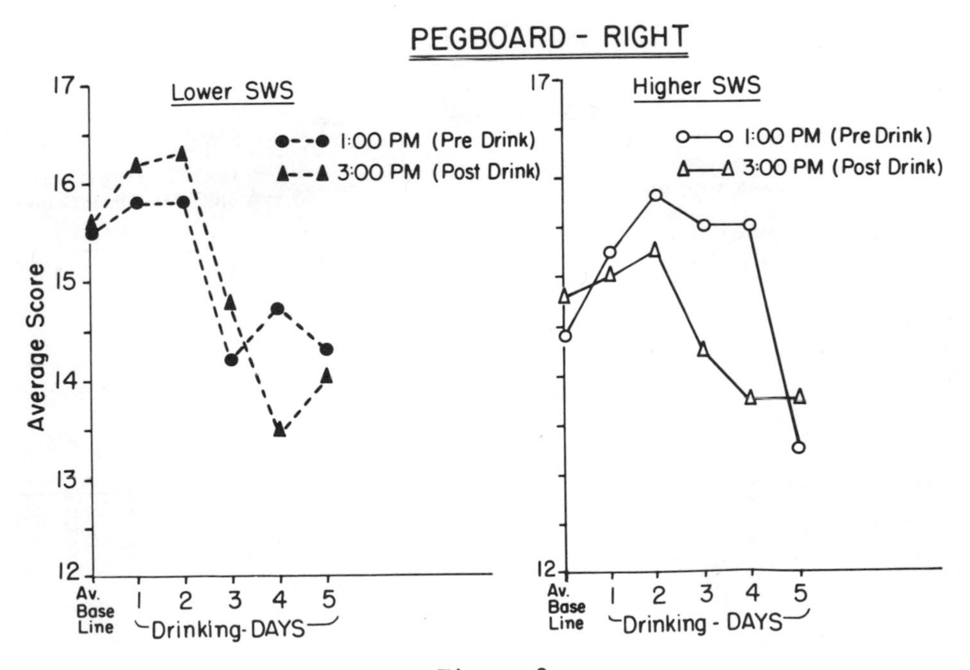

Figure 3

post-drink (3 PM) performance for the two SWS groups demonstrated a stronger post-drink reduction in the higher SWS group (Figure 3). This would be consistent with a lesser degree of tolerance in the higher SWS group. However, it is noteworthy that the higher SWS group appeared to have a greater improvement initially, and a lesser reduction (until the fourth heavy drinking day), of the pre-drink 1 PM performance on successive drinking days. This suggested that the rate of recovery from the drinking of the previous afternoon and evening was more rapid in the higher SWS group and, likewise, that the lower SWS group may have had more residual impairment carried over from the drinking of the preceding days.

The left-handed performance showed no trend of 3 PM post-drinking reduction for either SWS group (Figure 4). Differences in tolerance between the two SWS groups was therefore suggested only in right-handed performance.

As for indications of possible differences in dependence, a post-drink improvement of performance would suggest dependence. For right-handed and left-handed performance this was more consistently observed in the lower SWS group.

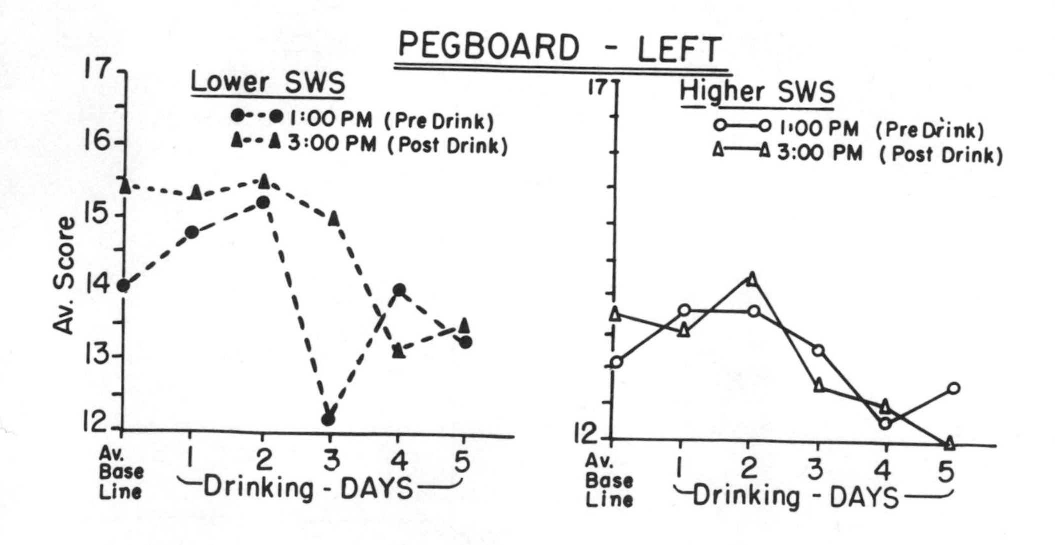

Figure 4

B. Withdrawal Period

In the comparison of the sleep findings during withdrawal, it was noted that the lower SWS group tended to have the peak % REM during withdrawal on the second withdrawal night and that the % REM peaked on the fourth withdrawal night for the higher SWS group. (Gross et al., this volume). It was suggested that this represented a difference in peak withdrawal hyperexcitability (since the REM reflects increased excitability) and that the withdrawal syndrome was a concomitant of the hyperexcitability. It was further proposed that the more rapid the shift to hyperexcitation, the more severe the withdrawal symptomatology. Consistent with this, two of the lower SWS group hallucinated during withdrawal while none of the higher SWS group did.

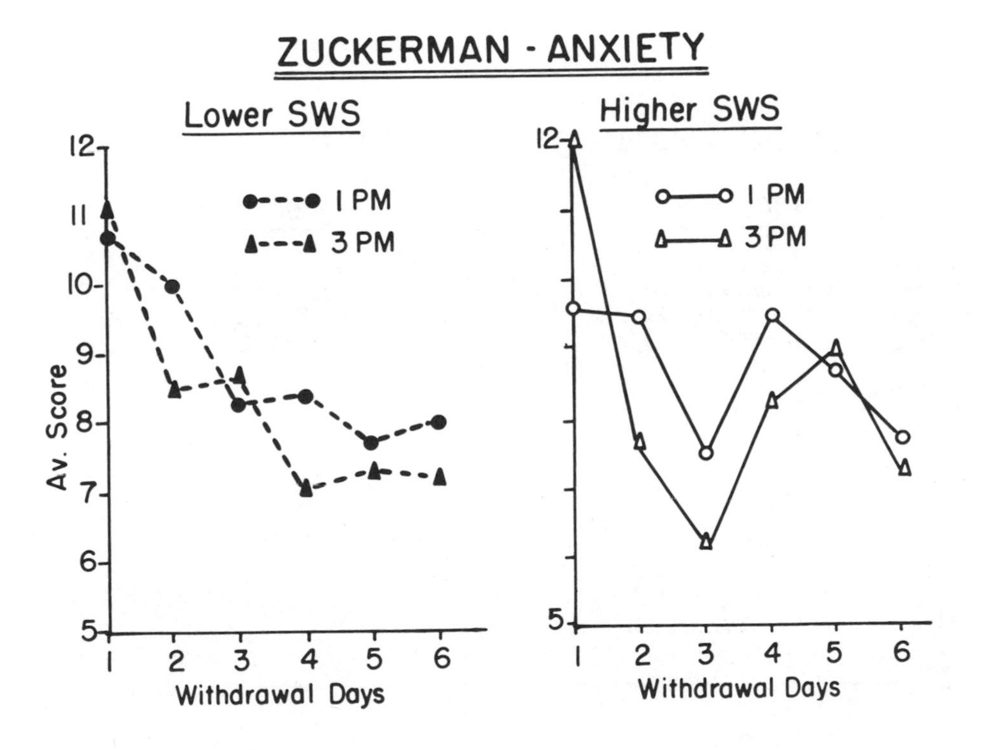

Figure 5

The average Zuckerman Anxiety scores of the higher SWS group demonstrated a secondary increase on Withdrawal Days 4 and 5 (Figure 5) which were the afternoons before and after the average % REM peak of this group. The lower SWS group did not demonstrate an early peak in anxiety to coincide with the earlier REM peak. However when the daily average of the 1 PM and 3 PM scores, and, even more, when the daily 3 PM scores were examined, the lower SWS group appeared to have a slower rate of decrease of anxiety than the higher SWS groups on Withdrawal Days 1-3.

The average Zuckerman Depression scores (Figure 6) of the higher SWS group showed a secondary increase on Withdrawal Day 5 but this was limited to the 1 PM testing (which also had a secondary increase in Withdrawal Day 3). The lower SWS group demonstrated neither a secondary increase at 1 PM, nor, as was the case with the Zuckerman Anxiety score, was there an indication of a reduced rate of decrease of severity early in withdrawal. The 3 PM depression scores were almost identical for both SWS groups.

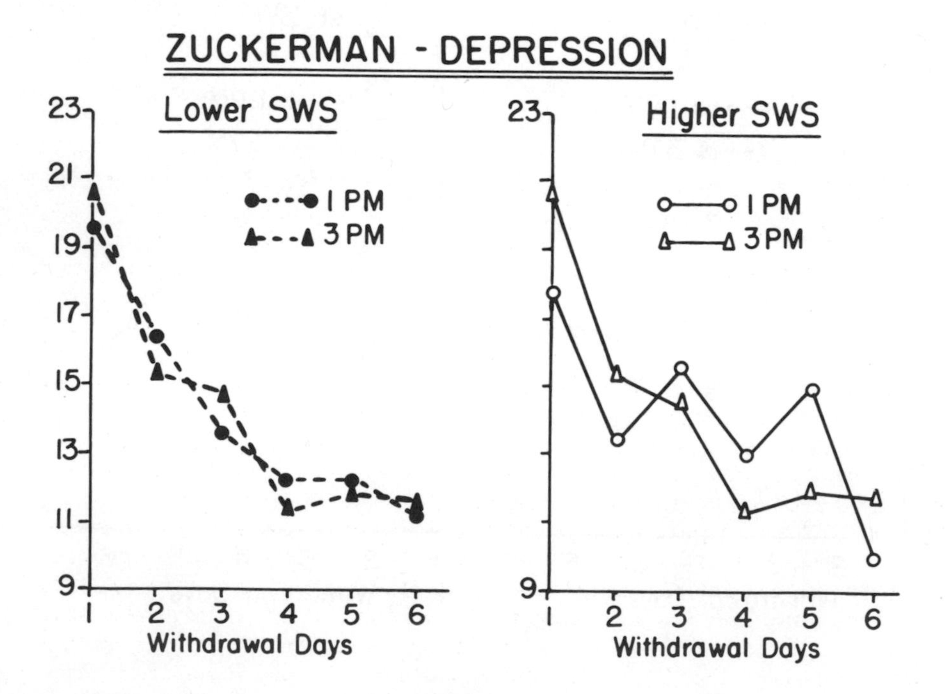

Figure 6

DISCUSSION

The preliminary evidence reported in this communication adds support to the hypothesis that the baseline level of SWS in alcoholics is negatively related to the level of carry-over of tolerance and dependence. This was observable in the comparison of the functional tolerance of the higher and lower SWS groups in the effect of the 2 PM dose of alcohol on the Zuckerman Anxiety scores but was not observable in the effect of alcohol on the Zuckerman Depression scores.

The possible relation between the SWS and carry-over of physical dependence was supported by the observation that on the second and third withdrawal days the average Zuckerman Anxiety scores appeared to be higher for the lower baseline SWS group and, on the fourth and fifth withdrawal days, it appeared to be higher for the higher baseline SWS group. The Zuckerman Depression levels gave no clear indication of differences.

The performance of the two SWS groups on the Purdue Pegboard during the drinking period was consistent with the hypothesis. It also suggested two questions about the nature of carry-over: 1) does carry-over involving functions of the cerebral hemispheres develop symmetrically or asymmetrically in relation to the dominant and non-dominant hemispheres?; and 2) is the effect of carry-over on certain functions related to impairment caused by alcohol intoxication, or delayed recovery from such impairment?

The Purdue Pegboard data suggested that the reacquisition of tolerance and dependence may involve the cerebral hemispheres asymmetrically. In examining the effects of the 2 PM drink (essentially the first drink of the day), the non-dominant hemisphere left-handed pegboard performance appeared to have greater functional tolerance and the tolerance of the two SWS groups did not appear to be different. Impaired performance following the 2 PM drink was more apparent in the right-handed performance (i.e., dominant hemisphere) and was greater in the higher SWS group (lesser carry-over). In contrast to this, evidence suggestive of dependence was more apparent in the performance involving the non-dominant hemisphere and was greater in the lower SWS group (greater carry-over). These findings would suggest that in functions which selectively involve the dominant or non-dominant hemispheres, the carry-over of tolerance and dependence may be greater in the non-dominant hemisphere. It is of interest to note that this appears to be related to the findings reported by Parsons (this volume) in which long term alcoholics, compared to short term alcoholics and normals, had a significant impairment of reaction time only when the stimulus was presented to the non-dominant hemispheres. Asymmetrical effects of alcohol on functions of the cerebral hemispheres were examined by Porjesz and Begleiter (this volume), in relation to visually evoked brain potentials. They observed that the visual evoked brain potential at the occipital site was more significantly reduced in the right hemisphere than the left.

What of the possible relationship between the carry-over of certain functions and the impairment caused by alcohol intoxication, delayed recovery from such impairment, or both? This possibility was suggested by comparison of the two SWS groups on right-handed pegboard performance (Figure 3). The higher SWS group continued to show what was, most likely, recovery from the alcohol effects of the preceding evening when tested at 1 PM during the first 3 days of heavy drinking. It was on those same days that the 2 PM drink reduced right-handed performance. The lower SWS group did not show the reduction by the 2 PM drink on those days but this group also failed to demonstrate the 1 PM recovery. When the higher SWS group showed a sharp reduction of performance at 1 PM on the fourth day of heavy drinking, the effect of the 2 PM drink was similar to what had been observed in the lower SWS group. This suggested that for certain functions there may be an interaction between the rate of reacquisition of tolerance and the rate of reduction of the capacity to recover from the effects of alcohol.

The limitations of this study are discussed elsewhere (Gross et al., this volume) and these results must be viewed as preliminary findings.

REFERENCES

Branchey, M., Rauscher, G. and Kissin, B., 1971. Modifications in the response to alcohol following the establishment of physical dependence. Psychopharmacologia 22:314-322.

Gross, M.M., Lewis, E. and Nagarajan, M., 1973. An improved quantitative system for assessing the acute alcoholic psychoses and related states (TSA and SSA). In: (Ed.) M.M. Gross, Alcohol Intoxication and Withdrawal: Experimental Studies, Advances in Experimental Medicine and Biology, Vol. 35, pp. 365-376, Plenum Press, New York.

Gross, M.M., Lewis, E. and Hastey, J., 1974. Acute alcohol withdrawal syndrome. In: (Eds.) B. Kissin and H. Begleiter, The Biology of Alcoholism, Vol. 3, pp. 191-264, Plenum Press New York.

Gross, M.M., Hastey, J.M., Lewis, E. and Young, N. (This volume) Slow wave sleep and carry-over of functional tolerance and physical dependence in alcoholics.

Kalant, H., 1973. Biological models of alcohol tolerance and physical dependence. In: (Ed.) M.M. Gross, Alcohol Intoxication and Withdrawal: Experimental Studies, Advances in Experimental Medicine and Biology, Vol. 35, pp. 3-13, Plenum Press, New York.

Mendelson, J.H., Stein, S. and McGuire, M.T., 1966. Comparative psychophysiological studies of alcoholic and nonalcoholic subjects undergoing experimentally induced ethanol intoxication. Psychosom. Med. 28:1.

Parsons, O.A. (This volume) Brain damage in alcoholics: Altered states of unconsciousness.

Porjesz, B. and Begleiter, H. (This volume) Alcohol and bilateral evoked brain potentials.

MARITAL INTERACTION DURING EXPERIMENTAL INTOXICATION AND THE RELATIONSHIP TO FAMILY HISTORY

Steven J. Wolin, M.D., Peter Steinglass, M.D., Paula Sendroff, M.A., Donald Davis, M.D., and David Berenson, M.D.

Department of Psychiatry, George Washington University School of Medicine, Washington, D.C. and Laboratory of Alcohol Research, NIAAA, St. Elizabeth's Hospital, Washington, D.C.*

INTRODUCTION

Over the past two years we have been collecting information on selected intact families in which alcoholism has persisted for many years, often over several generations. The orientation of the project is based upon an interactional model of chronic alcoholism in which problem drinking is viewed as an adaptive response to specific intrafamilial issues and is retained to provide interactional continuity.

Our group has now published several descriptions of this interactional model of alcoholism, focussing on dyadic transactions, entire family interactions and on the interface between the chronic alcohol abuser and the community.[6,7,8,9] In the current pilot study we have sought to further objectify our data collection methods and to generate specific testable hypotheses which hopefully will demonstrate the validity of this orientation. While there exists an extensive bibliography on the subject of interaction between the

*Drs. Wolin, Steinglass, Davis and Ms. Sendroff are now with the Center for Family Research, Ross Hall, George Washington University Medical Center, 2300 Eye Street, N.W., Washington, D.C. Dr. Berenson is now Director of the Alcohol Program, Bronx Psychiatric Center, Bronx, New York.

drinker and his spouse, his children, his community, etc., no systematic, "scientific" methodology has yet been developed. The project described in this report was initiated so that a fertile hypothesis-generating environment might be established in the difficult research area of experiential factors that might contribute to the maintenance of chronic alcoholism.

We have been interested in the behavioral consequences of alcohol abuse over several generations. While we are also impressed with the genetic possibilities and recent data described by Goodwin, et al.[1,2,3,5] some indications exist to make the study of interaction and the context in which abuse of alcohol occurs quite valuable. Specifically, there is little known about the impetus for daughters of alcoholics to select alcoholic husbands, an event occurring with considerable frequency. Similarly, the alcohol dependence of one particular child in a family with an alcoholic parent occurs repeatedly. The close examination of family life may yield important information on factors which determine these findings.

METHODS

A. Subject Selection:

Couples were selected who were married or living together at least five years without separation and in whom alcohol problems were self-diagnosed in at least one member of the couple. In these marriages we would expect that the patterns of interaction would demonstrate repetitious abuse patterns. The diagnosis of alcoholism was confirmed in addition by a positive response to the Brief Michigan Alcohol Screening Test.[4] The couples underwent extensive history taking, some interactional testing, and a period of ad lib alcoholization on a specially designed inpatient facility located on the grounds of St. Elizabeth's Hospital in Washington, D.C. and operated under the intramural program of the NIAAA. Two or three couples took part in each of four studies. This report comprises that data obtained from seven of these families.

During the nine day inpatient phase of the project, and following extensive physical examinations, alcohol was freely available from a self-service bar daytimes and early evenings. All subjects were encouraged to reproduce their typical home behavior patterns, drinking patterns and marital struggles. Twenty-four hour observations of behavior were made by trained staff who made no interventions. Daily multiple couple group therapy sessions, which had begun prior to hospitalization, were continued throughout the study. These sessions have been videotaped for future reference and analysis.

B. Family Trees:

Husband and wife were interviewed separately to derive an extended family tree. This genealogy was obtained by experienced interviewers and oriented towards (a) reports of alcoholism in other family members over four generations (children, subjects, their parents and their grandparents), (b) recalled patterns of interaction especially around drinking, which might have organized and influenced our subjects' own alcoholism patterns, and (c) character types and significant personality issues recollected about each family member. While information in some families was sketchy and each interviewee presented his own view and memory of past events, the skill of the interviewer, the focus upon interaction and the concurrent impression of the spouse taken separately, assisted us in obtaining as accurate a family tree as possible.

C. Interaction Recordings:

Several research tools were utilized to measure and rate individual performance and dyadic interaction of our "alcoholic couples." For example, the Ravich Interpersonal Game/Test (RIG/T) was administered. The RIG/T is an experimental game which has been developed for the purpose of studying decision-making, conflict, and conflict resolution under controlled and measurable conditions. The test requires husband and wife to each operate an electric train which, under most circumstances, will not reach a desired goal without significant negotiation and bargaining with the other game player.

Similarly, our couples were each given the Drewery Interpersonal Perception Test, an adaptation of the Edwards Personal Preference Schedule. This is a cross-matching task to demonstrate the differences between self-observation and observations by others. The results of all our test procedures are currently being analyzed and will be presented at a future date.

D. Inpatient Observations:

Regular recordings of interaction were obtained during the inpatient phase of the study. Multiple couple group therapy was held on a daily basis and was videotaped in order to record all interaction. In addition, various measurements were made by our staff throughout the day. Clinical descriptions of individual performance, drinking style and dyadic interaction were recorded. Behavioral observations focussed upon the comparison of sober and intoxicated states. For this report we will limit our observations to the clinical notes on several of the couples, in order that we might illustrate some relevant and recurrent issues in the dynamics of alcoholism in this population.

RESULTS

A. Family Histories:

The significance of the familial incidence of alcoholism is apparent in our seven couples. The frequency with which our alcoholics and their spouses have alcoholic relatives, far exceeded the (1-4%) in the general population. Many others have noted a familial factor in alcoholism. For us this suggests a linking process which transmits alcoholism from one generation to the next. We are aware of the recent studies which offer evidence to support a genetic mechanism. However, several factors impress upon us the importance of experiential family events:

First, genetic explanations appear insufficient to explain this continuity since much of the transmission occurred as the daughters of alcoholic men, non-alcoholic themselves, marry alcoholic men with surprising regularity. In five instances where this was a possible occurrence, it actually took place three times; i.e. wife of an alcoholic man would have an alcoholic father, although she had no alcohol problem herself.

Secondly, the processes of identification and imitation from one generation to the next seemed crucial from the personal and family histories. Couple members appeared to be inexorably repeating patterns of interactional style and drinking behavior carried on by parents and siblings with remarkable regularity. Two examples of this are presented below to illustrate this finding, since it is difficult to quantify but easy to comprehend:

B. Clinical Case #1:

In the S family (as illustrated in figure 1) is an alcoholic husband, aged 48. His father was an alcoholic. He was the only child in the family to abuse alcohol, and there were no other drinking problems in the extended family. His wife, Mrs. S, also had an alcoholic father. Her only sibling, a younger brother is alcoholic and there are several other alcoholic children among the grandmother's children. In addition, both husband and wife have clearly repeated the character pattern of their same sex parent. Mr. S describes himself as just like his alcoholic and dependent father of whom he is quite ashamed. When drunk his father was loud and abusive, constantly threatening violence. He rarely carried through on these threats, but they provided a central conflict in family living. When sober he was meek, illiterate and passively dependent upon his strong wife. Mr. S's mother was "of high standards," and she controlled the household. The wife, Mrs. S, matched her husband's family with exact complementarity. Her mother also married an alcoholic man, took care of him and reversed traditional roles by

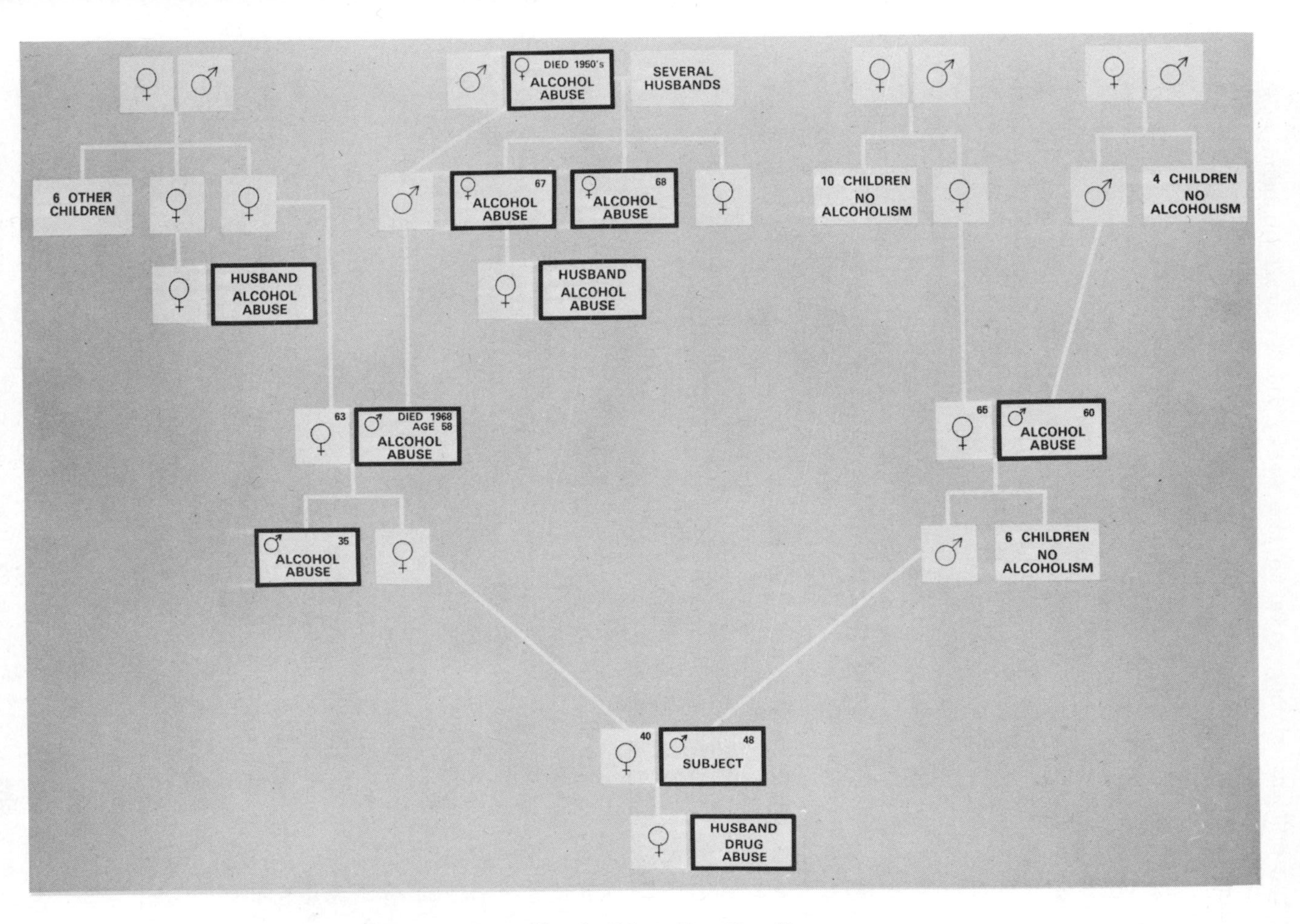

Figure 1. Alcoholism Family Tree

playing out the strong wife to the weak, ineffectual husband. She also had one younger brother, alcoholic, with whom this role reversal is similarly duplicated.

During the inpatient phase, and especially during drinking bouts an exaggeration occurred of the rigid role stereotypes that were described in the family history. Mr. S exhibited the role reversal by taking on many maternal and feminine tasks. He cooked and cared for the entire inpatient group of three couples. On the ward, Mrs. S, who is the major source of income at home, did not get involved in any of these activities. When drunk he became childish and foolish, appearing even more impotent and unmasculine. Mrs. S tolerated him, just as her mother had done, while becoming increasingly internally annoyed. Implied in her behavior was the notion that he was just a child with a bottle who couldn't follow through. As Mr. S drank further, he released more anger, but only towards his wife, never to the staff or other couples. The limitations on aggressive behavior were well known by this couple and what initially seemed angry and chaotic ended looking like a distorted and ritualized form of caring and intimacy. Consequently, in both the role reversal and in the expression of anger this couple repeated an alcoholism interaction system from their parental generation.

C. Clinical Case #2:

As shown in figure 2, the wife is the alcohol abuser in this family. She clearly identified herself as being like her father, an alcoholic, during our history-taking interviews. She sees getting drunk as providing the release and excitement that would otherwise be absent. She recalls the unpredictability and instability of her childhood home and clearly wishes to reproduce it. She married a man like her quiet-tempered mother and attempts to seduce him into taking care of her and joining her in drinking. She rigorously and consciously rebelled against her moralistic, teetotalling mother in modelling her behavior after that of her father. However, she married a man who would rely on her hysterical actions to give family life purpose and direction. Earlier generations of this family have behaved in a similar fashion with numerous other problem drinkers. Mr. D described his father as a gentle, soft-spoken man who never lost his temper, a description given to him by others. His own mother was a loud, outspoken woman not unlike his wife. A clear repetition was evident - husband acts like passive father and marries a strong woman; wife, acting strongly and primitively, identifies with her father and marries a man like her mother.

On our ward, Mrs. D drank daily with dramatic behavioral consequences. She would become excited, seductive and emotionally labile. She encouraged her husband to join her in the drinking, which he would refuse. Attempts at physical closeness on her part

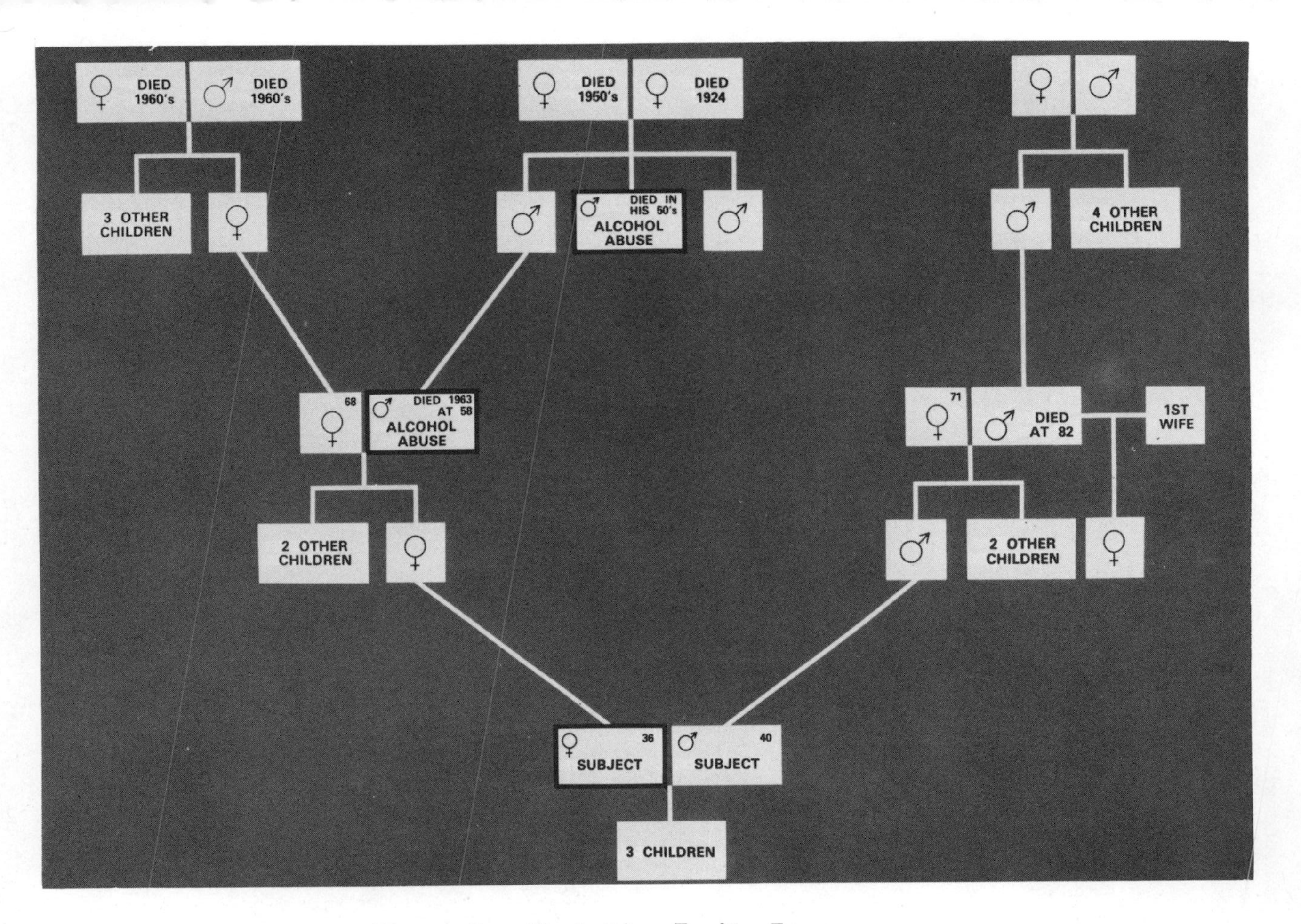

Figure 2. Alcoholism Family Tree

were matched with attempts, albeit impulsive, of emotional closeness. In general, Mrs. D acted out with remarkable accuracy, the dramatic performances of her father. Mr. D became flustered and anxious in the face of his wife's behavior. At first, he tried to care for her. In doing so he became overresponsible and tense. Then, when confronted with her great demands for attention and closeness he would panic and push her away. His wife would then see him as rigid and unloving, exactly as both she and her father saw her mother. As with the S family, alcohol provided the vehicle to express this behavioral script. Intoxicated states emphasized specific aspects of sober personality issues and demanded performance by the partner in a rigid stereotyped fashion. These patterns could well be termed "family alcoholism rituals."

DISCUSSION

In a careful examination of the family trees in these selected families a striking continuity is observable. Alcohol abuse is perpetuated over several generations in conjunction with the repetition of certain personality characteristics. The individual with these traits repeats interaction patterns which they had described for their parents and which utilized alcohol as a crucial vehicle.

In addition, observations made during naturalistic experimental intoxication have shown these interaction patterns to be quite stable. Individual behavior is reliably repeated in these intact marriages as alcoholism persists. This persistence of character styles and family interaction patterns demonstrates the potential for hypotheses which link chronic alcohol abuse in intact families with the transmission of alcoholism across generations.

Our increased understanding has led to the development of several hypotheses and to two important research designs in which they will now be tested. One project is oriented to the issue of transmission of alcoholism across generations. A focus on selected areas of family life, termed "family rituals," will be studied in those families where transmission has occurred and compared to families without transmission. By utilizing several novel data-gathering devices we hope to predict the transmittor family and to understand those family characteristics which encourage alcoholism transmission. Essentially this project intends to demonstrate that certain families through its ritualized use of alcohol actually encourage alcoholism in its children.

A second approach to the problem of the alcoholic family will be the study of its homeostatic systems. Here we will systematically measure specific variables which appear central in the maintenance of alcoholic family systems in given intact families. The research design in this project will be to observe selected families in their home environment. Groups of couples will meet at each other's homes and interaction will be recorded. The research team will become intimately involved with each of approximately 30 families. Hopefully those important behavioral events which precede and sustain abusive drinking will be recorded and measured. Together these two projects will potentially make a significant contribution to our understanding of alcoholism in the family.

REFERENCES

1. Goodwin, D.W. Is alcoholism hereditary? Arch. Gen. Psychiat. 25:545-549, 1971.

2. Goodwin, D.W., Schulsinger, F., Hermansen, L., Guze, S.B., and Winokur, G. Alcohol problems in adoptees raised apart from alcoholic biologic parents. Arch. Gen. Psychiat. 28:238-243, 1973.

3. Goodwin, D.W., Schulsinger, F., Moller, N., Hermansen, L., Winokur, G., and Guze, S.B. Drinking problems in adopted and nonadopted sons of alcoholics. Arch. Gen. Psychiat. 31:164-169, 1974.

4. Pokorny, A.D., Miller, B.A., and Kaplan, H.B. The brief MAST: a shortened version of the Michigan Alcoholism Screening Test. Amer. J. Psychiat. 129 (3) :118-121, 1972.

5. Schuckit, M., Goodwin, D., and Winokur, G. A study of alcoholism in half siblings. Amer. J. Psychiat. 128:1132-1136. 1972.

6. Steinglass, P., Weiner, S., and Mendelson, J. A systems approach to alcoholism: a model and its clinical application. Arch. Gen. Psychiat. 24:401-408, 1971.

7. Steinglass, P., Weiner, S., and Mendelson, J. Interactional issues as determinants of alcoholism. Amer. J. Psychiat. 128: 275-280, 1971.

8. Weiner, S., Tamerin, J., Steinglass, P., et al. Familial patterns in chronic alcoholism: a study of a father and son during experimental intoxication. Amer. J. Psychiat. 127: 1646-1651, 1971.

9. Wolin, S.J. and Steinglass, P. Interactional behavior in an alcoholic community. Med. Ann. of D.C. 43 (4) :183-187, 1974.

INDEX